Collins

CSEC®

MATHEMATICS

Raphael Johnson

Reviewers: Melissa Andrade, Joanne Baker, Marcus Caine, Lenore Dunnah

Collins

William Collins' dream of knowledge for all began with the publication of his first book in 1819. A self-educated mill worker, he not only enriched millions of lives, but also founded a flourishing publishing house. Today, staying true to this spirit, Collins books are packed with inspiration, innovation and practical expertise. They place you at the centre of a world of possibility and give you exactly what you need to explore it.

Collins. Freedom to teach.

Published by Collins
An imprint of HarperCollins*Publishers* Limited
The News Building
1 London Bridge Street
London
SE1 9GF

HarperCollins Publishers
Macken House,
39/40 Mayor Street Upper,
Dublin 1, D01C9W8, Ireland

Browse the complete Collins Caribbean catalogue at
www.collins.co.uk/caribbeanschools

© HarperCollins *Publishers* Limited 2019

10 9 8 7 6 5 4 3

ISBN 978-00-0-830446-1

Collins CSEC® Mathematics is an independent publication and has not been authorised, sponsored or otherwise approved by **CXC®**.

CSEC® is a registered trademark of the **Caribbean Examinations Council (CXC®)**.

British Library Cataloguing in Publication Data

A catalogue record for this publication is available from the British Library.

The publishers gratefully acknowledge the permission granted to reproduce the copyright material in this book. Every effort has been made to trace copyright holders and to obtain their permission for the use of copyright material. The publishers will gladly receive any information enabling them to rectify any error or omission at the first opportunity.

Author: Raphael Johnson
Reviewers: Melissa Andrade, Joanne Baker, Marcus Caine and Lenore Dunnah
Publisher: Dr Elaine Higgleton
Commissioning editor: Tom Hardy
Project leader: Julianna Dunn
Copy editors: Joan Miller, Steven Matchett and Mary Nathan
Proofreaders: Amanda Dickson and Sam Hartburn
Illustrator: Ann Paganuzzi
Production controller: Lyndsey Rogers
Typesetter: Siliconchips Services Ltd
Cover designers: Kevin Robbins and Gordon MacGilp
Cover image: Kudryashka/Shutterstock
Printed and bound by Grafica Veneta S. P. A.

MIX
Paper | Supporting responsible forestry
FSC™ C007454

This book is produced from independently certified FSC™ paper to ensure responsible forest management.

For more information visit: www.harpercollins.co.uk/green

Contents

Introduction

Mathematics has always been one of the most challenging of all subjects for students, and in the Caribbean less than half of students sitting the CXC® CSEC® Mathematics exam in the last 3 years have achieved a pass. Collins CSEC® Mathematics textbook has been designed to help you understand and practise the mathematical concepts you need to master, in order not only to pass your CSEC® examination, but to understand the importance of maths in everyday life.

So how will this textbook help you pass? What makes it different?

- As you would expect, the content covers in detail the entire, new CSEC® Mathematics syllabus (May–June 2018) including topics and concepts never dealt with before in other maths texts.

- It provides a simple and clear, detailed, step-by-step explanation of how to solve maths problems and precise and accurate definitions of various mathematical terms and concepts.

- You will find a list of objectives for each chapter, stating the expected changes in your knowledge, skills and attitudes by the end of the chapter. This information helps you to measure where you have a good understanding of a concept, and where you might need more practise.

- Importantly it includes the real-life application of maths as shown through many concrete examples, exercises, illustrations and pictures. We have endeavoured throughout to help you to understand how maths fits into everyday life and activities. Maths isn't a purely academic or theoretical subject, and we believe that showing its application in everyday life will help demystify the subject, help give greater context for maths topics, and that this understanding will give you the motivation to study, leading to a better grade in the exam.

- There are many varied worked examples to provide practice for each concept.

- We also provide a large number of graded exercises with answers. These exercises are presented in three levels of difficulty: **access** level, which provides practice of the concepts being taught and which should be achievable by the majority, if not all students; **CSEC® pass** level, which you need to be able to tackle comfortably to pass the exam; **top grade** level, which provides practice for those hoping to achieve a pass at grade 1 or 2. These three levels are colour-coded with purple, blue and green, respectively. We have done this so that you — and your teacher — know exactly where you are tracking in terms of the attainment requirements needed for a pass — or a top grade — in CSEC® maths. The majority of students should be able to do the access level questions; students who can comfortably answer the blue questions should be on course to pass with a grade 3; students who can tackle the green questions should be tracking for a pass at grade 1 or 2.

- This textbook is not organised like other textbooks – our content follows what we call a *spiral curriculum approach* in which earlier topics are revisited at a higher level at a later stage. This will allow you to build your skills step-by-step – what you have learned in each chapter prepares you for the next chapter.

- There is a summary of the main concepts at the end of every chapter.

- Each chapter includes examination type questions at the end.

- We have included special comment boxes containing: (i) *Did you know?* Items (ii) Definitions of *Key terms* (iii) *Hints and tips* (iv) *Important formulae* (v) *Worked examples* (vi) *Reminders*.

- There are three complete CSEC® Mathematics type exams (Multiple Choice and Long Answer) with answers provided separately.

- The School-Based Assessment (SBA) – which in essence is about maths in real life – is covered in each chapter where we show the real-life application of a maths concept, and there is also a complete chapter on the SBA in Mathematics where we explain what makes a good SBA topic (and what does not) and how you can measure if your SBA topic is appropriate. This chapter also includes guidance on the problem-solving activities that can help you develop your SBA and a sample SBA project.

- There is a special chapter on Using Your Calculator which includes a full guide to all keys, which you can refer to throughout your studies so that you always know how to complete calculations.

This textbook covers everything you need for CSEC® Mathematics. What is different – we believe – is our unique approach, devised to help more students pass, and to help more students achieve a top grade, with topics ordered following a spiral curriculum approach that will help you learn. All topics are fully contextualised to everyday life and to help with the SBA. You understand how you are tracking against CSEC® standards with graded practice exercises, which will give confidence for the exam, and will allow you to know where you need to seek assistance or put in more practice.

Raphael Johnson,
Grenada,
1st September 2019

How to use this book

Learning objectives clearly shown

Prior knowledge - what you need to know or revise before starting the chapter - clearly listed

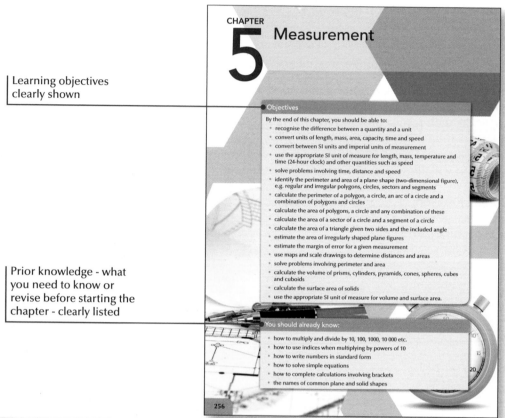

CHAPTER 5 Measurement

Objectives

By the end of this chapter, you should be able to:

* recognise the difference between a quantity and a unit
* convert units of length, mass, area, capacity, time and speed
* convert between SI units and imperial units of measurement
* use the appropriate SI unit of measure for length, mass, temperature and time (24-hour clock) and other quantities such as speed
* solve problems involving time, distance and speed
* identify the perimeter and area of a plane shape (two-dimensional figure), e.g. regular and irregular polygons, circles, sectors and segments
* calculate the perimeter of a polygon, a circle, an arc of a circle and a combination of polygons and circles
* calculate the area of polygons, a circle and any combination of these
* calculate the area of a sector of a circle and a segment of a circle
* calculate the area of a triangle given two sides and the included angle
* estimate the area of irregularly shaped plane figures
* estimate the margin of error for a given measurement
* use maps and scale drawings to determine distances and areas
* solve problems involving perimeter and area
* calculate the volume of prisms, cylinders, pyramids, cones, spheres, cubes and cuboids
* calculate the surface area of solids
* use the appropriate SI unit of measure for volume and surface area.

You should already know:

* how to multiply and divide by 10, 100, 1000, 10 000 etc.
* how to use indices when multiplying by powers of 10
* how to write numbers in standard form
* how to solve simple equations
* how to complete calculations involving brackets
* the names of common plane and solid shapes

256

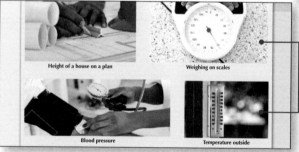

Height of a house on a plan   Weighing on scales

Blood pressure   Temperature outside

Photographs and diagrams show the application of maths in real-life situations

Key terms are defined where they are first used

Frequently used and important units and formulae are highlighted to promote learning and memorising

Worked examples provide practice for each concept

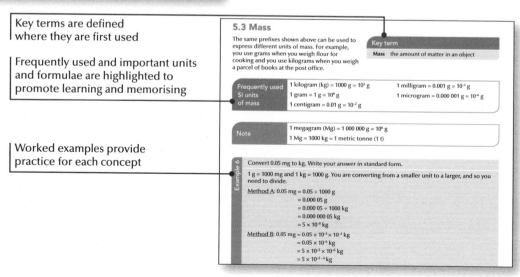

5.3 Mass

The same prefixes shown above can be used to express different units of mass. For example, you use grams when you weigh flour for cooking and you use kilograms when you weigh a parcel of books at the post office.

Key term

Mass the amount of matter in an object

| Frequently used SI units of mass | 1 kilogram (kg) = 1000 g = 10^3 g
1 gram = 1 g = 10^0 g
1 centigram = 0.01 g = 10^{-2} g | 1 milligram = 0.001 g = 10^{-3} g
1 microgram = 0.000 001 g = 10^{-6} g |

| Note | 1 megagram (Mg) = 1 000 000 g = 10^6 g
1 Mg = 1000 kg = 1 metric tonne (1 t) |

Example 6

Convert 0.05 mg to kg. Write your answer in standard form.

1 g = 1000 mg and 1 kg = 1000 g. You are converting from a smaller unit to a larger, and so you need to divide.

Method A: 0.05 mg = 0.05 ÷ 1000 g
= 0.000 05 g
= 0.000 05 ÷ 1000 kg
= 0.000 000 05 kg
= 5×10^{-8} kg

Method B: 0.05 mg = $0.05 \times 10^{-3} \times 10^{-3}$ kg
= 0.05×10^{-6} kg
= $5 \times 10^{-2} \times 10^{-6}$ kg
= $5 \times 10^{-2-6}$ kg

Graded exercises help teachers and students understand where they are tracking in terms of passing the CSEC® exam

Access level questions - highlighted with pink - provide practice of concepts at a level that should be achievable by the majority of students.

CSEC® pass level questions - highlighted with blue - are questions that students need to be able to tackle comfortably to pass the exam.

Top grade level questions - highlighted with green - show where students need to be working if they hope to achieve a pass at grade 1 or 2.

Did you know? boxes give extra information that students might find interesting but that is not essential for the exam.

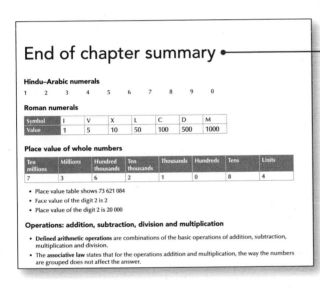

Exercise 5B

1. Convert these measurements to kg.
   a 2000 g   b 30 000 cg   c 5 Mg   d 100 mg

2. Convert these measurements. Use indices.
   a 2 000 000 µg to g   b 200 mg to Mg   c 70 cg to kg

3. Convert these measurements.
   a 0.000 023 kg to µg (in standard form correct to 1 decimal place)
   b 234 800 kg to mg (in standard form correct to two decimal places)
   c 2.55×10^{5} µg to Mg (in standard form)
   d 0.0098×10^{-3} µg to Mg (in standard form correct to 1 significant figure)

4. Convert these measurements. Express your answers in standard form correct to 1 significant figure.
   a 2ng to pg   b 2000 000 Tg to cg
   c 400 fg to Gg   d 0.000 008 2 Mg to ng

5.4 More units of length and mass

The SI is the official measuring system in most countries, including the Caribbean countries. In the Caribbean, the imperial system is still widely used in many countries, especially by builders, electricians, plumbers, technicians, etc.

Did you know?	The change from the imperial or English system began as early as 1795 in France and was completed as late as the 1970s in most British Commonwealth countries.

End of chapter summary

End-of-chapter summaries highlight the main concepts covered in the chapter.

Hindu–Arabic numerals

1   2   3   4   5   6   7   8   9   0

Roman numerals

Symbol	I	V	X	L	C	D	M
Value	1	5	10	50	100	500	1000

Place value of whole numbers

Ten millions	Millions	Hundred thousands	Ten thousands	Thousands	Hundreds	Tens	Units
7	3	6	2	1	0	8	4

- Place value table shows 73 621 084
- Face value of the digit 2 is 2
- Place value of the digit 2 is 20 000

Operations: addition, subtraction, division and multiplication

- **Defined arithmetic operations** are combinations of the basic operations of addition, subtraction, multiplication and division.
- The **associative law** states that for the operations addition and multiplication, the way the numbers are grouped does not affect the answer.

Examination-style practice questions are given at the end of each chapter.

Examination-type questions for Chapter 1

1. Using a calculator or otherwise, calculate the exact value of the following.

   a $\dfrac{10\frac{1}{2} - 2(3\frac{1}{2} + 1\frac{1}{2})}{3\frac{3}{4} \times \frac{2}{5}}$

   b $\dfrac{5.16 \div 1.2 - (1.5)^{2}}{1.3 + 1.2}$

2. Evaluate $\sqrt{125} - (1.4)^{2}$ and give your answer correct to:
   a two decimal places
   b three significant figures.

3. a In a class, the ratio of boys to girls is 3 : 2 respectively. There are 20 girls in the class.
       i   How many boys are in the class?
       ii How many students are there in total in the class?
    b A teacher wanted to share some sweets among 4 children. She decided to share the sweets in the ratio 2 : 3 : 4 : 6 based on the ages of the children. The oldest child got 12 sweets.
       i   What was the total amount shared?
       ii How much did each of the other 3 children receive?

4. a After 5 innings, a cricket team has scored an average of 280 runs. The runs scored for 4 of those innings were: 250, 300, 320 and 200. What was their score for the fifth innings?
    b A bus takes 5 hours for a trip, travelling at an average speed of 50 km/h.
       i   What is the length of the trip?

1 Number theory and computation

Objectives

By the end of this chapter, you should be able to:

- appreciate the development of different numeration systems
- understand and appreciate the decimal numeration system
- evaluate numerical expressions using any of the four basic operations on real numbers
- compute powers of real numbers of the form x^a, where $a \in Q$
- list the set of factors and multiples of a given integer
- compute the HCF or LCM of two or more positive integers
- convert between fractions, percentages and decimals
- calculate any fraction or percentage of a given quantity
- express one quantity as a fraction or percentage of another
- express a value to a given number of significant figures or decimal places
- write any rational number in scientific notation
- convert measurements from one set of units to another
- compare quantities through ratio, proportion and rates
- divide a quantity in a given ratio
- compute terms of a sequence following a given rule
- given the terms of a sequence, derive an appropriate rule
- distinguish numbers among sets of numbers
- order a set of real numbers
- state the value of a digit of a numeral in a given base
- solve problems involving concepts in number theory.

You should already know:

- how to use the decimal place-value number system consisting of the digits 0, 1, 2, 3, 4, 5, 6, 7, 8, 9
- how to count to at least 1 000 000
- how to add, subtract, multiply and divide whole numbers, decimals and fractions (these topics will be revised in this chapter)
- multiplication and division facts to 12×12.

Time shown by numbers on a digital clock

In this chapter, you will learn about the history of numbers and about the different properties of numbers. You will have opportunities to practise calculating and solving real-life problems that involve the many uses of numbers in everyday life. Examples include: counting, telling the time, checking speed, using phones, reading dates, measuring weight and other quantities, determining ages, choosing sizes of clothes, calculating amounts, making financial transactions.

Lengths on a tape measure

You will also use fractions, decimals, percentages, ratio and proportion to tackle everyday problems involving shopping, speed, comparison of sizes and quantities, sharing, etc.

Dialling phone numbers

Examples of number and computation in real life

The pictures show everyday examples of numbers. Which of them have you seen in real life?

Time shown by numbers on a clockface

Numbers on playing cards

Price of gas and diesel

Answers on a calculator

Roman numerals on a clock face

Account numbers

1.1 History of numbers

In ancient civilisations, people used to record events and successes (for example killings during a hunt) by making marks or drawings on cave walls and ceilings.

Over time and across the world, different civilisations developed their own unique methods of counting, writing numerals and keeping records.

Here are some examples of these number systems and sets of numerals.

1.1.1 Hindu–Arabic numerals

| 1 | 2 | 3 | 4 | 5 | 6 | 7 | 8 | 9 | 0 |

These are the symbols that are used today in the modern decimal system. You can use them to make numbers of different sizes:

| 10 | 37 | 400 | 5000 | 2 834 695 |

1.1.2 Roman numerals

These were used throughout the Roman Empire more than 2500 years ago, and have been used ever since. The table shows the symbols used today as Roman numerals and what their values are.

Symbol	I	V	X	L	C	D	M
Value	1	5	10	50	100	500	1000

Roman numerals are still used today. You may see them on buildings, on some clock faces and on some books.

You can combine the Roman numerals I, V, X, L, C and M in different ways to make different numbers.

When a numeral is written on the right of a numeral that has the same or greater value, you add the values together. For example:

XI = X + I = 10 + 1 = 11   VI = V + I = 5 + 1 = 6   LX = L + X = 50 + 10 = 60

When a numeral is written on the left of a numeral that has greater value, you subtract the lesser value from the greater. For example:

XL = L − X = 50 − 10 = 40   IX = X − I = 10 − 1 = 9   XC = C − X = 100 − 10 = 90

Roman numerals appear on old buildings in Italy

A bar placed over a Roman numeral shows that that number is multiplied by 1000. For example:

$\overline{IV}$ = 4000 $\overline{L}$ = 50 000 $\overline{M}$ = 1 000 000

Tip

When using Roman numerals, try to use as few numerals as possible to show a number. For example, for 60, write LX rather than XXXXXX.

However, when there's a choice between adding or subtracting, you usually add. For example, for 1700, write MDCC rather than MCCCM.

Example 1

Write these as Roman numerals.

(a) 8   (b) 400

(a) 8 is more than 5 and less than 10.

$8 = 5 + 3$

$= V + III$ (Since 5 = V and 3 = III)

$= VIII$ (You are adding so put the numerals with the lesser value on the right of the greater)

$8 = VIII$

(b) 400 is more than 100 and less than 500.

$400 = 500 - 100$

$= D - C$ (Since 500 = D and 100 = C)

$= CD$ (You are subtracting so put the numerals with the lesser value on the left of the greater)

$400 = CD$

Example 2

Write these in Hindu–Arabic numerals.

(a) XIV   (b) XLV

(a) Using the rule 'lesser numeral to the left of the greater numeral means subtract', work out IV first:

$IV = 5 - 1$ (Since I = 1 and V = 5)

$= 4$

Now using the rule 'lesser numeral to the right of the greater numeral means add', add:

$XIV = X + IV$

$= 10 + 4$

$= 14$

$XIV = 14$

(b) Using both rules, subtract 10 from 50 and add 5:

$XLV = 50 - 10 + 5$

$= 45$

$XLV = 45$

1.1.3 Brahmi numerals

These symbols were used in India nearly 2000 years ago. At that time, this system had no symbol for zero and it had different symbols for 10, 20, 30, etc. and 100, 200, 300, etc. The table shows the first nine symbols and their values.

Symbol	—	=	≡	Ψ	⊢	ϐ	?	ς	?
Value	1	2	3	4	5	6	7	8	9

1.1.4 Modern Arabic numerals

These are also called Eastern Arabic numerals. They are used today in Egypt, Sudan, Saudi Arabia, Oman and other countries in the Mashriq region of the world. The table shows the numerals and their values.

Symbol	٠	١	٢	٣	٤	٥	٦	٧	٨	٩
Value	0	1	2	3	4	5	6	7	8	9

Numbers are made from the numerals in the same way as in the decimal number system. For example:

٤٧ = 47

Number plates on cars in Abu Dhabi have numerals in Arabic and modern Hindu–Arabic styles

1.1.5 Ancient Egyptian hieroglyphic numerals

5000 years ago, the ancient Egyptians used pictures called hieroglyphs to represent numbers. The table shows the numerals and their values.

Symbol	𓏤	𓎆	𓍢	𓆼	𓂭	𓆐	𓁨
Value	1	10	100	1000	10 000	100 000	1 000 000

The ancient Egyptians repeated the numerals to make numbers. For example:

$= (5 \times 1) + (4 \times 10) + (2 \times 100) + (5 \times 1000) + (3 \times 10\ 000) + (2 \times 100\ 000)$

$= 5 + 40 + 200 + 5000 + 30\ 000 + 200\ 000$

$= 235\ 245$

Example 3

(a) Express the number 68 in modern Arabic numerals.

(b) This number is shown in ancient Egyptian hieroglyphic numerals.

Use Hindu–Arabic numerals to show the value of this number.

(a) 68 = ٦٨

(b) $= (4 \times 1) + (3 \times 10) + (0 \times 100) + (6 \times 1000) + (5 \times 10\ 000)$
$+ (1 \times 100\ 000)$

$= 4 + 30 + 6000 + 50\ 000 + 100\ 000$

$= 156\ 034$

$= 156\ 034$

---

1 Number theory and computation

Exercise 1A

1 Write these in Roman numerals.

    a 12   **b** 83

2 Write these in Hindu–Arabic numerals.

    a XIX   **b** XCV

3 Use Brahmi numerals to show the number 8.

4 Use modern Arabic numerals to show the number 145.

5 Use ancient Egyptian hieroglyphic numerals to show the number 890.

6 Write these in Roman numerals.

    a 29   **b** 150

7 Write these in Hindu–Arabic numerals.

    a CIII   **b** CXL

8 Use modern Arabic numerals to show 8731.

9 How will you write the year 2018 using Roman numerals?

10 Express the following Roman numerals in Hindu–Arabic numerals.

    a CLXXXVI   **b** MMDCX

11 This number is shown in modern Arabic numerals.

Express the number in:

    a Hindu–Arabic numerals

    b Roman numerals

    c ancient Egyptian hieroglyphic numerals.

1.2 Classification of numbers

There are different types of number.

Natural or counting numbers ($\mathbb{N}$) are the numbers: 1, 2, 3, 4, etc.

Whole numbers ($\mathbb{W}$) are the numbers: 0, 1, 2, 3, 4, 5, etc.

Whole numbers are therefore represented by 0 and all the natural numbers.

Integers ($\mathbb{Z}$) are all the positive and negative natural numbers and 0 or the negative and positive whole numbers.

Examples of integers: –5, –4, –3, –2, –1, 0, 1, 2, 3, 4, 5, 6, etc.

Rational numbers (ℚ) are the numbers that can be written as fractions, $\{\frac{p}{q}: p$ and q are integers, $q \neq 0\}$. These are:

- integers
- negative and positive fractions
- negative and positive decimals (terminating and recurring decimals).

Examples of rational numbers are: $–3, 0, 7, \frac{1}{2}, –\frac{3}{4}, 0.87, –0.5, 4$.

Irrational numbers (P) are numbers that cannot be written as fractions or terminating or recurring decimals. They can be expressed as decimals with an infinite number of digits to the right of the decimal point.

Examples of irrational numbers: $\pi, \sqrt{2}$ ($\pi = 3.141592654\ldots$; $\sqrt{2} = 1.414213562\ldots$)

Real numbers (ℝ) are all the rational and irrational numbers.

Examples of real numbers are: $\sqrt{9}, –\sqrt{3}, \frac{1}{4}, 0, 12$.

Imaginary numbers (A) are multiples of the square root of –1 and expressions that include the square root of –1.

Examples of imaginary numbers are: $\sqrt{-1}, \sqrt{-4}, 2\sqrt{-1}$.

1.2.1 Sets of numbers

Any group or collection of objects of the same kind is known as a set. (There is more about sets in Chapter 6.)

Use capital letters to name sets and curly brackets to show the members of the sets. You can show the relationships between sets in rectangles and circles called **Venn diagrams.** (Use rectangles for larger sets or the universal set and circles for smaller sets or subsets.)

You can use **set notation** to show the different types of numbers described above.

Set of natural numbers: (ℕ) = {1, 2, 3, 4, …}

Set of whole numbers: (𝕎) = {0, 1, 2, 3, 4, …}

Set of integers: (ℤ) = {–4, –3, –2, –1, 0, 1, 2, 3, 4, …}

Set of rational numbers: (ℚ) = $\{–3, 0, 7, \frac{1}{2}, –\frac{3}{4}, 0.87, –0.5, \sqrt{4}, \ldots\}$

Set of irrational numbers (P) = $\{\pi, \sqrt{2}, \sqrt{3}, \sqrt{5}, \ldots\}$

Set of real numbers (rational and irrational numbers) (ℝ) = $\{\sqrt{9}, \sqrt{3}, \frac{1}{4}, 0, 12, –4, \ldots\}$

Set of imaginary numbers (A) = $\{\sqrt{-4}, \sqrt{-1}, 2\sqrt{-1}, \ldots\}$

Any set that is included in another set is called a **subset.**

The symbol ⊂ (is contained in) is used to show subsets.

Based on the definitions above:

ℕ ⊂ 𝕎 ⊂ ℤ ⊂ ℚ ⊂ ℝ

Also $P \subset ℝ$ but $A \not\subset ℝ$

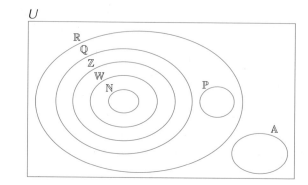

The above information can also be represented by a Venn diagram (right). The universal set is the set of real numbers.

Other sets of numbers are defined according to their properties.

Other types of numbers include:

A rectangular number is a number that can be represented by a number of objects or a series of dots arranged in the shape of a rectangle.

Examples of rectangle numbers: $6 = 2 \times 3$ and $8 = 4 \times 2$

A triangular number is a number that can be represented by a number of objects or a series of dots arranged in the shape of an equilateral triangle.

The first triangular number is 1 followed by 3, 6, 10 and 15. Each successive number is found by adding another row with one extra symbol.

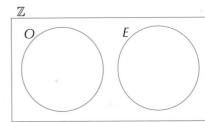

Prime numbers are numbers divisible only by themselves or 1. Examples of prime numbers are: 2, 3, 5, 7.

Composite numbers are numbers that are divisible by numbers other than themselves and 1.

Note	2 is the smallest and only even prime number.

Even numbers are integers that are exactly divisible by 2 (that is, with no remainder). Examples of even numbers are: 0, 2, 4, 6, –8. An even number can be negative or positive.

Odd numbers are all the integers that are not exactly divisible by 2. Examples are: –7, 1, 3, 5, 7.

These other types of numbers can also be represented using set notation.

The set of even numbers: $E = \{2, 4, 6, 8, \ldots\}$

The set of odd numbers: $O = \{1, 3, 5, 7, \ldots\}$

A Venn diagram can be used to represent odd and even numbers, with the universal set being integers.

1.2.2 Closure

A set of numbers is **closed** under an operation if, when the operation is carried out on any two numbers from the set, the result is also a member of the set. This is called the **law of closure**.

For example:

- the sum of 2 and 3 from the set of whole numbers is 5, which is a whole number
- the product of 5 and 7 equals 35, which is a whole number.

Similar results are produced for any two whole numbers. Hence, the set of whole numbers is closed under addition and multiplication.

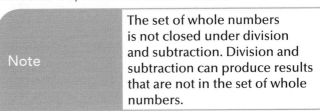

Note	The set of whole numbers is not closed under division and subtraction. Division and subtraction can produce results that are not in the set of whole numbers.

Prove whether or not the set of whole numbers is closed under division.

Take two whole numbers, e.g. 5 and 4, and divide 5 by 4.

$5 \div 4 = 1.25$

1.25 is not a whole number.

Therefore, the set of whole numbers is not closed under division.

Prove whether or not the set of odd numbers is closed under subtraction.

Consider the two odd numbers 3 and 7.

$7 - 3 = 4$

4 is an even number.

This means that the set of odd numbers is not closed under subtraction.

1.3 Ordering a set of real numbers

You can order a set of real numbers by arranging them from the smallest to the greatest (ascending order) or from the greatest to the smallest.

Order this set of numbers in ascending order: 5, –3, 2, 18, 7, 4.

–3, 2, 4, 5, 7, 18

Write these numbers in descending order: 1.4, $1\frac{1}{2}$, 3, 0.9

3, $1\frac{1}{2}$, 1.4, 0.9

Exercise 1B

1 Give one example of each type of number.

    a natural number   **b** whole number   **c** integer   **d** rational number

    e irrational number   **f** real number   **g** imaginary number

2 Write down all the natural numbers between –2 and 10.

3 Write down all the whole numbers from 0 to 10.

4 Write down all the integers from –10 to 5.

5 Order these numbers in ascending order: 0.001, $\frac{2}{100}$, 10^{-2}, $100^{\frac{1}{2}}$, 1.000 001

6 Arrange in descending order: $\sqrt{1000}$, $\sqrt[3]{3^3}$, 27 000, 3(10.2), $\frac{92}{3}$

7 Look at this set of numbers: -1, $\sqrt{9}$, 1.5, $1\frac{3}{4}$, $\sqrt{\frac{1}{4}}$, $\sqrt{3}$, 100, $\sqrt{-4}$, 0, -0.45, $\frac{14}{3}$

Write down all the:

a integers   b natural numbers   c real numbers

d irrational numbers   e imaginary numbers.

8 Name the integers between 1.7 and 3.2.

9 How many whole numbers are there from 0 to 100?

10 Show that the set of prime numbers is not closed under the operations subtraction and multiplication. Choose your own numbers as examples.

11 Represent the set of prime numbers and whole numbers from 0 to 20 in a Venn diagram.

12 Which of these sets of numbers is the largest?

a whole numbers   b natural numbers   c integers   d real numbers

13 Is there a smallest integer and a largest integer? If so, what are they? If not, why not?

14 Is there a smallest whole number and a largest whole number? If so, what are they? If not, why not?

15 Name all the triangular numbers between 1 and 20 and sketch each of them.

16 Write down all the rectangular numbers between 1 and 20 and sketch each of them.

17 a For the natural numbers between 1 and 20, write down the largest

     i prime number   ii composite number

     iii odd number   iv even number.

b Calculate the sum of your answers to **a**.

c Calculate the product of your answers to **a**.

1.4 Place value of whole numbers

The decimal system uses ten digits that are the Hindu–Arabic numerals: 0, 1, 2, 3, 4, 5, 6, 7, 8, 9. You can combine the digits to make different numbers. The position of a digit in a number shows its value.

Each position in a number has a specific name as shown in the place-value table below.

Ten thousands	Thousands	Hundreds	Tens	Units
5	9	2	1	7

The number shown in the place-value table is: 59 217.

It is made of:

5 ten thousands	$= 5 \times 10\,000$	$= 50\,000$
9 thousands	$= 9 \times 1000$	$= 9000$
2 hundreds	$= 2 \times 100$	$= 200$
1 ten	$= 1 \times 10$	$= 10$
7 units	$= 7 \times 1$	$= 7$
Adding these values		$= 59\,217$

Each digit in the number above has two values: a **place value** and a **face value**.

You can use the place values of digits to write 59 217 in **expanded form**.

In expanded form, $59\,217 = 50\,000 + 9000 + 200 + 10 + 7$.

> **Key terms**
>
> **Face value of a digit**   the actual value of a digit in a number
>
> **Place value of a digit**   given by its position in a number (units, tens, hundreds, etc.)

> **Key term**
>
> **Expanded form**   a number written as the sum of the place values of its digits

Example 8

Given the number 59 217 (the number above), write the face value and the place value of the digits 2 and the 7.

The face value of the digit 2 is 2.

The place value of the digit 2 is $2 \times 100 = 200$.

The face value of the digit 7 is 7.

The place value of the digit 7 is $7 \times 1 = 7$.

Any digit in a number can be written in terms of its place value and face value.

Example 9

Given the number 901 876, write down the place value of the digits 9, 0 and 1.

Place value of the digit $9 = 9 \times 100\,000$

$= 900\,000$

Place value of the $0 = 0 \times 10\,000$

$= 0$

Place value of the $1 = 1 \times 1000$

$= 1000$

Exercise 1C

1   What is the face value of the digit 3 in the number 53?

2   What is the place value of the digit 6 in the number 26?

3   Write down the face value and place value of the digit 3 in the whole number 4382.

4   Write down the face value and place value of the digit 9 in the number 19 013.

5   In the numbers 32 456 and 91 078, which digit has the greater place value: 4 or 7?

6 In the whole numbers 235 and 789, which number has the digit with the smallest place value? State the digit and its place value.

7 Given the numbers 9021 and 8734, write:

    a the digits with the highest and lowest place values

    b the sum of the digits in your answers to **a**.

8 Write down the place value of the digit 4 in the number 948 731 260.

9 Draw a place value table and write in the numbers 1 021 508 and 9 723 456.

10 Write the number 987 000 213 in expanded form.

11 Given two numbers 123 456 and 789 321, calculate the product of the place value of the digit 9 and the face value of the digit 4.

12 Given the number 3782, calculate the sum of the face value of each digit.

13 Given the number 8079, calculate the product of the place value of each digit.

1.5 Revision: Addition, subtraction, multiplication and division

The four basic operations used for calculation are addition, subtraction, multiplication and division.

1.5.1 Addition (sum)

<table>
<tr><td>

Example 10

What is the sum of 3, 7 and 9?

$3 + 7 + 9 = 19$

The sum of 3, 7 and 9 is 19.

</td><td>

Key terms

Addition   combining two or more numbers or quantities

Sum   result of an addition

Total   another word for the result of an addition

The symbol **+**   the plus sign shows addition

</td></tr>
</table>

1.5.2 Subtraction (difference)

<table>
<tr><td>

Example 11

What is $13 - 8$?

$13 - 8$ means 8 is removed from a total of 13.

$13 - 8 = 5$

</td><td>

Key terms

Subtraction   taking one number or quantity from another

Difference between   the numerical difference between two numbers is found by subtracting one number from the other or counting up from the smaller number to the larger

The symbol **−**   the minus symbol shows subtraction

</td></tr>
</table>

Note   You can use the result of Example 11 to say that the difference between 13 and 8 is 5.

1.5.3 Multiplication (product)

Example 12

What is 3×5?

$3 \times 5 = 5 + 5 + 5$

$\qquad = 15$

Alternatively:

$5 \times 3 = 3 + 3 + 3 + 3 + 3$

$\qquad = 15$

If you know your multiplication tables, you will know that $5 \times 3 = 15$.

Key terms

Multiplication   the same as repeatedly adding the same number or quantity

Product   the result of multiplication

The symbol ×   the multiplication symbol shows multiplication

Multiplication tables   tables that organise multiplication facts, also called times tables

1.5.4 Division (quotient)

Example 13

What is $12 \div 4$?

$12 \div 4 = 3$

Notes

$12 \div 4 = 3$ means:

- there are three fours in twelve
- when 12 items are grouped equally in 4 groups, there are 3 items in each group.

Key terms

Division   sharing or grouping numbers or items equally so that each share or group contains the same number

The symbol ÷   the division symbol shows division

Quotient   the result of division

Dividend   the number being divided in a calculation. For example, in $56 \div 8 = 7$, the dividend is 56.

Divisor   the number dividing into another in a calculation. For example, in $108 \div 12 = 9$, the divisor is 12.

Example 14

Share 9 mangoes equally among 3 people.

Write this word problem as a division calculation:

$9 \div 3 = 3$

Each person receives 3 mangoes.

Example 15

How many groups of 3 marbles can be made from 24 marbles?

$24 \div 3 = 8$

1.5.5 Operations with whole numbers

The basic operations of addition, subtraction, multiplication and division can be used with place value to solve problems involving whole numbers.

Example 16

Work out 987 + 864.

```
        9 8 7
987 + 864:   + 8 6 4
        ─────────
```

```
        9 8 7
      + 8 6 4
        ─────────
              1
              1
```

Start with the units.

4 units + 7 units = 11 units

11 units = 1 ten + 1 unit

Write 1 in the units place of the answer.

Write 1 under the tens place of the answer.

```
        9 8 7
      + 8 6 4
        ─────────
            5 1
          1   1
```

1 ten + 6 tens + 8 tens = 15 tens

15 tens = 1 hundred + 5 tens

Write 5 in the tens place of the answer.

Write 1 under the hundreds place of the answer.

```
        9 8 7
      + 8 6 4
        ─────────
          8 5 1
        1 1   1
```

1 hundred + 8 hundreds + 9 hundreds = 18 hundreds

18 hundreds = 1 thousand + 8 hundreds

Write 8 in the hundreds place of the answer.

Write 1 under the thousands place of the answer.

```
        9 8 7
      + 8 6 4
        ─────────
        1 8 5 1
        1 1   1
```

1 thousand + 0 thousands = 1 thousand

Write 1 in the thousands place of the answer.

987 + 864 = 1851

Example 17

Work out 1234 − 987.

```
        1 2 3 4
1234 − 987:   −   9 8 7
        ─────────────
```

Start with the units: 4 − 7

Because 7 is greater than 4, take 1 ten from the 3 tens in the tens column.

3 tens − 1 ten = 2 tens

```
        1 2 ²3̸ ¹4
      −   9 8 7
        ─────────────
                  7
```

Cross out the 3 tens and write 2 tens.

Change the 1 ten to 10 units.

1 ten + 4 units = 14 units

14 units − 7 units = 7 units

Write 7 in the units place in the answer.

Example 17 continues on next page

The tens column now contains 2 tens – 8 tens.

Because 8 is greater than 2, take 1 hundred from the 2 hundreds in the hundreds column.

$$1\ {}^1\cancel{2}\ {}^1\cancel{3}\ {}^2\ 4$$
$$-\quad\ 9\ 8\ 7$$
$$\overline{\qquad\quad 4\ 7}$$

2 hundreds – 1 hundred = 1 hundred

Cross out the 2 hundreds and write 1 hundred.

Change the 1 hundred to 10 tens.

10 tens + 2 tens = 12 tens

12 tens – 8 tens = 4 tens

Write 4 in the tens place in the answer.

The hundreds column now contains 1 hundred – 9 hundreds.

Because 9 is greater than 1, take 1 thousand from the 1 thousand in the thousands column.

$${}^0\cancel{1}\ {}^1\cancel{2}\ {}^1\cancel{3}\ {}^2\ 4$$
$$-\quad\ 9\ 8\ 7$$
$$\overline{\qquad\ 2\ 4\ 7}$$

1 thousand – 1 thousand = 0

Cross out the 1 thousand.

Change the 1 thousand to 10 hundreds.

10 hundreds + 1 hundred = 11 hundreds

11 hundreds – 9 hundreds = 2 hundreds

Write 2 in the hundreds place in the answer.

1234 – 987 = 247

Example 18

Work out 432 ÷ 24.

432 ÷ 24:   24⟌432

Divide 24 into each part of 432.

Try to divide 24 into 4.

This doesn't give a whole number answer so write 0 above the 4.

$$\begin{array}{r} 0\ 1 \\ 24\overline{)4\ 3\ 2} \\ -\ 2\ 4 \\ \hline 1\ 9 \end{array}$$

$1 \times 24 = 24$

Try to divide 24 into 43.

There is one 24 in 43.

Write 1 above the 3.

43 – 24 = 19

$$\begin{array}{r} 0\ 1\ 8 \\ 24\overline{)4\ 3\ 2} \\ -\ 2\ 4 \\ \hline 1\ 9\ 2 \\ -\ 1\ 9\ 2 \\ \hline 0 \end{array}$$

$1 \times 24 = 24$

$8 \times 24 = 192$

Bring the 2 down and write it next to the 19.

Try to divide 24 into 192.

$24 \times 8 = 192$

Write 8 above the 2.

192 – 192 = 0

432 ÷ 24 = 18

The method used here is called **long division**.

Example 19

Work out 389 × 75.

389 × 75:

```
    3 8 9
×     7 5
```

```
    3 8 9
×     7 5
─────────
     4  5
```

Start with the units digit of 75, which is 5.

Multiply 389 by 5.

First, multiply the 9 units of 389 by 5.

$5 \times 9 = 45$

Using place value, 45 units = 4 tens + 5 units.

Write 5 in the units place of the answer for 389 × 5.

Write 4 under the tens place of the answer for 389 × 5.

```
    3 8 9
×     7 5
─────────
    4 4 5
    4
```

Next, multiply the 8 tens of 389 by 5.

5×8 tens = 40 tens

Using place value, 40 tens = 4 hundreds + 0 tens.

0 tens + 4 tens = 4 tens

Write 4 in the tens place of the answer for 389 × 5.

Write 4 under the hundreds place of the answer for 389 × 5.

```
    3 8 9
×     7 5
─────────
  1 9 4 5
    4 4
```

Now multiply the 3 hundreds of 389 by 5.

5×3 hundreds = 15 hundreds

Using place value, 15 hundreds = 1 thousand + 5 hundreds.

5 hundreds + 4 hundreds = 9 hundreds

Write 9 in the hundreds place of the answer for 389 × 5.

Write 1 in the thousands place of the answer for 389 × 5.

```
      3 8 9
×       7 5
───────────
  1 9 4 5
    4 4
2 7 2 3 0
  6 6
```

Follow a similar procedure to multiply 389 by the tens digit of 75, which is 7.

However, because the place value of 7 is 70, first write 0 in the units place of the answer for 389 × 70.

```
      3 8 9
×       7 5
───────────
  1 9 4 5
    4 4
2 7 2 3 0
  6 6
───────────
2 9 1 7 5
  1
```

Finally, use place value to add both answers together.

389 × 75 = 29 175

The method used here is called **long multiplication**.

Exercise 1D

1 Given the numbers 8 and 3, calculate their:

    a sum   **b** difference   **c** product   **d** quotient.

2 40 mangoes are shared equally among 8 children. How many does each child receive?

3 Using place value, perform the following calculations.

    a 235 + 987   **b** 1234 – 999   **c** 345 × 87   **d** 20 648 ÷ 232

4 Solve the following problems.

    a 1876 + 887   **b** 98 723 – 68 759   **c** 4486 × 378   **d** 238 938 ÷ 5689

5 A man working for $8698.00 per month receives an increase of $1345.00 on his salary. What is his new monthly salary?

6 A total of 12 345 people were in a stadium watching a football match. 5679 of these were females. How many males were there?

7 A pharmacist was sorting tablets into bottles. She placed 28 tablets in each bottle. 1204 tablets were placed in bottles. How many bottles did she use?

8 Four top cricketers scored these numbers of runs during their career: 12 345, 10 987, 8794 and 15 678. Calculate the total number of runs scored by the four players.

9 An airplane flew a distance of 12 897 km on Monday and on Tuesday it flew 13 989 km. How many more kilometres did it fly on Tuesday than on Monday?

10 A company exporting fruits packed 35 crates of oranges, 27 crates of pineapples, 46 crates of grapefruits, and 57 crates of plums. Each orange crate contained 350 oranges; each grapefruit crate 284 grapefruits; each pineapple crate contained 178 pineapples; and each plum crate contained 784 plums. Find the total number of fruits exported.

11 Complete these calculations. Take care with place value.

    a 78 977 + 9989   **b** 100 000 – 98 721   **c** 57 896 × 6798   **d** 1 156 926 ÷ 9483

12 The populations of five countries are: 120 000; 234 567; 93 756; 333 459 and 186 987. What is the total population of the five countries?

13 In a census taken in a country with a population of 2 845 789, it was found that there were 783 942 men and 1 236 954 children. How many of the population were women?

14 A container contains 100 large cartons and each carton contains 10 small boxes containing 100 marbles. How many marbles are in the container?

15 The budgets of five Caribbean towns are as follows: $2 876 934; $5 326 657; $3 456 129; $7 890 567 and $10 009 288. What is the total amount of money in the budgets of these five towns?

16 A cup of sand contains about 2 million grains. One cup of sand can fill 16 tablespoons. How many grains can fill 11 tablespoons?

1.6 Defined arithmetic operations

In addition to the four basic arithmetic operations (addition, subtraction, multiplication and division) other operations can be created as a combination of these basic operations. These operations are known as **defined arithmetic operations**.

Example 20

Calculate $2 \triangle 3$ where $\triangle$ means double the first number and subtract the second.

$2 \triangle 3 = (2 \times 2) - 3$
$\qquad = 4 - 3$
$\qquad = 1$
$2 \triangle 3 = 1$

Example 21

Given that $x \, ! \, m = 2(m + x) - 3x + 2m$, find $3 \, ! \, 4$

$3 \, ! \, 4$ means $x = 3$ and $m = 4$
Therefore, $3 \, ! \, 4 = 2(4 + 3) - (3 \times 3) + (2 \times 4)$
$\qquad\qquad = (2 \times 7) - 9 + 8$
$\qquad\qquad = 14 - 9 + 8$
$\qquad\qquad = 13$
$\qquad 3 \, ! \, 4 = 13$

1.7 Associativity, commutativity and distributivity

1.7.1 Associative law

The associative law states that for the operations addition and multiplication, the way the numbers are grouped does not affect the answer.

For example, for addition:
$\qquad (4 + 6) + 8 = 10 + 8 = 18$
$\qquad 4 + (6 + 8) = 4 + 14 = 18$
$\qquad (4 + 6 + 8) = 18$

For example, for multiplication:
$\qquad (5 \times 6) \times 4 = 30 \times 4 = 120$
$\qquad 5 \times (6 \times 4) = 5 \times 24 = 120$
$\qquad (5 \times 6 \times 4) = 120$

The associative law does not apply for division and subtraction.

For example, for subtraction:
$\qquad (8 - 4) - 2 = 4 - 2 = 2$
$\qquad 8 - (4 - 2) = 8 - 2 = 6$

For example, for division:
$\qquad (72 \div 12) \div 3 = 6 \div 3 = 2$
$\qquad 72 \div (12 \div 3) = 72 \div 4 = 18$

1.7.2 Commutative law

The commutative law states that the order in which numbers are added or multiplied does not affect their total or product.

For example, for addition:

$$2 + 3 + 9 = 14$$
$$3 + 2 + 9 = 14$$
$$9 + 3 + 2 = 14$$
$$9 + 2 + 3 = 14$$

For example, for multiplication:

$$4 \times 5 \times 3 = 60$$
$$5 \times 4 \times 3 = 60$$
$$3 \times 5 \times 4 = 60$$

The commutative law does not apply for subtraction and division.

For example, for subtraction:

$$7 - 1 = 6$$
$$1 - 7 = -6$$

For example, for division:

$$12 \div 4 = 3$$
$$4 \div 12 = \frac{1}{3}$$

Both the associative and commutative laws can be used to simplify calculations involving addition and multiplication.

Example 22

Work these out.

(a) $23 + 50 + 47 =$

(b) $17 \times 20 \times 5 =$

(a) $23 + 50 + 47 = (23 + 47) + 50$
$\qquad\qquad\qquad = 70 + 50$
$\qquad\qquad\qquad = 120$

(b) $17 \times 20 \times 5 = (5 \times 20) \times 17$
$\qquad\qquad\qquad = 100 \times 17$
$\qquad\qquad\qquad = 1700$

1.7.3 Distributive law

The distributive law states that multiplication is distributive over addition and subtraction.

This means that for a set of numbers p, q and r:

$p(q + r) = pq + pr$ and $p(q - r) = pq - pr$

Example 23

Work out $5(2 + 3)$.

$5(2 + 3) = (5 \times 2) + (5 \times 3)$
$\qquad\quad = 10 + 15$
$\qquad\quad = 25$

Example 24

Work out $3(8 - 5)$.

$3(8 - 5) = (3 \times 8) - (3 \times 5)$
$\qquad\quad = 24 - 15$
$\qquad\quad = 9$

1.8 Identity for addition and multiplication

1.8.1 Additive identity

When 0 is added to any number, the answer is always the number you started with.

0 is called the **additive identity** because the identity of the number remains unchanged when 0 is added.

Example 25

Work these out.

(a) $0 + 12 =$

(b) $2 + 0 =$

(c) $-8 + 0 =$

(a) $0 + 12 = 12$

(b) $2 + 0 = 2$

(c) $-8 + 0 = -8$

1.8.2 Multiplicative identity

When any number is multiplied by 1, the product is the original number.

The number 1 is called the **multiplicative identity** because the number remains unchanged when multiplied by 1.

Example 26

Work these out.

(a) $2 \times 1 =$

(b) $1 \times -3 =$

(a) $2 \times 1 = 2$

(b) $1 \times -3 = -3$

1.9 Additive and multiplicative inverses

1.9.1 Additive inverse

The inverse of a number under addition combines with the number to give the identity for addition, which is 0.

Key term

Inverse of a number   under a particular operation, the inverse combines with the number to give the identity for that operation.

Example 27

What is the additive inverse of 4?

The identity for addition is 0.

The number that has to be added to 4 to give 0 is -4:

$4 + -4 = 4 - 4$

$\qquad = 0$

Note   Similarly, the additive inverse of -4 is 4: $-4 + 4 = 0$

1.9.2 Multiplicative inverse

The inverse of a number under multiplication combines with the number to give the identity for multiplication, which is 1.

What is the multiplicative inverse of each of these?

(a) 2

(b) −3

(a) Since $2 \times \frac{1}{2} = 1$, the multiplicative inverse of 2 is $\frac{1}{2}$.

(b) Since $-3 \times -\frac{1}{3} = 1$, the multiplicative inverse of −3 is $-\frac{1}{3}$.

1.10 Multiplication and division involving zero

1.10.1 Multiplication by zero

Zero multiplied by any number equals zero.

Work these out.

(a) $5 \times 0 =$

(b) $0 \times -7 =$

(a) $5 \times 0 = 0$

(b) $0 \times -7 = 0$

1.10.2 Division by zero

If you divide any number by progressively smaller and smaller values the result becomes larger and larger. Remember that division may be compared to repeated subtraction, and no matter how many times you subtract zero from a number, the result will be unchanged. So any number divided by zero results in a value so large that it is undefined. The value 'tends to infinity'.

Work these out.

(a) $10 \div 0 =$

(b) $-8 \div 0 =$

(a) $10 \div 0 = \infty$

(b) $-8 \div 0 = \infty$

1.10.3 Zero divided by a number

Zero divided by any number equals zero.

Work these out.

(a) $0 \div 100 =$

(b) $0 \div -1 =$

(a) $0 \div 100 = 0$

(b) $0 \div -1 = 0$

1.11 Order of arithmetic operations (BODMAS)

When a calculation contains more than one operation, perform the operations in a particular order. The order of the operations is known as BODMAS because you perform the operations in that order:

B brackets

O orders/of (powers of numbers)

D division

M multiplication

A addition

S subtraction

Solve: $3(4 - 1) + 4^2 - 8 \div 2 \times 7$

$$3(4 - 1) + 4^2 - 8 \div 2 \times 7 = 3 \times 3 + 4^2 - 8 \div 2 \times 7 \quad \text{(First, work out what's inside the brackets.)}$$
$$= 3 \times 3 + 16 - 8 \div 2 \times 7 \quad \text{(Work out the power.)}$$
$$= 3 \times 3 + 16 - 4 \times 7 \quad \text{(Divide.)}$$
$$= 9 + 16 - 28 \quad \text{(Multiply.)}$$
$$= 25 - 28 \quad \text{(Add.)}$$
$$= -3 \quad \text{(Subtract.)}$$

$3(4 - 1) + 4^2 - 8 \div 2 \times 7 = -3$

Exercise 1E

1 Solve the following, and write the laws or identities you have used.

   a $100 + 0$ **b** 40×1

2 Find the multiplicative inverse of $\frac{3}{4}$.

3 Solve the following.

   a $25 \times 3 \times 0$ **b** $2 + 8 \div 3 \times 0$ **c** $3 + 2 - 5 \div 9 \times 11$

4 Find the additive inverse of −12.

5 Solve the following problem: $2(5 - 1) + 8 \div 4 - 3 \times 1$

6 Calculate 2 # 5, where # means: double the first number and add the result to the triple of the second number.

7 Using the associative law, show all the different arrangements for carrying out the following calculations and give the answers.

   a $3 + 5 + 17 + 10$ **b** $3 \times 6 \times (-2) \times 1$

8 Show why the commutative law does not apply to these calculations.

   a $20 \div 5$ **b** $15 - 6$

9 Given that $L * M = 2L \div L^2M$, find 4 * 3. Give the answer in its simplest form.

10 Simplify: $-3 - 2(5 - 1) + 3 \times 4 - 1 \times 0$

11 Using the distributive law, solve:

    a $3(5-4)+6(8-7)$ **b** $7(9-12)-(2-8)$

12 Given that $a \# b = 2a - 3b + ab^2$ and $p * q = 3(p - q)$, simplify $(3 \# 5 \# 7) \div (2 * 4 * 6)$.

13 Using the distributive law, solve the following. Express your answer in its lowest term:

$$\frac{3(5-8)-(9-2)\times -2+8}{17(1-3)+8\div(16-2\times 9)}$$

1.12 Powers of numbers

Any number multiplied by itself once or more times can be written using **power notation** (also called **index notation**). The power or index indicates how many 'lots' of the base are multiplied together.

power or index

$$n^a$$

base

For example: $3 \times 3 = 3^2$ where 3 is the base and 2 is the power.

1.12.1 Square numbers

The square of a number is the number multiplied by itself.

Examples of square numbers are: $2 \times 2 = 4$, $3 \times 3 = 9$, $4 \times 4 = 16$, $5 \times 5 = 25$, etc.

> **Key term**
>
> **Square number**   number of the form $n \times n$ where n is any whole number

If you use a set of objects to represent a square number, they can be arranged in a square.

Example 33

These footballs have been arranged in a square. What square number does this represent?

$4 = 2 \times 2$

They represent the square number 4.

Example 34

These flowers have been arranged in a square. What square number does this represent?

$9 = 3 \times 3$

They represent the square number 9.

Example 35

What is the square of 10?

$10 \times 10 = 100$

The square of 10 is 100.

1.12.2 Square root

The square root of a number x is that number which when multiplied by itself gives x.

Example 36

What is the square root of 64?

When 8 is multiplied by itself (8×8) it gives 64.

Therefore, the square root of 64 equals 8.

Also $-8 \times -8 = 64$

Therefore, $\pm\sqrt{64} = \pm 8$

You can also write $64^{\frac{1}{2}} = \pm 8$

Key terms

The symbol $\sqrt{n}$ shows the positive square root of the number n within the symbol. For example, $\sqrt{112}$ is the positive square root of 112.

The symbols $n^{\frac{1}{2}}$ also show the square root of the number n. The power $\frac{1}{2}$ shows the square root. For example, $49^{\frac{1}{2}}$ is the square root of 49.

Example 37

What is the value of $\pm\sqrt{36}$?

Since $6 \times 6 = 36$ and $(-6) \times (-6) = 36$ then:

$\pm\sqrt{36} = \pm 6$

1.12.3 Other powers of numbers

Powers can be whole numbers or fractions.

Example 38

What is the value of 5^4?

$5^4 = 5 \times 5 \times 5 \times 5$

$= 625$

Example 39

What is the value of $8^{\frac{1}{3}}$?

$8^{\frac{1}{3}}$ is the cube root of 8. This is the number a such that:

$a \times a \times a = 8$

$2 \times 2 \times 2 = 8$

Therefore, the cube root of 8 equals 2.

$8^{\frac{1}{3}} = 2$

Notes

You can also write $\sqrt[3]{8} = 2$.

The small 3 shows that it is the cube root.

In the same way, $n^{\frac{1}{5}} = \sqrt[5]{n}$ is the fifth root of n. This is the number a where $a^5 = n$. This means that $a \times a \times a \times a \times a = n$.

The symbol $\sqrt{}$ is called a radical and you can use it with different numbers to show different roots of numbers.

- Any number raised to the power of 1 is the number itself.
- Any number raised to the power of 0 is 1.

Example 40

What is the value of 8^1?

$8^1 = 8$

Example 41

What is the value of 99^0?

$(99)^0 = 1$

Example 42

What is the value of $1 \times 1 \times 1 \times 2 \times 2 \times 2 \times 3 \times 3 \times 4 \times 4$?

$1 \times 1 \times 1 \times 2 \times 2 \times 2 \times 3 \times 3 \times 4 \times 4 = 1^3 \times 2^3 \times 3^2 \times 4^2$

$\qquad\qquad = 1 \times 8 \times 9 \times 16$

$\qquad\qquad = 1152$

1.13 Reciprocal of a number

The reciprocal of a number n is 1 divided by the number.

You can write this as: $\frac{1}{n}$

For example, the reciprocal of $4 = \frac{1}{4}$.

Since $\frac{1}{0}$ is undefined, the reciprocal is found only of non-zero numbers.

When the reciprocal of a number is multiplied by the original number, the answer is 1. Therefore, the reciprocal of a number is the same as the multiplicative inverse of a number.

Did you know?

Another way of writing the reciprocal of n is $n^{-1} = \frac{1}{n}$

$n^{-2} = \frac{1}{n^2}$

$n^{-3} = \frac{1}{n^3}$ etc.

For example, $10^{-4} = \frac{1}{10^4} = \frac{1}{10\ 000}$

Example 43

State the reciprocal of the following.

(a) 9 (b) $\frac{1}{2}$

(a) Reciprocal of $9 = \frac{1}{9}$

(b) Reciprocal of $\frac{1}{2} = \frac{2}{1}$

$\qquad\qquad\qquad = 2$

Exercise 1F

1. Evaluate: $2^3 \times 5^1 - 4^0$

2. Simplify $\sqrt{16 \times 5^2}$

3 Find the product of: the reciprocal of $\frac{3}{4}$, half the square root of 64, the square of 2, and 8 to the power of 0.

4 Write as a single power of 2:
$128 \times \sqrt[3]{8} \times (64)^{\frac{1}{2}}$

5 Express the following as a power of 10: $(1000)^0 \times (1000)^1 \times (1000)^{-1} \times (10\,000)^{\frac{1}{2}} \div (100)^{-\frac{1}{2}}$

6 Write this as a product of powers of 2 and 3: $8 \times 16 \times 36^2$

7 Solve this: the reciprocal of 4 which is then multiplied by half the square root of 64.

8 Make a sketch, using circles, to represent the following square numbers.
  a 36   **b** 49

9 Simplify: $32 \times 9 \times \sqrt[3]{91 - 64}$. Write your answer as a product of powers of 2 and 3.

10 The operation ϕ means the square root of the first number added to the cube of the second number all divided by 10 times the cube root of their product. Find the value of $9 \phi 3$.

1.14 Factors and multiples

Example 44

What are the factors of 8?

$8 \div 1 = 8$	$8 \div -1 = -8$
$8 \div 2 = 4$	$8 \div -2 = -4$
$8 \div 4 = 2$	$8 \div -4 = -2$
$8 \div 8 = 1$	$8 \div -8 = -1$

Factors of 8 are 1, –1, 2, –2, 4, –4, 8 and –8.

Key term

Factor of a number   any positive or negative integer that can divide exactly into the number (without a remainder).

Did you know?   The number 1 is a factor of all numbers.

Example 45

What is the lowest prime factor of 8?

Looking at the answers to Example 40, you can see that the lowest prime factor of 8 is 2.

Key term

Prime factor   a factor that is also a prime number.

Key term

Prime number   a number that has exactly two factors, itself and 1.

Example 46

What are the first four multiples of 3?

Multiples of 3 are: 3, 6, 9, 12, etc.

Key term

Multiple of a number   any number that the number is a factor of. For example, 4 divides exactly into 16, 72 and 100. 16, 72 and 100 are all multiples of 4.

Generally, the multiples of any number a are $a \times n$ where n is an integer.

What are the multiples of 4 between 0 and 15?

$4 \times 1 = 4$

$4 \times 2 = 8$

$4 \times 3 = 12$

The multiples of 4 between 0 and 15 are 4, 8 and 12.

Remember

- 0 can be a multiple of a number n but it cannot be a factor.

    $0 \times n = 0$

    $0 \div n = 0$

    $n \div 0$ is undefined

- Factors and multiples can be negative numbers or positive numbers.
- All numbers are factors and multiples of themselves.

1.14.1 Highest common factor (HCF)

Some factors will divide into more than one number. These factors are called **common factors**. For example:

- factors of 12: 1, 2, 3, 4, 6, 12
- factors of 18: 1, 2, 3, 6, 9

The factors common to both 12 and 18 are 1, 2, 3 and 6.

The largest of these common factors is 6.

A quicker way to find HCF is by successive division.

You can use successive division by prime factors common to both numbers to find the HCF of 12 and 18, as shown in the diagram.

Starting with the smallest prime numbers, divide by **2**, and then by **3**.

The remaining two numbers to be divided are 2 and 3. These are **prime numbers** and their common factor is 1.

The HCF is therefore given by: $2 \times 3 = 6$ (the product of the common factors of 12 and 18).

Key term

Highest common factor (HCF) of two or more numbers
the largest number that is a common factor of the numbers

2	12	18
3	6	9
1	2	3

Remember

A prime number is a number that has only itself and 1 as factors.

For example, 2, 17, 53 are prime numbers.

Note

1 is not a prime number, since the definition of a prime number says that it has exactly two factors (the number itself and 1).

Example 48

Find the HCF of 10, 15 and 20.

5	10	15	20
1	2	3	4

5 is the only common factor of 10, 15 and 20.

5 is the HCF of 10, 15 and 20.

Tip

When finding the HCF by the division method, remember that the numbers that are used to divide into the group of numbers must be factors of all the numbers in the group.

1.14.2 Lowest common multiple (LCM)

Some numbers are multiples of more than one number. These are common multiples.

Multiples of 4 are: 4, 8, **12**, 16, 20, **24**, etc.

Multiples of 6 are: 6, **12**, 18, **24**, 30, etc.

Therefore, two of the multiples common to 4 and 6 are 12 and 24.

The lowest common multiple is 12.

Key term

Lowest common multiple (LCM) of two or more numbers   the lowest number that is a multiple of each of the numbers.

A quicker method to find the LCM is to use successive division. For example, to find the LCM of 4 and 6:

2	4	6
	2	3

First, divide by 2 (the first prime number) and write the answer under each number. Then find the LCM by multiplying the bold numbers together (the prime number (2) and the answers (2 and 3) from the division):

LCM = 2 × 2 × 3

= 12

Example 49

Find the LCM of 10, 15 and 20.

2	10	15	20
2	5	15	10
3	5	15	5
5	5	5	5
	1	1	1

2 does not divide exactly into 15 so write 15 here.

Multiply the bold numbers together.

2 × 2 × 3 × 5 × 1 × 1 × 1 = 60

The LCM of 10, 15 and 20 is 60.

Notes

- In making successive divisions it is good practice to use prime numbers.
- Unlike when finding the HCF, the successive divisors do not have to be factors of all the numbers in the group.

1 Identify all the factors of 12.

2 Write down all the prime factors of 20.

3 Identify all the multiples of 3 between 10 and 20.

4 What is the HCF of this set of numbers: 4, 8, 12 and 20?

5 What is the LCM of 5, 10 and 15?

6 Determine the square of each of the following.

    a 15   **b** 20   **c** 32

7 Calculate the square root of the following numbers.

    a 81   **b** 121   **c** 225

8 What is the reciprocal of the number $-\frac{7}{8}$?

9 State the largest and smallest prime numbers between 0 and 10 and calculate their product.

10 Express 105 as the product of prime numbers.

11 Write down all the multiples of 10 less than 100.

12 Divide the smallest multiple of 4 by the largest factor of 8. Write down the result.

13 What is the least number of mangoes that can be shared equally between groups of 8, 12 and 16 children?

14 An alarm goes off on three cell phones after every 5, 10 and 15 seconds. The phones are activated at the same time. After how many seconds will the alarms on all the phones go off at the same time?

15 A board measuring 6 cm by 40 cm is to be painted with black and white squares.

    a What is the largest square (in cm²) that can be shown on the board?

    b Determine the number of squares of the size found in **a** that can be shown on the board.

1.15 Fractions: revision

You can find fractions of numbers, objects or groups of objects. In each case, the numbers, objects and groups are the wholes. Here are some examples.

> **Key term**
>
> **Fraction** part of a whole

Fractions get their names according to the number of equal parts in the whole. The line in A has been divided into thirds. The bag of objects in C has been divided into halves. The rectangle in D has been divided into quarters.

In B, the circle has been divided into 8 equal parts. Each equal part is called one-eighth ($\frac{1}{8}$). If five of the equal parts were shaded, then five-eighths ($\frac{5}{8}$) of the circle is shaded.

A. One **whole length** divided into 3 equal parts

B. One **whole circle** divided into 8 equal parts

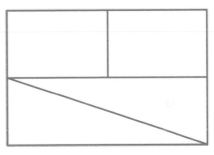

C. One **whole set** of 14 objects divided into 2 equal sets of 7 objects

D. One **whole rectangle** divided into 4 equal areas (different shapes)

You write fractions like this:

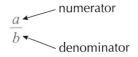

numerator

denominator

Key terms

Denominator   number of equal parts that a whole has been divided into

Numerator   number of equal parts being considered

$\frac{3}{4}, \frac{12}{19}$

All unit fractions are proper or vulgar fractions. Their value is less than 1.

Key terms

Unit fractions   fractions with a numerator of 1. For example: $\frac{1}{4}, \frac{1}{2}, \frac{1}{17}$

Proper or vulgar fractions   fractions with a numerator less than the denominator. For example: $\frac{5}{7}, \frac{3}{4}, \frac{12}{19}$

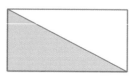

$\frac{1}{2}$ is shaded   unit fraction

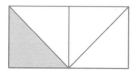

$\frac{3}{4}$ is unshaded   proper fraction

Improper fractions are greater than a whole. Their value is greater than 1.

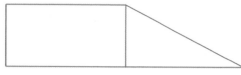

$\frac{3}{2}$ rectangles = $1\frac{1}{2}$ rectangles

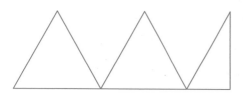

$\frac{5}{2}$ equilateral triangles = $2\frac{1}{2}$ equilateral triangles

Example 50

Write the improper fraction $\frac{11}{5}$ as a mixed number.

$11 \div 5 = 2\frac{1}{5}$

1.15.1 Equivalent fractions

Equivalent fractions are fractions that have different numerators and different denominators but have equal value. You can make equivalent fractions by multiplying or dividing the numerator and denominator of a fraction by the same number. This does not change the value of the fraction.

Example 51

Use division to calculate a fraction that is equivalent to $\frac{4}{16}$.

Divide the numerator and denominator by 4:

$$\frac{4}{16} = \frac{4 \div 4}{16 \div 4}$$

$$= \frac{1}{4}$$

$$\frac{4}{16} = \frac{1}{4}$$

Example 52

Use multiplication to calculate a fraction that is equivalent to $\frac{3}{5}$.

Multiply the numerator and denominator by 3:

$$\frac{3}{5} = \frac{3 \times 3}{5 \times 3}$$

$$= \frac{9}{15}$$

$$\frac{3}{5} = \frac{9}{15}$$

1.15.2 Reducing a fraction

A fraction can be reduced to its lowest term by dividing the numerator and denominator by the same number.

Key terms

Lowest terms a fraction in which the numerator and denominator have no common factors is in its lowest terms

Simplest form a fraction in its lowest terms

Example 53

Reduce the fraction $\frac{24}{40}$ to its lowest terms.

24 and 40 are even numbers so start by dividing the numerator and denominator by 2:

$$\frac{24}{40} = \frac{24 \div 2}{40 \div 2}$$

$$= \frac{12}{20}$$

Divide again by 2:

$$\frac{12}{20} = \frac{12 \div 2}{20 \div 2}$$

$$= \frac{6}{10}$$

And again:

$$\frac{6}{10} = \frac{6 \div 2}{10 \div 2}$$

$$= \frac{3}{5}$$

$\frac{24}{40}$ is $\frac{3}{5}$ in its lowest terms.

Example 53 continues on next page

Alternative method:

8 is the highest common factor of 24 and 40. Divide the numerator and denominator by 8:

$$\frac{24}{40} = \frac{24 \div 8}{40 \div 8}$$

$$= \frac{3}{5}$$

Exercise 1H

1 Identify the fraction shown in each picture.

   a Each colour of the rope is the same length. What fraction of the total length is yellow?

   b What fraction of the set of marbles shown is red?

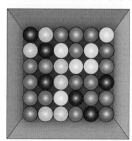

> **Hint**
> Count the total number of marbles. Count the red marbles.

2 What fraction of the rectangle shown is shaded?

3 What type of fraction is each of these?

   a $\frac{3}{4}$ **b** $\frac{5}{3}$ **c** $1\frac{1}{2}$

4 Make a sketch to represent each fraction.

   a $\frac{7}{8}$ **b** $\frac{9}{5}$ **c** $3\frac{5}{8}$

5 Name four fractions that are equivalent to $\frac{3}{5}$.

6 Reduce the following fractions to their simplest form.

    a $\frac{8}{12}$ **b** $\frac{20}{60}$

7 In a class of 40 students, 15 like Mathematics, 10 like English, 5 like History, 8 like Spanish and 2 like French. What fraction of the class like Mathematics and English?

8 Each small triangle in the pattern has the same area.

    a What fraction of the entire pattern is shaded?

    b What is the name given to that fraction when expressed in its lowest terms?

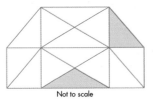

Not to scale

9 What fraction of the circle shown below is not shaded? (Sector angle = 15°)

Not to scale

10 Write 30 seconds as a fraction of 60 weeks.

1.16 Operations with fractions

1.16.1 Addition and subtraction of fractions

A Addition and subtraction of fractions with the same denominator

When adding or subtracting fractions that have the same denominator, add or subtract the numerator and use the same denominator.

(a) $\frac{1}{4} + \frac{3}{4} = \frac{(1+3)}{4} = \frac{4}{4} = 1$

 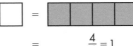

    $\frac{1}{4}$ + $\frac{3}{4}$ = $\frac{4}{4} = 1$

(b) $\frac{5}{8} - \frac{3}{8} = \frac{(5-3)}{8} = \frac{2}{8} = \frac{1}{4}$

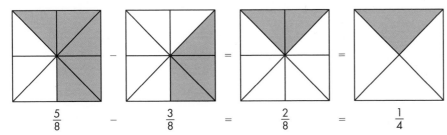

    $\frac{5}{8}$ − $\frac{3}{8}$ = $\frac{2}{8}$ = $\frac{1}{4}$

Example 54

What is $\frac{3}{7} + \frac{5}{7}$?

$$\frac{3}{7} + \frac{5}{7} = \frac{3+5}{7}$$
$$= \frac{8}{7}$$
$$= 1\frac{1}{7}$$

Example 55

What is $\frac{9}{16} - \frac{5}{16}$?

$$\frac{9}{16} - \frac{5}{16} = \frac{9-5}{16}$$
$$= \frac{4}{16}$$
$$= \frac{1}{4}$$

B Addition and subtraction of fractions with different denominators

It is easier to add and subtract fractions when they have the same denominator. Hence, begin by making the denominators equal. To do this, write both fractions as equivalent fractions with the same denominator.

Example 56

What is $\frac{1}{3} + \frac{1}{4}$?

To find equivalent fractions for $\frac{1}{3}$ and $\frac{1}{4}$ that have the same denominator, use the multiplying method from Example 48.

The denominator of $\frac{1}{4}$ is 4 so multiply the numerator and denominator of $\frac{1}{3}$ by 4:

$$\frac{1}{3} = \frac{1 \times 4}{3 \times 4}$$
$$= \frac{4}{12}$$

The denominator of $\frac{1}{3}$ is 3 so multiply the numerator and denominator of $\frac{1}{4}$ by 3:

$$\frac{1}{4} = \frac{1 \times 3}{4 \times 3}$$
$$= \frac{3}{12}$$

Therefore, $\frac{1}{3} + \frac{1}{4} = \frac{4}{12} + \frac{3}{12}$

$$= \frac{7}{12}$$

You can use diagrams to show this is true.

Divide a whole into thirds.

Now, divide the whole into quarters.

The whole is now divided into twelfths.

$$\frac{1}{3} = \frac{4 \text{ equal parts}}{12 \text{ equal parts}}$$

$$= \frac{4}{12}$$

$$\frac{1}{4} = \frac{3 \text{ equal parts}}{12 \text{ equal parts}}$$

$$= \frac{3}{12}$$

$$\frac{1}{3} + \frac{1}{4} = \frac{4}{12} + \frac{3}{12}$$

$$= \frac{7}{12}$$

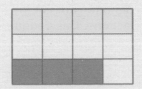

Example 57

What is $\frac{3}{4} - \frac{1}{5}$?

Find equivalent fractions for $\frac{3}{4}$ and $\frac{1}{5}$ that have the same denominators:

$$\frac{3}{4} = \frac{3 \times 5}{4 \times 5}$$

$$= \frac{15}{20}$$

$$\frac{1}{5} = \frac{1 \times 4}{5 \times 4}$$

$$= \frac{4}{20}$$

Now subtract:

$$\frac{3}{4} - \frac{1}{5} = \frac{15}{20} - \frac{4}{20}$$

$$= \frac{11}{20}$$

Example 58

Solve these.

(a) $\frac{2}{3} + \frac{3}{4}$ (b) $\frac{3}{4} - \frac{1}{7}$

(a) $\frac{2}{3} = \frac{2 \times 4}{3 \times 4}$

$\quad = \frac{8}{12}$

$\frac{3}{4} = \frac{3 \times 3}{4 \times 3}$

$\quad = \frac{9}{12}$

Therefore, $\frac{2}{3} + \frac{3}{4} = \frac{8}{12} + \frac{9}{12}$

$\qquad\qquad = \frac{8 + 9}{12}$

$\qquad\qquad = \frac{17}{12}$

$\qquad\qquad = 1\frac{5}{12}$

(b) $\frac{3}{4} = \frac{3 \times 7}{4 \times 7}$

$\quad = \frac{21}{28}$

$\frac{1}{7} = \frac{1 \times 4}{7 \times 4}$

$\quad = \frac{4}{28}$

Therefore, $\frac{3}{4} - \frac{1}{7} = \frac{21}{28} - \frac{4}{28}$

$\qquad\qquad = \frac{17}{28}$

Sometimes, the fractions to be added or subtracted have denominators that have common factors. In these instances, use the LCM of the denominators of the fractions you are adding or subtracting as the denominator of the equivalent fractions. This denominator is called the **lowest common denominator of the fractions (LCD)**.

Example 59

Solve these.

(a) $\frac{3}{8} + \frac{5}{6}$ (b) $\frac{7}{12} - \frac{4}{9}$

(a) $\frac{3}{8} + \frac{5}{6}$

The lowest common multiple of 8 and 6 is 24.

Therefore, the lowest common denominator of $\frac{3}{8}$ and $\frac{5}{6}$ is 24.

$$8 \times 3 = 24$$

Therefore, multiply the numerator and denominator of $\frac{3}{8}$ by 3 to make an equivalent fraction with a denominator of 24.

$$\frac{3}{8} = \frac{3 \times 3}{8 \times 3}$$
$$= \frac{9}{24}$$

$$6 \times 4 = 24$$

Therefore, multiply the numerator and denominator of $\frac{5}{6}$ by 4 to make an equivalent fraction with a denominator of 24.

$$\frac{5}{6} = \frac{5 \times 4}{6 \times 4}$$
$$= \frac{20}{24}$$

Therefore, $\frac{3}{8} + \frac{5}{6} = \frac{9}{24} + \frac{20}{24}$

$$= \frac{29}{24}$$
$$= 1\frac{5}{24}$$

(b) The LCM of 12 and 9 is 36.

Therefore, the LCD of $\frac{7}{12}$ and $\frac{4}{9}$ is 36.

$$12 \times 3 = 36$$

Therefore, $\frac{7}{12} = \frac{7 \times 3}{12 \times 3}$

$$= \frac{21}{36}$$

$$9 \times 4 = 36$$

Therefore, $\frac{4}{9} = \frac{4 \times 4}{9 \times 4}$

$$= \frac{16}{36}$$

$$\frac{7}{12} - \frac{4}{9} = \frac{21}{36} - \frac{16}{36}$$
$$= \frac{5}{36}$$

C Addition and subtraction of mixed numbers

(a) $2\frac{3}{4} + 1\frac{3}{8}$ (b) $4\frac{2}{5} - 2\frac{5}{6}$

(a) $2\frac{3}{4} + 1\frac{3}{8} = (2+1) + (\frac{3}{4} + \frac{3}{8})$ (Group the whole numbers and the fraction parts.)

$\qquad = 3 + (\frac{6}{8} + \frac{3}{8})$ (Use the LCM to make the denominators the same.)

$\qquad = 3 + \frac{9}{8}$

$\qquad = 3 + 1\frac{1}{8}$

$\qquad = 4\frac{1}{8}$

(b) $4\frac{2}{5} - 2\frac{5}{6} = (4-2) + (\frac{2}{5} - \frac{5}{6})$

$\qquad = 2 + (\frac{12}{30} - \frac{25}{30})$ (LCD of $\frac{2}{5}$ and $\frac{5}{6}$ is 30.)

$\qquad = 1 + (\frac{30}{30} + \frac{12}{30} - \frac{25}{30})$ (Because $\frac{25}{30}$ is larger than $\frac{12}{30}$, change a whole one to $\frac{30}{30}$.)

$\qquad = 1 + (\frac{42}{30} - \frac{25}{30})$

$\qquad = 1 + \frac{17}{30}$

$\qquad = 1\frac{17}{30}$

1.16.2 Multiplication of fractions and mixed numbers

A Multiplying a fraction or mixed number by a whole number

Work out $4 \times \frac{3}{4}$.

This problem can be solved by using different methods.

(i)   Using diagrams:

$\frac{3}{4}$ can be shown as 3 shaded sectors out of 4.

$4 \times \frac{3}{4}$ makes 12 shaded sectors.

You can rearrange the shaded sectors to form 3 wholes.

(ii)   Repeated addition: $\frac{3}{4} + \frac{3}{4} + \frac{3}{4} + \frac{3}{4} = \frac{12}{4}$

$\qquad\qquad\qquad\qquad\qquad = 3$

(iii)   Multiplying the numerator by the whole number:

$\qquad \frac{3}{4} \times 4 = \frac{3 \times 4}{4}$

$\qquad\qquad = \frac{12}{4}$

$\qquad\qquad = 3$

(iv)   Multiplying the fraction by the whole number and dividing by common factors where possible. This is called cancelling:

$\qquad {}^1 4 \times \frac{3}{4_1} = 3$

Cancelling   dividing numerators and
denominators by common factors

A full crate of water contains 24 bottles. How many bottles of water are there in a case that is $\frac{3}{4}$ full?

To solve this problem, work out $\frac{3}{4}$ of 24. This means calculating $\frac{3}{4} \times 24$.

$$\frac{3}{4} \times 24 = \frac{3}{4^1} \times 24^6$$
$$= 3 \times 6$$
$$= 18$$

There are 18 bottles in a crate that is $\frac{3}{4}$ full.

Use a diagram to check this.

The 24 bottles are split into four equal groups.

Each group contains $\frac{1}{4}$ of the bottles.

There are 6 bottles in each group.

Since $\frac{1}{4}$ of the bottles = 6,

then $\frac{3}{4}$ of the bottles = 3 × 6 bottles

$$= 18 \text{ bottles}$$

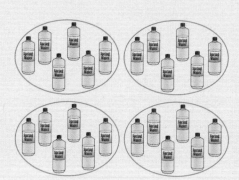

B   Multiplying a fraction by a fraction

Tip

When multiplying fractions, check whether there are common factors. If there are, you can use cancelling to make the calculations easier.

Tom owns $\frac{3}{4}$ of an acre of land and $\frac{5}{6}$ is planted with trees. Calculate what area, in acres, is planted with trees.

Work out $\frac{3}{4} \times \frac{5}{6}$.

$$\frac{3}{4} \times \frac{5}{6} = \frac{\overset{1}{3}}{4} \times \frac{5}{6_2}$$ (Divide the numerator and denominator by 3.)

$$= \frac{1}{4} \times \frac{5}{2}$$

$$= \frac{5}{8}$$

Example 64

What is $\frac{5}{7} \times \frac{3}{8}$?

$$\frac{5}{7} \times \frac{3}{8} = \frac{5 \times 3}{7 \times 8}$$
$$= \frac{15}{56}$$

C Mixed number times mixed number

Example 65

Calculate the area in square yards of a piece of cloth measuring $2\frac{1}{4}$ yards by $1\frac{1}{2}$ yards.

To solve this problem, you have to work out $2\frac{1}{4} \times 1\frac{1}{2}$.

Convert the mixed numbers to improper fractions and then follow the rules of fraction × fraction:

$2\frac{1}{4} = \frac{9}{4}$ and $1\frac{1}{2} = \frac{3}{2}$

Therefore, $2\frac{1}{4} \times 1\frac{1}{2} = \frac{9 \times 3}{4 \times 2}$
$$= \frac{27}{8}$$
$$= 3\frac{3}{8}$$

The area of the cloth is $3\frac{3}{8}$ square yards.

Example 66

What is $3\frac{3}{5} \times 6\frac{2}{3}$?

$3\frac{3}{5} = \frac{18}{5}$ and $6\frac{2}{3} = \frac{20}{3}$

Therefore, $3\frac{3}{5} \times 6\frac{2}{3} = \frac{18}{5} \times \frac{20}{3}$

$$= \frac{\overset{6}{\cancel{18}} \times \overset{4}{\cancel{20}}}{\underset{1}{\cancel{5}} \times \underset{1}{\cancel{3}}}$$

(Divide the numerator and denominator by 3 and by 5.)

$$= \frac{6 \times 4}{1 \times 1}$$

$$= 24$$

1.16.3 Division of fractions

As with addition, subtraction and multiplication of fractions, real-life problems sometime require the division of fractions.

Example 67

Share $\frac{3}{4}$ of a pizza equally between five children.
How much does each child receive?

The calculation is $\frac{3}{4} \div 5$.

Dividing by 5 is the same as finding $\frac{1}{5}$ of an amount.

As you saw in Example 62, to find a fraction of an amount, multiply the amount by the fraction.

Therefore, $\frac{3}{4} \div 5 = \frac{3}{4} \times \frac{1}{5}$

$$= \frac{3 \times 1}{4 \times 5}$$

$$= \frac{3}{20}$$

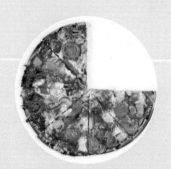

Use diagrams to show that this is true.
Divide the pizza into quarters.

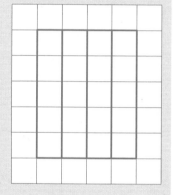

There is $\frac{3}{4}$ of the pizza to share between the children.

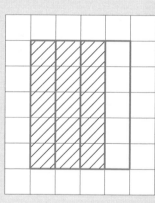

There are 5 children so divide the pizza into fifths.

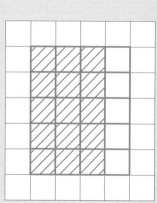

Each child will receive $\frac{1}{5}$ of the $\frac{3}{4}$ of the pizza.

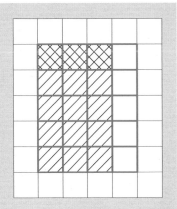

This is $\frac{3}{20}$ of the pizza.

Example 68

What is $\frac{3}{5} \div \frac{7}{8}$?

$\frac{3}{5} \div \frac{7}{8} = \dfrac{\frac{3}{4}}{\frac{5}{8}}$ (Set out the division like this.)

$= \dfrac{\frac{3}{5} \times \frac{8}{7}}{{}_1\frac{7}{8} \times \frac{8}{7}{}_1}$ (Multiply the numerator and the denominator by $\frac{8}{7}$. This doesn't change the value of the fraction.)

(In the denominator, cancel the common factors.)

$= \dfrac{\frac{3}{5} \times \frac{8}{7}}{1}$

$= \frac{3}{5} \times \frac{8}{7}$ (You now have a calculation that involves $\frac{3}{5}$ multiplied by the reciprocal of $\frac{7}{8}$.)

$= \frac{24}{35}$

Tip

When dividing by a fraction $\frac{a}{b}$, multiply by its reciprocal $\frac{b}{a}$. For example:

$\frac{5}{8} \div \frac{4}{9} = \frac{5}{8} \times \frac{9}{4}$

$25 \div \frac{7}{12} = 25 \times \frac{12}{7}$

1.16.4 Mixed operations with fractions

Example 69

Simplify $\dfrac{15\frac{1}{2} - \frac{3}{4}\left(4\frac{1}{4} + 5\frac{1}{2}\right)}{8\frac{1}{4} + 3\frac{1}{2} \times 6\frac{1}{2}}$.

Follow BODMAS and use the methods for adding, subtracting, multiplying and dividing fractions.

For the numerator:

Bracket first gives:

$$4\frac{1}{4} + 5\frac{1}{2} = 9 + \left(\frac{1}{4} + \frac{1}{2}\right)$$

$$= 9 + \left(\frac{2}{8} + \frac{4}{8}\right)$$

$$= 9 + \frac{6}{8}$$

$$= 9\frac{3}{4}$$

For the denominator:

Multiplication gives:

$$3\frac{1}{2} \times 6\frac{1}{2} = \frac{7}{2} \times \frac{13}{2}$$

$$= \frac{7 \times 13}{2 \times 2}$$

$$= \frac{91}{4}$$

$$= 22\frac{3}{4}$$

$$\frac{15\frac{1}{2} - \frac{3}{4}\left(4\frac{1}{4} + 5\frac{1}{2}\right)}{8\frac{1}{4} + 3\frac{1}{2} \times 6\frac{1}{2}} = \frac{15\frac{1}{2} - \frac{3}{4} \times 9\frac{3}{4}}{8\frac{1}{4} + 22\frac{3}{4}}$$

$$= \frac{15\frac{1}{2} - \frac{3}{4} \times \frac{39}{4}}{8 + 22 + \frac{1}{4} + \frac{3}{4}}$$

$$= \frac{15\frac{1}{2} - \frac{117}{16}}{30 + \frac{4}{4}}$$

$$= \frac{15\frac{1}{2} - 7\frac{5}{16}}{31}$$

$$= \frac{(15 - 7) + \left(\frac{1}{2} - \frac{5}{16}\right)}{31}$$

$$= \frac{8 + \frac{3}{16}}{31}$$

$$= \frac{8\frac{3}{16}}{31}$$

$$= \frac{\frac{131}{16}}{31}$$

$$= \frac{131}{16} \times \frac{1}{31}$$

$$= \frac{131}{496}$$

Exercise 1I

1 Add the fractions $\frac{2}{9}$ and $\frac{1}{9}$.

2 Solve $\frac{4}{5} - \frac{2}{5}$.

3 Solve $\frac{3}{5} + \frac{2}{3}$.

4 Solve $\frac{5}{8} - \frac{3}{4}$.

5 Solve $\frac{5}{16} \times \frac{4}{5}$.

6 Solve $\frac{5}{9} \div \frac{2}{3}$.

7 Evaluate $2\frac{3}{4} + 5\frac{1}{2}$.

8 Solve $5\frac{2}{3} - 4\frac{1}{6}$.

9 Solve $6 \times 1\frac{1}{3}$.

10 Evaluate $3\frac{3}{4} \times 2\frac{1}{2}$.

11 Evaluate $12 \div 1\frac{2}{3}$.

12 Solve $8\frac{4}{5} \div 4\frac{2}{5}$.

13 Solve $\dfrac{1\frac{1}{2} - 2\frac{3}{4} + 4\frac{1}{4}}{\frac{1}{3} \times 1\frac{1}{5}}$.

14 A landowner owned $5\frac{3}{4}$ acres of land. She used $2\frac{1}{2}$ acres for farming, $1\frac{3}{5}$ acres for buildings and she decided to sell the remainder. How much did she sell?

15 Solve $\dfrac{2\frac{4}{5} \times \frac{5}{7} - 1\frac{2}{3}}{1\frac{5}{6} + 2\frac{2}{5}}$.

16 A man built a path to his house. On Monday, he constructed a length of 78 ft. On Tuesday, he built $2\frac{4}{5}$ times Monday's length. He completed the road on Wednesday with a length that was $3\frac{7}{8}$ times the total from Monday and Tuesday. What was the final length of the complete road?

17 A man bought $7\frac{3}{4}$ yards of cloth to make trousers and a jacket. He used $2\frac{1}{2}$ yards to make the trousers and $3\frac{2}{3}$ yards to make the jacket. He decided to make 4 fancy pockets on his jacket with the remainder of the cloth. How much cloth was used for each pocket?

18 Solve $\dfrac{7\frac{1}{4} + 2\frac{1}{2} \times 3\frac{3}{4}}{1\frac{1}{4}(5 - 2\frac{2}{3})}$.

19 Miss Jane bought $2\frac{1}{2}$ cakes from the grocery and she was given $3\frac{1}{4}$ cakes by her neighbour as a donation for a children's party. During the party, $4\frac{3}{4}$ cakes were eaten. How much of the cake remained?

20 John, Mary and Peter had $3\frac{1}{4}$ pizzas to share among themselves. John ate $\frac{3}{4}$ of a pizza. Mary took $\frac{2}{5}$ of a pizza. Peter was very hungry and he ate $1\frac{1}{5}$ pizzas. The remainder was then shared equally among 3 small children. How much did each child receive?

21 Simplify $\dfrac{5\frac{5}{9} - 3\frac{1}{2}(2\frac{1}{2} - 1\frac{7}{8})}{9\frac{1}{4}(8\frac{3}{4} + 2\frac{5}{7}) - (1\frac{1}{2} + \frac{3}{8})}$.

22 The diagram shows a rectangular field of dimensions $1\frac{3}{8}$ km by $\frac{4}{5}$ km. A path just inside the field has a width of $\frac{1}{200}$ km. Use the formula area = length × width to calculate:

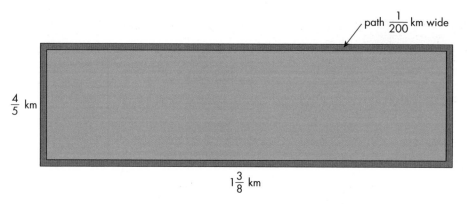

$\frac{4}{5}$ km

path $\frac{1}{200}$ km wide

$1\frac{3}{8}$ km

a total area of the field

b area of the path

c area of land available for use.

1.17 Decimal fractions: revision

1.17.1 Decimals and place value of decimals

You have looked at the place value of whole numbers (see Section 1.2 on page 7). Numbers can have three digits, four digits, etc. and you can show them in a place-value table, where the digit with the smallest place value is in the units column on the right.

You can also set up a column to the right of the units column. This column contains digits that are fractions with a denominator of 10. You can add other columns to the right of this, for fractions of denominators 100, 1000, etc. Each column has a specific name, as shown in the place-value table below. These are **decimals**.

Hundreds	Tens	Units	Decimal point	tenths	hundredths	thousandths	ten thousandths
		0	.	1			
		0	.	0	0	5	
		6	.	4			
	4	0	.	0	0	0	7

Hence, decimals are another way of representing fractions and mixed numbers. They are shown as tenths, hundredths, and other fractions that are powers of 10. For example, looking at the numbers in the place value table:

$0.1 = \frac{1}{10}$

$0.005 = \frac{5}{1000}$

$6.4 = \frac{64}{10}$

$40.0007 = \frac{400\,007}{10\,000}$

Look at this place-value table.

Ten thousands	Thousands	Hundreds	Tens	Units	Decimal point	tenths	hundredths	thousandths	ten thousandths
5	9	2	1	7	.	3	4	6	8

The number represented above in the place-value table is 59 217.3468

It has:

5 ten thousands	$5 \times 10\,000$	=	50 000
9 thousands	9×1000	=	9000
2 hundreds	2×100	=	200
1 ten	1×10	=	10
7 units	7×1	=	7
3 tenths	$3 \times \frac{1}{10}$	=	0.3
4 hundredths	$4 \times \frac{1}{100}$	=	0.04
6 thousandths	$6 \times \frac{1}{1000}$	=	0.006
8 ten thousandths	$8 \times \frac{1}{10\,000}$	=	0.0008
Adding these values:			59 217.3468

Just as in a whole number, each digit in the number above has two values: a place value and a face value.

Remember

- Face value refers to the actual value of the digit.
- Place value is based on the digit's position in the place-value chart (unit, tens, tenth, etc.).

Example 70

Look at the number 59 217.3468 shown in the place-value table above.

(a) What are the face value and place value of the digit 8?

(b) What are the face value and place value of the digit 3?

(a) Face value of the 8 is 8 and place value of the 8 is $\frac{8}{10\,000} = 0.0008$.

(b) Face value of the 3 is 3 and place value of the 3 is $\frac{3}{10} = 0.3$.

1.17.2 Converting between decimals and fractions

Any fraction can be written as a decimal and vice versa.

Generally, to convert any fraction (including improper fractions) to a decimal, divide the numerator by the denominator.

Example 71

Convert these fractions to decimals.

(a) $\frac{1}{2}$ (b) $\frac{3}{4}$ (c) $\frac{7}{5}$

(a) $\frac{1}{2} = 1 \div 2$

     $= 0.5$

(b) $\frac{3}{4} = 3 \div 4$

     $= 0.75$

(c) $\frac{7}{5} = 1\frac{2}{5}$

     $= 1 + (2 \div 5)$

     $= 1 + 0.4$

     $= 1.4$

To convert a decimal to a fraction, count the digits on the right of the decimal point. This shows what power of 10 you should use for the denominator.

Example 72

Convert 0.35 to a fraction.

0.35 has two digits on the right of the decimal point.

Therefore, the denominator will be $10^2 = 100$.

$0.35 = \frac{35}{100}$

Example 73

Convert 3.002 to a fraction.

3.002 has three digits on the right of the decimal point.

Therefore, the denominator will be $10^3 = 1000$.

$3.002 = \frac{3002}{1000}$

You can also write this as a mixed number:

$3.002 = 3 + 0.002$

$= 3 + \dfrac{2}{1000}$

$= 3\dfrac{1}{500}$

Exercise 1J

1 Give five of your own examples of a decimal number.

2 Given this set of decimal numbers, state:

a the smallest **b** the largest.

0.001   0.9   1.00001   0.009   0.1   1.1

3 Copy the square below. Shade it so that 0.75 of the square is shaded.

4 Draw rectangles to represent the decimal 2.35.

> **Hint**   Use three identical rectangles. Divide one of the rectangles into 100 and shade appropriately.

5 Convert these fractions to decimals.

a $\dfrac{1}{10}$ **b** $\dfrac{1}{100}$ **c** $\dfrac{1}{50}$ **d** $\dfrac{3}{1000}$ **e** $\dfrac{150}{10\,000}$ **f** $\dfrac{8}{20}$ **g** $\dfrac{4}{25}$ **h** $\dfrac{7}{5}$ **i** $\dfrac{3}{8}$

6 Convert these decimals to fractions or mixed numbers.

a 0.7   **b** 0.95   **c** 1.009   **d** 0.0003   **e** 157.06

7 Expand the decimal number 123.56 to show the place value of each digit.

8 Given the number 98 763.01245, state:

a the place value of the digits 2, 8, 0, 6 and 5.

b the face value of the digits 2, 8, 0, 6 and 5.

9 Using a drawing of a geometrical figure (e.g. rectangle), represent the decimal number 1.03.

10 Convert the fraction $\frac{3}{40}$ to a decimal.

11 Convert the decimal 0.000 008 08 to a fraction.

12 Expand the decimal number 4350.892, showing how its value can be obtained by adding the digits according to their place value.

13 What decimal number does this expansion represent? 8000.09 + 70.1 + 0.005

14 Expand the number 356 872.0194, showing how the digits can be combined according to their place value.

15 Explain how you can use a diagram to represent the decimal number 1.001.

16 Convert the fraction $\frac{1}{40}$ to a decimal.

17 What is 2 000 000 + 9985.01 + 887.006 + 0.000 003?

18 Explain how you can use a diagram to represent the decimal number 5.005.

1.18 Operations with decimals

1.18.1 Addition and subtraction of decimals

Solve 29.457 + 18.596

To add decimals, use the same method that you use for whole numbers. Align the decimal points to ensure that you add thousandths to thousandths, hundredths to hundredths, etc. Start with the digits with lowest place values.

$$29.457 + 18.596 \qquad \begin{array}{r} 2\,9.4\,5\,7 \\ +\,1\,8.5\,9\,6 \\ \hline 4\,8.0\,5\,3 \\ \hline {\scriptstyle 1\ 1\ \ 1\ 1} \end{array}$$

Start with the thousandths.

6 thousandths + 7 thousandths = 13 thousandths

$\qquad\qquad\qquad\qquad$ = 1 hundredth + 3 thousandths

Write 3 thousandths in the thousandths place in the answer.

Write 1 hundredth under the hundredths place in the answer.

Then add the hundredths and so on.

29.457 + 18.596 = 48.053

Example 75

Solve 4.123 – 2.289

To subtract decimals, use the same method that you use for whole numbers. Align the decimal points to ensure that you subtract thousandths from thousandths, hundredths from hundredths, etc. Start with the digits with lowest place values.

4.123 – 2.289

$$
\begin{array}{r}
{}^{3}\,{}^{1}0\,{}^{1}1\,{}^{1} \\
4.\cancel{1}\,\cancel{2}\,3 \\
-\;2.2\;8\;9 \\
\hline
1.8\;3\;4
\end{array}
$$

Start with the thousandths: 3 thousandths – 9 thousandths

Because 9 is greater than 3, take 1 hundredth from the 2 hundredths in the hundredths column.

2 hundredths – 1 hundredth = 1 hundredth

Cross out the 2 hundredths and write 1 hundredth.

Change the 1 hundredth to 10 thousandths.

10 thousandths + 3 thousandths – 9 thousandths = 4 thousandths.

Write 4 in the thousandths place in the answer.

Then subtract the hundredths and so on.

4.123 – 2.289 = 1.834

1.18.2 Multiplication of decimals

To multiply decimals, use similar methods to those you use for whole numbers. You can use long multiplication.

Example 76

Solve 23.4 × 0.54

Write the question vertically, and align the decimal points.

Ignore the decimal places and perform the multiplication as if they were two whole numbers.

Begin with the lowest value digit.

Here, start by multiplying 23.4 by the 4 of 0.54.

$$
\begin{array}{r}
2\;3.4 \\
\times\quad 0.5\,4 \\
\hline
9\;3\;6 \\
1\,1\;7\;0\;0 \\
\hline
1\,2.6\;3\;6 \\
{}_{1}
\end{array}
$$

After completing the multiplication, count the total number of decimal places in the two numbers in the calculation. This is the number of decimal places in the answer.

In 23.4 × 0.54, there are three decimal places. Count three places from the digit with the smallest value (the digit furthest on the right), and write in the decimal point between the 6 and the 2.

23.4 × 0.54 = 12.636

1.18.3 Division of decimals

Example 77

Solve $159.33 \div 5.64$

First, convert the division to one involving whole numbers. Write the division as a fraction and multiply to change the denominator to a whole number.

$$159.33 \div 5.64 = \frac{159.33}{5.64}$$

$$= \frac{159.33 \times 100}{5.64 \times 100} \quad \text{(Multiply by } 100 = 10^2 \text{ because there are two decimal places in both 159.33 and 5.64)}$$

$$= \frac{15\,933}{564}$$

Remember	When you multiply the numerator and the denominator of a fraction by the same number, the value of the fraction remains the same.

You can now solve $15\,933 \div 564$ by using the long division method in Example 18 on page 16 for dividing whole numbers.

```
              2 8.2 5
    5 6 4 )1 5 9 3 3.0 0
          − 1 1 2 8
            4 6 5 3
          − 4 5 1 2
            1 4 1 0
          − 1 1 2 8
              2 8 2 0
            − 2 8 2 0
                    0
```

Write the decimal point in the answer after you've reached the last digit of the dividend if it's a whole number or at the decimal point if it's a decimal.

$159.33 \div 5.64 = 28.25$

1.19 Approximating numbers

1.19.1 Approximating numbers to a given number of decimal places

Example 78 shows another division of decimals. Look at the answer.

Example 78

Solve $128.5 \div 6.21$. Give your answer to two decimal places.

$$128.5 \div 6.21 = \frac{128.5}{6.21}$$

$$= \frac{128.5 \times 100}{6.21 \times 100} \quad \text{(Multiply by } 100 = 10^2 \text{ because there are two decimal places in 6.21.)}$$

$$= \frac{12\,850}{621}$$

$$\begin{array}{r} 20.692 \\ 621\overline{)12850.000} \\ -1242 \\ \hline 4300 \\ -3726 \\ \hline 5740 \\ -5589 \\ \hline 1510 \\ -1242 \\ \hline 268 \end{array}$$

You could go on with the division, taking down more zeros. However, this question asks for a specific number of decimal places. The number of decimal places determines the accuracy of the answer – the more decimal places the more accurate the answer.

To write a decimal number correct to a certain number of decimal places, first note the digit in the decimal place you are interested in. Then look at the next digit. If the next digit is 5 or greater, then round up the digit by 1. If the next digit is less than 5, then the digit before is left unchanged.

Here, the answer must be given to two decimal places, so look at the answer when it has three decimal places: 20.692

2 is less than 5 so to two decimal places, the answer is 20.69.

$128.5 \div 6.21 = 20.69$ to two decimal places

Example 79

Round the number 0.0897 to three decimal places.

The digit in the position of the third decimal place (thousandths) is 9.

The next digit is 7, which is greater than 5.

Therefore, round up to the next thousandth.

Rounding 9 thousandths up to the next thousandth results in 10 thousandths.

10 thousandths = 1 hundredth + 0 thousandths

Add the 1 hundredth to the 8 hundredths.

Therefore, 0.0897 rounded to three decimal places is 0.090.

1.19.2 Approximating numbers to a given number of significant figures

Writing numbers to a specific number of significant figures is another way of showing the accuracy of a measurement or value.

When you write an answer to a given number of decimal places, you only count the number of digits after the decimal point.

However, when you write answers to a given number of significant figures, you start with the first non-zero digit in the number. This can include the digits before the decimal point as well as after the decimal point, as long as the digits are not zero. Look at these examples.

- The number 28.09 has four significant figures and two decimal places. Include the third digit 0 as a significant figure as it holds the tenths place.

- The number 0.000 768 4 has four significant figures and seven decimal places. Here, don't include the zeros before the digit 7. The digit 7 is the first non-zero digit so begin with that.

- The number 6500 has two significant figures and no decimal places. Here, don't include the zeros when counting significant figures as they aren't holding places between non-zero digits.

Example 80

Write the number 130.069 to five significant figures.

130.069 to five significant figures will include the digits 1, 3, 0, 0 and the fifth significant figure will be in the hundredths place.

There is a 6 in the hundredths place but as the thousandths digit is 9, which is greater than 5, round the digit 6 up to 7.

Therefore, 130.069 to five significant figures is 130.07.

Example 81

Write the number 873 to two significant figures.

Counting from left to right, the second significant figure is 7.

Since the next digit is 3, which is less than 5, the 7 remains unchanged.

Replace the 3 with 0 as a place holder.

Therefore, 873 to two significant figures is 870.

Example 82

Write the number 0.002 01 to two significant figures.

The first significant figure is 2.

The next digit is 0 but it's followed by 1.

Therefore, the number 0.002 01 to two significant figures is 0.0020

1.20 Standard form (scientific notation)

Writing a number in standard form is a way to present numbers so that you can compare them easily.

Numbers in standard form are shown by $A \times 10^n$, where A is a number between 1 and 10 ($1 \leq A < 10$) and n is a positive or negative integer.

To write a number in standard form, multiply or divide the number by 10 until it is between 1 and 10. Then write the power of 10 that is needed to restore it to its original size: n is positive for numbers greater than 1 and negative for numbers less than 1.

Example 83

Write 2001.05 in standard form.

$2001.05 \div 10 = 200.105$

$200.105 \div 10 = 20.0105$

$20.0105 \div 10 = 2.00105$

$1 \le 2.00105 < 10$

You have divided 2001.05 by 10 three times. You have divided by 1000, which is 10^3.

Therefore, $2001.05 = 2.00105 \times 10^3$ (Use positive 3 because the starting number was more than 1.)

Example 84

Express the number 0.00509 in standard form.

$0.00509 \times 10 = 0.0509$

$0.0509 \times 10 = 0.509$

$0.509 \times 10 = 5.09$

$1 \le 5.09 < 10$

Therefore, $0.00509 = 5.09 \times 10^{-3}$ (Use negative 3 because the starting number was less than 1.)

Standard form can also be used together with decimal places and significant figures to express a number even more concisely.

Example 85

Write 298.7 in standard form correct to 1 decimal place.

$298.7 = 2.987 \times 10^2$

Writing 2.987 to one decimal place, round 9 tenths to 10 tenths, which is 1 unit.

Therefore, 2.987 to one decimal place is 3.0.

298.7 is 3.0×10^2 in standard form to one decimal place.

1.21 Mixed operations with decimals

Two or more operations can be included in a calculation involving decimals. Use BODMAS to help you work out the answer.

Example 86

Simplify $\frac{7.32(1.5 - 0.8) + 13.46}{9.1 - 6.2 \times 1.05}$, giving your answer correct to two decimal places.

Look at the numerator, using BODMAS:

First, the brackets: $1.5 - 0.8 = 0.7$

Multiplication: $7.32 \times 0.7 = 5.124$

Addition: $5.124 + 13.46 = 18.584$

Now look at the denominator, again using BODMAS:

Multiplication: $6.2 \times 1.05 = 6.51$

Subtraction: $9.1 - 6.51 = 2.59$

Example 86 continues on next page

$$\frac{\text{Numerator}}{\text{Denominator}} = \frac{18.584}{2.59}$$

$$= \frac{1858.4}{259} \quad \text{(Use long division to work this out.)}$$

$$= 7.175 \quad \text{(Work this out to three decimal places.)}$$

$$= 7.18 \quad \text{correct to two decimal places}$$

Exercise 1K

1 Solve these.

   a $1.23 + 5.46$ **b** $6.789 - 5.462$

2 Solve these.

   a 3.21×4 **b** $84.6 \div 2$

3 Evaluate $45.7 + 9.98 + 0.76$

4 Solve $231.34 - 98.57$

5 Solve 0.0876×1000.

6 Solve $4.03 \div 100$.

7 Evaluate 236.19×2.3

8 Solve $98.2 \div 0.31$

9 Solve $34.78 + 9.81$. Give your answer correct to one decimal place.

10 Evaluate $92.11 - 8.99$. Give your answer correct to three significant figures.

11 Solve 40.58×2.9. Give your answer correct to two decimal places.

12 Evaluate $235.7 \div 1.1$, giving your answer correct to three significant figures.

13 Evaluate $1987.867 + 98.885$.

14 Solve 0.833×0.045. Give your answer correct to one significant figure.

15 Perform the following long division, giving your answer in standard form correct to two decimal places: $1876.213 \div 0.023$

16 Find the value of $2.2534 - 1.8976$, giving your answer in standard form correct to three significant figures.

17 Solve $\frac{14.2 - 5.1 \times 2.3}{0.6 + 1.2}$. Write your answers in standard form and correct to:

   a one decimal place   **b** three significant figures.

18 Solve the following:

$$\frac{0.3 - 8.2 + 3.4\,(2.5 - 1.8)}{5.8 - 1.1 \times 3.9 + 4.8}$$

Express your answer in standard form:

a correct to two decimal places

b correct to three significant figures.

19 The floor of a room measures 15.3 m by 18.6 m. The furniture in the room takes up an area of 80.7 m^2. There are five people in the room.

a How much standing area is allotted to each person? (Assume that the remaining area is divided equally among the five people.)

b The height of the room is 3.85 m. Calculate the three-dimensional space given by area × height of the room.

c How many tiles, each one measuring 30 cm^2, would have been used to tile the entire floor?

20 Mike shops for groceries with $200.00. He buys:

- 7 lbs flour at $1.50 per lb
- 14 lbs meat at $5.75 per lb
- 6 bottles of wine at $13.40 a bottle
- 12 lbs sugar at $2.20 per lb.

How many sweets can he purchase with the change if each sweet cost 75¢?

1.22 Percentages

A percentage, like a decimal, is also another type of fraction. It is a fraction with a denominator of 100. Therefore, percentage really means 'out of a 100'.

For example: a student gets 75 out of 100 in her math exam. Her result can be given as 75 percent.

Key term

The symbol % shows that this is a percentage. It can even be seen as the digits 100 written in a unique way.

1.22.1 Converting fractions to percentages

Convert a fraction to a percentage by simply multiplying the fraction by 100%.

Example 87

Convert $\frac{1}{2}$ to a percentage.

$\frac{1}{2} = \frac{1}{2} \times 100\% = 50\%$

1.22.2 Converting decimals to percentages

Convert a decimal to a percentage by multiplying by 100%.

Example 88

Express 0.7 as a percentage.

$0.7 = \frac{7}{10}$

$\qquad = \frac{7}{10} \times 100\%$

$\qquad = 70\%$

1.22.3 Calculating a percentage of a quantity

Example 89

Calculate 10% of 140.

$10\% \text{ of } 140 = \frac{10}{100} \times 140$

$\qquad\qquad = \frac{1}{10} \times 140$

$\qquad\qquad = 14$

1.22.4 Express one quantity as a percentage of another

Example 90

Mary has 80 cents and John has $1.20. What is Mary's amount as a percentage of John's?

Before calculating, both quantities must be in the same unit of measure.

John's amount is in dollars. Convert it to cents as Mary's is.

$1.20 \times 100 = 120$ cents

Mary's amount as a percentage of John's $= \frac{80}{120} \times 100\%$

$\qquad\qquad\qquad = \frac{\overset{2}{\cancel{80}}}{\underset{3}{\cancel{120}}} \times 100\%$

$\qquad\qquad\qquad = 0.666\ldots \times 100\%$

$\qquad\qquad\qquad = 66.666\ldots\%$

$\qquad\qquad\qquad \approx 67\%$

1.22.5 Converting percentages to fractions

Example 91

Convert 20% to a fraction in its lowest terms.

$20\% = \frac{20}{100}$

$\qquad = \frac{\overset{1}{\cancel{20}}}{\underset{5}{\cancel{100}}}$

$\qquad = \frac{1}{5}$ (using equivalent fractions)

1.22.6 Converting percentages to decimals

Convert 73% to a decimal.

$73\% = \frac{73}{100}$

$= 73 \div 100$

$= 0.73$

Exercise 1L

1 Convert to percentages.

  a $\frac{1}{2}$ **b** $\frac{3}{4}$ **c** $4\frac{1}{5}$ **d** 0.75   **e** 1.82   **f** 0.0063   **g** $\frac{7}{5}$ **h** 2

2 Convert the percentages to **i** fractions or mixed numbers **ii** decimals.

  a 40%   **b** 120%   **c** 0.2%   **d** 1.05%   **e** 565%

3 Calculate 30% of $40.00.

4 In a certain school, 60% of the students are girls. There are 200 boys at the school. How many students attend the school?

5 Tom ate 80% of a pizza. What fraction remained?

6 Make a simple sketch of a rectangular garden that is 60% planted with vegetables.

7 Harry gave his sister $1.50 and kept $6.00 for himself. What was his sister's share as a percentage of his share?

8 Arrange the following values in ascending order: 72%, $\frac{3}{4}$, 1.2, 0.78, $1\frac{1}{2}$, $\frac{8}{5}$, 150%.

9 A man's salary of $10 000 was increased by 20%. What is his new salary?

10 15% of the length of a string is 20 cm. What is the whole length of the string?

11 A boy had 100 marbles and gave away 5%. How many marbles did he have left?

12 $140.00 was shared between three boys: James, Tommy and Phillip. James received 20%, Tommy got 60% and Phillip received the remainder. How much did each boy receive?

13 A farmer uses 24% of his agricultural lands for vegetables, $\frac{2}{5}$ for fruits and 30% for grazing his animals. What fraction of his land remains unused?

14 Ten seconds is what percentage of 10 weeks?

15 Mike went shopping with some money. He was going to spend 60% of his money on clothes. He used 40% of his clothes money to buy sportswear. He spent 25% of the sportswear money on football kit. He spent $200.00 on the football kit. How much money did he start off with?

16 A survey of the population of a country showed that 0.005% of the population were musicians, 0.0085% were dancers and 0.01% were singers.

a Were there more musicians, dancers or singers?

b There were 50 singers. How many musicians were there?

c What was the population of the country?

17 In an election to choose the president of an organization, $\frac{1}{4}$ of the voters voted for candidate A, 20% voted for candidate B, $\frac{1}{3}$ voted for candidate C, 16% voted for candidate D, and the rest of the voters did not vote for anyone.

a What percentage of the voters did not vote for anyone?

b 6 voters did not vote for anyone. What was the approximate number of voters in the election?

c Which candidate actually won the election?

18 A pair of shoes were originally priced at $120.00. They were put on sale at a price 47% less than the original price. A few months later, the reduced price was increased by 245%. What was the final price of the shoes?

1.23 Conversion between units of measure

People measure many different things, for example, the width of a road, the mass of a stone, the capacity of a container, etc. You use units to say how long or heavy things are or how much they hold.

In mathematics, you usually use metric units. Metric units are based on powers of 10.

Common metric units of length

Name of unit	Symbol	Value in metres	Value in metres (indices)
kilometre	km	1000	10^3
metre	m	1	10^0
centimetre	cm	0.01	10^{-2}
millimetre	mm	0.001	10^{-3}

Common imperial units of length

The metric system replaced the imperial or English system. This system is still used in some places.

Name of unit	Symbol	Value
mile	mi	1 mi = 1760 yd
yard	yd	1 yd = 3 ft
foot	ft	1 ft = 12 in
inch	in	_____

> **Note** 1 inch = 2.5 cm

Common metric units of mass

Name of unit	Symbol	Value in grams	Value in grams (indices)
kilogram	kg	1000	10^3
gram	g	1	10^0
centigram	cg	0.01	10^{-2}
milligram	mg	0.001	10^{-3}

Common imperial units of mass

Name of unit	Symbol	Value
UK ton	t	1 UK ton = 2240 lbs
US ton	t	1 US ton = 2000 lbs
pound	lb	1 lb = 16 oz
ounce	oz	_____

> **Note** 1 kg = 2.2 lbs

Common units of time

Name of unit	Symbol	Value
year	yr	1 yr = 12 mths
month	mth	1 mth = 4 wks
week	wk	1 wk = 7 dys
day	dy	1 dy = 24 h
hour	h	1 h = 60 mins
minute	min	1 min = 60 s
second	s	_____

Common units of capacity and volume

Name of capacity unit	Symbol	Value in litres	Value in litres (indices)
gallon	gal	4.54	4.54×10^0
litre	l	1	10^0
cubic centimetre	cc	0.001	10^{-3}
millilitre	ml	0.001	10^{-3}

Name of volume unit	Symbol	Value in cubic metres	Value in cubic metres (indices)
cubic metre	m³	1	10^0
cubic centimetre	cm³	0.000001	10^{-6}
cubic millimetre	mm³	0.000 000 001	10^{-9}

> **Tip**
>
> When you convert from one unit to another, follow these rules:
> - from larger to smaller units – multiply
> - from smaller to larger units – divide.

Example 93

Convert 200 mm to km.

First, convert from mm to m.

200 mm to m

1 m = 1000 mm

Divide 200 mm by 1000 (10^3) because you're converting from a smaller unit to a larger unit.

200 mm = 0.2 m

Now convert m to km.

1 km = 1000 m

Divide 0.2 m by 1000 (10^3).

$\quad$ 0.2 m = 2×10^{-4} km

200 mm = 2×10^{-4} km

Example 94

Convert 2.5 kg to cg.

First, convert from kg to g.

1 kg = 1000 g

2.5 kg = 2.5×10^3 g (Larger to smaller unit so multiply.)

Now, convert from g to cg.

1 g = 100 cg

2.5×10^3 g = $2.5 \times 10^3 \times 10^2$ cg

$\qquad\qquad$ = 2.5×10^5 cg

$\quad$ 2.5 kg = 2.5×10^5 cg

Example 95

Convert 4 hours to seconds.

$\quad$ 4 hours = 4×60 mins

$\qquad\qquad$ = 240 mins

240 mins = 240×60 s

$\qquad\qquad$ = 14 400 s

$\quad$ 4 hours = 14 400 s

Example 96

A container can hold 4000 cc of liquid. Find the capacity of the container in litres.

4000 cc ÷ 1000 = 4 l

1 Number theory and computation

Exercise 1M

1 Convert the units of length.

   a 14.5 km to m   **b** 200 mm to m

2 Convert to mg.

   a 20.5 cg   **b** 100 g

3 A bucket contains 2.58 litres. What is this equivalent to in cubic centimetres?

4 How many minutes are the same as 3600 seconds?

5 A 400 m runner takes 0.8 minutes to complete a race. What distance did she run in 1 second?

6 A seed is 0.85 mm long. What is the total length of 2000 seeds? Express your answer in metres.

7 A farmer collected 7 bags of grain, each bag containing 64.8 kg. What is the total mass in grams?

8 The capacity of a rectangular container is given by the internal area of the base (length × breadth) times the depth. The dimensions of the base are 20 m by 30 m and the depth is 400 cm. What is the capacity of the container in litres?

9 An object travels a distance of 0.00082 cm in 200 milliseconds. How many metres does it travel in 1 second? (1 millisecond = 10^{-3} seconds)

10 Convert 60 mm³ to m³.

11 A tank can hold 5000 gallons of water. How many plastic bottles of water of capacity 350 ml can be filled from the tank?

12 A boy weighs 98 lbs. His brother is 11 kg lighter. What is their combined mass in lbs?

1.24 Ratio

In order to compare quantities, you may use loose terms like greater than, less than, bigger than, smaller than, etc. The term **ratio** is used to make more specific and direct comparisons between quantities.

Here is an example:

There are three types of fruits in a basket: 3 mangoes, 4 plums and 10 oranges.

Rather than saying there are more oranges than plums, or fewer mangoes than oranges, you can state:

the ratio of mangoes : plums : oranges = 3 : 4 : 10

Paul used a mixture of cement, sand and gravel to make concrete. He used twice the amount of sand as cement and three times the amount of gravel as cement.
What is the ratio cement : sand : gravel?

Let the quantity of cement be 1 unit.

Therefore: the quantity of sand will be 2 × 1 unit = 2 units

the quantity of gravel will be 3 × 1 unit = 3 units

Therefore, the ratio of cement : sand : gravel = 1 : 2 : 3

In a class, 5 children are working on Mathematics, 10 children on English, 15 children on History and 20 children on Physics. What is the ratio of the number of children working on the different subjects?

Mathematics : English : History : Physics = 5 : 10 : 15 : 20

In general, all fractions are ratios. The numerator compares the number of parts under consideration to the denominator, which represents the total number of parts. For example, the fraction $\frac{3}{4}$ compares 3 parts with the total number of parts, 4.

Other examples of ratios

- π ($\approx \frac{22}{7}$): ratio of the circumference of a circle to its diameter
- Probability: number of favourable outcomes : number of possible outcomes
- Boyle's law (Physics): $\frac{P_1}{V_2} = \frac{P_2}{V_1}$
- Number of beats : measure (music)

1.24.1 Reducing a ratio to its lowest terms

In Example 98, the ratio of Mathematics : English : History : Physics = 5 : 10 : 15 : 20

The HCF of the four numbers in the ratio is 5.

Divide each number by the HCF.

5 ÷ 5 : 10 ÷ 5 : 15 ÷ 5 : 20 ÷ 5 = 1 : 2 : 3 : 4

The ratio has been reduced to its lowest terms. This should be done whenever possible.

A ratio is reduced to its lowest term by dividing each number in the ratio by the HCF of the numbers.

1.24.2 Equivalent ratios

Ratios that can be reduced to the same lowest term are called equivalent ratios.

Investigate whether or not these ratios are equivalent ratios.

A : B : C = 2 : 3 : 4   P : Q : R = 10 : 15 : 20   L : M : N = 16 : 24 : 32

Divide each number in each ratio by its respective HCF.

A : B : C = 2 : 3 : 4

P : Q : R = 10 ÷ 5 : 15 ÷ 5 : 20 ÷ 5

= 2 : 3 : 4

L : M : N = 16 ÷ 8 : 24 ÷ 8 : 32 ÷ 8

= 2 : 3 : 4

Therefore, the ratios are all equivalent ratios.

1.24.3 Dividing quantities in a given ratio

Example 100

Share $100.00 among David, Deborah and Winston in the ratio 2 : 3 : 5 respectively.

Total all the values in the ratio.

$2 + 3 + 5 = 10$

1 share = $100 ÷ 10

       = $10

1 share = $10

> **Hint**
>
> First, find out how much 1 share is. Then you can work out all the other shares.

Therefore:

2 shares = $10 × 2

       = $20

3 shares = $10 × 3

       = $30

5 shares = $10 × 5

       = $50

David receives $20.

Deborah receives $30.

Winston receives $50.

1.25 Direct and indirect proportion

1.25.1 Direct proportion

When two or more quantities increase or decrease by the same ratio, they are in **direct proportion**.

For example:

The cost of one orange is $2.00. Therefore, two oranges cost $4.00 and three oranges cost $6.00. Similarly, the cost of 5 lbs of flour is $10.00. The cost of 1 lb flour is $2.00.

Example 101

Sam buys an exercise book for $2.50. How much does he pay for 4 exercise books?

1 book costs $2.50

Using direct proportion, cost of 4 books = $2.50 × 4

                           = $10.00

1.25.2 Indirect (inverse) proportion

Two or more quantities are in **indirect** or **inverse proportion** when, as one quantity increases, the other decreases.

Example 102

5 men build a wall in 10 days. How long will it take 10 men working at the same rate to build the same wall?

More men will take less time.

Doubling the men will halve the time.

$$\text{Time taken} = \frac{10 \text{ days}}{2}$$
$$= 5 \text{ days}$$

1.26 Rates

Examples of rates are hotel rates (prices), service rates (cleaning) and speed. A rate is a measure of how one quantity changes compared to another, for example, bread and money (all prices are rates and ratio because a price is a ratio of money to a measure of quantity).

All rates are therefore good examples of ratios.

Example 103

A hotel charges $40.00 per night. A guest pays $320.00. How many nights can the guest stay for?

cost per night : 1 night = total cost : number of nights

$40.00 : 1 night = $320.00 : number of nights

$$\frac{40}{1} = \frac{320}{\text{number of nights}}$$

$$\text{number of nights} = \frac{320}{40}$$
$$= 8$$

The guest can stay for 8 nights.

Example 104

A cleaner charges $20.00 per hour for his services. How much does he receive after 8 hours of work?

Charge : time = $20 : 1 hour

$x : 8 hours = $20 : 1 hour

Therefore, $20 × 8 : 8 hours

Therefore, $160 : 8 hours

The cleaner receives $160.00 after 8 hours.

Example 105

A vehicle travels at 60 km/h. How far has it travelled in 5 hours?

Distance travelled : time = 60 km : 1 h

5 × 60 km : 5 × 1 h (Multiply both sides by 5.)

300 km : 5 h

The vehicle has travelled 300 km.

1.27 Arithmetic mean (average)

The arithmetic mean or average of a set of numbers is the sum of the numbers divided by the number of values.

Calculate the arithmetic mean of this set of numbers: 4, 5, 6, 7, 8.

$$\text{Arithmetic mean} = \frac{4+5+6+7+8}{5}$$

$$= \frac{30}{5}$$

$$= 6$$

Exercise 1N

1. Four cricketers have scores of 40, 60, 80 and 100. Write the ratio of their scores – from the lowest to the highest – in its simplest form.

2. $60.00 is shared between 3 children in the ratio 1 : 2 : 3. How much does each child receive?

3. A boy buys a ball for $1.50. How much does he pay for 6 balls?

4. Five men build a hut in 20 days. How long will it take 10 men of similar ability to build the same hut?

5. Calculate the average age of five children of ages 10, 8, 6, 4 and 2 years.

6. A boat travels at 150 km/h. How far does it travel in 1 h and 20 minutes?

7. How long does a vehicle travelling at 20 km/h take to cover 18 metres?

8. Three men shared a sum of money in the ratio 3 : 5 : 7. The man who got the largest share got $21.00. What were the other shares?

9. A cricketer purchased cricket gear that cost $540.00. He purchased:
   - 2 bats that cost $75.00 each
   - 2 pairs of gloves at $40.00 a pair
   - 3 cricket balls that cost $60.00 each
   - a helmet.

   He received $30.00 change. What was the price of the helmet?

10. Some mangoes were shared between 3 boys in the ratio 3 : 5 : 7. The boy who received the biggest share got 14 mangoes. How many mangoes were shared?

11. Five students are 4, 7, 3, 6 and 5 years old. What is their average age?

12. A school used 2352 gallons of water in two weeks. Calculate the average amount of water used per day.

13 A platform can support a maximum load of 450 lbs. Some people step onto the platform. The weights of the people are:

- 3 children of average mass $45\frac{3}{4}$ lbs
- 2 men of masses $94\frac{1}{2}$ lbs and $147\frac{1}{4}$ lbs
- 2 women of average mass 107 lbs

  a Will the platform be able to withstand the total mass?

  b What is the difference between the total mass and the maximum load?

14 The average of six numbers is 10. Five of the numbers are 8, 9, 12, 14 and 6. What is the sixth number?

15 A chef creates a recipe in which he uses: 5 lbs flour, $\frac{1}{2}$ lb sugar, $\frac{1}{4}$ lb butter and 1 oz salt. What is the ratio of the ingredients in whole numbers? Use 16 oz = 1 lb.

16 Two tailors take three days to sew five shirts. How many days will six tailors of similar ability take to sew fifteen shirts?

17 The average of 20 numbers is 40. The numbers 35, 40, 45, 50 and 55 are added to the 20 numbers. What is the new average?

18 John wrote 4 tests in English and got an average mark of 78. He scored an average of 81 in 3 Mathematics tests and had an average of 92 marks from 5 history tests. Calculate his average for all the tests.

19 The average height of 15 students in a class is 1.6 m. A new student joined the class and the average height decreased to 1.59 m. What was the height of the new student?

20 A student researched into the average number of corn seeds on a cob. He found out that there was an average of 750 grains in each ear of corn. How many ears will produce a total of 49 500 grains of corn?

21 A document has 126 pages with an average of 274 words per page. How many words does the document have?

22 A school has 20 classes. The average number of students in each class is 40. The average age of students for ten classes is 12 years and the average age in the other ten classes is 14. Write the total number of years of all the students in the school in the form $a \times bn$, where a, b and n are whole numbers

23 Five long jumpers (group A) jumped distances of $7\frac{1}{2}$ m; $8\frac{3}{4}$ m; $6\frac{1}{4}$ m; $9\frac{1}{4}$ m and 8 m. Another five jumpers (group B) jumped an average of $8\frac{1}{4}$ m.

  a Comparing both sets of jumps, which group had the longest combined jump?

  b What is the difference in combined distances between the two groups A and B?

24 There are 40 math students in a class. The top 5 students have an average of 94.8 marks; the next 19 students have an average of 76.3; and the next 9 students have an average of 63.7. The average score for the whole class is 65. Calculate the average score for the other students in the class.

1.28 Sequences

A **sequence** is a set of numbers that follow a pattern or obey a particular rule. Each number in the sequence is called a **term** of the sequence, denoted by the letter *T*.

Sequences can be formed through the application of arithmetic operations: addition, subtraction, multiplication and division or a combination of operations. These form the sequence rules.

Look at these examples.

- The sequence 2, 4, 6, 8,… is formed by adding 2 to each successive term.

    The rule for this sequence is 'add 2'.
- The sequence 2, 4, 8, 16, 32… is formed by multiplying each successive term by 2.

    The rule for this sequence is 'multiply by 2'.

A combination of operations can lead to more complicated 'rules'. For example, the sequence 1, 4, 9, 16, 25… is formed by squaring each successive natural number.

Example 107

Write the sequence with the rule.

(a) Cube each successive natural number.

(b) Start with 25 and subtract 7 from the previous term.

(c) Multiply each successive natural number by 10.

(a) $1^3, 2^3, 3^3, 4^3, 5^3$… = 1, 8, 27, 64, 125…

(b) 25, 25 − 7, 25 − 14, 25 − 21, 25 − 28… = 25, 18, 11, 4, −3…

(c) 1 × 10, 2 × 10, 3 × 10, 4 × 10, 5 × 10 = 10, 20, 30, 40, 50…

Example 108

Create a sequence based on these triangular numbers, which are represented by dots arranged in the shape of a triangle.

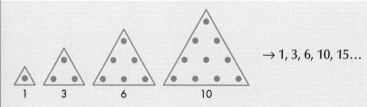

→ 1, 3, 6, 10, 15…

1   3   6   10

Example 109

Given the sequence 1, 3, 5, 7, identify:

(a) the rule

(b) the next two terms in the sequence.

(a) Each term is obtained by adding 2 to the previous term.

    The rule is 'add 2'.

(b) Fifth term = 7 + 2

        = 9

    Sixth term = 9 + 2

        = 11

Example 110

Given the sequence 1, 4, 16, 64, identify:

(a) the rule

(b) the next two terms in the sequence.

(a) Each term is obtained by multiplying the previous term by 4.

The rule is 'multiply by 4'.

(b) Fifth term = 64×4

$= 256$

Sixth term = 256×4

$= 1024$

Example 111

Given the sequence 8, –4, 2, –1, identify:

(a) the rule

(b) the next two terms in the sequence.

(a) 1st term = 8

2nd term = $8 \div -2$

$= -4$

3rd term = $-4 \div -2$

$= 2$

4th term = $2 \div -2$

$= -1$

The rule is 'divide by –2'.

(b) Fifth term = $-1 \div -2$

$= \frac{1}{2}$

Sixth term = $\frac{1}{2} \div -2$

$= -\frac{1}{4}$

Exercise 10

1   What are the next two terms in this sequence? 3, 6, 9, 12, …

2   Look at this sequence: 0.8, 1.0, 1.3, 1.7, 2.2, …

  a What is the rule for this sequence?

  b Identify the next two terms.

3   Write a sequence for the rule: 'the square of each consecutive whole number plus 1'.

4   Write a sequence for the rule: 'half of each consecutive whole number'.

5   Identify the missing term x in the sequence 1, 11, 31, x, 101, 151…

6   Look at this sequence: 1, 10, 100, 1000, 10 000, …

  a What is the rule for this sequence?

  b Identify the next two terms.

7 Identify the next two terms in the sequence $\frac{1}{2}$, 2, $3\frac{1}{2}$, 5, …

8 The first term of a sequence is 1 and the rule is 'the square root of consecutive square numbers'. What are the first five terms in the sequence?

9 What is the rule for this sequence? 1, $\frac{1}{2}$, 3, $\frac{1}{4}$, 5, $\frac{1}{6}$, …

10 Identify the next two terms in this sequence: 4, 0.25, 2, 0.5, 0, 0.75, …

11 What is the rule for this sequence? 20, 19, 15, 6, –10, …

12 What is the first term, *A*, in the following sequence? *A*, 5, 17, 53, 161, …

13 What are the next three terms in the sequence? 1, $\frac{1}{2}$, 3, $\frac{1}{4}$, 5, …

14 Given the rule 'natural number × –0.01' write down the first 5 terms of the sequence in which the first term is –0.01.

1.29 Number bases

The **decimal** or **denary** system (also called the base 10 system) is the most widely used number system today. It consists of ten digits: 0, 1, 2, 3, 4, 5, 6, 7, 8, 9. As you saw on page 7, digits in numbers have different place values, based on their positions.

In addition to the decimal or base 10 number system, there are other groups of numbers or bases. They consist of different digits.

1.29.1 Base 2 (binary system)

The **binary system**, also known as base 2, is used in computers. It consists of two digits: 0 and 1. The largest digit is 1, which is 1 less than the base of 2.

This is a place value table for base 2. Moving from right to left in the place-value table, the powers of 2 increase by 1.

2^4	2^3	2^2	2^1	2^0
	1	0	1	1
1	1	1	0	1

Note The small numbers written after the number indicate the base in which the number is written. Generally, decimal numbers do not include the small 10 except where it is used for clarification.

Look at the binary number 1011_2 shown in the table. To convert it to base 10:

$(1 \times 2^3) + (0 \times 2^2) + (1 \times 2^1) + (1 \times 2^0) = (1 \times 8) + (0 \times 4) + (1 \times 2) + (1 \times 1)$

$$= 8 + 0 + 2 + 1$$

$$= 11_{10}$$

Example 112

Look at the binary number 11101_2 shown in the table. Convert it to base 10.

$(1 \times 2^4) + (1 \times 2^3) + (1 \times 2^2) + (0 \times 2^1) + (1 \times 2^0) = (1 \times 16) + (1 \times 8) + (1 \times 4) + (0 \times 2) + (1 \times 1)$

$$= 16 + 8 + 4 + 0 + 1$$

$$= 29_{10}$$

Example 113

Convert 20 in base 10 to binary.

Divide the number repeatedly by 2 and write down the remainder for each division. The answer will be the remainders, taken from the bottom to the top.

2	20
2	10 remainder 0
2	5 remainder 0
2	2 remainder 1
2	1 remainder 0
	0 remainder 1

20 in base 10 = 10100_2

1.30 Operations with binary numbers

These basic operations consist of: addition, subtraction, multiplication and division.

1.30.1 Addition of binary numbers

Example 114

Solve $1010_2 + 1111_2$.

$$1010_2 + 1111_2: \quad \begin{array}{r} 1010 \\ +1111 \\ \hline 11001 \\ 11 \end{array}$$

Start with 2^0: $1 + 0 = 1$

2^1: $1 + 1 \quad = (1 \times 2^2) + 0$

2^2: $1 + 0 + 1 = (1 \times 2^3) + 0$

2^3: $1 + 1 + 1 = (1 \times 2^4) + 1$

2^4: 1

$1010_2 + 1111_2 = 11001_2$

Example 115

Solve $1011_2 + 1101_2 + 111_2$.

$$1011_2 + 1101_2 + 111_2: \quad \begin{array}{r} 1011 \\ 1101 \\ + 111 \\ \hline 11111 \\ 111 \end{array}$$

$1011_2 + 1101_2 + 0111_2 = 11111_2$

Tip	For addition: $0 + 1 = 1$ $1 + 1 = 0$ carry 1 $1 + 1 + 1 = 1$ carry 1.

1.30.2 Subtraction of binary numbers

Example 116

Solve $111_2 - 11_2$.

$$111_2 - 11_2: \quad \begin{array}{r} 111 \\ - 11 \\ \hline 100 \end{array}$$

$111_2 - 11_2 = 100_2$

1 Number theory and computation

Example 117

Solve $1100_2 - 11_2$.

$$1100_2 - 11_2: \quad -\begin{array}{r} 1\overset{0}{1}\overset{1\!\!\!1}{1}\overset{1}{0}0 \\ 1\,1 \\ \hline 1\,0\,0\,1 \end{array}$$

Start with 2^0: $0 - 1$

Because 1 is greater than 0, take 1×2^1 from the 2^1 column.

Change 1×2^1 to 2×2^0.

$2 - 1 = 1$

Write 1 in the 2^0 place in the answer.

Continue with 2^1 and so on.

$1100_2 - 11_2 = 1001_2$

1.30.3 Multiplication of binary numbers

Example 118

Solve $1101_2 \times 111_2$.

Use the long multiplication method, and also follow the rules of binary addition.

$$\begin{array}{r} 1\,1\,0\,1 \\ \times \quad 1\,1\,1 \\ \hline 1\,1\,0\,1 \\ 1\,1\,0\,1\,0 \\ 1\,1\,0\,1\,0\,0 \\ \hline 1\,0\,1\,1\,0\,1\,1 \\ \hline {\scriptstyle 1\ 1\ 1} \end{array}$$

$1101_2 \times 111_2 = 1011011_2$

1.30.4 Division of binary numbers

Use the long division method, and also follow the rules of binary subtraction.

Example 119

Solve $11100_2 \div 100_2$.

$$\begin{array}{r} 1\,1\,1 \\ 1\,0\,0\,\overline{)1\,1\,1\,0\,0} \\ -\,1\,0\,0 \\ \hline 1\,1\,0 \\ -\,1\,0\,0 \\ \hline 1\,0\,0 \\ 1\,0\,0 \\ \hline 0 \end{array}$$

$11100_2 \div 100_2 = 111_2$

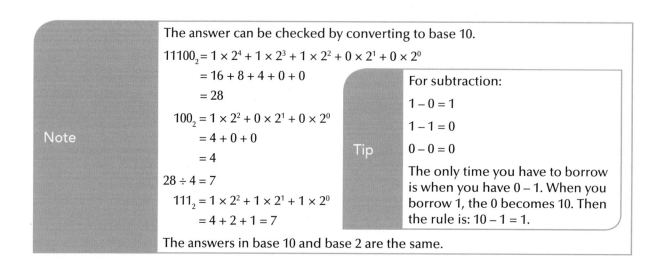

Note

The answer can be checked by converting to base 10.

$11100_2 = 1 \times 2^4 + 1 \times 2^3 + 1 \times 2^2 + 0 \times 2^1 + 0 \times 2^0$

$\quad = 16 + 8 + 4 + 0 + 0$

$\quad = 28$

$100_2 = 1 \times 2^2 + 0 \times 2^1 + 0 \times 2^0$

$\quad = 4 + 0 + 0$

$\quad = 4$

$28 \div 4 = 7$

$111_2 = 1 \times 2^2 + 1 \times 2^1 + 1 \times 2^0$

$\quad = 4 + 2 + 1 = 7$

The answers in base 10 and base 2 are the same.

Tip

For subtraction:

$1 - 0 = 1$

$1 - 1 = 0$

$0 - 0 = 0$

The only time you have to borrow is when you have $0 - 1$. When you borrow 1, the 0 becomes 10. Then the rule is: $10 - 1 = 1$.

1.31 Base 4 and base 8

Two other number bases commonly used are base 4 and base 8 (octal).

1.31.1 Base 4

The digits that make up base 4 are: 0, 1, 2, and 3.

You can convert from base 10 to base 4 and vice versa (following the procedure established with binary).

Example 120

Convert 30_{10} to base 4.

4	30
4	7 remainder 2
4	1 remainder 3
	0 remainder 1

$30_{10} = 132_4$

A number written in base 4 can be expanded to give a number in base 10.

Example 121

Convert 1023_4 to base 10.

$(3 \times 4^0) + (2 \times 4^1) + (0 \times 4^2) + (1 \times 4^3) = 3 + 8 + 0 + 64$

$\quad = 75$

$1023_4 = 75_{10}$

1.31.2 Operations with base 4 numbers

Addition in base 4

Example 122

$123_4 + 102_4$

$$123_4 + 102_4 : \begin{array}{r} 1\,2\,3 \\ +\,1\,0\,2 \\ \hline 2\,3\,1 \\ {\scriptstyle 1} \end{array}$$

$123_4 + 102_4 = 231_4$

Subtraction in base 4

Example 123

Solve $231_4 - 133_4$.

$$231_4 - 133_4 : \begin{array}{r} {\scriptstyle 1\,1\,2\ \ 1} \\ 2\,3\,1 \\ -\,1\,3\,3 \\ \hline 3\,2 \end{array}$$

$231_4 - 133_4 = 32_4$

Multiplication in base 4

Example 124

Solve $213_4 \times 23_4$.

$$213_4 \times 23_4 : \begin{array}{r} 2\,1\,3 \\ \times\ \ \ 2\,3 \\ \hline 1\,3\,1\,1 \\ 1\,0\,3\,2\,0 \\ \hline 1\,2\,2\,3\,1 \\ {\scriptstyle 1} \end{array}$$

Use long multiplication and the rules for addition in base 4.

Start with the 4^0: $3 \times 3 = 9$

$$= 2 \times 4^1 + 1$$

Write 1 in the 4^0 column of the answer.

Write 2 under the 4^1 column of the answer.

4^1: $(3 \times 1) + 2 = 5$

$$= (1 \times 4^1) + 1$$

Write 1 in the 4^1 column of the answer.

Write 1 under the 4^2 column of the answer.

Continue with the 4^3 column and so on.

$213_4 \times 23_4 = 12231_4$

Division in base 4

Solve $320_4 \div 32_4$.

Use long division.

$$320_4 \div 32_4 : \quad 32 \overline{)\begin{array}{r} 10 \\ 320 \\ -32 \\ \hline 00 \end{array}}$$

$320_4 \div 32_4 = 10_4$

1.31.3 Base 8 (octal)

You can convert from base 10 to base 8 and vice versa.

Convert 105_{10} to base 8.

8	105
8	13 remainder 1
8	1 remainder 5
	0 remainder 1

$105_{10} = 151_8$

Convert 567_8 to base 10.

$$567_8 = (7 \times 8^0) + (6 \times 8^1) + (5 \times 8^2)$$
$$= (7 \times 1) + (6 \times 8) + (5 \times 64)$$
$$= 7 + 48 + 320$$
$$= 375$$

$567_8 = 375_{10}$

1.31.4 Operations with base 8 (octal) numbers

Addition in base 8

Solve $567_8 + 246_8$.

$$567_8 + 246_8 : \quad \begin{array}{r} 5\,6\,7 \\ +2\,4\,6 \\ \hline 1\,0\,3\,5 \\ {\scriptstyle 1\ \ 1} \end{array}$$

$567_8 + 246_8 = 1035_8$

Subtraction in base 8

Solve $1211_8 - 732_8$.

$$1211_8 - 732_8 : \quad \begin{array}{r} 1\ {}^1\!2\ {}^{10}\!\cancel{1}\ {}^1\!1 \\ \ \ 7\ 3\ 2 \\ \hline 2\ 5\ 7 \end{array}$$

$1211_8 - 732_8 = 257_8$

Multiplication in base 8

Example 130

Solve $672_8 \times 45_8$.

$$672_8 \times 45_8 : \quad \begin{array}{r} 6\,7\,2 \\ \times \quad\ 4\,5 \\ \hline 4\,2\,4\,2 \\ 3\,3\,5\,0\,0 \\ \hline 3\,7\,7\,4\,2 \end{array}$$

Start with 8^0: $5 \times 2 = 10 = (1 \times 8^1) + 2$

Write 2 in the 8^0 place in the answer.

Write 1 under the 8^1 column in the answer.

8^2: $(5 \times 7) + 1 = 36$

$\qquad = (4 \times 8^1) + 4$

Write 4 in the 8^1 place in the answer.

Write 4 under the 8^2 column in the answer.

Continue for 8^2 and so on.

$672_8 \times 45_8 = 37742_8$

Division in base 8

Example 131

Solve $132_8 \div 12_8$.

$$132_8 \div 12_8 : \quad 12\,\overline{\smash{)}\,\begin{array}{l}1\,1 \\ 1\,3\,2 \\ \end{array}}$$
$$\begin{array}{r} -1\,2 \\ \hline 1\,2 \end{array}$$

$132_8 \div 12_8 = 11_8$

Exercise 1P

1 Expand the number 87569_{10} to show the place value of each digit. Add them to obtain the number.

2 Convert 21_{10} to binary.

3 Give four examples of binary numbers.

4 Look at the numbers below. State which number base each number is most likely to belong to.

    a 789   **b** 11211   **c** 1004333   **d** 101116   **e** 1234567

5 Convert 10110_2 to base 10.

6 Solve $101001_2 + 10110_2$.

7 Solve $101001_2 - 100001_2$.

8 Solve $1011_2 \times 111_2$.

9 Solve $1111_2 \div 101_2$.

10 Convert 235_{10} to base 4.

11 Convert 333_4 to base 10.

12 Solve $1230_4 + 233_4$.

13 Solve $1211_4 - 322_4$.

14 Solve $1122_4 \times 222_4$.

15 Solve $2310_4 \div 33_4$.

16 Convert 988_{10} to base 8.

17 Convert 726_8 to base 10.

18 Solve $257_8 + 456_8$.

19 Solve $1117_8 - 222_8$.

20 Solve $2356_8 \times 67_8$.

21 Solve $666_8 \div 222_8$

22 Solve $78_{10} + 1110_2 - 233_4 + 25_8$, giving your answer in base 10.

23 Solve $123_4 \times 675_8$, giving your answer in base 10.

24 Solve $10101_2 \times 897_{10}$, giving your answer in base 4.

25 Solve $9_{10} \times 7_8 \times 3_4 \times 11_2$, expressing your answer in base 2,

26 Solve $123_4 \div 27_8$, giving your answer in base 8.

1.32 The use of calculators in number theory and computation

This chapter best lends itself to the use of calculators. Using a calculator can reinforce your conceptual knowledge. You will learn how to carry out the different mathematical procedures and use the algorithms to solve mathematical problems and not rely totally on calculators. For example, calculators can be very useful in performing operations with large whole numbers. However, you should know how to manipulate small numbers without the use of calculators and be able to apply formulae and rules like the BODMAS rule.

Some applications of calculators in number theory and computation:

- The four basic operations (addition, subtraction, multiplication and division): the keys (+) (×) (−) and (÷) together with the (=) sign are used.
  - Example: 40 + 60 = 100 (answer appearing on the screen).
- Squares and powers in general: The keys (x^2) or (x^y) or (shift x^3) can be used.
  - Example: $(14)^2$: the keys (1) (4) (x^2) = 196
- Square root and cube root
  - Example: $\sqrt[3]{1000}$
- Reciprocal: key ($\frac{1}{x}$)
  - Example: reciprocal of 4
- Fraction to percentage
  - Example: $\frac{4}{5}$ converted to percentage
- Reducing a fraction to its lowest terms
  - Example: $\frac{42}{60}$ to its lowest terms and LCM/HCF of 42 and 60

Notes

- The calculator is very useful for checking your answers, especially when calculating with large numbers. However, it is very important that conceptual understanding is established before using calculators.
- Some calculators have more features and are able to perform more operations.

End of chapter summary

Hindu–Arabic numerals

1   2   3   4   5   6   7   8   9   0

Roman numerals

Symbol	I	V	X	L	C	D	M
Value	1	5	10	50	100	500	1000

Place value of whole numbers

Ten millions	Millions	Hundred thousands	Ten thousands	Thousands	Hundreds	Tens	Units
7	3	6	2	1	0	8	4

- Place value table shows 73 621 084
- Face value of the digit 2 is 2
- Place value of the digit 2 is 20 000

Operations: addition, subtraction, division and multiplication

- **Defined arithmetic operations** are combinations of the basic operations of addition, subtraction, multiplication and division.
- The **associative law** states that for the operations addition and multiplication, the way the numbers are grouped does not affect the answer.

  $(a + b) + c = a + (b + c)$

  $(a \times b) \times c = a \times (b \times c)$

- The **commutative law** states that the order in which numbers are added or multiplied does not affect their total or product.

  $a + b + c = a + c + b = c + b + a$

  $a \times b \times c = a \times c \times b = c \times b \times a$

- The **distributive law** states that multiplication is distributive over addition and subtraction.

  $a(b + c) = ab + ac$

  $a(b - c) = ab - bc$

- The **additive identity is 0** because a number remains unchanged when 0 is added.
- The **multiplicative identity is 1** because a number remains unchanged when multiplied by 1.
- The **additive inverse** of a number n is $-n$.
- The **multiplicative inverse** of a number n is $\frac{1}{n}$.
- **Multiplication by zero:** $n \times 0 = 0$
- **Division by zero:** $n \div 0$ is undefined
- **Division of zero:** $0 \div n = 0$

- Order of operations is given by BODMAS:

  B brackets

  O orders/of (powers of numbers)

  D division

  M multiplication

  A addition

  S subtraction

Powers of numbers

- The power or index indicates how many times the base is multiplied by itself.
- $a \times a \times a \times a \times \dots \times a = a^b$ where a is the base and b is the power.
- The square number $n^2 = n \times n$
- The cube number $n^3 = n \times n \times n$
- The square root of a number $n = \sqrt{n} = n^{\frac{1}{2}}$
- The reciprocal of a number n is $\frac{1}{n} = n^{-1}$

Factors, prime numbers and multiples

- A **factor** of a number n is any positive or negative that can divide exactly into n (without a remainder).
- The number 1 is a factor of all numbers.
- A **prime number** is a number that has only two factors: itself and 1.
- A **prime factor** is a factor that is also a prime number.
- A **multiple** of a number n is any number that n is a factor of.
- The **highest common factor (HCF)** of two or more numbers is the largest number that is a common factor of the numbers.
- The **lowest common multiple (LCM)** of two or more numbers is the lowest number that is a multiple of all of the numbers.

Fractions

- A **fraction** is a part of a whole that has been divided into equal parts.
- The **denominator** of a fraction shows the number of equal parts that a whole has been divided into.
- The **numerator** of a fraction shows the number of equal parts being considered.
- A **unit fraction** is a fraction with numerator 1. For example: $\frac{1}{3}, \frac{1}{5}, \frac{1}{20}$
- A **proper or vulgar fraction** is a fraction with a numerator less than the denominator. For example: $\frac{2}{3}, \frac{4}{5}, \frac{12}{15}$
- An **improper fraction** (or top-heavy fraction) is a fraction with a numerator greater than the denominator. For example: $\frac{7}{6}, \frac{5}{4}, \frac{18}{15}$
- A **mixed number** is a number that consists of a whole number and a fractional part. It is another way of writing an improper fraction.
- Equivalent fractions are fractions that have different numerators and denominators but have equal value.

  $\frac{a}{b} = \frac{na}{nb}$

- A fraction **in its lowest terms** has a numerator and a denominator that have no common factors.
- A fraction in its **simplest form** is in its lowest terms.

Operations and fractions

- $\dfrac{a}{b} + \dfrac{c}{b} = \dfrac{a + c}{b}$
- $\dfrac{a}{b} - \dfrac{c}{b} = \dfrac{a - c}{b}$
- $\dfrac{a}{b} + \dfrac{c}{d} = \dfrac{(a \times d) + (c \times b)}{b \times d}$
- $\dfrac{a}{b} - \dfrac{c}{d} = \dfrac{(a \times d) - (c \times b)}{b \times d}$
- **Cancelling of fractions** involves dividing numerators and denominators by common factors.
- $\dfrac{a}{b} \times \dfrac{c}{d} = \dfrac{ac}{bd}$
- $\dfrac{a}{b} \div \dfrac{c}{d} = \dfrac{a}{b} \times \dfrac{d}{c} = \dfrac{ad}{bc}$

Place value of decimals

Hundreds	Tens	Units	Decimal point	tenths	hundredths	thousandths	ten thousandths
4	0	7	.	3	0	1	9

- Place value table shows 407.3019
- Face value of the digit 1 is 1
- Place value of the digit 1 is 0.001

Decimals

- $\dfrac{1}{10} = 0.1$, $\dfrac{1}{100} = 0.01$, $\dfrac{1}{2} = 0.5$, $\dfrac{1}{4} = 0.25$, $\dfrac{3}{4} = 0.75$
- $\dfrac{a}{b} = a \div b$
- $0.abcd = \dfrac{abcd}{10\,000}$
- A **terminating decimal** is a decimal with a fixed number of places. For example: 1.2345, 0.4, 15.678
- A **recurring decimal** is a decimal where the digits repeat infinitely. For example: 0.33333…, 0.66666…, 0.14141…
- 2504.35 has two decimal places and six significant figures.
- 0.00007102 has eight decimal places and four significant figures.
- Numbers in **standard form** are shown by $A \times 10^n$, where A is a number between 1 and 10 ($1 \leq A < 10$) and n is a positive or negative integer.
- $789.34 = 7.8934 \times 10^2$
- $0.0012 = 1.2 \times 10^{-3}$

Percentages

- A **percentage** is a type of fraction with a denominator of 100.
- The **symbol %** shows that this is a percentage.
- To convert a fraction or decimal to a percentage, multiply by 100%.
- $x\%$ of $n = \dfrac{x \times n}{100}$
- A quantity a as a percentage of $b = \dfrac{a}{b} \times 100\%$.

Converting between units of measure

- To convert from a larger to a smaller unit, multiply.
- To convert from a smaller to a larger unit, divide.

Ratio and proportion

- **Ratio** shows specific and direct comparison between quantities.
- When quantities A and B are in the ratio 1 : 3, there is three times as much B as A.
- When quantities A and B are in the ratio 2 : 5, there are two lots of A for every 5 lots of B.
- When two or more quantities are in **direct proportion**, then when one quantity increases or decreases the others increase or decrease in the same ratio.
- When two or more quantities are in **indirect** or **inverse proportion**, then when one quantity increases the other decreases.

Arithmetic mean (average)

- The **arithmetic mean** or **average** of a set of numbers is the sum of the numbers divided by the number of values.

Sequences

- A **sequence** is a set of numbers that follow a pattern or obey a particular rule.
- Each number in the sequence is called a **term**, denoted by the letter t.

Classification of numbers

- **Natural or counting numbers** ($\mathbb{N}$) are the numbers: 1, 2, 3, 4, etc.
- **Whole numbers** ($\mathbb{W}$) are the numbers: 0, 1, 2, 3, 4, 5, etc.
- **Integers** ($\mathbb{Z}$) are all the positive and negative natural numbers and 0.
  Examples of integers: –5, –4, –3, –2, –1, 0, 1, 2, 3, 4, 5, 6, etc.
- **Rational numbers** ($\mathbb{Q}$) are the numbers that can be written as fractions:
  - integers
  - negative and positive fractions
  - negative and positive decimals (terminating and recurring decimals).
    Examples of rational numbers are: $-3, 0, 7, \frac{1}{2}, -\frac{3}{4}, 0.87, -0.5, 4$.
- **Irrational numbers (P)** are numbers that cannot be written as fractions or terminating or recurring decimals.
  Examples of irrational numbers: $\pi, \sqrt{2}$ ($\pi = 3.141592654\ldots$; $\sqrt{2} = 1.414213562\ldots$)
- **Real numbers** ($\mathbb{R}$) are all the rational and irrational numbers.
  Examples of real numbers are: $\sqrt{49}, -\sqrt{5}, \frac{1}{4}, 0, 12$.
- **Imaginary numbers (A)** are multiples of the square root of –1 and expressions that include the square root of –1.
  Examples of imaginary numbers are: $\sqrt{-1}, \sqrt{-9}, 2\sqrt{-1}$
- A **rectangle number** is a number that can be represented by a number of objects or a series of dots arranged in the shape of a rectangle.
  Examples of rectangle numbers: $12 = 3 \times 4$ and $56 = 7 \times 8$

- A **triangle number** is a number that can be represented by a number of objects or a series of dots arranged in the shape of an equilateral triangle.

  The first five triangle numbers are: 1, 3, 6, 10, and 15.

- **Composite numbers** are numbers that are divisible by numbers other than themselves and 1.

- **Even numbers** are integers that are exactly divisible by 2 (that is, with no remainder). Examples of even numbers are: 0, 2, 4, 6, −8. An even number can be negative or positive.

- **Odd numbers** are all the integers that are not exactly divisible by 2. Examples are: −7, 1, 3, 5, 7.

Closure

- A set of numbers is **closed** under an operation if, when the operation is carried out on any two numbers from the set, the result is also a member of the set. This is called the **law of closure**.

Number bases

- The **decimal** or **denary** system (also called the **base 10 system**) consists of ten digits: 0, 1, 2, 3, 4, 5, 6, 7, 8, 9.

- The **base 2** or **binary system** (also known as base 2) consists of two digits: 0 and 1.

2^4	2^3	2^2	2^1	2^0
1	0	1	0	1

- The number shown in the base 2 place value table is 10101_2.
- The face value of the first 1 in 10101_2 is 1.
- The place value of the first 1 in 10101_2 is $1 \times 2^4 = 1 \times 16 = 16$.
- The **base 4** system consists of four digits: 0, 1, 2, and 3.

4^4	4^3	4^2	4^1	4^0
2	0	1	3	1

- The number shown in the base 4 place value table is 20131_4.
- The face value of the 3 in 20131_4 is 3.
- The place value of the 3 in 20131_4 is $3 \times 4^1 = 3 \times 4 = 12$.
- The **base 8** or **octal system** consists of seven digits: 0, 1, 2, 3, 4, 5, 6 and 7.

8^4	8^3	8^2	8^1	8^0
5	3	4	7	2

- The number shown in the base 8 place value table is 53472_8.
- The face value of the 4 in 53472_8 is 4.
- The place value of the 4 in 53472_8 is $4 \times 8^2 = 4 \times 64 = 256$.

Examination-type questions for Chapter 1

1 Using a calculator or otherwise, calculate the exact value of the following.

a $\dfrac{10\frac{1}{2} - 2(3\frac{1}{2} + 1\frac{1}{2})}{3\frac{3}{4} \times \frac{2}{5}}$

b $\dfrac{5.16 \div 1.2 - (1.5)^2}{1.3 + 1.2}$

2 Evaluate $\sqrt{125} - (1.4)^2$ and give your answer correct to:

  a two decimal places

  b three significant figures.

3 **a** In a class, the ratio of boys to girls is 3 : 2 respectively. There are 20 girls in the class.

    i How many boys are in the class?

    ii How many students are there in total in the class?

  b A teacher wanted to share some sweets among 4 children. She decided to share the sweets in the ratio 2 : 3 : 4 : 6 based on the ages of the children. The oldest child got 12 sweets.

    i What was the total amount shared?

    ii How much did each of the other 3 children receive?

4 **a** After 5 innings, a cricket team has scored an average of 280 runs. The runs scored for 4 of those innings were: 250, 300, 320 and 200. What was their score for the fifth innings?

  b A bus takes 5 hours for a trip, travelling at an average speed of 50 km/h.

    i What is the length of the trip?

    ii How much time will the bus take if its average speed is increased to 70 km/h?

5 **a** Given $2302_4 = 2K + (3 \times 4^2) + 0 + (2 \times 4^0)$ find the value of K.

  b Convert $(5 \times 8^4) + (6 \times 8) + (7 \times 8^0)$ to a number in base 8.

  c Express $4(4^4 + 4^3 + 1)$ as a number in base 4.

6 A contractor discovered that, over the last 5 years, the cost of building a two-bedroom house has increased from $140 000.00. This was due to the increase in the cost of materials, labour and overheads. Previously, costs were divided between materials, labour and overheads in the ratio 14 : 10 : 4. Over the last 5 years, these costs have increased by 5%, 10% and 15% respectively. Calculate the percentage increase in the cost of the house as a result.

2 Consumer arithmetic

Objectives

By the end of this chapter, you should be able to:

- calculate discount, sales tax, profit and loss
- calculate percentage profit and percentage loss
- express profit, loss, discount, mark up and purchase tax as a percentage of a basic value
- solve problems involving marked price, selling price, cost price, profit, loss or discount
- solve problems involving payments by instalments, such as hire purchase and mortgages
- solve problems involving simple interest, compound interest (use of formulae), appreciation and depreciation, measures and money (currency conversion), rates and taxes, utility bills, invoices and shopping bills, salaries and wages, insurance and investments.

You should already know:

- the basics of number theory and computation.

Introduction

In this chapter you will learn how to perform calculations relating to normal, everyday business transactions, including calculations associated with personal budgets, personal shopping, payment of bills and salaries.

The chapter will also help you to develop an appreciation for speed and accuracy in everyday financial transactions; it addresses some advantages and disadvantages of particular investments and models and solutions to real-world problems.

Additionally, the chapter clarifies a number of financial calculations that many people find challenging, such as simple and compound interest, taxes, hire purchase, mortgages and utilities payments.

Examples of consumer arithmetic in real life

Black Friday deals offer massive discounts on various items.

Some shops offer discount cards to regular customers.

Prices of groceries may be displayed electronically.

Profit sharing means that employees have a share in profits.

In duty-free shops the prices of goods are dramatically reduced due to the removal of duties.

The bar graph shows a steady percentage increase in workers' salaries.

Money in the form of notes and coins: an important ingredient in the financial process.

Financial success – your mortgage loan is approved!

A mortgage is an important financial undertaking.

Cash may be deposited or withdrawn at a bank.

Money may be changed from one currency to another.

Money may be transferred electronically, in online purchasing.

2.1 Discount, sales tax, profit and loss

2.1.1 Discount

Shops usually display their goods showing the price they want you to pay. This is the marked price. However, they often offer a reduction in the price, or a **discount**, so you may pay less than the marked price.

Discount offered on a TV set

Calculation of discount

You can calculate the discounted price of an item by subtracting the amount of the discount from the marked price:

discounted price = marked price – discount

The discount is often expressed as a percentage of the marked price. To calculate the percentage, divide the discount by the original (marked) price and multiply by 100:

$$\text{percentage discount} = \frac{\text{discount (\$)}}{\text{marked price (\$)}} \times 100\%$$

> **Key term**
>
> **Discount** is the amount, expressed as a quantity ($) or percentage (%), that the seller takes off the marked price so that the customer pays less for the item.

When you know the percentage discount, you can determine the discounted price by first calculating the discount and then subtracting that amount from the marked price.

discounted price = marked price – discount percentage of marked price

Alternatively:

$$\text{discounted price} = \frac{(100 - \text{percentage discount}) \times \text{marked price}}{100}$$

Example 1

A student was given a discount of $35.00 on a book priced at $80.00. How much did she pay for the book?

Discounted price = marked price – discount

$\qquad\qquad = \$80 - \$35$

$\qquad\qquad = \$45.00$

Example 2

A shirt priced at $60.00 was sold at $45.00. What was the percentage discount?

$\dfrac{60 - 45}{60} \times 100 = \dfrac{15}{60} \times 100$

$\qquad\qquad = 25\%$

Example 3

During a Christmas sale a television marked at $2000.00 was offered with a 20% discount. What was the final price of the television?

20% of $2000 = \dfrac{20}{100} \times 2000$

$\qquad\qquad = \$400.00$

$\Rightarrow$ final cost $= 2000 - 400$

$\qquad\qquad = \$1600.00$

Alternatively, since a 20% discount is offered, the price of the television becomes $100\% - 20\% = 80\%$ of its original value.

80% of $\$2000.00 = \dfrac{80}{100} \times 2000$

$\qquad\qquad\quad = \$1600.00$

Example 4

Tom bought his cell phone at a discounted price of $2000.00. If the discount was 10%, what was the original or marked price?

Using the formula:

discounted price = marked price – discount percentage of marked price

$\qquad 2000 =$ marked price $- 10\%$ of marked price

$\qquad\qquad = (1 - 0.1)$ marked price

$\qquad\qquad = 0.9$ marked price

$\qquad$ Marked price $= \dfrac{2000}{0.9}$

$\qquad\qquad\qquad = \$2222.22$

Exercise 2A

1. A radio marked for $200.00 is sold at a 5% discount. What is the discounted price?

2. A barber offers his customers a discount of 5¢ in the dollar. How much does a trim priced at $15.00 cost?

3. A guitar originally priced at $240.00 was sold for a discounted price of $200.00. What is the percentage discount?

4. A shopkeeper sells envelopes at $2.00 each. He offers a discount of 5% on each envelope. Calculate the discounted cost of 45 envelopes.

5 A boy buys a book at a discounted price of $80.00. If the discount given is 8%, what is the marked price?

6 A store owner gives a 5% discount on all his bicycles. For a bicycle priced at $240.00 (without discount) he decides to give an additional discount of 3%. What is the final price of the bicycle?

2.1.2 Sales tax

On almost everything you buy, the government charges a **sales tax** to help them raise money to pay for the services they provide for the country.

Example 5

A boy buys a bicycle priced at $300.00 and sales tax of 10% is added. How much does he pay for the bicycle?

$$10\% \text{ of } 300 = \frac{10}{100} \times \$300$$

$$= \$30.00$$

So the amount to be paid = $300 + $30

$$= \$330$$

Exercise 2B

1 A girl buys a dress originally priced at $60.00, with an additional sales tax of 7%. What is the final price of the dress?

2 A refrigerator originally priced at $2000.00 is sold with an additional sales tax of 14%. What is the final price of the refrigerator?

3 The cost of a KFC kid's meal increases from $10.00 to $15.00 with sales tax. What is the percentage sales tax?

4 The cost of a shirt, after a 4% sales tax is added, is $50.00. What was the original price of the shirt?

2.1.3 Profit and loss (per cent)

Whenever goods and services are bought and sold, there is likely to be a difference between the **cost price** (CP) and the **selling price** (SP). Then the person offering the goods or services will make a **profit** or a **loss**.

Profit (*P*) is selling price – cost price
(*P* = SP – CP)

Loss (*L*) = cost price – selling price
(*L* = CP – SP)

Example 6

A man bought a used vehicle for $18000.00 and sold it for $20000.00. What was the profit?

Profit = SP – CP

$\quad\quad$ = 20000 – 18000

$\quad\quad$ = $2000

Example 7

A student pays $120.00 for a textbook and sells it to another student for $80.00. Did he make a profit or loss and what was the amount?

CP = $120.00; SP = $80.00 ⇒ CP > SP ⇒ loss

The loss is 120 – 80 = $40.

2.1.4 Percentage profit and loss

Profit and loss may be expressed as percentages of the cost price.

Example 8

A student bought a laptop for $1200.00 and sold it for $1400.00. What was his **percentage profit**?

$$\text{Percentage profit} = \frac{1400 - 1200}{1200} \times 100\%$$

$$= \frac{200}{1200} \times 100\%$$

$$= 16.7\%$$

Example 9

Allana bought a cell phone for $600.00 and sold it to Mary for $580.00. Determine her **percentage loss**.

$$\text{Percentage loss} = \frac{\text{cost price} - \text{selling price}}{\text{cost price}} \times 100\%$$

$$= \frac{600 - 580}{600} \times 100\%$$

$$= \frac{20}{600} \times 100\%$$

$$= 3.3\%$$

Exercise 2C

1 A boy sells his toy train to his friend for $5.60. Given that he paid $6.25 for the toy train how much was his profit or loss?

2 One bottle of soft drink costs the shopkeeper $2.00 to buy. The shopkeeper sold a case of 24 bottles of soft drink at $1.50 each. What was the loss or profit he incurred?

3 A shopkeeper buys 80 exercise books for $200.00 and sells them to schoolchildren at $2.00 each. Does he make a profit or loss and how much is it?

4 A boy bought a cricket bat for $80.00 and sold it for $60.00. Calculate the percentage loss.

5 James bought a bus for $80 000 and sold it for $90 000. Determine the percentage profit.

6 Andrea purchased a pen for $8.95 and then sold it to her friend for $7.90. Calculate the percentage loss.

7 A woman made a profit of 5% when she sold a household item for $25.50. What was the cost price of the item?

8 A householder bought a new washing machine for $4850.00 and sold it to a neighbour, to make a profit of 7%. What was her selling price?

2.1.5 Percentage change

Generally, when any quantity increases or decreases, you can calculate the percentage change in the numerical value in a similar way as you find percentage profit and loss.

Example 10

The number of boys in class 2J increased from 8 to 10. What is the percentage increase in the number of boys?

$$\text{Percentage increase} = \frac{10 - 8}{8} \times 100$$

$$= \frac{2}{8} \times 100$$

$$= 25\%$$

Example 11

When a woman reduced the number of hours she worked, the amount she was paid was reduced from $2800.00 to $2400.00 per month. Calculate the percentage decrease in the amount she was paid.

$$\text{Percentage decrease} = \frac{2800 - 2400}{2800} \times 100$$

$$= \frac{400}{2800} \times 100$$

$$= 14.3\%$$

Tips	Take cost price as 100%.
	A price that leads to a loss will be less than 100% of the CP.
	A price that leads to a profit will be more than 100% of the CP.
	When calculating percentage profit or loss, cost price is always the denominator.

Exercise 2D

1 John's overall performance as a cricketer was enhanced when his tally of wickets moved from 16 to 20. What was his percentage increase in wickets?

2 The students' performance in mathematics increased from 24 to 30 passes. What was the percentage increase in passes?

3 The number of accidents in a country made a big jump from 12 to 20 during the course of one year. What was the percentage increase?

4 According to one statistician, the number of people using social media over the last 9 years increased by 95% to 9.5 billion. What was the original number?

5 The amount a man weighed decreased by 0.5% after 3 days of exercise. He now weighs 220 lbs. What did he originally weigh?

6 After receiving a 7.2% increase in pay, an electrician was paid $2045.00 per month. What would his pay be if it had been increased by 11.3%? Calculate the difference between the two increases.

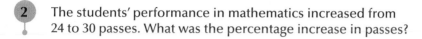

2.2 Simple and compound interest

When someone borrows money from a bank, they are expected to pay the bank a charge, for borrowing the money. Likewise, when you put money into a savings account, the bank generally pays a fee because they can use the money you deposit to finance other transactions. This charge, or fee, is called **interest** and is paid at regular intervals, based on a percentage set by the bank.

Key term
Interest is the profit earned from investing or saving a sum of money in a bank or company, or the amount paid for borrowing a sum of money from a financial institution or lender.

Key terms
The amount borrowed or deposited is the **principal**, the percentage rate is the rate at which interest is calculated and the total of the principal and interest, at the end of the period of the loan or investment, is the **amount**.

The money invested or borrowed is the **principal**. The percentage earned on the investment, or paid on the loan, is the **percentage rate** and is usually calculated annually. The interest earned (or paid) is calculated as this percentage of the principal. The total obtained by adding the interest to the principal is called the **amount**.

2.2.1 Simple interest

With **simple interest**, the amount steadily increases over time, while the principal remains unchanged. The amount, therefore, is the total sum of money accumulated after a period of time or the total amount that should be repaid after the agreed time.

Interest, I, payable or earned depends on:

- the principal (amount borrowed or invested), P
- the percentage rate (rate of interest paid or charged), R
- the period or time of the loan or investment, T

$$I = \frac{PRT}{100} \qquad A = P + I$$

P, R and T can be found by transposition of the formula:

$$P = \frac{100I}{RT} \quad R = \frac{100I}{PT} \quad T = \frac{100I}{PR}$$

Example 12

What is the simple interest on $400.00, borrowed for four years at 5% per annum?

$$
\begin{aligned}
I &= \frac{PRT}{100} \\
&= \frac{400 \times 5 \times 4}{100} \\
&= 4 \times 5 \times 4 \\
&= 80
\end{aligned}
$$

The simple interest is $80.

Example 13

A student invested $200.00 at 10% per annum. How long will it take for his savings to amount to $400.00?

This question is asking how long it will take for the interest paid on the investment to increase the value of the whole investment from $200 to $400. The interest paid is the difference between the final amount of the investment, A, and the original principal, P.

$$
\begin{aligned}
I &= A - P \\
&= 400 - 200 \\
&= 200
\end{aligned}
$$

Using the simple interest formula, $I = \dfrac{PRT}{100}$

$$\Rightarrow 200 = \frac{200 \times 10 \times T}{100}$$

$$\Rightarrow \quad T = 10$$

It will take 10 years.

2.2.2 Compound interest

Compound interest differs from simple interest in that the interest is continuously added to the principal on a regular basis, usually annually, resulting in the principal continuously increasing.

The amount obtained from investing or borrowing at compound interest is:

$$A = P\left(1 + \frac{r}{100}\right)^n$$

where A is the amount, P is the principal, R is the rate% and n is the number of years.

Example 14

A man invests $600.00 for two years at 5% compound interest. Calculate the amount of money he receives at the end of the two years.

At the end of year 1:

$$I = \frac{PRT}{100}$$

$$= \frac{600 \times 5 \times 1}{100}$$

$$= 30$$

$\Rightarrow$ new principal = $600 + $30

$$= \$630.00$$

At the end of year 2:

$$I = \frac{PRT}{100}$$

$$= \frac{630 \times 5 \times 1}{100}$$

$$= 31.5$$

$\Rightarrow$ final amount $= \$630 + \$31.50$

$$= \$661.50$$

Alternatively, you could have used the compound interest formula:

$$A = P\left(1 + \frac{r}{100}\right)^n$$

$$\Rightarrow A = 600\left(1 + \frac{5}{100}\right)^2$$

$$= \$661.50$$

where A is the amount, P is the principal, R is the rate% and n is the number of years.

2.2.3 Depreciation and appreciation

Depreciation is a reduction in value of an asset. It is calculated annually as a percentage of the book value (starting value) of the asset at the beginning of the year. The book value at the beginning of the next year is found by subtracting the depreciation from the previous year's book value (the reducing balance method).

Depreciation is calculated as:

$$A = P\left(1 - \frac{r}{100}\right)^n$$

where A is the amount, P is the principal, R is the percentage rate and n is the number of years.

Tips

The formula for compound interest differs from that of depreciation in that the operator symbol in the brackets is a plus (+) instead of a minus (–).

For compound interest, as time progresses, the amount increases.

For depreciation, as time increases, the amount decreases.

Example 15

The owner of a machine shop purchases a drill press for $4000.00. He estimates that depreciation will occur at 15% of the book value every year. What will be the value after 3 years?

After the first year the book value $= 4000 - 4000 \times \dfrac{15}{100}$

$$= 4000 - 600$$
$$= \$3400$$

After the second year the book value $= 3400 - 3400 \times \dfrac{15}{100}$

$$= 3400 - 510$$
$$= \$2890$$

After the third year the book value $= 2890 - 2890 \times \dfrac{15}{100}$

$$= 2890 - 433.5$$
$$= \$2456.50$$

Alternatively, using the formula:

$$A = P\left(1 - \frac{r}{100}\right)^n$$

$$= 4000\left(1 - \frac{15}{100}\right)^3$$

$$= 4000 \times 0.85^3$$

$$= \$2456.50$$

Some things can increase in value over time. This is called **appreciation**. Fine paintings by old masters increase in value over time. The Mona Lisa by Leonardo da Vinci was valued at about $100m in the 1960s, but today it would be worth more than $600m.

Houses kept in good condition can also appreciate in price, especially in a city. A house in Montego Bay in Jamaica worth $100 000 in 2001 could be worth $160 000 twenty years later. Its value has appreciated.

The Mona Lisa is on display at the Louvre museum in Paris.

Key term

Appreciation is the increase in value over time of an asset such as a house or a valuable antique.

Example 16

A house purchased in 2008 for $120 000 was sold in 2018. If the average annual rate of appreciation on this house is 5%, what was its value in 2018?

The value in 2018 can be found using the compound interest formula.

$$A = P\left(1 + \frac{r}{100}\right)^n$$

$$A = 120\,000\left(1 + \frac{5}{100}\right)^{10}$$

$$A = 120\,000(1.05)^{10}$$

$$A = \$195\,500 \text{ to 4 sig. fig.}$$

In 2018 the house was worth about $195 500.

Exercise 2E

1. Calculate the simple interest on $200.00 borrowed at 5% per annum for three years.

2. A man invests $2000.00 for five years at a rate of 10% per annum simple interest. How much does he receive at the end of the five years?

3. Find the principal that will produce simple interest of $100.00 in three years at 2% per annum.

4. A student has to pay simple interest of $150.00 on a sum of money borrowed for six years at a rate of 3%. What was the amount borrowed?

5. Find the simple interest on $400.50 invested for $6\frac{1}{4}$ months at $3\frac{1}{2}$% per annum.

6. How many years will it take a sum of $3000.00 invested at 5% per annum to generate simple interest of $200.00?

7. The simple interest on $10 000.00 for 28 months is $500.00. Find the percentage rate.

8 A businessman invests $5000.00 for five years at 4% compound interest. How much does he receive at the end of the five years?

9 Mrs Davis bought a house for $75 000 three years ago. It is in a good location and its value has appreciated at a rate of 10% for the last three years. What is the value of the house now to 3 significant figures?

10 The value of a sewing machine depreciates each year by 10% of its value at the beginning of the year. The value when new is $1000.00. Find its value after five years.

11 A teacher borrows $6000.00 from a bank at 5% compound interest. In how many years will the debt amount to $10 000.00?

12 A father invests $10 000.00 in the bank for 10 years at compound interest, to pay for his son's education. What rate will ensure that this investment amounts to $20 000.00 at the end of the 10 years?

13 A man buys a brand new SUV vehicle for $120 000.00 and decides to trade it in after four years. The estimated depreciation is 12% for each year.

    a Calculate the value of the vehicle after four years.

    b Write the value of the SUV after the four years as a percentage of its value when new.

2.3 Hire purchase and mortgages

2.3.1 Hire purchase

Hire purchase is a way of buying goods that allows you to pay over an extended period. It entails:

- making a down payment or deposit
- paying the balance of the purchase price plus the interest in a number of instalments.

The components of hire purchase include:

- the **deposit**, which is a specific percentage of the marked price
- the **outstanding balance**, which is marked price − deposit
- the **balance payable**, which is monthly instalment × number of months
- the **hire purchase price**, which is deposit + balance payable
- the **interest**, which is a charge based on a percentage of the outstanding balance after the deposit is paid
- the **total interest charged**, which is hire-purchase price − marked price
- the **percentage interest rate**, which is $\frac{\text{interest charged}}{\text{outstanding balance}} \times 100$.

> **Key term**
>
> **Hire purchase** is a way of buying goods or services by making an initial payment, called a deposit, followed by a number of smaller payments called instalments.

Example 17

A house owner buys a refrigerator on hire purchase for $1500.00. He pays a deposit of 20% of the selling price and is charged interest at 10% on the outstanding balance. If he pays the balance in 12 monthly instalments, how much will he pay for each instalment?

Cost price of refrigerator = $1500.00

$$\text{Deposit} = 20\% \text{ of } 1500$$

$$= \frac{20}{100} \times 1500$$

$$= \$300.00$$

Outstanding balance is 1500 − 300 = $1200.00

| Note | Calculate the interest as simple interest for questions like this. |

10% interest on outstanding balance is $\frac{10}{100} \times 1200 = \$120.00$

Total amount to be repaid is outstanding balance + interest = 1200 + 120

$$= \$1320.00$$

Since there are 12 instalments, the amount per instalment is 1320 ÷ 12 = $110.00

Example 18

Harry bought his stereo set for $800.00 on hire purchase, paying a deposit of 30% of the purchase price. He was told that there would be 7% interest on the outstanding balance and that the interest and outstanding balance must be repaid in instalments of $100.00. How many instalments will be required to make the full repayment?

Cost price of stereo = $800.00

$$\text{Deposit} = 30\% \text{ of } \$800.00$$

$$= \frac{30}{100} \times 800$$

$$= \$240.00$$

Outstanding balance = 800 − 240

$$= \$560.00$$

7% interest on outstanding balance = $\frac{7}{100} \times 560$

$$= \$39.20$$

Total amount to be repaid = outstanding balance + interest

$$= \$560.00 + \$39.20$$

$$= \$599.2$$

Since each instalment will be $100.00, the number of instalments will be 599.2 ÷ 100 and this will be rounded up to 6.

Exercise 2F

1   A stove, priced at $2000.00, is sold on hire purchase with a deposit of $600.00 and 20 monthly instalments. Interest at 5% is charged on the outstanding balance. Find the full cost of buying the stove on hire purchase.

2   A householder purchases a chair for $800.00. She pays a deposit of 10% of the marked price and interest at 10% is added to the outstanding balance. The total (balance plus interest) is to be paid in 10 monthly instalments. Calculate how much each instalment will be.

3   A computer priced at $3680.00 is sold on hire purchase with a deposit of $400.00 and 12 monthly instalments of $300.00. Calculate the rate of interest charged on the outstanding balance.

4   A lady decides to purchase a suite of furniture on hire purchase. She makes a deposit of $500.00 and then interest of 8% is charged on the outstanding balance of $2400.00. Find the marked price of the furniture and the period of repayment of the loan, given that monthly instalments of $324.00 are to be made.

5   A student purchases a cell phone priced at $800.00 by paying 15% interest and making six monthly instalments. What is the amount of interest paid and how much is each instalment?

6   A vehicle is offered for sale at a cash price of $25 000.00. It can also be purchased on hire purchase with a deposit of 20% and monthly instalments of $200.00 per month over a period of 10 years. Determine:

     a   the amount of the deposit   **b**   the outstanding balance

     c   the number of monthly instalments   **d**   the balance payable

     e   the hire purchase price   **f**   the interest charged

     g   the percentage interest charged.

2.3.2 Mortgages

When you need to make a big purchase, such as a house, you may need to spread the payments over several years. In this case, you take out a **mortgage** on the house.

Like hire purchase, taking out a mortgage requires making a small deposit (such as 10% of the full purchase price) with the financial institution providing the balance (in this case 90%) in the form of a loan.

> **Key term**
>
> A **mortgage** is a loan given by a bank or other financial institution to assist in the purchase of property such as a house or land. The purchaser makes a small down payment (usually ⩽ 20% of the total value) and then borrows the rest, repaying it in monthly instalments.

The period of repayment for a mortgage is generally quite long, typically 10 to 25 years, during which time the borrower makes monthly instalments. The monthly instalments include interest, which results in the borrower having to repay much more than the amount borrowed.

The components of a mortgage include:

- the **deposit**, which is a percentage of the cost of the property
- the **mortgage amount**, which is the outstanding balance after payment of the deposit
- the **interest payable**, which is the amount of interest to be paid on the mortgage amount

- the **total amount repayable**, which is the mortgage amount plus the interest payable
- the **monthly amount repayable**, which is the total amount repayable ÷ number of months
- the **full amount paid**, which is the deposit + (monthly amount repayable × number of months)

Example 19

A house priced at $200 000.00 is offered with a 90% mortgage repayable over a 15-year period. Calculate:

(a) the deposit

(b) the mortgage

(c) the total amount paid to the bank at a monthly instalment of $2000.00

(d) the full amount paid for the house.

(a) Deposit = 10% of 200 000

$$= \frac{10}{100} \times 200\,000$$

$$= \$20\,000.00$$

(b) Mortgage amount = price − deposit

$$= 200\,000 - 20\,000$$

$$= \$180\,000.00$$

(c) Total amount paid to the bank in instalments = 2000 × 12 × 15

$$= \$360\,000.00$$

(d) Full amount paid = amount paid + deposit

$$= 360\,000 + 20\,000$$

$$= \$380\,000.00$$

Example 20

A vehicle originally priced at $20 000.00 is offered for a 5% deposit followed by monthly instalments of $200.00 over a period of 20 years. Find:

(a) the amount of the deposit

(b) the amount to be repaid over the 20-year period

(c) the interest paid

(d) the total amount to be paid, including interest.

(a) The deposit is 5% of cost price = $\frac{5}{100} \times 20\,000$

$$= \$1000.00$$

(b) The amount less the deposit is 20 000 − 1000 = $19 000.00

(c) The total amount is monthly instalment × 12 × 20 = 200 × 12 × 20

$$= \$48\,000.00$$

Interest = total amount − amount less deposit

$$= 48\,000 - 19\,000$$

$$= \$29\,000$$

(d) The total amount paid for the vehicle = deposit + total amount to be repaid

$$= 1000 + 48\,000$$

$$= \$49\,000$$

Exercise 2G

1 A summer cottage is priced at $100 000.00. The lending agency offers a 90% mortgage. Calculate the deposit and mortgage amount.

2 A flat is on sale for $140 000.00. What is the percentage deposit, given that the mortgage amount is $133 000?

3 A beach-front property is on sale for a cash price of $350 000.00. It can also be bought via a mortgage arrangement requiring a 13% deposit and monthly payments of $3500 to be made over a period of 20 years.

Calculate:

a the deposit   **b** the mortgage amount   **c** the total amount repayable.

4 A boat costing $80 000.00 can be purchased by first paying a deposit of 12% of the selling price and taking out a bank loan for the remainder. Given that interest is also charged and the monthly instalment required is $400.00 for 18 years, calculate:

a the deposit   **b** the mortgage amount

c the interest payable   **d** the total amount payable.

5 A three-bedroom house is put up for sale for a cash price of $220 000.00. It can also be bought by taking out a mortgage of 92% of the market price and paying monthly instalments over a 15-year period. Given that the total amount repayable is $320 000.00, calculate:

a the deposit   **b** the mortgage amount   **c** the monthly amount repayable

d the interest payable   **e** the total amount repayable.

6 A house can be purchased with an 80% mortgage of $80 400.00, paying monthly instalments of $500.00 over a 15-year period. Calculate:

a the deposit   **b** the cost of the house   **c** the number of monthly instalments

d the interest payable   **e** the total amount payable.

2.4 Rates and taxes

Governments need to raise money for the maintenance of services and community programmes. One way of doing this is by charging **rates** on buildings. So every property in a town or city is assigned a **rateable value**, R_v, by the valuation department, based on the size, location and condition and actual value of the property.

Rates payable are based on a percentage of the rateable value. They are levied (charged) at an agreed amount, the **rate percentage**, R_p, such as 50¢ or 75¢, 'in the dollar' of the rateable value.

The rates, R, to be paid over a year, are calculated as:

R = rateable value × percentage charge set for the year

$\quad = (R_v) \times (R_p)$

> ### Key terms
>
> **Rates** are charges made on property owners, to raise money to maintain community services.
>
> Each property is assigned a **rateable value**, based on its size, location and condition and actual value.
>
> The **rate percentage** is the percentage or cent in the dollar charge, set by the local authority, to determine the **rates payable**, which is the amount to be paid in the year.

The rateable value of a building is $200.00 and the charge set for the year is 80¢ in the dollar. What is the rate payable for the year?

Using the formula:

rate payable in 1 year = rateable value × rate charged

$$R = R_v \times R_p$$
$$= 200 \times 0.8$$
$$= \$160.00$$

The total rateable value of all the properties in a village is $12 000.00. Determine the rate that will result in the valuation department receiving $4000.00 this year.

$$\text{Rate} = \frac{\text{rates payable in one year}}{\text{total rateable value}}$$

$$= \frac{4000}{12\,000}$$

$= 0.33$ This may be expressed as 33% or 33 cents in the dollar.

The rate charged by the valuation department for a building is 60% (60¢ in the $). If the rate payable per year is $500.00, find the rateable value of the building.

Rates payable $R = 500$

$$\text{So } R_v \times \frac{60}{100} = 500$$

$$R_v = \frac{500 \times 100}{6}$$

$$R_v = \$833.33$$

Exercise 2H

1. The rateable value of all the business places in a village is $2400.00 and the rates are 75 cents in the dollar. Calculate the rates payable.

2. The rateable value of a property is $800.00. If the rate payable in one year is $140.00, find the rate charged.

3. The rate charged by the valuation department on a certain business premises is 55%. Given that the rate payable in one year is $800.00, find the rateable value.

4 In a shopping district the total rateable value of all the businesses is $200 456 800.00. Calculate the rate needed to cover the expenses for that district, which amounts to $100 000 582.00 in a particular year.

5 In one year a supermarket owner pays $3246.00 in rates, charged at 30 cents in the dollar. Find the rateable value of the supermarket.

6 The rateable value of a house is $1600.00. Find the rates payable when the rate is 69%. How much more will the home owner have to pay if the rate increases to 75 cents in the dollar?

7 The rateable value of a hotel in the city is $20 000. Calculate the income from a rate of 8%. If the income required increases by $500.00 what will the new rate be?

2.5 Utilities payments

Utilities are the services that are important in everyday life, such as electricity, water and telephone. Customers receive regular **utility bills**, listing all the charges and the amount they need to pay.

> **Key term**
>
> A **utility bill** is a document used to charge customers for the use of utilities such as electricity, water and telephone.

2.5.1 Electricity bills

The price of electricity in homes or businesses is based on the quantity of electricity (or number of **kilowatt-hours**) used in a certain time period (generally one month) at a stated rate per unit. The bill generally includes a charge for fuel used and other non-fuel charges, such as a charge for provision of the service, government charges (value-added tax, environmental levy) and a fixed meter rental charge. The rates per unit used may vary, based on the time of the day.

> **Key term**
>
> A unit of electricity is a **kilowatt-hour** (kWh) and 1 kWh = 1000 Wh.

You can calculate the number of units of electricity used by an appliance by multiplying the power rating of the appliance (in watts, W) by the number of hours it is in use. So:

amount of electricity used by an appliance = power × number of hours in use.

Look at the sample electricity bills.

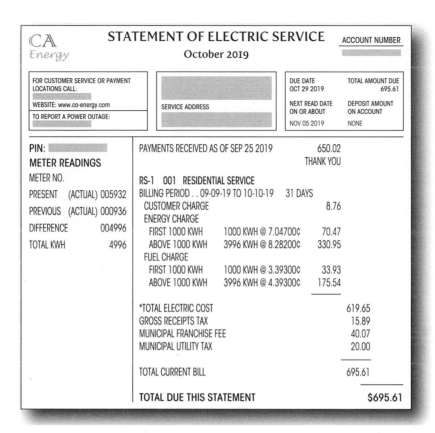

STATEMENT OF ELECTRIC SERVICE

CA Energy

October 2019

ACCOUNT NUMBER

FOR CUSTOMER SERVICE OR PAYMENT LOCATIONS CALL:		DUE DATE OCT 29 2019	TOTAL AMOUNT DUE 695.61
WEBSITE: www.ca-energy.com	SERVICE ADDRESS	NEXT READ DATE ON OR ABOUT	DEPOSIT AMOUNT ON ACCOUNT
TO REPORT A POWER OUTAGE:		NOV 05 2019	NONE

PIN:

METER READINGS

METER NO.

PRESENT (ACTUAL) 005932

PREVIOUS (ACTUAL) 000936

DIFFERENCE 004996

TOTAL KWH 4996

PAYMENTS RECEIVED AS OF SEP 25 2019   650.02
                                            THANK YOU

RS-1 001 RESIDENTIAL SERVICE
BILLING PERIOD . . 09-09-19 TO 10-10-19 31 DAYS
  CUSTOMER CHARGE   8.76
  ENERGY CHARGE
    FIRST 1000 KWH   1000 KWH @ 7.04700¢   70.47
    ABOVE 1000 KWH   3996 KWH @ 8.28200¢   330.95
  FUEL CHARGE
    FIRST 1000 KWH   1000 KWH @ 3.39300¢   33.93
    ABOVE 1000 KWH   3996 KWH @ 4.39300¢   175.54

*TOTAL ELECTRIC COST   619.65
GROSS RECEIPTS TAX   15.89
MUNICIPAL FRANCHISE FEE   40.07
MUNICIPAL UTILITY TAX   20.00

TOTAL CURRENT BILL   695.61

TOTAL DUE THIS STATEMENT $695.61

| CUSTOMER NO. | ACCOUNT NO. | ACCOUNT TYPE Domestic |

Actual   ELECTRICITY CHARGES

METER READINGS		NO. OF DAYS	USAGE THIS PERIOD (kWh)	TYPE OF SERVICE	NON-FUEL	FUEL	DEMAND / FLOOR AREA	DUE DATE	CURRENT ELECTRICITY CHARGES
31-Aug-18	01-Oct-18	31	181	Metered	$73.43	$84.48		09-Nov-18	$157.91
22133	22314								

RATES / kWh (unit)		ELECTRICAL USAGE HISTORY				BILLING DETAILS	
FUEL	$0.46672	PERIOD	DAYS	USAGE (kWh)	kWh / DAY	PREVIOUS BALANCE	$327.01
NON-FUEL	$0.40570	01-Oct-18	31	181	6	LESS PAYMENT	$167.25 CR
GOVERNMENT CHARGES ECS		31-Aug-18	28	168	6	ADJUSTMENTS	$0.00
ENVIRONMENTAL LEVY	$10.00	03-Aug-18	31	177	6	BROUGHT FORWARD	$159.76
VAT (non-fuel) 15%	$4.99	03-Jul-18	32	183	6		
VAT (other) 15%	$0.00	01-Jun-18	30	165	6	ELECTRICITY CHARGES	$157.91
		02-May-18	28	155	6	GOVERNMENT CHARGES	$14.99
		04-Apr-18	33	184	6	TOTAL CURRENT CHARGES	$172.90
		02-Mar-18	28	134	5	**TOTAL AMOUNT DUE**	**$332.66**

NOTES:
We remind you of balance due. This is our first notice, please pay balance within 10 days of bill date shown below.

Example 24

How many units of electricity are used when a 40 W bulb remains lit for 24 hours?

$$\text{Number of units} = \frac{40}{1000} \times 24$$

$$= \frac{960}{1000}$$

$$= 0.96 \text{ kWh}$$

$$= 0.96 \text{ units}$$

A house owner's present meter reading is 376 and his previous reading was 188. Given that the fixed rate per unit is $0.355 49 for fuel and $0.4057 for non-fuel, VAT is $5.42 and the environmental levy is $10.00, calculate the amount due on the bill.

Number of units used is 376 − 188 = 188

Fuel charge = 188 × 0.355 49

$\qquad$ = $66.83

Non-fuel charge = 188 × 0.4057

$\qquad$ = $76.27

Government charges = 5.42 + 10.00

$\qquad$ = $15.42

Total amount on the bill = 66.83 + 76.27 + 15.42

$\qquad$ = $158.52

The charge for electricity used by a 1000 W appliance running continuously over 5 days is calculated based on a fixed fuel charge of $0.60 per kWh and an energy charge of 20 cents per kWh.

Calculate the total amount of the bill and the actual amount paid, given that a 5% discount was allowed.

The amount of power used = 1000 W × 24 h × 5

$\qquad$ = 120 000 Wh

$\qquad$ = 120 kWh

Fuel charge = number of kWh × rate

$\qquad$ = 0.6 × 120

$\qquad$ = $72.00

Energy charge = 0.2 × 120

$\qquad$ = $24.00

Total bill = 72 + 24

$\qquad$ = $96.00

5% discount = $\dfrac{5}{100}$ × 96

$\qquad$ = $4.80

Total bill after discount = 96.00 − 4.80

$\qquad$ = $91.20

Exercise 2I

1 How many units of electricity are used when a 60 W bulb is left on continuously for 2 days?

2 A 200 W refrigerator operates for one month constantly. How many units of electricity will it consume?

3 A 10 kW electric stove operates for 6 hours. How many units of electricity will it consume?

4 A 20 kW appliance operates for 14 hours non-stop. How many units of electricity are used in that time period?

5 Calculate the number of kWh a 350 W freezer will consume in 20 days, working constantly.

6 A householder receives an electricity bill for $250.00 after using 800 kWh of electricity. What is his electricity cost per kWh?

7 The charges on a property owner's bill for electricity are:

      fixed charge of $30.00 for meter use

      tax of $30.00

      20 cents per kWh used.

Find his total bill when he uses 600 kWh of electricity.

8 The quarterly electricity bill for a consumer in one of the Caribbean islands is made up of:

      a fixed charge of $10.00 per month

      the first 100 kWh at $0.2 per kWh and

      25¢ per kWh after that.

      VAT $15.00

      a discount of 15% on the total bill.

The consumer uses a total of 1050 kWh in a quarter (three months). Determine how much he will have to pay to settle his bill.

9 These are details of an electricity bill for a consumer on one of the Caribbean islands.

Name of customer: John Doe	Service for the period: 1/2/2016 – 31/3/2016
Previous meter reading:100982	Current meter reading: 101522
kWh used:	
Fuel charge: 40¢ per kWh	Energy charge: 17¢ per kWh
VAT: $15.00	

Using the information given in the bill, determine:

a the number of kWh used   **b** the energy charge

c the fuel charge   **d** the total amount payable on the bill

e the actual amount paid, given that a discount of 12% was earned for early payment.

2.5.2 Water bills

Users of water, which is normally provided by the state, are required to pay **water rates**, which are fees or charges for the service provided. Water rates are calculated according to the volume of water used by the consumer and different rates may apply, depending on how the water is used. The period of payment may be monthly, quarterly, half-yearly or yearly.

Key term

Water rates are charges made for the supply and disposal of water and are calculated according to the volume of water used by the consumer.

Water rates are calculated from a basic charge for the consumption of 1 m³ of water.

Here is a sample water bill.

					SERVICE ADDRESS		
OFFICE: ▓▓▓▓▓▓▓					BELLEVUE		

METER NUMBER	READ DATES		BILLING DAYS	METER READINGS			USAGE
	PRESENT	PREVIOUS		CODE	PRESENT	PREVIOUS	
▓▓▓▓▓▓	11/22/2018	10/23/2018	30	MR	0421	0396	5,500 GAL

USAGE HISTORY				BILLING DETAILS	
Month	Days	Water use (Gals)	Water use per day (Gals)		
				PREVIOUS BALANCE	85.22
				Less payments	
10/2018	29	4,619	159	Adjustments	
09/2018	27	4,179	155	BALANCE BROUGHT FORWARD	85.22
08/2018	35	4,399	126	CONSUMPTION	59.12
07/2018	34	4,839	142	BASIC	10.80
06/2018	28	4,399	157	SEWER	
				OTHER	
				VAT	
				TOTAL AMOUNT DUE	**$155.14**

A householder uses 60 m³ of water in three months at the rate of $5.00 per m³. The water rate then dropped in the next three months and consumers were charged $3.50 per m³ for the first 50 m³ and $4.00 per m³ for any additional amount consumed. The householder used 80 m³ of water over the next three months. There is an overall discount of 10% on all water bills. How much did the consumer have to pay for water for the six-month period?

Cost for the first 60 m³ $= 5 \times 60$

$\qquad\qquad\qquad\quad = \$300.00$

Cost for the next 50 m³ $= 3.5 \times 50$

$\qquad\qquad\qquad\quad = \$175.00$

Cost for the last 30 m³ $= 4 \times 30$

$\qquad\qquad\qquad\quad = \$120.00$

Total cost $= 300 + 175 + 120$

$\qquad\quad = \$595.00$

Discount of 10% of 595 $= \$59.5$

Final bill $= 595 - 59.5$

$\qquad\quad = \$535.50$

Exercise 2J

 1. The Harris family used 24 m³ of water at the rate of $4.00 per m³ in one month. How much did they pay for water for that month, given that they also received a 5% discount?

 2. A secondary school used 542 m³ of water in the first quarter and 632 m³ in the second quarter of a particular year. The water rates in the first quarter were $3.50 per m³ but changed to $4.25 per m³ in the second quarter. How much did the school pay for water in the first half of the year?

3 The Phillips family used 140 m³ of water in 2016 and 168 m³ in 2017. In 2016 the water rates were $2.00 for the first 50 m³ and $3.00 for every cubic metre in excess of 50 m³. There was a discount of 10% on all bills. In 2017, however, the rates had changed to $3.40 for the first 60 m³, $2.80 for the next 30 m³ and $2.00 for every cubic metre in excess of 90 m³. Consumers in 2017 were given a discount of 8% on all bills. In each year, a 5% tax was charged on all final bills. In which year was the Phillips' water bill higher and by how much?

4 A guest house consumed 1500 m³ of water for the first half of the year, during which time the rates were $3.15 per m³ for the first 400 m³, $4.10 for the next 400 m³ and $4.85 for amounts over 800 m³. A 5% water tax was charged on all bills and then a 9% discount was given on all bills paid in full. What was the total amount paid by the guest house owner?

2.5.3 Telephone bills

The amounts that telephone service providers charge their customers are based on the duration of calls made. Different rates may apply to local and international calls. The period of payment is generally monthly and the **telephone bill** includes:

- a charge based on the duration of each call, with different rates for local and international calls
- a fixed charge for the rental of the telephone
- an internet usage charge
- VAT.

Here are some sample telephone bills.

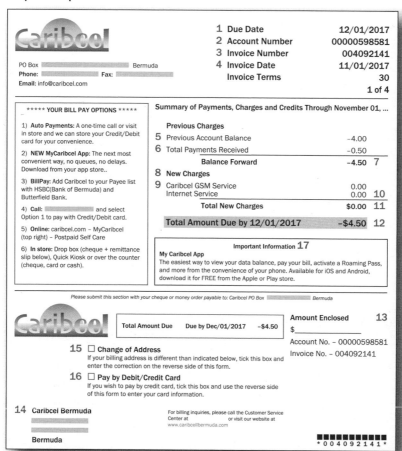

Your November Bill

Bill Date:
26 Nov 2018

Account # ▮▮▮▮▮▮
VAT Invoice # ▮▮▮▮▮▮
Primary Service # ▮▮▮▮▮▮

glow

How much do I owe?

Your payment is

EC$141.71

DID YOU KNOW that you can call us 24 hours a day, 7 days a week, for information on your bill balance, our products, your internet problems or to report a fault. Call Tollfree on ▮▮▮▮▮▮ .

What did I use?

Digital Service	$66.52
Telephone	$56.70
Account Charges	$0.00
Discounts	$0.00
Taxes	$18.49
Current Total	EC$141.71
Outstanding Charges	EC$0.00
Total Due	**EC$141.71**

Example 28

This is a summary of Mr Johnson's telephone bill.

Usage charges

Date	Type of call/charge	Quantity	Number of minutes	Cost ($)
26 October–25 November	local	45	24 005	64.50
26 October–25 November	international	3	45	32.00
26 October–25 November	internet usage (fixed)			50.00
26 October–25 November	Rental charge (fixed monthly)			25.00
	Rental and usage charges			171.50
	Value added tax (VAT)	15% VAT on charges		25.73

Find the final amount of the bill.

To find the final amount on the bill, add up all the various charges.

Final bill amount = 171.50 + 25.73

$\qquad$ = $197.23

Exercise 2K

1 Use this table to calculate the monthly telephone bill for the John household.

Name	First 5 minutes ($)	Additional minutes ($)	Service charge ($)	Government tax ($)
John	5.25	20.50	30.00	14.00

2 The information in this table is used to calculate the telephone bill for the James family for the month of January, 2017.

Telephone usage record

Calls	Number of minutes	Fixed charge/charge per minute ($)	Total ($)
Local calls	10 009	40.00 (fixed)	40
Land line to cell phone	40	2.00	
Long distance	50	1.50	
Government tax (15% of total)			

What is the total January bill for the James family?

3 This is John Doe's telephone bill for January.

> Telephone bill: John Doe
>
> Account number 101022304   Billing date: 01/01/19–31/01/19
>
> Reading: 01/01/19: 1340
>
> Reading: 31/01/19: 1540   Number of local calls.................
>
> Landline to cell phone calls: 35 minutes @ $2.00 per minute ....
>
> Oversea calls $65.20
>
> Monthly service fee ....
>
> Local calls ....
>
> VAT ....
>
> Tax on oversea calls ....

The bill includes:

- a fixed monthly service fee of $20.00 for local calls (not exceeding 50)
- a charge of 50¢ per call for local calls (more than 50 up to 120)
- a charge of $0.75 per call for local calls (in excess of 120)
- a variable charge based on destination country for overseas calls
- a charge of $2.00 per minute on landline to cell phone calls
- a tax of 45% of the cost of all overseas calls
- VAT of 15% on all calls (local and overseas).

Calculate:

a the cost of landline to cellphone calls   **b** the monthly service charge

c the cost of local calls   **d** the VAT

e the tax on overseas calls   **f** the total charge on the bill.

2.6 Currency and foreign exchange

2.6.1 Currency

The countries of the world use difference types of **money,** or **currency,** for financial transactions. In most English-speaking Caribbean countries, the basic currency is the dollar ($) which may be subdivided into 100 smaller units called cents (¢).

> **Key term**
>
> **Currency** is the **money** used by a country for its financial transactions.

Monetary amounts can therefore be expressed in dollars ($) or cents (¢) or both.

Example 29

Write each amount in two different ways.

(a) Five dollars   (b) Ten dollars   (c) Three dollars and five cents

(d) Five thousand and five cents

(a) Five dollars = $5.00 or 500¢

(b) Ten dollars and twenty cents = $10.20 or 1020¢

(c) Three dollars and five cents = $3.05 or 305¢

(d) Five thousand and five cents = 5005¢ or $50.05

Denominations

To simplify financial transactions, different **denominations** of coins and notes are used.

Coins available are: five cents (5¢); ten cents (10¢); twenty-five cents (25¢); one dollar ($1.00).

Notes available are: one dollar, five dollar, ten dollar, twenty dollar, fifty dollar, one hundred dollar, and so on.

> **Key term**
>
> **Denomination** is the name given to a specific currency amount, usually in the form of coins or notes (such as dollars, pound, dime).

Example 30

Mary is given $45.40 to make some purchases for her mother. Give two possible combinations of dollar notes and coins that her mother could have given Mary.

Possible combinations for $45.40:

- 2 × $10.00 ($20), 1 × $20.00 ($20), 1 × $5.00 ($5), 25 cents (25¢), 10 cents (10¢) and five cents (5¢)
- 5 × $5.00 ($25), 2 × $10.00 ($20) and 4 × 10 cents (10¢).

You need to be able to carry out **basic financial operations** to solve problems.

Key term

Basic financial operations involve the use of the four basic operations: addition, subtraction, multiplication and division in calculating financial transactions.

Example 31

These sums of money were given out to five children:

$6.75; $10.00, $4.35, $8.40, $15.50.

Find the total amount given out.

6.75 + 10.00 + 4.35 + 8.40 + 15.50 = $45.00

Example 32

A housewife went to the market with $200.00. Her purchases came to $160.50. How much did she have left?

200.00 − 160.50 = 39.50

She had $39.50 left.

Example 33

Twelve students in a class were each given $2.50.

What was the total amount given out to the students?

2.5 × 12 = $30.00

Example 34

$1840 was used to purchase 8 identical lamps. What was the cost of each lamp?

1840 ÷ 8 = 230

Each lamp cost $230.00.

Exercise 2L

1 Convert:

    a $5.03 to ¢   **b** 20 085¢ to $.

2 Using different denominations (5¢, 10¢, 25¢, 50¢, $1, $5, $10, $20, $50, $100), give three different ways of presenting $125.80.

3 These sums of money were given out to five people:

    $23.50, $124.10, $77.00, $9.05, $214.60

    a What was the total sum given out?

    b What was the difference between the greatest and smallest sums?

4 A sum of $248.08 was shared equally among 8 students. How much money did each student receive?

5 Fourteen pensioners were each given $342.00 by the government. What was the total sum of money given out to these pensioners?

6 A housewife went shopping to buy:

        5 lbs flour @ $2.50 per lb

        6 lbs sugar @ $3.75 per lb

        4 lbs potato @ $2.40 per lb

        5 boxes matches @ $1.50 per box

        10 mangoes @ 1.25 each

        5 lbs fish @ $4.50 per lb

    She paid with a $100.00 note. How much change did she receive?

2.6.2 Foreign exchange

Different countries have their own monetary systems. To facilitate **foreign exchange**, which involves trade and travel, there are conversion or exchange rates between the currencies.

Exchange rates fluctuate as a result of changing global demand and supply, which in turn can be affected by political and economic stability, natural disasters, government and business international trading, travel and tourism.

> **Key term**
>
> **Foreign exchange** is the exchange of currencies between different countries, using an agreed rate of conversion.

Tourism

The US dollar is the most traded currency on the planet, for many reasons. It is considered to be the unofficial global reserve currency – the one most widely accepted by other countries. It also serves as a benchmark or target rate and the standard currency for many commodities, such as crude oil and precious metals.

Other popular currencies include the euro (European countries), the Japanese yen, the British pound, the Swiss franc and the Canadian dollar.

Here are some images of currencies from different countries, with a list of exchange rates for different currencies (as at January 2019) compared with the US dollar.

Barbados 2 dollars

US dollar

Guyana 20 dollars

Jamaica 1000 dollars

Trinidad & Tobago 100 dollars

British pound coin

Exchange rate table	
US dollar (US$)	1.000 USD
Euro (€)	0.879
British pound (£)	0.776
Trinidadian dollar	6.774
Japanese yen	109.638
Canadian dollar	1.330
Eastern Caribbean dollar	2.700
Chinese yuan renminbi	6.798
Argentine peso	37.806
Jamaican dollar	131.87
Guyana dollar	236.3
Barbados dollar	2.000

Example 35

A tourist from the United States changes US$400.00 into Barbados dollars at the exchange rate of US$1.00 = BDS$2.00. She spends BDS$300.00 and travels to Dominica where she converts BDS$180.00 to $EC at the rate of US$1.00 = EC$2.7. She spends EC$200.00 in Dominica and on leaving she changes her money back to US currency. How many US$ is she left with?

US$400 = 400 × 2

＝ BDS$800

800 – 300 = BDS$500.00 remaining.

BDS$180 = US$90.00 = 90 × 2.70

＝ EC$243.00

If she spends EC$200.00 she is left with EC$43.00.

500 – 180 = BDS$320

She is therefore left with BDS$320.00 + EC$43.00

$$= \frac{320}{2} + \frac{43}{2.7}$$

$$= US\$(160 + 15.93)$$

$$= US\$175.93$$

Exercise 2M

1 Use the conversion table above to convert:

a EC$250.00 to US$ b £54 to US$ c US$221 to BD$

d CAN$102 to US$ e JAM$62.5 to US$ f US$40 to GUY$

g US$150 to Chinese yuan.

(2) Use the conversion table above to convert:

   a EC$25.50 to TT$ **b** GUY$400 to BDS$ **c** JAM$405.50 to EC$

   d £75.75 to EC$ **e** EC$1428.80 to US$.

(3) A British tourist travelling to the Caribbean changed £200 to EC$ when she arrived in St Lucia and spent EC$150.00 while she was there. She later travelled to Trinidad where she changed EC$200 to TT$ and spent half of her TT$. Her final stop was in Jamaica where she purchased JAM$2000.00 and spent three-quarters of that money. She then changed all her money to US$. How many US$ did she receive?

(4) A visitor to the US from the Caribbean changed EC$3000.00 to US$ at the rate of US$1 = EC$2.70. After spending 75% of her US dollars on shopping, she used the remainder to purchase BDS$ (US$1 = BDS$2.00). She changed $\frac{1}{4}$ of the BDS$ to EC$ and $\frac{5}{8}$ to TT$. How much of each currency was she left with? (US$1 = TT$6.73)

2.7 Salaries and wages

2.7.1 Salaries

People such as civil servants, teachers, secretaries and company managers are normally paid monthly. The amount they earn over a month is their monthly **salary**.

A person's **gross annual salary** is found by multiplying their monthly salary by 12.

A person's **net monthly salary** is what they are paid after taxes, union dues, etc. are deducted from the gross salary.

> **Key term**
>
> A **salary** is the monthly or annual payment workers receive for services rendered.

A teacher will receive a salary

> **Key terms**
>
> **Gross annual salary**   is the gross monthly salary × 12
>
> **Net monthly salary**   is the gross monthly salary − deductions

Example 36

A civil servant receives a monthly salary of $4400.00. What is his annual salary?

Annual salary = 4400 × 12

                = $52 800.00

Example 37

The principal of a secondary school receives a gross annual salary of $72 000.00. Deductions of $1200.00 are made every month. Find the net monthly salary and the net annual salary.

Gross monthly salary = 72 000 ÷ 12

                      = $6000.00

Net monthly salary = gross monthly salary − deduction

                  = 6000 − 1200

                  = $4800.00

Net annual salary = 4800 × 12 = $57 600.00

2.7.2 Wages

People such as factory or construction workers or transport drivers are likely to be paid more frequently. The weekly amount they are paid is their **wages**.

- **Overtime** rates depend on the number of hours, or when they are normally worked, and are calculated as:
  - time and a quarter, which is $1\frac{1}{4}$ times the basic rate
  - time and a half, which is $1\frac{1}{2}$ times the basic rate
  - double time, which is twice the basic rate.

Key terms

Wages are payments received by workers who are normally paid hourly, daily, weekly or fortnightly for their work.

- The amount received per hour is the **basic rate**.
- The number of hours worked per week is called the **basic week** (for example, 5 days a week for 8 hours a day = 40 hours)
- The amount of money earned in a basic week is the **basic wage**.
- Two consecutive basic weeks constitute a **basic fortnight**.
- Hours worked more than the basic week are called **overtime**.

Example 38

A seamstress works a **basic week** of 40 hours at a **basic rate** of $8.50 per hour. What is her fortnightly wage?

Weekly wage = 40×8.5

= $340.00

Fortnightly wage = 340×2

= $680.00

Example 39

A dock worker works a basic week of 35 hours (7 hours each day for 5 days) and is paid $280.00 per week. If he works 6 hours' overtime on Saturday at time and a half and 5 hours on Sunday at double time, what will be his gross weekly wage?

His basic rate is $280 \div 35 = 8$

He earns $8.00 per hour.

The time and a half rate is $1\frac{1}{2} \times 8 =$ $12.00 per hour.

On Saturday his overtime pay is $12 \times 6 =$ $72.00

The double time rate is $8 \times 2 =$ $16.00 per hour.

On Sunday his overtime pay is $16 \times 5 =$ $80.00

His gross weekly wage is $280 + 72 + 80 =$ $432.00

2.7.3 Commission

Commission is a bonus payment. It may be a reward for selling more than a basic amount. It is normally calculated as a percentage of the total value of products sold above the fixed basic amount.

Gross wage = basic wage + commission

Key term

Commission is an amount paid to a salesperson in addition to their basic wage.

Example 40

A salesman is paid a basic wage of $420 a week and earns commission of 5% of the value of goods sold. During the week he sells goods worth $6000.00. Find his total wage for the week.

Commission = 5% of 6000

$$= \frac{5}{100} \times 6000$$

$$= \$300.00$$

Total wage = basic wage + commission

$$= 420 + 300$$

$$= \$720.00$$

Exercise 2N

1. A teacher receives a monthly salary of $2800.00. What is her annual salary?

2. A civil servant gets an annual salary of $48 000.00. Calculate his monthly salary.

3. A government statistician gets a gross monthly salary of $3400.00. Deductions amounting to $314.00 are made on a monthly basis. Find:

   a the net monthly salary

   b the gross annual salary

   c the annual deductions

   d the net annual salary.

4. A seamstress works a basic week of 40 hours at a basic rate of $8.50 per hour. Calculate her basic wage.

5. A machinist works for a basic wage of $540.00. Given that his basic week consists of 40 hours, calculate his basic rate.

6. A barber's basic rate is $10 per hour. He receives a basic wage of $500.00. How many hours constitute the basic week?

7. An insurance agent is given 5% commission on her sales of policies. If in one week she sells $4000.00 worth of policies, what is her commission?

8. A salesman is paid a basic wage of $200.00 a week. He is also given commission of 4.5% on the sale of goods from $500.00 to $900.00 and 6.5% commission on sales greater than $900.00.

   a How much commission will he receive for sales amounting to $2500.00?

   b What is his total wage for that week?

9. A man's basic rate for a 35-hour week is $6.50 per hour. Calculate his fortnightly wage.

10. A secretary works a basic week of 45 hours and is paid a basic rate of $8.00 per hour. What is his gross wage if he works an additional 4 hours on Saturday, for which he is paid time and a half, and 5 hours on Sunday, for which he is paid double time?

11 The workers of an oil company work a basic week of 40 hours at a basic rate of $12.60 per hour. They can also work overtime from Monday to Friday, at time and a quarter, and on Saturday, at time and a half, as well as Sunday at double time. Calculate the gross wage of a worker who, in addition to his basic week, works 17 hours' overtime from Monday to Friday, 7 hours' overtime on Saturday and 5 hours' overtime on Sunday.

12 A tailor works a basic week of 40 hours, earning a basic wage of $440.00. He also earns an overtime wage of $300.00 on Saturday, at time and a half, and $400.00 on Sunday, at double time. Calculate:

    a his basic rate   **b** the number of overtime hours.

2.8 Income tax

Income tax is one of the major sources of income for governments. Any worker earning an income greater than a specified minimum amount is expected to contribute by the payment of income tax.

Every worker is granted specific non-taxable allowances, so part of their income is tax free.

Non-taxable allowances include:

- personal allowance
- spouse allowance
- child allowance
- dependant allowance
- national insurance allowance
- mortgage interest allowance
- credit union shares allowance
- government bonds allowance.

Key terms

Income tax is a tax on worker's earnings above a certain minimum and is calculated as a specific percentage of the worker's taxable income (gross income – non-taxable allowances).

The income before tax is **gross income.** Income after deduction of taxes is **net income.**

Example 41

A civil servant's salary is $30 000.00 per year. Her taxable income is found by deducting:

- personal allowance of $150.00 per month
- spouse allowance of $100.00 per month
- child allowance of $40.00 per month
- national insurance of $50.00 per month.

She then pays tax at 30%.

Calculate her taxable income and the amount she has to pay in income tax.

The total amount to be deducted from the gross annual salary is:

- personal allowance of 150 × 12 = 1800
- spouse allowance of 100 × 12 = 1200
- child allowance of 40 × 12 = 480
- NIS of of 50 × 12 = 600

giving a total of $4080.

Taxable income = 30 000 − 4080

$$= \$25\,920.00$$

Total tax payable is 30% of 25 920 = $\dfrac{30}{100} \times 25\,920$

$$= \$7776.00$$

2.9 Insurance

Life is full of risks such as losing one's laptop, having a road traffic accident or damage to one's house. We cannot avoid these risks altogether, but we can insure against them. **Insurance** is an agreement with a company that if an accident happens they will pay out an agreed amount to help repair your car or replace your laptop and so on. They do this in return for an insurance **premium** which is paid to them on a monthly or an annual basis. The premium may be calculated as a percentage of the value of the item being insured.

Key terms

Insurance is a contract between an individual and an insurance company which states that the company will pay an agreed amount in the event of an accident or loss.

Premium is the monthly or annual payment which the insured person must make to the insurance company to ensure that the insurance policy remains in force.

Example 42

Joanna owns a 4 × 4 car which is valued at $25 000. Her insurance company has agreed to insure the car against fire and theft for 2% of the value of the car. What is the insurance premium that Joanna will have to pay?

Insurance premium = 2% of the value of the car

$$= \frac{2}{100} \times 25\,000$$

$$= \$500$$

Joanna's insurance premium will be $500 per year

Exercise 20

 1 A worker's taxable monthly income is $2480.00. If tax is paid at 15% per annum, find the amount paid in income tax.

 2 When income tax is levied at 5.5%, a civil servant pays $150.00. Calculate his taxable income.

3 A single mother has two children aged 12 and 8 and earns $30 000.00 per year. She also supports one dependent relative.

She is entitled to allowances of:

- single person allowance $1200
- child under 10 years $200
- child 10–14 years $300
- dependent relative $350
- NIS $250

a Calculate the amount she pays in income tax, levied at 15% per annum.

b Find her monthly taxable income.

4 Mrs Elliot wants to insure her house against hurricane damage. Her house is worth $250 000. One company quotes an insurance premium of $600 to cover her house against hurricane damage each year. Another company says that its premium will be 0.4% of the value of the house each year. Which is the cheaper option?

5 James has to insure his car, which is worth $10 000. An insurance company offers to insure it for a premium calculated as 1% of the value of the car, each year. What would be James's annual insurance premium?

6 An engineer has a monthly gross salary of $8000.00. He makes a 2% donation of his salary to charity. He contributes $175 to NIS and is allowed a $400.00 per month personal allowance, $600 per month as children allowance and $300 per month for a dependent relative.

His income tax rates are 15% on the first $300 and 20% on the remainder. Calculate:

a the total non-taxable annual income   **b** the total monthly taxable income

c the annual tax.

End of chapter summary

Discount, sales tax, profit and loss

- **Discount** is the amount, expressed as a quantity ($) or percentage (%), that the seller takes off the marked price so that the customer pays less for the item.
- **Sales tax** is the amount of money that the government adds to the basic price of goods or services; it is normally calculated as a percentage of the basic price.
- When the selling price is more than the cost price the dealer makes a **profit**.
- When the selling price is less than the cost price the dealer makes a **loss**.
- **Cost price** is the amount paid to purchase the goods and services.
- **Selling price** is the amount the goods and services are sold for.
- **Percentage profit** $= \dfrac{\text{profit}}{\text{cost price}} \times 100\%$
- **Percentage loss** $= \dfrac{\text{loss}}{\text{cost price}} \times 100\%$

Interest and depreciation

- **Interest** is the profit earned from investing or saving a sum of money in a bank or company, or the amount paid for borrowing a sum of money from a financial institution or lender.
- The amount borrowed or deposited is the **principal**, the percentage rate is the rate at which interest is calculated and the total of the principal and interest, at the end of the period of the loan or investment, is the **amount**.
- For **simple interest**, the principal is unchanged over the period of the loan or investment, interest is added regularly.
- For **compound interest**, the principal increases at the end of each period as the interest is added to it and itself attracts interest.
- **Depreciation** is the yearly reduction in value of assets such as equipment, motor vehicles and machinery.
- **Appreciation** is the increase in value over time of an asset such as a house or a valuable antique.

Hire purchase and mortgages

- **Hire purchase** is a way of buying goods or services by making an initial payment, called a deposit, followed by a number of smaller payments called instalments.
- A **mortgage** is a loan given by a bank or other financial institution to assist in the purchase of property such as a house or land. The purchaser makes a small down payment (usually $\leqslant 20\%$ of the total value) and then borrows the rest, repaying it in monthly instalments.

Rates and taxes

- **Rates** are charges made on property owners, to raise money to maintain community services.
- Each property is assigned a **rateable value**, based on its size, location and condition and actual value.
- The **rate percentage** is the percentage or cent in the dollar charge, set by the local authority, to determine the **rates payable**, which is the amount to be paid in the year.

Utility payments

- A **utility bill** is a document used to charge customers for the use of utilities such as electricity, water and telephone.
- A unit of electricity is a **kilowatt-hour** (kWh) and 1 kWh = 1000 Wh.
- **Water rates** are charges made for the supply and disposal of water and are calculated according to the volume of water used by the consumer.
- A **telephone bill** is a document listing the charges to be paid by customers for telephone services provided by telephone companies.
- **Currency** is the **money** used by a country for its financial transactions.
- **Denomination** is the name given to a specific currency amount, usually in the form of coins or notes (such as dollars, pound, dime).
- **Basic financial operations** involve the use of the four basic operations: addition, subtraction, multiplication and division in computing financial transactions.
- **Foreign exchange** is the exchange of currencies between different countries, using an agreed rate of conversion.
- **Insurance** is a contract to pay a customer in the event of an accident or loss. It requires the regular payment of an insurance **premium**.

Salaries, wages, commission and income tax

- A **salary** is the monthly or annual payment workers receive for services rendered.
- **Gross annual salary** is the gross monthly salary × 12 and **net monthly salary** is the gross monthly salary less deductions
- **Wages** are payments received by workers who are normally paid hourly, daily, weekly or fortnightly for their work. The amount paid per hour is the **basic rate** and the number of hours worked per week is the **basic week**. The amount earned in a basic week is the **basic wage**. Two consecutive basic weeks constitute a **basic fortnight**. Hours worked more than the basic week are called **overtime**.
- **Commission** is an amount paid to a salesperson in addition to their basic wage.
- **Income tax** is a tax on worker's earnings above a certain minimum and is calculated as a specific percentage of the worker's taxable income (gross income – non-taxable allowances).
- The income before tax is **gross income**. Income after deduction of taxes is **net income**.

Examination-type questions for Chapter 2

1   Jasmine borrowed $12 400 at 10% per annum compound interest.

    a   Calculate the interest on the loan for the first year.

    b   At the end of the first year she was able to repay $6050.00. How much did she owe at the beginning of the second year?

    c   Calculate the interest on the remaining balance for the second year.

2   A bank pays 12% per annum compound interest on all fixed deposits. A customer deposits $14 540.00 in his account. Calculate the total amount in the account after five years.

3   A salesperson working for a fixed salary of $4280.00 per month is also given a 5% commission on the value of commodities sold above $5000.00 annually. In one year he sells $30 000.00 worth of commodities. Calculate his total salary for that year.

4   A department store imports 20 electrical fans for sale. The fans cost $1200.00 each and the customs department charges $140.00 duty on each fan.

    a   Calculate the total cost of all the fans, inclusive of duty.

    b   If the department store sells each fan at $1900.00, determine the total profit made on the sale of the fans.

    c   Calculate the percentage profit, giving your answer correct to the nearest whole number.

5   The table below shows part of Mrs James' stationery bill for the school's bookstore.

Item	Quantity	Unit price ($)	Total price ($)
Note book	20	3.50	P
Pen	Q	1.25	62.50
Ruler	100	R	75.00
Eraser	120	0.65	S
Dictionary	W	6.55	91.70
Total			Y
VAT (20%)			Z

    a   Calculate the values of: P, Q, R, S, W, Y and Z

    Mrs James sold note books at $3.00 each, pens at $1.00 each, rulers at 90¢ each, erasers at 75¢ each and dictionaries at $7.00 each.

    b   Given that she sells all of her stock, calculate whether she made a profit or loss overall and work out the percentage profit or loss.

6   A scooter is advertised at a cash price of $675.00. It is also offered on hire purchase for a down payment of $102.00 and 12 monthly instalments of $82.00.

    a   Calculate the total hire purchase price of the scooter.

    b   How much would you save by paying cash for the scooter?

    c   Write down the cash price as a percentage of the hire purchase price.

7   These are the exchange rates offered by a bank.

    a   A tourist changes US$820.00 to BDS$ at the rate offered by the bank. How much did she receive, in BDS$?

    b   She spent BDS$1200.00 then she changed half of the remainder to TT$ and the other half to EC$. How many TT$ and how many EC$ did she receive?

| EC$1.00 = TT$3.50 |
| US$1.00 = BDS$2.00 |
| US$1.00 = TT$6.50 |

8   This is John Peters' telephone bill.

Name: John Peters	Address: Riverville	Account number: 9540012	Arrears ($)
Previous reading date: 31/03/19	Present reading date: 30/04/19	# local calls	76.90
# calls: 7717	# calls:7822	?	
Monthly service fee for calls not exceeding 50	$45.00		
Charge on calls exceeding 50 (35¢ per call)	?		
Overseas calls (based on usage/overseas rate)	$68.50		
65% tax on all overseas calls	?		
Total charge	?		

Calculate, for the period under review:

a   the number of local calls   b   the charge for local calls

c   the tax on overseas calls   d   the total charge.

CHAPTER 3 Algebra

Objectives

By the end of this chapter you should be able to:

- explain directed numbers and their use in arithmetic operations
- describe the use of symbols in algebra to represent numbers, operations, variables and relations
- explain and give examples of algebraic notation
- state the laws of indices and use these laws to simplify expressions with integral indices
- simplify algebraic expressions, using the four basic operations
- perform substitution on various algebraic expressions
- perform binary operations in algebra
- carry out simple factorisation and expansion of algebraic expressions, using the distributive law
- define and solve algebraic fractions
- define and solve simple linear equations in one unknown
- give examples of and solve simultaneous equations in two unknowns
- define and solve simple linear inequalities in one unknown
- perform a transposition on algebraic equations – change the subject of a formula
- find the highest common factor (HCF) of algebraic terms
- find the product of two binomials
- factorise a four-term algebraic expression by grouping
- factorise quadratic expressions of the type:
  - coefficient of $x^2 = 1$
  - coefficient of $x^2 \neq 1$
  - difference of two squares
  - a perfect square
- rewrite a quadratic expression into the form $a(x + h)^2 + k$ (completing the square)
- solve quadratic equations by:
  - factorising
  - completing the square
  - using the quadratic formula

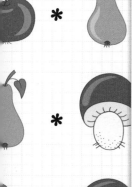

	TUE	WED	THU	FRI
	+19 °C	+15°C	+10°C	+8°C

- solve word problems based on linear equations, linear inequalities, simultaneous equations, quadratic equations
- solve quadratic and linear simultaneous equations
- prove two algebraic expressions to be identical (equations as compared to identities)
- explain direct and inverse variation and solve problems, using equations of the form:
  - $y \propto x \rightarrow y = kx$ (direct variation)
  - $y \propto \dfrac{1}{x} \rightarrow y = \dfrac{k}{x}$

You should already know:

- concepts of number theory and computation

Introduction

This chapter on algebra will help you to appreciate algebra, both as a language and as a form of communication. You will learn to appreciate the role of symbols and algebraic strategies in solving problems, in mathematics and in related fields, and to work with abstract concepts.

Use of algebra can help solve many everyday problems in areas such as finance, measurement, planning and organisation, because many ideas can be expressed in terms of algebraic expressions and equations. Through using algebra, householders can, for example, manage their budgets, pay bills, take care of health care costs and even do their shopping.

On a more professional level, algebra supports the development of critical thinking skills, logic, deductive and inductive reasoning as well as laying the foundation for higher learning and career success in many disciplines such as mathematics, economics, engineering, physics, programming, medicine and science in general. In fact, many scientific laws are expressed as algebraic equations.

Examples of algebra in real life

The change in vertical height of the road, compared to the horizontal distance along the same stretch of road, is its gradient and is calculated in the same way as the gradient of a line.

The price increases in direct proportion to the amount of gas pumped into the fuel tank, so can be expressed algebraically, as a direct variation.

The path of the ball after it is thrown by the player can be expressed using algebra.

3.1 Directed numbers and arithmetic operations on them

Directed numbers, as the name suggests, are numbers that have 'direction': they may be positive (denoted by the + sign) or negative (denoted by the – sign). While a negative number must always have its sign in front of it, a number without a sign is taken to be positive.

Directed numbers can be shown on a number line.

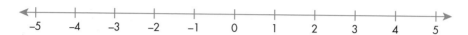

Conventionally, numbers to the left of 0 are negative, those to the right of 0 are positive. So, starting from 0, numbers increase in size positively to the right and negatively to the left.

Examples of directed numbers are +3 (which may also be written simply as 3) and –5.

> **Key term**
>
> **Directed numbers** are numbers with a positive (+) or negative (–) sign attached.

3.1.1 Addition and subtraction of directed numbers

Adding and subtracting two numbers with the same sign

For small numbers, you can use the number line.

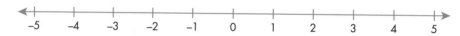

Example 1

2 + 3 = ?

2 + 3 = 5   From 0, move 2 places to the right, then jump 3 more places right, ending up on 5.

Example 2

–1 – 2 = ?

–1 – 2 = –3   From 0, move 1 place to the left, then jump 2 more places left, ending up on –3.

These two examples lead to a general rule.

Example 3

(a) 272 + 23 = ?   (b) –68 – 42 = ?

Applying the rule above:

(a) 272 + 23 = 295

(b) –68 – 42 = –(68 + 42)

= –110

> **Key facts**
>
> To **add or subtract** two directed numbers with the **same sign**, ignore the signs, add the numbers and then reinstate the sign.

Adding and subtracting two numbers with different signs (+ and –)

Again, for small numbers, you can use the number line.

Example 4

5 – 3 = ?

5 – 3 = 2   From 0, move 5 places to the right, then move 3 places to the left, ending up on 2.

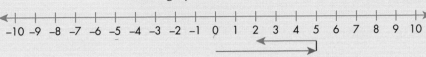

Example 5

–7 + 4 = ?

–7 + 4 = –3   From 0, move 7 places to the left to –7, then move 4 places to the right, ending up on –3.

These two examples lead to a general rule.

> ### Key facts
>
> To **add or subtract** two directed numbers with **different signs**, ignore the signs, subtract the smaller number from the larger and then reinstate the sign of the larger number.

Example 6

Evaluate: (a)   123 – 89   (b)   –35 + 24

           (c)   –65 + 87.

Applying the rule above:

(a)   123 – 89 = 34   (b)   –35 + 24 = –11

(c)   –65 + 87 = 22

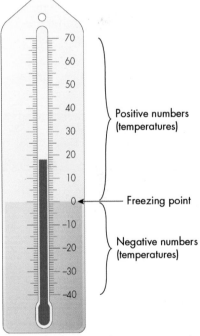

The numbers on a thermometer are directed numbers.

Exercise 3A

1   Use the number line to add or subtract.

   a   3 + 3   **b**   –5 – 4   **c**   –11 + 2   **d**   9 – 1   **e**   –8 + 8

2   Find the value of 37 + 45.

3   Find the value of –89 – 27.

4   Evaluate 67 – 59.

5   Evaluate –113 – 37.

6   The temperature indicated by a thermometer rose from –10 °C to 23 °C. By how many degrees did the temperature increase?

7 Solve –14 – 34 + 16 – 27.

8 Find the value of 20 + 60 + 77 – 114.

9 Solve 234 + 3567 – 982 – 2238.

10 Find the value of 90 289 – 45 678 + 36 009.

11 Solve 28 329 758 – 6 488 849 – 192 394.

3.1.2 Multiplication of directed numbers

Multiplication of two directed numbers with different signs (+ × –)

Example 7

Evaluate: (a) 2×-4 (b) -3×3.

You can treat multiplication as repeated addition.

(a) Following on from examples 1 and 2, you can represent 2×3 on a number line as 3 + 3 or 2 + 2 + 2 in the positive direction (to the right of zero). In the same way, to represent 2×-4 on the number line think of a boy standing on the zero and making two jumps to the left, each jump equal to 4 units in the negative direction, so he ends on –8.

(b) Using a similar approach, you can find -3×3 by thinking of the boy starting on zero and making three jumps of 3 units each in the negative direction, ending on –9.

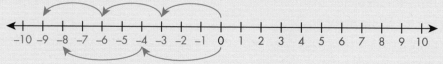

Therefore: $2 \times -4 = -8$ and $-3 \times 3 = -9$.

These two examples lead to this general rule.

Key fact

When **multiplying** two directed numbers with **different signs** (negative and positive), multiply as normal – the answer obtained is always negative.

Multiplication of two directed numbers with the same signs (+ × +) or (– × –)

Once again, for small numbers, you can use the number line.

From the previous examples you may conclude that when multiplying **two positive** numbers, you can use repeated addition on the number line and the result is always positive, for example, $4 \times 2 = 4 + 4$ or 2 + 2 + 2 + 2 = 8.

Example 8

Evaluate: 2×3.

Using the number line, start from 0 and move 2 places to the right (+2). Do this 3 times, treating multiplication as repeated addition, and ending on 6. So $2 \times 3 = 6$.

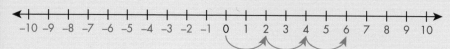

The next example demonstrates how to use the number line to multiply **two negative numbers**.

Example 9

Evaluate -4×-2.

Following on from example 7, -4×2 would be represented by starting from 0 and making two jumps of 4 units each, to the left (in the negative direction) on the number line ending on -8. However -4×-2 is represented by backward jumps from 0, due to the presence of the negative sign in -2.

Generally, representing a negative number on the number line implies a movement to the left, in the negative direction; however, the introduction of a second negative indicates a movement in the opposite direction, which is to the right or positive. Therefore, on the number line, -4×-2 translates into the boy making two backward jumps, each of 4 units, ending on 8.

$-4 \times -2 = 8$

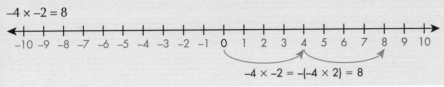

$-4 \times -2 = -(-4 \times 2) = 8$

These two examples of multiplication of two directed numbers of the **same** sign lead to this general rule.

Key fact

When **multiplying** two directed numbers with the **same sign** (negative or positive), multiply as normal – the answer obtained is always positive.

Exercise 3B

1 Use the number line to work these out.

   a 3×4 **b** -5×-3 **c** 1×-9 **d** -6×2

2 Calculate 23×7.

3 Calculate -17×-7.

4 Find the value of 15×-14.

5 Evaluate -23×12.

6 Calculate 45×37.

7 A tailor cut 22 pieces of cloth, each of length 35 cm. Find the total length of the cloth.

8 Calculate $-78 \times -22 \times 98$.

9 Find the value of $89 \times -92 \times -12$.

10 Evaluate $62 \times 89 \times -45 \times -23$.

11 Solve $67 \times -99 \times 46 \times 87$.

3.1.3 Division of directed numbers

Division of two directed numbers with the same sign (+ ÷ + or – ÷ –)

Example 10

Evaluate: (a) 6 ÷ 3 (b) –8 ÷ –2.

(a) The number 6, on the number line, can be divided into three equal parts, each equivalent to 2.

6 ÷ 3 = 2

(b) The number –8 can be divided into four equal parts, each equivalent to –2.

–8 ÷ –2 = 4

Again, the examples of division of two directed numbers of the same sign lead to this general rule.

Key fact

When **dividing** two numbers with the **same sign** (+ ÷ + or – ÷ –) the answer is always positive.

Division of two directed numbers with different signs (+ ÷ – or – ÷ +)

Example 11

Evaluate: (a) 10 ÷ –5 (b) –6 ÷ 2.

Think back to the boy on the number line.

(a) Imagine him standing on 10, facing in the positive direction because 10 is a positive number.

He is going to jump 5 places in a negative direction, because –5 is a negative number.

How many jumps will he make?

He makes 2 jumps, both backwards, so the answer is –2.

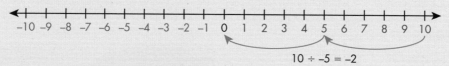

10 ÷ –5 = –2

10 ÷ –5 = –2

(b) Now he stands on –6, facing in a negative direction because –6 is a negative number.

He makes 3 jumps, each of 2 units, but again he is moving backwards, so the answer is –3.

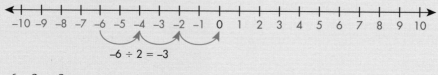

–6 ÷ 2 = –3

–6 ÷ 2 = –3

These two examples of division of two directed numbers of different signs lead to this general rule.

Key facts

When **dividing two directed numbers** with **different signs** (+ and –) the result is always negative.

Did you know? The popular statement: 'Two negatives give a positive,' can be quite misleading, since it is only true for multiplication and division of two directed numbers.

Exercise 3C

1. Use the number line to solve these.

   a $12 \div 3$ **b** $-10 \div -2$ **c** $-8 \div 8$ **d** $15 \div -5$

2. Solve $-24 \div -6$.

3. Solve $52 \div 13$.

4. Solve $70 \div -14$.

5. How many pieces of thread, all of equal length 600 cm, can a boy cut from a total length of 8000 cm? What length remains?

6. Find the value of $-156 \div 12 \div (-26)$.

7. Solve $-22\,440 \div -220 \div (-28 \div 14)$.

8. Evaluate $(-88\,220 \div 11) \div (-120 \div 60)$.

9. Solve $(59\,768 \div 482) \div (-4464 \div 36)$.

3.1.4 Mixed operations with directed numbers

Mixed operations relate to combinations of addition, subtraction, multiplication and division. These four operations may all appear in a single problem involving directed numbers.

Tip	When there are more than two directed numbers in the same calculation, it is advisable to work with two numbers at a time.

Remember	Always apply the rules of BODMAS to mixed operations with directed numbers.

Example 12

Evaluate $-2 + 3 \times 4(-6 \div -3) + 8 - 5$.

Apply the rules of BODMAS (see Chapter 1, section 1.9): for mixed operations, complete the calculations inside brackets (B) first, followed by 'of' (O), division (D), multiplication (M), addition (A) and subtraction (S).

At the same time, follow the rules of directed numbers.

$-2 + 3 \times 4(-6 \div -3) + 8 - 5 = -2 + 3 \times 4 \times 2 + 8 - 5$

$\qquad\qquad = -2 + 24 + 8 - 5$

$\qquad\qquad = 22 + 3$

$\qquad\qquad = 25$

Alternatively, you could put all the positive numbers together and all the negative numbers together during the penultimate step, to give:
$-2 - 5 + 24 + 8 = -7 + 32 = 25$.

Summary: working with directed numbers	To add or subtract two directed numbers with the same sign, add and keep the sign.
	If the signs are different, subtract the number with the smaller magnitude (size) from the larger and keep the sign of the number with the larger magnitude.
	For the multiplication and division of two directed numbers: same signs give a positive answer and different signs give a negative answer.

Exercise 3D

1 Solve $-2 + 3 - 5$.

2 Solve $6 \times -3 \times 4$.

3 Find the value of $20 \div 4 \times -3$.

4 Evaluate $5 - 8 \times 3 + 6 \div 3$.

5 Solve $2(4 - 5) - 7(3 - 9) + 1$.

6 Evaluate:
$$\frac{3 - 7 \times 8 + 6 \div 3(5 - 4)}{9 + 6 - 3 \times 5 + 8(2 - 1)}$$

7 Write in its simplest form:
$$\frac{2(3 - 7) \times 6 - 4(5 - 6)}{10 \times 2 - 1 + 8(3 - 4)}$$

8 Solve:
$$\frac{2 + 0 \times 3 - 5 + 8 - 10}{8 - 0(2 + 4) - 9 + 0 + 5}$$

3.2 The meaning and importance of algebra

3.2.1 What is algebra?

Algebra is a branch of mathematics that uses numbers, letters and other symbols to carry out various arithmetic and other mathematical operations. Algebra is really a very effective way of solving real-life problems.

Think about a boy admiring a number of juicy, yellow mangoes on a branch when suddenly a strong gust of wind blew, knocking 5 of these mangoes to the ground and leaving only six on the branch.

How many mangoes were on the branch before the wind blew?

You can use algebra to solve real-life problems like this one. If you take the original number of mangoes on the tree as x, then $x - 5 = 6$. You have now expressed an everyday problem as a simple algebraic equation which you can easily solve to find that there were originally 11 mangoes on the branch.

In algebra, letters, called **variables**, are used to replace unknown numbers.

Arithmetic operations and relations can be written as combinations of letters or numbers and letters, for example, the sum of two numbers represented by x and y can be written as $x + y$ and three times a number represented by k can be written as $3 \times k$ or $3k$ instead of 3×6.

Numbers and variables combined in this way form **terms**.

Examples of **algebraic terms** include $2x$, $9y^3$, $(z + 2)^2$. The number in front of the letter is called the **coefficient**, the letters are variables. Both variable and coefficient can be raised to a power or index.

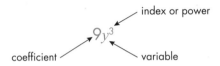

If you had three apples and five bananas in one bag, and four bananas and two apples in another, how many would you have? You can add the apples and add the bananas, to find that you have five apples and nine bananas. In algebra, you can make similar calculations, adding $3a + 5b$ to $4b + 2a$, to get $5a + 9b$, because $3a$ and $2a$ are **like terms**, as are $5b$ and $4b$. However, $5a$ and $9b$ are **unlike** terms.

Key term

A **variable** is an unknown number or quantity that can be represented by a letter in algebra.

Key term

An **algebraic term** is a combination of letters and numbers or variables and coefficients, such as $3x$.

Key term

In algebraic terms such as $3x$, x is the variable and 3 is the **coefficient** of x. It tells you how many xs you have.

Key terms

Like terms are made up of exactly the same letters and powers, for example, $3x$, $10x$, $100x$. **Unlike terms** do not have the same variable or power, for example, $3y$ and $3x$ are unlike terms.

Key term

An **algebraic expression** is a set of algebraic terms linked by signs, such as + or –. An algebraic expression consisting of two terms is a **binomial expression**.

Exercise 3E

1. Give two examples of an algebraic term.

2. From your own example of an algebraic term, identify the coefficients, variable(s) and powers.

3. Give one example of a **binomial expression**.

4. Sort these terms into like terms and unlike terms.

   $2ax$, $3x^2$, $5x$, $12xy$, $10x^2$, $12ax$, $100x$, $16xy$, $24xy^2$, $3ax^2$, $2x^2y$, $6xy^2$

5. A girl went shopping for school supplies at a stationery outlet. She purchased:

   x exercise books at \$2.00 each, y pens at \$1.00 each, $2x$ erasers at 50 cents each, $3y$ notebooks at \$4.00 each, $\frac{1}{2}x$ rulers at \$1.00 each, x packs of typing paper at \$18.00 a pack and y drawing sets at \$2.00 per set. Write down an expression in terms of x and y for the total cost of all the items purchased.

3.3 Algebraic notation

To use algebra, you need to extract the information from the words in a problem and then express it in **algebraic notation**.

Example 13

Use algebraic notation to write down ten times the product of two numbers.

Let the two numbers be w and y.

Then ten times their product is $10 \times w \times y = 10wy$.

Example 14

Use algebraic notation to express the product of three numbers divided by their sum.

Let the three numbers be L, M and N.

Then the product is $L \times M \times N = LMN$ and their sum is $L + M + N$.

The product of the three numbers divided by their sum $= \dfrac{LMN}{L + M + N}$

Exercise 3F

1. Use algebraic notation to state the sum of three numbers.

2. Express the difference of two numbers in algebraic notation.

3. Use algebraic notation to state the product of four numbers.

4. Express the quotient of two numbers in algebraic notation.

5. Use algebraic notation to state the sum of two numbers divided by twice their difference.

6. Use algebraic notation to write down four times the sum of four numbers, multiplied by the square of their product.

7. Use algebraic notation to express three-quarters of the quotient of two numbers multiplied by half their product.

8. Express in algebraic notation the square root of the sum of five numbers divided by the cube of their product.

9. Use algebraic notation to write down the square of the quotient of two numbers multiplied by the square root of half their difference, added to five times the difference of their product and sum.

3.4 Laws of indices

Indices or powers indicate a quantity being multiplied by itself one or more times.

For numbers:

$3 \times 3 = 3^1 \times 3^1$

$\quad = 3^{1+1}$

$\quad = 3^2$

$\quad = 9$

and in algebra:

$p \times p \times p = p^{1+1+1}$

$\quad = p^3$

The first law of indices – multiplication

When multiplying powers of the same number, called the base, add the powers.

Example 15

Simplify $a^2 \times a^5$.

$a^2 \times a^5 = a^{5+2}$

$\quad = a^7$

The second law of indices – division

When dividing powers of the same base, subtract the indices.

Example 16

Simplify $m^8 \div m^6$.

$m^8 \div m^6 = m^{8-6}$

$\quad = m^2$

The third law of indices – raising a power to another power

When raising a power of a base to another power, multiply the indices.

Example 17

Simplify $(5x^2)^3$.

$(5x^2)^3 = 5^{1\times3} \times x^{2\times3}$

$\quad = 5^3 \times x^6$

$\quad = 125x^6$

The fourth law of indices – negative indices

When a quantity is raised to a negative power, invert it and change the sign of the power to positive (negative indices are equivalent to finding the reciprocal of the quantity), for example, $2^{-1} = \frac{1}{2}$.

Example 18

Simplify $10w^{-3}k^2t^{-4}$.

Each negative index is written as a reciprocal. Positive indices remain unchanged.

$10w^{-3}k^2t^{-4} = \dfrac{10k^2}{w^3t^4}$

The fifth law of indices – fractional indices

If a base is raised to a fractional power, the numerator of the fraction indicates a power and the denominator indicates a root.

Example 19

Simplify: $\left(9x^6y^9\right)^{\frac{2}{3}}$.

Everything in the brackets is raised to the power of $\frac{2}{3}$.

Find the cube root and raise the result to the power of 2.

$$\left(\sqrt[3]{\left(8x^6y^9\right)}\right)^2 = \left(2x^2y^3\right)^2 = 4x^4y^6$$

Find the cube root of each element of the product: 8, x^6 and y^9.

The sixth law of indices – zero index

Any base raised to the power of zero is 1.

Example 20

Evaluate   (a) $(10\,000x)^0$ (b) $\left(\dfrac{24p}{13\,546acy}\right)^0$.

(a) $(10\,000x)^0 = 1$

(b) $\left(\dfrac{24p}{13\,546acy}\right)^0 = 1$

Tip	Any quantity raised to the power of zero is equal to 1 because: $1 = \dfrac{a}{a} = \dfrac{a^1}{a^1} = a^{1-1} = a^0$

Exercise 3G

1 Simplify these powers.

    a $p^4 \times p^5$ **b** $q^2 \div q^3$ **c** $r^5 \times r^{-2}$

2 Simplify $(2x^2)^3 + 5(x^3)^4$.

3 Simplify $4(x^{-3})^3 - 5x^6(2x)^{-2}$.

4 Simplify $\dfrac{10x^2y^{-3} \times 4}{20x^{-1}y}$.

5 Write in its simplest form $\left(8a^3b^6\right)^{\frac{1}{3}} \times (16)^{-\frac{3}{4}}$.

6 Simplify $2.5(a^2b^3)^0 + 7.5(x^3y^2)^0$.

3.5 Using the four basic operations to simplify algebraic expressions

3.5.1 Addition and subtraction of algebraic terms

You have already seen that you can add or subtract only like terms.

Key fact

You can only **add or subtract** terms if they are like terms, for example, $2x - 3x + 5x = 4x$.

Example 21

Simplify: (a) $9x + 12x$
(b) $7y + 9y + 2y^2$.

(a) $9x + 12x = 21x$

(b) $7y + 9y + 2y^2 = 16y + 2y^2$

Example 22

Simplify $12a - 6ab - 15a + 7ab$.

$12a - 15a + 7ab - 6ab = -3a + ab$

Exercise 3H

1 Add these algebraic terms: $2x + 3x + 5x$.

2 Find the sum of $3x^2 + 6x + 20x + 40x^2$.

3 Subtract: **a** $50x - 22x$ **b** $100ay - 75ay$

4 Write an expression for the total area of a square of side x cm and a rectangle measuring $3x$ cm by $4x$ cm.

5 Simplify: $2x + 5x - 7x - 2x + 4x$.

6 Simplify: $2ab + 3ab^2 - 4a^2b - ab$.

7 Simplify: $12 + 2x^2 - 3x + bx - ax$.

8 Simplify: $2.7n^2 - 1.8n + 3.3n^2 - 5.2n + 2n$.

9 The volume of a cylinder is given by the expression: $V = \pi r^2 h$, where r represents the radius of the base and h is the height. A smaller cylinder has radius $\frac{3}{8}$ of the radius of the original cylinder and its height is $\frac{1}{4}$ the height of the original cylinder. Write an expression for the volume of the smaller cylinder.

3.5.2 Multiplication of algebraic terms

Although you can add and subtract only like algebraic terms, you can multiply terms by like and unlike terms. When you multiply two different variables together, you simply write the letters as a product, for example, $a \times b \times c = abc$.

Key fact

You can **multiply and divide** terms by like and unlike terms, for example, $5y \times 2y^3 = 10y^4$, $12a^3 \div 4a^2 = 3a^{3-2} = 3a$.

When multiplying terms containing the same variable or letter you must add the powers of the variables. For example:

$$r \times r = r^{1+1}$$
$$= r^2$$

$$y^3 \times y^4 = y^{3+4}$$
$$= y^7$$

$$3p^5 \times 2p^3 = 3 \times 2 \times p^5 \times p^3$$
$$= 6p^8$$

The first law of indices covers the multiplication of powers, stating, for example, that $x^2 \times x^3 = x^{3+2} = x^5$ and the second law deals with the division of powers, for example, $10y^6 \div 5y^4 = \frac{10}{5}y^{6-4} = 2y^2$.

Example 23

Simplify: (a) $5p \times 12p^3$ (b) $12k^8 \times 8k^{-3}$.

(a) $5p \times 12p^3 = 5 \times 12 \times p^{1+3}$

$= 60p^4$

(b) $12k^8 \times 8k^{-3} = 12 \times 8 \times k^{8-3}$

$= 96k^5$

Exercise 3I

1 Determine the product of x, y and z.

2 Determine the product of a, $2a$ and $3a$.

3 Simplify $2a \times 3b \times 4c$.

4 A box has length y, breadth $2y$ and height $3y$. Write an expression for the volume of the box in terms of y.

5 Simplify $ab \times bc \times ac \times acb$.

6 Multiply these algebraic terms: $5pq$, $10pq^2$, $4p^2q$, $2p$.

7 Simplify $10r^2 \times 14r^3 \times 5r^5 \times 8r^{-6}$.

8 The total surface area of a cuboid block of wood is given by the expression: $2(2a \times 3a) + 2(3a \times 4a) + 2(2a \times 4a)$. Write a simplified expression for the volume of the block.

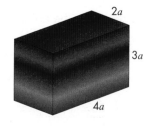

9 Remove the brackets and simplify.

a $11x(3x - 2) + 4(2x^2 - 6) - x^2(5 - x)$

b $3(ab^2cd^2 + ab^2c) - 5ac^2b(3a - 5b^2c) - (bc + abc)$

3.5.3 Division of algebraic terms

When dividing terms in the same variable or letter, you need to subtract the powers; for example, $M^4 \div M^3 = M^{4-3} = M$. Coefficients are divided as normal.

This is covered by the laws of indices (section 3.4) as relating to negative indices. The fourth law states that when a number or variable is raised to a negative power, the result is the reciprocal of the number or variable and the power changed to the equivalent positive power, for example: $2k^{-3} = \frac{2}{k^3}$.

Example 24

Simplify $\dfrac{20p^3r^6}{4p^2r^2}$.

$\dfrac{{}^5 20p^3r^6}{{}_1 4p^2r^2} = 5p^{3-2}r^{6-2}$

$= 5pr^4$

Example 25

Simplify $\dfrac{12p^8 \times 4p^{-3}}{6p^{-2}}$.

Apply the laws of indices.

$\dfrac{12p^8 \times 4p^{-3}}{6p^{-2}} = \dfrac{48p^{8-3} \times p^2}{6}$

$= \dfrac{48}{6} \times p^{8-3+2}$

$= 8p^7$

Exercise 3J

1 Simplify $4ab \div 2a$.

2 Simplify $pqrs \div pqs$.

3 Simplify $6a^2b^3c^4 \div 3abc$.

4 Simplify $\dfrac{x^3y^2z + xyz}{2xyz + 3xyz}$.

5 Simplify this expression to its simplest form.

$$\frac{10pq \times 4p^2q}{20pq^2}$$

6 A cuboid container of dimensions l by b by h contains water up to a height of $\frac{1}{4}h$. Derive an expression in terms of l, b and h for the quantity of empty space in the container. Note: volume of cuboid $= l \times b \times h$.

7 Simplify $\dfrac{5xy\left(2x^2y^3 + 3x^2y^3\right)}{8xyz - 2xyz - 4xyz}$.

8 Simplify $\left(\dfrac{4y^3z^{-2}}{12y^{-5}z^{-3}}\right)^3$.

3.5.4 Mixed operations in algebra

You need to be able to add, subtract, multiply and divide algebraic terms just as you do for numbers.

Example 26

Simplify $\dfrac{80v^3 + 20v^2 \times v - 10v^3}{10v \times 4v^2 \times 3}$.

$$\frac{80v^3 + 20v^2 \times v - 10v^3}{10 \times 4 \times 3v^3} = \frac{90v^3}{120v^3}$$

Collect like terms in the numerator and simplify the denominator.

$$= \frac{3}{4}$$

Simplify to lowest terms.

Example 27

Simplify $\dfrac{12a^3b^4}{10h^2p^5} \div \dfrac{4a^2b^4}{8h^3p^4}$.

$$\frac{12a^3b^4}{10h^2p^5} \div \frac{4a^2b^4}{8h^3p^4} = \frac{^6\cancel{12a^3b^4}}{_5\cancel{10h^2p^5}} \times \frac{^2\cancel{8h^3p^4}}{\cancel{4a^2b^4}}$$

Follow the rules for dividing fractions, cancelling the numbers.

$$= \frac{6 \times 2 \times a \times h}{5 \times p}$$

Simplify to lowest terms.

$$= \frac{12ah}{5p}$$

1 Simplify $2y - 3y + 5y \times 6y + 12y$.

2 Simplify $8a - 2b \times 5a + 12a \div 2$.

3 Remove the bracket and simplify. $2p(p - 2) - 3p(2p - 5p)$.

4 Simplify $\dfrac{10pqr + pr^2 - 2pq(r + 1)}{5pqr}$.

5 Simplify $\dfrac{100mn}{80v^2w^3} \times \dfrac{40v^3w^4}{50m^{-1}n^{-2}}$.

6 Simplify $\dfrac{10ab^2cd}{8x^2y^2z^2} \div \dfrac{15a^3bc^2d}{24xyz} \times \dfrac{ab}{xy}$.

3.6 Algebraic fractions

Examples of **algebraic fractions** include
$\dfrac{2}{x^2}$, $\dfrac{3y}{3b^3}$, $\dfrac{2a^2b}{y-3}$, $\dfrac{x+2}{5x^3}$ and $\dfrac{2x^2 + 4x - 5}{-1}$. You have
already been simplifying them by reducing
them to their simplest terms. You can also apply
the four operations of addition, subtraction,
multiplication and division to algebraic fractions
in a similar way to how you manipulate simple fractions.

> **Key term**
>
> An **algebraic fraction** is a fraction with an algebraic expression in the numerator or denominator or both.

3.6.1 Multiplication of algebraic fractions

When you multiply two or more algebraic fractions, start by exploring the possibility of cancelling (as you did in a previous section). If you can't cancel, then:

- multiply the terms in the numerators to find the numerator of the product
- multiply the terms in the denominators to find the denominator of the product.

Example 28

Simplify: (a) $\dfrac{10pq}{5mn} \times \dfrac{12mn^2}{8p^2q^4}$ (b) $\dfrac{12ab}{17xyz} \times \dfrac{8b^2c}{5x^2yz^3}$

(a) $\dfrac{^2\cancel{10pq}}{_1\cancel{5mn}} \times \dfrac{^3\cancel{12mn^2}}{_2\cancel{8p^2q^4}\,^3}$ (cancelling) $\rightarrow \dfrac{6n}{2pq^3} \rightarrow \dfrac{3n}{pq^3}$

(b) $\dfrac{12ab}{17xyz} \times \dfrac{8b^2c}{5x^2yz^3} = \dfrac{96ab^3c}{85x^3y^2z^4}$ Cancelling is not possible.

3.6.2 Division of algebraic fractions

The rules for division of algebraic fractions are the same as for division of simple fractions.

When dividing algebraic fractions:

- invert the divisor (the second fraction) and write down its reciprocal
- change the sign from division to multiplication.

Then proceed as for multiplication, cancelling where possible.

Example 29

Simplify: $\dfrac{6ab^4c}{10a^2b^3c^2} \div \dfrac{8b^2c}{5abc}$

$$\dfrac{6ab^4c}{10a^2b^3c^2} \div \dfrac{8b^2c}{5abc} = \dfrac{{}^36ab^4c}{{}_210a^2b^3{}^2c^2} \times \dfrac{{}^15abc}{{}_48b^2c}$$

Invert the divisor and cancel.

$$= \dfrac{3ab^4c}{8ab^4c^2}$$

Cancel again if you can.

$$= \dfrac{3}{8c}$$

3.6.3 Addition and subtraction of algebraic fractions

When adding or subtracting fractions, follow the same procedure as for ordinary fractions, first finding the lowest common denominator, and so on.

Example 30

Simplify: (a) $\dfrac{2}{a} + \dfrac{3}{2a}$ (b) $\dfrac{4x}{p} + \dfrac{5y}{q}$ (c) $\dfrac{6}{x^2} - \dfrac{5}{x^3}$ (d) $\dfrac{a}{l+m} - \dfrac{b}{x+y}$.

(a) The LCD is $2a$ (since a is a factor of $2a$).

$$\dfrac{2}{a} + \dfrac{3}{2a} = \dfrac{2 \times 2 + 3 \times 1}{2a}$$

$$= \dfrac{4+3}{2a}$$

$$= \dfrac{7}{2a}$$

(b) The LCD is $p \times q = pq$ (since p and q have no common factors).

$$\dfrac{4x}{p} + \dfrac{5y}{q} = \dfrac{4x \times q + 5y \times p}{pq}$$

$$= \dfrac{4qx + 5py}{pq}$$

(c) The LCD of x^2 and x^3 is x^3 (since x^2 is a factor of x^3).

$$\dfrac{6}{x^2} - \dfrac{5}{x^3} = \dfrac{6x-5}{x^3}$$

(d) The LCD of $(l+m)$ and $(x+y)$ is $(l+m)(x+y)$.

$$\dfrac{a}{l+m} - \dfrac{b}{x+y} = \dfrac{a(x+y) - b(l+m)}{(l+m)(x+y)}$$

Exercise 3L

1 Give five examples of an algebraic fraction.

2 Simplify $\dfrac{2x^2y}{3xy} \times \dfrac{5xy}{4x^3y^4} \times \dfrac{2x}{4}$.

3 Simplify $\dfrac{9a^2b^3}{6ab^2} \div \dfrac{12ab^4}{20a} \times \dfrac{10ab}{5b}$.

4 Simplify $\dfrac{1}{x} + \dfrac{2}{y} + \dfrac{3}{z}$.

5 Simplify $\dfrac{2a}{x} + \dfrac{3b}{x^2} - \dfrac{4c}{x^3}$.

6 Simplify $\dfrac{5a}{4y} - \dfrac{4b}{3y} + \dfrac{c}{5y}$.

7 Simplify $\dfrac{5x}{2y - 3} + \dfrac{2x}{y + 2}$.

8 Simplify $\dfrac{a + 2}{y + 1} - \dfrac{b + 3}{y + 2} + \dfrac{2b + 5}{y + 3}$.

3.7 Substitution

If you are given values for the variables in an algebraic expressions, you can find a numerical value for it by **substituting** these values and then doing the arithmetic calculation.

Key fact

You can find a numerical value for an algebraic expression by **substituting** values to replace the letters or symbols that make up the expression; for example, to evaluate $10rs^2 + 5r$ when $r = 3$ and $s = 5$: $10 \times 3(5)^2 + 5 \times 3 = 750 + 15 = 765$.

Example 31

Given that $x = 3$ and $y = 5$, evaluate $3x - 2y + 5$.

Replace x in the expression by 3 and y by 5.

$3x - 2y + 5 = 3(3) - 2(5) + 5$

$= 9 - 10 + 5$

$= 14 - 10$

$= 4$

Example 32

Given that $a = -4$, $b = 7$ and $c = 1$, evaluate $\dfrac{2a^3 - 3b^2 + c}{5abc - a}$

$\dfrac{2(-4)^3 - 3(7)^2 + 1}{5(-4)(7)(1) - (-4)} = \dfrac{2 \times (-64) - 3 \times 49 + 1}{-140 + 4}$

$= \dfrac{-128 - 147 + 1}{-136}$

$= \dfrac{-274}{-136}$

$= \dfrac{137}{68}$

$= 2\dfrac{1}{68}$

Tips

When substituting it is good practice to use brackets or multiplication signs, for example, $abc = (-4)(7)(1)$ or $-4 \times 7 \times 1$ to avoid computation errors. Do **not** just write -471.

Exercise 3M

1 Given that $a = 2$ and $b = 3$, evaluate $3a + b$.

2 Given that $x = -3$, $y = 4$ and $z = -1$, evaluate $2x + 3y - 5z$.

3 Given that $p = 7$, $q = -6$ and $r = 5$, evaluate $2p(3q - r) - (p - q - r)$.

4 Given that $x = 2$, $y = -3$ and $z = 4$, evaluate $\dfrac{x + y + z}{x - y - z}$.

5 Calculate the value of the expression $\dfrac{2a + 3b - 4c}{a + b - 3c}$ when $a = 6$, $b = 2$ and $c = -3$.

6 Given that $e = 0$, $f = -2$ and $g = 3$, evaluate $\dfrac{10e^3 - 3f^3 + fg^e - fg}{f + g + e}$.

3.8 Binary operations

In algebra, a **binary operation** is a new operation based on a combination of the four basic algebraic operations. This new operation can be represented by any symbol that you choose, apart from +, −, × or ÷.

Key term

A **binary operation** is an operation that is made from a combination of the four basic algebraic operations. This new operation can be represented by any symbol other than: +, −, × or ÷.

Example 33

Given that $a \blacklozenge b = 2a - 2b + 1$, find $2 \blacklozenge 3$.

$2 \blacklozenge 3$ means that $a = 2$ and $b = 3$.

Substituting for a and b:

$$2 \blacklozenge 3 = 2(2) - 2(3) + 1$$
$$= 4 - 6 + 1$$
$$= -1$$

Example 34

Given that $v \diamond w = \dfrac{2v^3 - 4w^2}{3vw}$ find $-3 \diamond -2$.

$$(-3) \diamond (-2) = \frac{2(-3)^3 - 4(-2)^2}{3 \times (-3) \times (-2)}$$

$$= \frac{2(-27) - 4(4)}{18}$$

$$= \frac{-54 - 16}{18}$$

$$= -\frac{70}{18}$$

$$= \frac{35}{9}$$

$$= 3\frac{8}{9}$$

Exercise 3N

1 Given that $a \circledcirc b = 2a - b$, find $2 \circledcirc 3$.

2 If $p \circledast q = 2(p - q)^2$, evaluate $3 \circledast 5$.

3 Given that $r \blacktriangle s$ means $(r + 2s)^{\frac{1}{2}}$, find $9 \blacktriangle 8$.

4 Given that $w \diamondsuit n = \sqrt{w^2 - n^3}$, evaluate $4 \diamondsuit 3$.

5 In a certain binary operation $(p \circledast q)$ the symbol $\circledast$ means 'double the first number and add the square root of the second number'. Calculate $8 \circledast 9$.

6 Given that $m \, \text{☯} \, n$ means $m^3 + mn^2 - 2m^2n$, evaluate $2 \, \text{☯} \, 4$.

7 If $h \, \clubsuit \, k$ means $h^3 + k^2 - 2hk$, evaluate $2 \, \clubsuit \, (3 \, \clubsuit \, - 4)$.

8 Given that $a \blacklozenge b = \dfrac{a^2 - b^3}{a^3 + b^2}$ find $\dfrac{2 \blacklozenge - 3}{1 \blacklozenge 4}$.

3.9 Using the distributive law to expand algebraic expressions

In an algebraic expression such as $a(b + c)$, the number or variable in front of the bracketed expression multiplies each term inside the brackets: $a(b + c) = ab + ac$.

3.9.1 Multiplication of binomials

When you multiply two binomials, for example, $(a + b)(c + d)$, each term in the first set of brackets multiplies both terms in the second set.

$$(a + b)(c + d) = a \times c + a \times d + b \times c + b \times d$$

$$= ac + ad + bc + bd$$

Tip

When removing brackets and carrying out all the operations in algebra, you must follow the rules of directed numbers, for example, a minus outside a term in brackets will change the sign of every term inside.

Example 35

Simplify each expression.

(a) $3y(2y - 5y^2)$ (b) $(3p - 2r)(6p - r)$

(a) $3y(2y - 5y^2) = 3y \times 2y + 3y \times -5y^2$

$\qquad = 6y^2 - 15y^3$

(b) $(3p - 2r)(6p - r) = (3p) \times (6p) + (3p) \times (-r) + (-2r) \times (6p) + (-2r) \times (-r)$

$\qquad\qquad = 18p^2 - 3pr - 12pr + 2r^2$

$\qquad\qquad = 18p^2 + 2r^2 - 15pr$

Note

The arrows indicate which terms you need to multiply.

1 Simplify $2a(a - 2b)$.

2 Simplify $4w(3w - 2w^2)$.

3 Simplify $3x(2x^2 - 4x - 5)$.

4 Find the product of the binomials $(2x - 3)$ and $(x + 4)$.

5 Simplify $(x - 4)(2x^2 - 5x + 2)$.

6 Simplify $(3x^2 - y^2)(4x^2 + 5xy)$.

7 Remove the brackets and simplify this expression.

$$\frac{(3x - 5y)(2x + 4y)}{(x - 2y)(x + 2y)}$$

3.10 Simple factorisation of an algebraic expression

Factorisation can be considered as the reverse of expansion. In simple factorisation the common factor or term is removed and placed outside the rest of the term, which is set inside a pair of brackets, for example, $ax + ay = a(x + y)$.

Key term

Factorisation means splitting an algebraic expression into two or more factors.

If the common factor has more than one term (for example, a binomial) then two brackets are used to separate the factors.

Example 36

Factorise $3abc + 6bcd - 9acd$.

Find the largest number factor that will divide exactly into the coefficients of each term (3).

Find the letter or group of letters that appear in each term (c).

Now place the combined common factor (number and letters, $3c$) outside the brackets and divide each term by it.

$$3abc + 6bcd - 9acd = 3c\left(\frac{^{}\cancel{3}ab\cancel{c}}{\cancel{3}\cancel{c}} + \frac{^{2}\cancel{6}b\cancel{c}d}{\cancel{3}\cancel{c}} - \frac{^{3}\cancel{9}a\cancel{c}d}{\cancel{3}\cancel{c}}\right)$$
$$= 3c(ab + 2bd - 3ad)$$

The method you have seen is equivalent to finding the **highest common factor** (HCF) of all the terms and placing it outside the bracket, then dividing each term by that HCF.

Key terms

The **highest common factor (HCF)** of a group of numbers is the highest number that can divide exactly into each number in the group. The **HCF of a set of monomials** (single-term algebraic expressions) is the product of the HCF of the coefficients and the lowest powers of the common letters in the monomials.

Example 37

Factorise $8p^2m^3 - 6pm^2 + 12p^3m^4$.

The HCF of 8, 6 and 12 (coefficients) is 2, the HCF of p^2, p and p^3 is p and the HCF of m^3, m^2 and m^4 is m^2.

So $8p^2m^3 - 6pm^2 + 12p^3m^4 = 2pm^2(4pm - 3 + 6p^2m^2)$

Exercise 3P

1 Find the HCF of $3ax^2$, $12a^2x^2$ and $6ax$.

2 Find the HCF of $20a$, $15ab$ and $10a^2b$.

3 Factorise each expression by first finding the HCF.

  a $18x^3 + 24x - 12x^2$ **b** $5p^3r^4 + 10p^2r^3$

4 Factorise this expression.

$$\frac{2ab^5}{x^3y^3} + \frac{6a^2b}{5x^2y^2} - \frac{9ab}{10}$$

5 Factorise this expression.

$atp(x^2 - y^3) - tp^2(x^2 - y^3)$

6 Factorise this expression.

$$\frac{2a + b}{3x - 5} - \frac{6ax + 3bx}{9x^2 - 15}$$

3.11 Simple linear equations in one unknown

A simple **linear equation** in one unknown can be compared to a scale, as shown in the diagram. In the left pan there is an unknown mass $(x + 5)$ kg and in the right pan there is a mass of 9 kg. For the scale to remain balanced, whatever is added or removed on one side must be added or removed on the other side.

> **Key term**
>
> A **linear equation** is an equation in one or two unknowns, where the highest power in any term is 1, such as $x = 5$ or $2x + 3y = 17$. Its graph is a straight line.

If 5 kg is removed from the left pan, x kg remains and, when the same is done on the right, 4 kg remains, therefore $x = 4$ kg.

$$x + 5 = 9 \Rightarrow x + 5 - 5 = 9 - 5$$

$$\Rightarrow x = 4$$

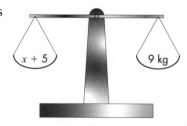

If three identical packages were placed on the left pan, balancing the 9 kg on the right, you could use equations to find the mass of each package.

Let the mass of each package be x kg. The mass of the three packages will be $3x$ kg.

mass in left pan = mass in right pan

Therefore: $3x$ kg = 9 kg. Dividing each side of the scale by 3:

$$\frac{3x}{3} = \frac{9}{3} \rightarrow x = 3 \text{ kg}$$

Using the model of the scale, you can identify different types of equation, based on the various ways of balancing the scale. At least seven types of simple equation can be identified. Some are described here.

1 Equations requiring multiplication

Example 38

Solve the equation $\frac{y}{2} = 5$.

Multiply both sides by 2:

$\frac{y}{2} \times 2 = 5 \times 2 \Rightarrow y = 10$

Quicker method: Cross-multiply.

$\frac{y}{2} \times \frac{5}{1} \Rightarrow y \times 1 = 5 \times 2$

$\Rightarrow y = 10$

2 Equations requiring division

Example 39

Solve the equation $4x = 12$.

Divide both sides by 4:

$\frac{4x}{4} = \frac{12}{4} \Rightarrow x = 3$

3 Equations requiring addition and subtraction

Example 40

Solve these equations.

(a) $x - 7 = 3$ (b) $x + 9 = 1$

(a) $x - 7 = 3$ (requiring addition)

$\Rightarrow x - 7 + 7 = 3 + 7$

$\Rightarrow x = 10$

(b) $x + 9 = 1$ (requiring subtraction)

$\Rightarrow x + 9 - 9 = 1 - 9$

$\Rightarrow x = -8$

Tip

Adding or subtracting from both sides can prove to be quite tedious. A quicker method is to place all the numbers on one side of the equation and all the variables on the other, remembering to change the signs when moving across the equals sign from one side to the other. This procedure is referred to as **transposition**.

$x + 4 = 11$

$x = 11 - 4 = 7$ (bringing across the 4 from left to right)

4 Equations with the unknown (variable) on both sides

Example 41

Solve the equation $2x - 7 = 12x + 3$.

Arrange the variables and numbers on different sides (not forgetting to change the signs).

$-7 - 3 = 12x - 2x$

$\Rightarrow -10 = 10x$

$\Rightarrow -1 = x$

5 Equations containing brackets (on one or both sides)

Example 42

Solve the equation $5(2x - 3) = 2 - 4(3x - 1)$.

Remove the brackets by using the distributive law.

$10x - 15 = 2 - 12x + 4$

$\Rightarrow 10x + 12x = 2 + 15 + 4$

$\Rightarrow 22x = 21$

$\Rightarrow x = \dfrac{21}{22}$

6 Equations involving fractions on one or both sides

Example 43

Cross-multiplying:

$\dfrac{5}{3x - 2} = \dfrac{3}{4} \Rightarrow 4 \times 5 = 3(3x - 2)$

$\Rightarrow 20 = 9x - 6$

$\Rightarrow 20 + 6 = 9x$

$\Rightarrow x = \dfrac{26}{9}$

Key fact

When you need to solve an equation such as $\dfrac{3x}{2} = \dfrac{x + 14}{3}$ you can **cross-multiply**, to get $3x \times 3 = 2 \times (x + 14)$ and simplify from there. It is the same as multiplying both sides by the product of the denominators.

Tip

When cross-multiplying it is advisable to use brackets so that the distributive law will be correctly applied.

7 Equations with roots

Example 44

$\sqrt{3x - 1} = 4$

Square both sides to get rid of the square root, since $\left(\sqrt{a}\right)^2 = a$.

This is covered by the fifth law of indices, on fractional indices (see section 3.4).

$\left(\sqrt{a}\right)^2 = \left(\sqrt{a}\right) \times \left(\sqrt{a}\right) = a^{\frac{1}{2}} \times a^{\frac{1}{2}} = a^1 = a$

$\left[\sqrt{3x - 1}\right]^2 = 4^2$

$\Rightarrow 3x - 1 = 16$

$\Rightarrow 3x = 17$

$\Rightarrow x = \dfrac{17}{3}$

Exercise 3Q

1 Solve the equation $\dfrac{x}{2} = 6$.

2 Solve the equation $5x = 10$.

3 Solve the equation $x - 7 = 3$.

4 Solve the equation $x + 8 = 5$.

5 Solve the equation $2x - 4 = x + 7$.

6 Solve the equation $3(2x - 6) = 4x + 2$.

7 A square has an area of 64 cm². Given that the length of one side is x, find the value of x.

8 A rectangle is 4 cm longer than it is wide. Given that the width is w and the perimeter is 20 cm, find its area.

9 Solve the equation $5(4 - 2x) = 8(2x - 1)$.

10 Solve the equation $\dfrac{2x}{7} = \dfrac{5}{3}$.

11 Solve the equation $\dfrac{6x}{4} + \dfrac{3}{5} = \dfrac{2}{3} - \dfrac{5x}{2}$.

12 The angles in a triangle are $x°$, $2x°$ and $40°$. Given that all the angles add up to $180°$, find the value of x.

13 A man decided to share a sum of money among his three sons: Adam, Bill and Chris. Adam was given $\$x$, Bill was given twice that amount and Chris received $1\frac{1}{2}$ times as much as Adam. Given that Chris received $\$60.00$, how much did the other sons receive and what was the total amount shared?

14 Solve the equation $\sqrt{3(2x - 8)} = \dfrac{3}{4}$.

15 Solve the equation $\dfrac{2x - 3}{4} + \dfrac{5x + 2}{3} = \dfrac{x + 1}{2} - \dfrac{4x + 6}{3}$.

16 A householder decided to do some shopping. She took $\$500.00$ and went to the fish market, where she bought fish costing $\$y$. She then went to the vegetable market, where she spent three times that amount. At the clothes store she spent $\$80.00$ more that she spent at the vegetable market and realised that she had only $\$30.00$ left.

   a Use algebraic notation to state the amounts spent at the vegetable market and store.

   b Write and solve an equation in y and determine how much was spent in each area.

3.12 Inequalities

Look again at the scales. This time they are unbalanced – one side has a greater mass than the other. This is an example of an **inequality** (or inequation).

In this example, the amount on the right-hand side weighs more, or is heavier than the amount on the left-hand side, causing the scales to be unbalanced.

This inequality can be written as: $12 > x + 5$ (12 is greater than $x + 5$).

Other inequality symbols used are:

- $<$ meaning 'less than'
- $\leq$ meaning 'less than or equal to'
- $\geq$ meaning 'greater than or equal to'.

There are also seven types of inequality, matching the seven types of equation, and they are solved with similar strategies. There is one exception, however: multiplication by a negative number changes the sign on both sides and reverses the inequality symbol.

This means that if there is a negative sign in the coefficient of x and it is made positive by multiplying (or dividing) both sides of the equation by -1, this results in a change of sign on the other side of the equation as well, **and** a reversal of the inequality symbol ($<$ to $>$ or $>$ to $<$). For example:

$$-2x < 3 \Rightarrow x > -\frac{3}{2}$$

> **Key term**
>
> An **inequality** shows the relationship between two expressions that are not equal to each other, using a sign such as $\neq$ (not equal to), $>$ (greater than), $\geq$ (greater than or equal to), $<$ (less than) or $\leq$ (less than or equal to).

$x + 5$

12 kg

Example 45

Simplify $3(2x - 1) + 4 \geq 6 + 2(4x - 5)$.

$6x - 3 + 4 \geq 6 + 8x - 10$

$\Rightarrow 6x - 8x \geq 3 - 4 + 6 - 10$

$\Rightarrow -2x \geq -5$

$\Rightarrow x \leq \dfrac{5}{2}$ (The sign changes on both sides therefore the symbol is reversed.)

Example 46

Simplify $\dfrac{2y}{3} - \dfrac{2}{5} > \dfrac{6y}{4} - \dfrac{1}{2}$.

When there are two fractions on one or both sides, you must combine both fractions before cross-multiplying.

Finding the LCD on both sides: Cross-multiplying:

$\dfrac{5 \times 2y - 3 \times 2}{15} > \dfrac{2 \times 6y - 4 \times 1}{8}$ $8(10y - 6) > 15(12y - 4)$

$\Rightarrow \dfrac{10y - 6}{15} > \dfrac{12y - 4}{8}$ $\Rightarrow 80y - 48 > 180y - 60$

$\Rightarrow 80y - 180y > 48 - 60$

$\Rightarrow -100y > -12$

$\Rightarrow y < \dfrac{12}{100}$ (The symbol is reversed.)

Example 47

The area of a triangle must not be more than 60 cm². If the base of the triangle is 10 cm what is its greatest height?

Area of triangle $= \frac{1}{2}$ base × height

So $\frac{1}{2} \times 10 \times h \leqslant 60$

$\qquad h \leqslant 60 \div 5$

$\qquad h \leqslant 12$

Example 48

Solve the inequality $\sqrt{\dfrac{x-4}{2}} > \dfrac{\sqrt{x}}{3}$.

Eliminate the square root by squaring both sides of the inequality.

$\dfrac{x-4}{2} > \dfrac{x}{9}$

Multiplying both sides by 18: $\quad 9(x-4) > 2x$

$\qquad\qquad\qquad\qquad \Rightarrow 9x - 36 > 2x$

$\qquad\qquad\qquad\qquad \Rightarrow x > 5\frac{1}{7}$

Exercise 3R

1 Solve the inequality $2x - 3 < 1$.

2 Solve $3(2x - 4) > 4x + 6$.

3 The area of a square must not be greater 40 cm². Calculate the greatest possible length of the sides.

4 The area of a rectangle must not be more than 24 cm². Given that the length of the rectangle is 6 cm, what is the greatest width it can have?

5 Solve the inequality $12(3x - 5) \leqslant 4(2x - 8)$.

6 Solve the inequality $\dfrac{9x}{8} \geqslant \dfrac{5}{7}$.

7 Solve the inequality $\dfrac{3x - 1}{5} < \dfrac{7}{9}$.

8 The area of a circle needs to be greater than 12 cm². What is the smallest radius possible?

9 Solve: $\sqrt{\dfrac{2x-5}{5}} \leqslant \dfrac{2}{3}$.

10 Solve: $\dfrac{12x}{5} + \dfrac{3}{4} < \dfrac{11}{6} - \dfrac{5x}{2}$.

11 The volume of a cylinder should be smaller than 28 cm³. Given that the radius is 5 cm, what is the maximum height of the cylinder?

3.13 Transposition – changing the subject of a formula

You will often need to work with **formulae** in mathematics. For example, $F = ma$ is a formula used to calculate force (F) where m is mass and a is acceleration. If $m = 20$ kg and $a = 2$ m s^{-2}, $F = 20$ kg $\times$ 2 m s$^{-2} = 40$ N.

The **subject** of a formula is the single variable that appears alone, on one side (usually the left-hand side); for example, in the formula $F = ma$, F is the subject. You can calculate its value by substituting values for all the other variables.

Key terms	Key term
A **formula** is an equation that describes the relationship between two or more quantities. The **subject** of the formula is the variable that stands alone on one side.	You can change the subject of the formula through the process of **transposition**. For example, to make m the subject of the formula $F = ma$, divide both sides of the equation by a to give $\dfrac{F}{a} = m$.

Example 49

Given the formula $s = \dfrac{(v + u)t}{2}$, make u the subject.

First: cross-multiply: $2s = (v + u)t$

Divide by t: $\dfrac{2s}{t} = v + u$

Bring v over to the left-hand side, leaving u alone on the right (remembering to change the sign).

$$\frac{2s}{t} - v = u$$

Then swap sides: $u = \dfrac{2s}{t} - v$

Example 50

Change the subject of the formula $x = py - 2ry$ so that y is the subject.

$$x = py - 2ry$$
$$x = y(p - 2r)$$
$$\frac{x}{(p - 2r)} = y$$
$$y = \frac{x}{(p - 2r)}$$

Example 51

Express the equation $V = \dfrac{nRT}{P}$ so that P is the subject.

$$V = \frac{nRT}{P}$$
$$VP = nRT$$
$$P = \frac{nRT}{V}$$

Exercise 3S

1 Given the formula $F = ma$, name the subject of the formula, then make m the new subject.

2 Given the formula $v = u + at$, make t the subject.

3 Given the formula $C = 2m^2 + 5$ make m the subject.

4 Make n the subject of the formula $\dfrac{F}{L} = \dfrac{k(mn)}{P}$.

5 $M_1 + M_2 = \dfrac{4\pi^2 r^2}{GT^2}$. Transpose for T.

6 Make m the subject of the formula $\dfrac{h}{g} = R\dfrac{a}{b}\left[\dfrac{p}{q} \times \dfrac{L}{m}\right]$.

3.14 Simultaneous linear equations in two unknowns

To solve **simultaneous equations**, you need to find values of the variables that satisfy both equations. For example, given the equations:

$$x + y = 27$$
$$x - y = 3$$

it is possible to find the values of x and y for which both are true. In this case, $x = 15$ and $y = 12$.

Simultaneous equations may be solved by:

- the elimination method
- the substitution method.

> **Key term**
>
> **Simultaneous equations** are sets of equations in two or more variables that are simultaneously satisfied by the same set of values of these variables.

3.14.1 The elimination method

To use this method, you manipulate the equations to eliminate one variable and convert the two equations into one simple equation in one unknown. You can find the value of the other variable by substitution.

If one of the variables has the same sign and the same coefficient in both equations, you can eliminate it by subtraction. If one of the variables has different signs but the same coefficient in both equations, you can eliminate it by addition.

Example 52

Solve these equations.

$$x + y = 27 \quad \textbf{1}$$

$$x - y = 3 \quad \textbf{2}$$

Eliminate y by addition in this case, because the coefficients are equal but of opposite sign in the equations.

1 + 2: $\qquad (x + x) + (y - y) = 27 + 3 = 30$

$$2x + 0 = 30$$

$$x = 15$$

Substituting in **1** for x: $\quad 15 + y = 27$

$$y = 27 - 15$$

$$= 12$$

> **Tip**
>
> You can check the correctness of the answers by substituting the values obtained back into **one** of the equations and verifying that the left-hand side equals the right-hand side.

Check the answers by substituting in equation **2** for x and y.

LHS: $x - y = 15 - 12 = 3 =$ RHS

3.14.2 The substitution method

In the substitution method, you express one variable in terms of the other by transposing one of the equations, then substitute into the other equation to obtain one simple linear equation in one variable.

Referring back to the previous example:

$$x + y = 27 \qquad \textbf{1}$$

$$x - y = 3 \qquad \textbf{2}$$

From equation **2**, $x = y + 3$ (transposing for x)

Substituting in **1** for x: $\quad y + 3 + y = 27$

$$2y = 27 - 3 = 24$$

$$y = \frac{24}{2}$$

$$= 12$$

Substituting in **2** for y: $\quad x - 12 = 3$

$$x = 3 + 12$$

$$= 15$$

3.14.3 Simultaneous equations that have variables with different coefficients

If neither variable has the same coefficient in both equations, you need to make them equal by multiplying each equation by some number, then using elimination by subtraction or addition.

Example 53

Solve these equations.

$x + y = 3$ **1**

$3x - 2y = 4$ **2**

Using the elimination method to eliminate x or y.

Make the coefficients (of x or y) equal by multiplying equation **2** by the coefficient in equation **1** and vice versa, multiplying equation **1** by the coefficient in equation **2**.

To eliminate x in the above example, multiply equation **1** by 3 (the coefficient of x in equation **2**) thereby producing new equation **3**.

Equation **1** × 3 → $3x + 3y = 9$ **3**

It is essential to achieve the same numerical value of the coefficients, irrespective of the sign, because when the coefficient is the same for either variable the equations may be added or subtracted to eliminate that particular variable.

Note	One of the equations can be multiplied by a negative value to make the sign different.

Now, since the coefficients of x have the same sign and same numerical value in equations **2** and **3**, subtract equation **2** from equation **3**. Do this by subtracting corresponding terms.

$3x + 3y = 9$

$- (3x - 2y = 4)$

$\overline{0 + 5y = 5}$

When subtracting, the sign of each term in the equation being subtracted is changed, so

$3x - 3x = 0;$

$3y - (-2y) = 3y + 2y = 5y; 9 - 4 = 5$

So $5y = 5$

$y = \dfrac{5}{5}$

$= 1$

Substituting this back into one of the original equations (the simpler one) produces:

$x + 1 = 3$

$x = 3 - 1$

$x = 2$

The solution is $x = 2$, $y = 1$.

If instead of x you choose to eliminate the variable y first, the procedure is the same.

We can use linear equations to help solve real life problems. These are often presented just as words, not as a formula with symbols, and the first task is to take the words and to develop an algebraic equation from them which we can then solve in the normal way.

Example 54

Karen is shopping for hair extensions and sees that three packets of braids will cost $15 more than a small wig. If the wig is priced at $45, what is the price of a packet of braids?

$3 \times \text{braids} = \text{wig} + \$15$

$3b = w + 15$

$b = \dfrac{w + 15}{3}$

Since w is $45, we can calculate b.

$b = \dfrac{45 + 15}{3} = \dfrac{60}{3} = 20$

The packets of braids are $20 each.

1 Solve these simultaneous equations.

   a $x + y = 4$ **b** $2x - 2y = 3$ **c** $4a + 5b = 10$

      $x - y = 2$ $3x + 2y = 2$ $2a + 3b = 4$

2 A student bought 2 exercise books and 3 pencils for $5.00. If he had purchased 4 exercise books and 5 pencils the cost would be $9.00. What is the cost of 1 pencil and 1 exercise book?

3 Solve these simultaneous equations.

   $3x + 2y = 10$

   $2x - 4y = 4$

4 Solve these simultaneous equations.

   $4x + 5y = 1$

   $3x - 2y = 5$

5 Solve these simultaneous equations.

   $\dfrac{2x}{5} - \dfrac{3y}{4} = \dfrac{1}{2}$ **1**

   $\dfrac{1}{2}x + \dfrac{1}{4}y = 4$ **2**

6 Solve these simultaneous equations.

   $1.2x - 3.4y = 2.4$ **1**

   $0.4x - 1.2y = 1.2$ **2**

7 The cost of 2 bananas and 3 mangoes is $5.50. However, the cost of 5 bananas and 2 mangoes is $7.50. What's the cost of 1 banana and 1 mango?

3.15 Factorising by grouping

You have already learned about simple factorisation and the product of two binomials. These are two very important prerequisites for factorising by grouping.

Factorising by grouping is only possible when there are four algebraic terms. It is performed as two consecutive, simple factorisations of two pairs of algebraic terms.

You can always check your answer by multiplying the two binomials comprising your answer.

Example 55

Factorise $ax + by + bx + ay$.

First, group the four terms into two pairs, each with a common factor.

$$ax + bx + by + ay \quad \text{(common factors are } x \text{ and } y\text{)}$$

Factorise each pair: $x(a + b) + y(a + b)$

Factorise again, using the terms in brackets as the common factor.

$$(a + b)(x + y)$$

As a check, multiply the two binomials obtained from factorisation to obtain the original four terms.

Example 56

Factorise the expression $x^2 + 3x - 2x - 6$.

$$x^2 + 3x - 2x - 6 = x(x + 3) - 2(x + 3)$$
$$= (x + 3)(x - 2)$$

Exercise 3U

Factorise each expression.

1 $ak + bk + ra + rb$

2 $a(m^3 - n^2) - b(m^3 - n^2)$

3 $my^2 + ny^2 - mx^2 - nx^2$

4 $p^2lm - p^2ln + m - n$

5 $6x^2 - 12xy + 3xy - 6y^2$

6 $x^2 + xy - y^2 - xy$

7 $16a^2 + 20ab + 25b^2 + 20ab$

8 $(2a + b)(3x + 4y) - 6x - 8y$

9 $6x^2 - 3xy + 6x + 2xy - y^2 + 2y$

3.16 Factorising quadratic expressions

Quadratic expressions are written in the form $ax^2 + bx + c$, where a is the coefficient of x^2, b is the coefficient of x and c is the constant term. Examples of quadratic expressions include $2x^2 + 3x + 1$, $x^2 - 5x - 4$, $9x^2 - 4$, $2x^2 + 2x$ and $25x^2$.

Key term

A **quadratic expression** is an algebraic expression in which the highest power of the variable is 2.

3.16.1 Factorising quadratic expressions with coefficient of $x^2 = 1$

Example 57

Factorise $x^2 + 5x + 6$.

Find the factors of c that add up to the coefficient of x, which is b.

Factors of 6: 3×2 and $3 + 2 = 5$	(coefficient of x)
1×6 and $1 + 6 = 7$	(not the coefficient of x)
-3×-2 and $-3 + -2 = -5$	(not the coefficient of x)
-1×-6 and $-1 + -6 = -7$	(not the coefficient of x)

Write the factors in two brackets: $(x + 3)(x + 2)$.

You can find the product of the binomials to verify the answer.

3.16.2 Factorising quadratic expressions with coefficient of $x^2 \neq 1$

Example 58

Factorise: $2x^2 - 8x + 6$.

Find the product of a and c (ac).

$a = 2$, $c = 6$ so $ac = 2 \times 6 = 12$.

Find the factors of ac that will add to give the coefficient of x (b) = -8.

The factors of ac: 1×12 $1 + 12 = 13$ -1×-12 $-1 + -12 = -13$

2×6 $2 + 6 = 8$ -2×-6 $-2 + -6 = -8$

3×4 $3 + 4 = 7$ -3×-4 $-3 + -4 = -7$

Choose -2×-6 as b, the sum of the factors, is -8.

Substitute the factors for the middle term ($-8x$).

The expression becomes $2x^2 - 6x - 2x + 6$

Now factorise by grouping. $2x(x - 3) - 2(x - 3) = (x - 3)(2x - 2)$

3.16.3 Special cases

Factorising the difference of two squares

Example 59

Factorise: $25x^2 - 16$.

Find the square root of each term.

$25x^2 - 16 = (5x)^2 - (4)^2$

Write as an expression with two brackets, as:

$(5x - 4)(5x + 4)$

You can find the product of the binomials to verify the answer.

Factorising perfect squares

A perfect square is characterised by three terms:

- the first term is a perfect square
- the third term is a perfect square
- the second term = 2 × square root of first term × square root of third term.

Example 60

Factorise $100y^2 + 100y + 25$.

Find the square root of the first term: $\sqrt{100y^2} = 10y$.

Find the square root of the third term: $\sqrt{25} = 5$.

Show that: $2 \times \sqrt{1\text{st term}} \times \sqrt{3\text{rd term}} = 2\text{nd term}$: $2 \times 10y \times 5 = 100y = 2\text{nd term}$

Write the square roots in one set of brackets and square it: $(10y + 5)^2$

Factorising quadratic expressions of the type $ax^2 + bx$

Factorise the expression: $6x^2 + 12x$.

This turns out to be a simple factorisation.

$6x^2 + 12x = 6x(x + 2)$

3.16.4 Writing a quadratic expression in the form $a(x + h)^2 + k$ (completing the square)

Completing the square when the coefficient of $x^2 = 1$

A general expression for a perfect square is $(x + h)^2$, which is the perfect square of a binomial.

If this expression is expanded it gives $x^2 + 2hx + h^2$.

If you examine the expanded expression carefully you can observe that the constant term (h^2) is actually the square of half the coefficient of x ($2h$).

This is true for the square of any binomial when the coefficient of x^2 (a) is equal to 1, for example,

$(x + 4)^2 = x^2 + 8x + 16$ and $16 = (\frac{1}{2} \times 8)^2$.

You can draw a similar conclusion for the perfect square: $(x - h)^2 = x^2 - 2hx + h^2$. Note the significance of the change in sign.

You can use this property to **complete the square** for a quadratic expression when you are given the first two terms, or to rewrite a quadratic expression as a perfect square in the form:

$a(x + h)^2 + k$

> ### Key fact
>
> A quadratic expression, with coefficient of $x^2 = 1$ can be written as a perfect square by adding ($\frac{1}{2}$ coefficient of x)^2 to the first two terms. The entire expression can be written in the form $a(x + h)^2 + k$, for example,
>
> $x^2 + 4x + 3 = (x^2 + 4x + 2^2) + 3 - 4 = (x + 2)^2 - 1$.

Write the expression: $x^2 + 2x + 3$ in the form $a(x + h)^2 + k$.

Write the first two terms ($x^2 + 2x$) in a set of brackets with a space after the $2x$, leaving the 3 outside:

$(x^2 + 2x + \quad) + 3$

Now insert ($\frac{1}{2}$ coefficient of x)^2 in the space in the brackets: $(\frac{1}{2} \times 2)^2 = (1)^2 = 1$

The expression becomes $(x^2 + 2x + 1^2) + 3$ which is equal to $(x + 1)^2 + 3$.

The brackets now represent a perfect square but you must subtract the extra value added inside the brackets from the number outside the brackets, so that the original value of the expression is not changed.

$x^2 + 2x + 3 = (x + 1)^2 + 3 - 1$

$\qquad = (x + 1)^2 + 2$

> **Tip**
>
> You can check the correctness of the expression obtained after completing the square by expanding it and comparing it with the original expression – they should be the same.

Completing the square when the coefficient of $x^2 \neq 1$

First, factorise the expression by simple factorisation so that the coefficient of $x^2 = 1$; then follow the steps explained above, for when the coefficient of x^2 is equal to 1.

Complete the square for the quadratic expression $3x^2 - 12x + 1$.

Make the coefficient of $x^2 = 1$ by taking the coefficient of x^2 as a factor, to factorise the first two terms, leaving the constant term outside the brackets.

$3x^2 - 12x + 1 = 3(x^2 - 4x) + 1$

Now complete the square inside the brackets by adding the square of half the coefficient of x, remembering to subtract what is being added, in order to complete the square.

$3(x^2 - 4x + (-2)^2) + 1 - 12 = 3(x^2 - 4x + 4) + 1 - 12$

$$= 3(x - 2)^2 - 11$$

Tip

When subtracting the amount added in to complete the square, you need to take into account the factor outside the brackets, which multiplies the quantity inside the brackets. In the example, $3 \times 4 = 12$.

Exercise 3V

1 Give one example of a quadratic expression satisfying the given condition.

   a The coefficient of $x^2 = 1$ ($a = 1$)   **b** The coefficient of $x^2 \neq 1$

   c The coefficient of $x = 0$ ($b = 0$)   **d** The constant term $= 0$ ($c = 0$)

   e It is a perfect square.   **f** It represents a difference of two squares.

2 Factorise each of these quadratic expressions.

   a $x^2 + 6x + 5$ **b** $2x^2 + 3x + 1$ **c** $4x^2 + 8x + 4$ **d** $25a^2 - 9$

3 Factorise the quadratic expression $12x^2 - 24x$.

4 Write the quadratic expression $x^2 + 4x + 3$ in the form: $a(x + h)^2 + k$.

5 The product of two consecutive numbers is 12. Devise an algebraic expression to represent the product and then find the numbers.

6 Complete the square for the expression $4x^2 - 8x + 13$.

7 Complete the square for the expression $-5x^2 + 10x + 3$.

8 Write the quadratic expression $8.4x^2 - 4.2x - 5.2$ in the form $a(x + h)^2 + k$.

3.17 Solving quadratic equations

What is the difference between an **equation** and an **expression**? An **algebraic expression** may be a combination of variables (letters), numbers and operators (addition, subtraction, multiplication and division) but it does not contain an equals sign. For example, $2x^2y^3 + 4ab^4$ is an expression.

An equation is a statement that two quantities are equal and is made up of two algebraic expressions separated by an equals sign. For example, $(x + y)^2 – 2xy = 2y + x$ is an equation.

The highest power of the variable in a **quadratic equation** is 2.

Depending on the values of a, b and c, there are several types of quadratic equation, such as:

- $9x^2 = 16$
- $4x^2 + 8x = 0$
- $x^2 + 3x = –2$
- $4a^2 = 4a + 3$

- $4p^2 – 8p + 4 = 0$
- $7x^2 + 4x – 3 = 0$
- $5x^2 – 10x – 9 = 0$

3.17.1 Methods of solving quadratic equations

There are three main methods for solving quadratic equations:

- factorisation: after finding the factors, equate each of them to 0 and solve
- completing the square: equate the completed square to zero and solve
- using the quadratic formula: $x = \dfrac{-b \pm \sqrt{b^2 - 4ac}}{2a}$ when it's not possible to find factors.

3.17.2 Choosing the appropriate method for solving a quadratic equation

Example 64

Use the appropriate method of factorisation to solve these equations.

(a) $9x^2 = 16$ (b) $4x^2 + 8x = 0$ (c) $x^2 + 3x = –2$ (d) $4a^2 = 4a + 3$ (e) $4p^2 – 8p + 4 = 0$

(a) The two terms in this equation are exact squares, so we can use factorisation of the difference of two squares.

$9x^2 – 16 = 0 \quad\quad \Rightarrow (3x)^2 – 4^2 = 0$

$\Rightarrow (3x – 4)(3x + 4) = 0$

$3x – 4 = 0 \Rightarrow 3x = 4$

$\Rightarrow x = \dfrac{4}{3}$

$3x + 4 = 0 \Rightarrow 3x = –4$

$\Rightarrow x = -\dfrac{4}{3}$

Tip

After factorising the expression, equate each factor to 0 because if $a \times b = 0$ (a and b being the factors) then one or both factors must be 0 for the equation to be valid. Before factorising, you need to rearrange the equation to get 0 on one side, in the form $ax^2 + bx + c = 0$.

(b) This equation has no constant term and is in terms of x and x^2 only, so we can use simple factorisation.

$4x^2 + 8x = 0 \Rightarrow 4x(x + 2) = 0$

$4x = 0 \Rightarrow x = 0$

$x + 2 = 0 \Rightarrow x = –2$

Example 64 continues on next page

(c) In this example, the coefficient of x^2 is one so we choose the method of factorisation when the coefficient of $x^2 = 1$.

$x^2 + 3x = -2 \Rightarrow x^2 + 3x + 2 = 0$

Factorising: $(x + 2)(x + 1) = 0 \Rightarrow x + 2 = 0$ or $x + 1 = 0$

$\Rightarrow x = -2$ or $x = -1$

(d) Here the coefficient of x^2 is 4 so it is necessary to use the method of factorisation when the coefficient of $x^2 \neq 1$.

$4a^2 = 4a + 3 \Rightarrow 4a^2 - 4a - 3 = 0$

Factorising: $ac = 4 \times -3$

$\qquad\qquad\quad = -12$

Choose the factors of -12: $\quad 12 = -6 \times 2$ and $-6 + 2 = -4$

Substituting these factors: $\qquad 4a^2 - 6a + 2a - 3 = 0$

Now factorise by grouping: $\quad 2a(2a - 3) + 1(2a - 3) = 0$

$\qquad\qquad\qquad\qquad\qquad\quad \Rightarrow (2a - 3)(2a + 1) = 0$

$2a - 3 = 0 \Rightarrow a = \dfrac{3}{2}$ and $2a + 1 = 0 \Rightarrow a = -\dfrac{1}{2}$

(e) Looking at this equation note that it is a perfect square.

$\qquad 4p^2 - 8p + 4 = 0$

$(2p)^2 - 8p + (-2)^2 = 0$ and $2 \times 2p \times (-2) = -8p$

Rewrite this equation as a square: $(2p - 2)^2 = 0$

$2p - 2 = 0 \Rightarrow 2p = 2 \Rightarrow p = 1$

Example 65

Use the quadratic formula to solve the equation $7x^2 + 4x - 3 = 0$.

The quadratic formula gives $x = \dfrac{-b \pm \sqrt{b^2 - 4ac}}{2a}$.

Substituting $a = 7$, $b = 4$, $c = -3$:

$x = \dfrac{-4 \pm \sqrt{16 - 4(7)(-3)}}{2(7)}$

$ = \dfrac{-4 \pm \sqrt{100}}{14}$

$ = \dfrac{-4 \pm 10}{14}$

$ = \dfrac{6}{14}$ and $-\dfrac{14}{14}$

$x = \dfrac{3}{7}$ and $x = -1$

Example 66

Solve the equation $5x^2 - 10x - 9 = 0$ by completing the square.

Factorise the first two terms with the constant term outside the brackets:

$5(x^2 - 2x) - 9$

Complete the square inside the brackets by adding the square of half the coefficient of x.

$5(x^2 - 2x + 1^2) - 9$

Subtract the additional amount that was required to complete the square.

$5(x - 1)^2 - 9 - 5 = 5(x - 1)^2 - 14$

Set the completed square = 0.

$5(x - 1)^2 - 14 = 0$

$$\Rightarrow (x - 1)^2 = \frac{14}{5}$$

$$\Rightarrow x - 1 = \pm\sqrt{\frac{14}{5}}$$

$$x = 1 \pm 1.67$$

Example 67

A certain positive number x, when added to its square gives the sum 6. Devise an algebraic equation and solve it to find the number x.

$x^2 + x = 6$ (based on information given)

Rewriting the equation and equating to $0 \Rightarrow x^2 + x - 6 = 0$

Factors of -6: 3×-2 and $3 - 2 = 1$

$\Rightarrow (x + 3)(x - 2) = 0$

$x = -3$ or $x = 2$

x must be positive so $x = 2$.

The number is 2.

Exercise 3W

1. Solve these quadratic equations by factorisation.

   a $x^2 + 10x + 9 = 0$ b $6x^2 + 10x + 4 = 0$ c $64x^2 - 81 = 0$ d $9x^2 - 18x + 9 = 0$

2. A certain number, y, added to half its square gives the answer 4. Find the value of y.

3. Solve the equation $x^2 = 6x - 8$.

4. Solve the equation $5x^2 - 15x = -10$ by an appropriate method.

5. Use the quadratic formula to solve the equation $7x^2 - 4x - 5 = 0$.

6. The area of a square added to its perimeter gives the value 5. Find the length of a side of the square.

7. The length of one side of a rectangle is 3 cm more than the other. Given that the area of the rectangle is 18 cm², find the dimensions of the rectangle.

8 Use the method of completing the square to solve the equation $7x^2 - 28x + 2 = 0$.

9 Use the method of completing the square to solve the equation $-4x^2 + 6x + 9 = 0$.

10 A certain number subtracted from $\frac{3}{4}$ of its square gives the value 1. Find the number.

3.18 Simultaneous equations: one quadratic and one linear

Sometimes a pair of simultaneous equations can comprise one linear equation and one quadratic equation. In this case, the most appropriate method of solution would be substitution, where the linear equation is rearranged to give one variable written in terms of the other; the result of this rearrangement is substituted into the quadratic equation.

Example 68

Solve these simultaneous equations.

$2x - y = 3$ **1**

$3x^2 + y = 1$ **2**

From equation **1**: $2x - 3 = y$ (transposing)

Substituting in **2** for y: $3x^2 + 2x - 3 = 1$

$\Rightarrow 3x^2 + 2x - 4 = 0$

> **Note** Making y the subject is simpler than making x the subject, since this would introduce a fractional term that can be challenging when working with quadratic expressions.

The expression: $3x^2 + 2x - 4$ is not readily factorised by any of the five methods discussed above (see 3.16) so use the quadratic formula.

$$x = \frac{-b \pm \sqrt{b^2 - 4ac}}{2a}$$

$$= \frac{-2 \pm \sqrt{4 - 4(3)(-4)}}{6}$$

$$= \frac{-2 \pm \sqrt{52}}{6}$$

$$= \frac{-2 \pm 7.2}{6}$$

$$x = \frac{5.2}{6}, x = \frac{-9.2}{6}$$

$$x = 0.9, x = -1.5$$

Now substitute each value of x in **1** to find the values of y.

When $x = 0.9$, $y = 2x - 3$ When $x = -1.5$, $y = -1.5 \times 2 - 3$

$= 2 \times 0.9 - 3$ $= -6$

$= -1.2$

Alternatively, when the quadratic does not easily factorise, you can use the method of completing the square to solve the quadratic equation (see example 63). This same method could have been used in example 65 but would have resulted in more complex calculations.

3.19 Identities and equations – what is the difference?

An **identity** may sometimes be mistaken for an equation, and vice versa, but they are not the same.

Whereas an identity is always true for all values of the variable, an equation has to satisfy certain conditions and is only true for certain values of the variable.

Both identities and equations contain an equals sign (=) which separates the left-hand side from the right-hand side; however, in the case of an identity, the left-hand side can actually be derived from the right, and vice versa. In the case of an equation, this is not necessarily the case.

It can be concluded therefore that an identity is always an equation but an equation is not always an identity.

Examples of equations include:

- $2x + 3 = 5$ (only true when $x = 1$)
- $y = 2x + 4$ (the equation of a straight line, only true for certain values of x and y)
- $y = 5x^2 + 3x + 2$ (a quadratic equation, produces a parabolic graph for certain values of x and y).

Examples of identities (one side can be derived from the other) include:

- $(x + y)^2 = x^2 + 2xy + y^2$
- $a - b + c = a - (b - c)$
- $x^2 + 2x + 1 = (x + 1)^2$

Example 69

Show that $5(2x + 3) = 10x + 15$ is an identity.

LHS: $5(2x + 3) = 10x + 15$

RHS = $10x + 15$

   $15 = 15$

This is an identity.

Exercise 3X

1 Solve these simultaneous equations.

$x = 1$

$x^2 + y = 2$

2 Solve these simultaneous equations.

$x - y = 2$

$xy = -1$

3 Look at these equations.

- $7x - 3 = 2$
- $(3x + y)^2 = 9x^2 + 6xy + y^2$

   a State which is an identity and which is an equation only, explaining why.

   b Give two additional examples of equations and two of identities.

4 Solve these simultaneous equations.

$x - 2y = 1$

$x + y^2 = 0$

5 Solve these simultaneous equations.

$x + y = 4$

$x^2 + y^2 = 26$

6 Show that the equation: $4(2x - 3) - 2(5x - 7) = -2x + 2$ is an identity.

7 Solve these simultaneous equations.

$2x + y = 2$

$2x^2 - 3y^2 = 2$

8 Given that $(3x + 2)(4x - 3) - 5(2x - 1) = ax^2 + bx + c$ is an identity in x, find the value of a, b and c.

9 Solve these simultaneous equations.

$2x - y = 3$

$3x^2 + 5y^2 = 20$

3.20 Direct and inverse variation

3.20.1 Direct variation

Two quantities **vary directly** if they both increase or decrease at the same rate; for example, the distance travelled by a vehicle travelling at constant speed and the time taken vary directly. Distance is directly **proportional** to time. **Direct variation** indicates proportionality.

If a is directly proportional to b you can write $a \propto b$ where $\propto$ is the proportion sign. The proportion sign may be removed and replaced with a constant of proportionality. This results in the equation $a = kb$ where k is a constant of proportionality. Plotting a graph of a against b will result in a straight line, passing through the origin and with gradient k.

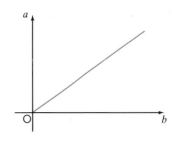

Looking at a table, based on proportionality, you can see that a and b increase proportionately, thus leading to a straight-line graph.

a	1	2	3	4	5
b	5	10	15	20	25

Examples of direct proportionality include:

- voltage and current
- force and mass
- total price of goods and quantity of goods bought.

Example 70

If A varies directly as B, and $A = 2$ when $B = 3$, find the value of A when $B = \dfrac{3}{4}$.

Find the constant of proportionality:

$$A = kB \Rightarrow k = \frac{A}{B}$$

$$= \frac{2}{3}$$

Substituting for k: $A = \dfrac{2B}{3}$

When $B = \dfrac{3}{4}$:

$$A = \frac{{}^{1}2}{{}_{1}3} \times \frac{3^{1}}{4_{2}}$$

$$= \frac{1}{2}$$

Example 71

Given that y varies directly as the square of x, and $y = 2$ when $x = 5$, find the value of y when $x = 3$.

$$y \propto x^2 \Rightarrow y = kx^2$$

$$\Rightarrow k = \frac{y}{x^2}$$

$$= \frac{2}{25}$$

$$= 0.08$$

When $x = 3$, $y = 0.08 \times 3^2$

$$= 0.08 \times 9$$

$$= 0.72$$

3.20.2 Inverse variation

Inverse proportion is the opposite of direct proportion. Two quantities are inversely proportional to each other if when one decreases the other increases and vice versa. For example, if the number of workers doing a job increases the time taken to do the job decreases.

Indirect variation indicates inverse proportionality.

If p varies inversely as q, then $p \propto \frac{1}{q}$. Removing the proportionality sign and replacing it with the constant of proportionality (k) results in the equation $p = \frac{k}{q}$.

This is a graph of p against q.

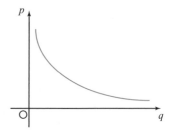

Example 72

L is inversely proportional to M and is equal to –5 when $M = 3$. Find the value of M when $L = -1$.

Finding the constant of proportionality, k:

$$L \propto \frac{1}{M} \rightarrow L = \frac{k}{M}$$

$$\Rightarrow LM = k$$

$$\Rightarrow k = -5 \times 3$$

$$= -15$$

Substituting for k: $L = -\frac{15}{M}$

Therefore when $L = -1$, $M = \frac{-15}{-1} = 15$

Example 73

Given that a is inversely proportional to b^2 and $a = 2$ when $b = 4$, find the positive value of b when $a = 9$.

$$a \propto \frac{1}{b^2}$$

$$\Rightarrow a = \frac{k}{b^2}$$

$$\Rightarrow k = ab^2$$

$$= 2 \times 4^2$$

$$= 32$$

$$\Rightarrow 9 = \frac{32}{b^2}$$

$$b^2 = \frac{32}{9}$$

$$\Rightarrow b = \frac{\sqrt{32}}{3}$$

Exercise 3Y

1   *A* varies directly as *B* and *A* equals 4 when *B* equals 3. Find the constant of proportionality and hence the value of *B* when *A* equals 10.

2   *y* is inversely proportional to *x* and *y* is equal to 10 when *x* is 2. Find the constant of proportionality and the value of *x* when *y* is 8.

3   *P* varies directly as the square of *q*. When $P = 5$, $q = 20$. Find the value of *P* when $q = 8$.

4   The volume (*V*) of a sphere varies directly as the cube of its radius. When *r* is equal to 2, *V* is 34. Find the value of *V* when *r* is equal to 0.5.

5   A bacterium multiplies according to the equation: $M = 4^{T+2}$ where *M* is the number of bacteria after a time *T* hours. Draw up a table showing the number of bacteria and hours taken and draw a graph to show that information. Use the graph to determine the number of bacteria after 3 hours.

6   Given that *U* varies inversely as H^4 and *U* equals $\frac{3}{4}$ when *H* equals $\frac{1}{4}$, find the value of *H* when *U* equals $\frac{5}{8}$.

End of chapter summary

Directed numbers and arithmetic operations on them

- **Directed numbers** are numbers with a positive (+) or negative (–) sign attached, for example, 5 and –3.
- To **add or subtract** two directed numbers with the **same sign**, ignore the signs, add the numbers and then reinstate the sign.
- To **add or subtract** two directed numbers with **different signs**, ignore the signs, subtract the smaller number from the larger and then reinstate the sign of the larger number.
- For the **multiplication and division** of two directed numbers: **same sign gives a positive answer and different signs give a negative answer.**

The meaning and importance of algebra

- A **variable** is an unknown number or quantity that can be represented by a letter in algebra.
- An **algebraic term** is a combination of letters and numbers or variables and coefficients, such as $3x$.
- In algebraic terms such as $3x$, x is the variable and 3 is the **coefficient** of x. It tells you how many xs you have.
- **Like terms** are made up of exactly the same letters and powers, for example, $3x$, $10x$, $100x$. **Unlike terms** do not have the same variable or power, for example, $3y$ and $3x$.
- An **algebraic expression** is a set of algebraic terms linked by signs, such as + or –. An algebraic expression consisting of two terms is a **binomial expression**.

Algebraic notation

- **Algebraic notation** is the use of letters and algebraic symbols to express mathematical word problems; for example, twice the sum of two numbers = $2(a + b)$.

The laws of indices

- Law 1: when multiplying powers of the same quantity, add the powers: $3x^2 \times 4x^3 = 12x^5$.
- Law 2: when dividing powers of the same quantity, subtract the powers: $12y^7 \div 6y^5 = 2y^{7-5} = 2y^2$.
- Law 3: when a power is raised to another power, multiply the two powers: $9(p^4)^3 = 9p^{12}$.
- Law 4: when a quantity is raised to a negative power, invert it and change the sign of the power to positive: $\dfrac{10k^{-3}}{5k^{-2}} = \dfrac{10k^2}{5k^3} = \dfrac{2}{k}$.
- Law 5: when a quantity is raised to a fractional power, the numerator of the fraction gives the power and the denominator indicates a root, for example, $9^{\frac{3}{2}} = \left(\sqrt{9}\right)^3 = 3^3 = 27$.
- Law 6: any quantity raised to the power zero equals 1, for example, $(10\,000\,000)^0 = 1$.

Using the four basic operations to simplify algebraic expressions

- **Addition and subtraction:** you can add or subtract terms only if they are like terms, for example, $2x - 3x + 5x = 4x$.
- **Multiplication and division:** you can multiply and divide like terms by like and unlike terms, for example, $5y \times 2y^3 = 10y^4$, $12a^3 \div 4a^2 = 3a^{3-2} = 3a$.
- You must apply the **laws of indices** when multiplying or dividing powers.

- Mixed operations in algebra involve problems that require more than one operation to solve them, for example, $\dfrac{2a + 5ab \times 3a^2b - 2ab}{7ab^3}$

Algebraic fractions

- An **algebraic fraction** is a fraction with an algebraic expression in the numerator or denominator or both, for example, $\dfrac{9x^2y}{ab + pq}$.
- You can apply the four operations of addition, subtraction, multiplication and division to algebraic fractions in a similar way to how you manipulate simple fractions.

    Addition: $\dfrac{3}{4x} + \dfrac{5}{2x} = \dfrac{3 + 10}{4x}$

    Subtraction: $\dfrac{6}{ab^2} - \dfrac{4}{ab} = \dfrac{6 - 4b}{ab^2}$

    Multiplication: $\dfrac{2a}{3b} \times \dfrac{5a}{4b} = \dfrac{10a^2}{12b^2} = \dfrac{5a^2}{6b^2}$

    Division: $\dfrac{4p^2}{5q} \div \dfrac{6p^3}{10q^3} = \dfrac{4p^2}{5q} \times \dfrac{10q^3}{6p^3} = \dfrac{40q^2}{30p} = \dfrac{4q^2}{3p}$

Substitution

- You can find a numerical value for an algebraic expression by **substituting** values to replace the letters or symbols that make up the expression; for example, to evaluate $10rs^2 + 5r$ when $r = 3$ and $s = 5$: $10 \times 3(5)^2 + 5 \times 3 = 750 + 15 = 765$.

Binary operations

- A **binary operation** is an operation that is made from a combination of the four basic algebraic operations. This new operation can be represented by any symbol other than: $+$, $-$, $\times$ or $\div$, for example, $a * b = 2a - 3b$.
- Multiplying out a term that comprises a number or variable in front of a bracketed term is **expanding the brackets**.

Using the distributive law to expand algebraic expressions

- The number or variable in front of the bracketed expression multiplies each term inside the brackets: $a(b + c) = ab + ac$.
- When you multiply two binomials, for example, $(a + b)(c + d)$, each term in the first set of brackets multiplies both terms in the second set, for example, $(a + b)(c + d) = a \times c + a \times d + b \times c + b \times d = ac + ad + bc + bd$.

Simple factorisation of an algebraic expression

- **Factorisation** means splitting an algebraic expression into two or more factors, for example, $ax + ay = a(x + y)$.
- The **highest common factor (HCF)** of a group of numbers is the highest number that can divide exactly into each number in the group. The **HCF of a set of monomials** (single-term algebraic expressions) is the product of the HCF of the coefficients and the lowest powers of the common letters in the monomials.

Simple linear equations in one unknown

- A **linear equation** is an equation in one or two unknowns, where the highest power in any term is 1, such as $x = 5$ or $2x + 3y = 17$. Its graph is a straight line.
- There are seven different types of equation that may need to be solved:

- equations requiring multiplication
- equations requiring division
- equations requiring addition and subtraction
- equations with the unknown (variable) on both sides
- equations involving roots
- equations involving fractions on one or both sides of the equation
- equations containing brackets on one or both sides of the equation.

- When you need to solve an equation such as $\frac{3x}{2} = \frac{x+14}{3}$ you can **cross-multiply**, to get $3x \times 3 = 2 \times (x+14)$ and simplify from there. It is the same as multiplying both sides by the product of the denominators.

Inequalities (or inequations)

- An **inequality** shows the relationship between two expressions that are not equal to each other, using a sign such as $\neq$ (not equal to), $>$ (greater than), $\geq$ (greater than or equal to), $<$ (less than) or $\leq$ (less than or equal to).
- There are seven types of inequality, requiring similar solutions to the seven types of equation.
- When solving inequations remember that multiplying through by a negative changes the inequality sign, for example, $-2x + 3 < 7 \Rightarrow -2x < 7 - 3 \Rightarrow x > -2$.

Transposition – changing the subject of the formula

- A **formula** is an equation that describes the relationship between two or more quantities. The **subject** of the formula is the variable that stands alone on one side.
- **Transposition** is the process of rearranging a formula so that a different letter or symbol becomes the subject; for example, $F = ma$ can be transposed to $m = \frac{F}{a}$.

Simultaneous linear equations in two unknowns

- **Simultaneous equations** are sets of equations in two or more variables that are simultaneously satisfied by the same set of values of these variables.
- Simultaneous equations may be solved by elimination or substitution. For example:

  $2x + y = 3$ **1**

  $3x - y = 7$ **2**
- By elimination: **1** + **2** $\Rightarrow 5x = 10 \Rightarrow x = 2$; $y = 3x - 7 = 6 - 7 = -1$
- By substitution, from **1**: $y = 3 - 2x$ and substituting for y in **2** $\Rightarrow 3x - (3 - 2x) = 7 \Rightarrow 3x - 3 + 2x = 7$ $\Rightarrow 5x = 10 \Rightarrow x = 2$, $y = -1$.

Factorising by grouping

- When factorising an expression with four terms, factorise as two pairs then combine to produce one pair of terms, for example, $ax + bx + ay + by = x(a + b) + y(a + b) = (a + b)(x + y)$.

Factorising quadratic expressions

- A **quadratic expression** is an algebraic expression in which the highest power of the variable is 2. Quadratic expressions are written in the form $ax^2 + bx + c$.
- There are five types of factorisation of quadratic expressions:
  - when the coefficient of $x^2 = 1$

- when the coefficient of $x^2 \neq 1$
- the difference of two squares
- a perfect square
- the form $ax^2 + bx$.
- A quadratic expression, with coefficient of $x^2 = 1$, can be written as a perfect square by adding $(\frac{1}{2}$ coefficient of $x)^2$ to the first two terms. The entire expression can be written in the form $a(x + h)^2 + k$, for example, $x^2 + 4x + 3 = (x^2 + 4x + 2^2) + 3 - 4 = (x + 2)^2 - 1$.
- If the coefficient of $x^2 \neq 1$, the expression is first factorised before completing the square, for example, $4x^2 + 8x + 1 \rightarrow 4(x^2 + 2x + 1^2) + 1 - 4 \rightarrow 4(x + 1)^2 - 3$.

Solving quadratic equations

- A **quadratic equation** is an equation of the type: $ax^2 + bx + c = 0$.
- A quadratic equation may be solved by factorising, after first taking all the terms to one side and equating to zero. Each factor is then equated to zero.
- There are five methods for factorising quadratic equations, matching those for factorising quadratic expressions, for example, to solve $x^2 + 4x + 3 = 0 \Rightarrow (x + 3)(x + 1) = 0 \Rightarrow x = -3$ or -1 (factorisation method – coefficient of $x^2 = 1$).
- Quadratic equations may also be solved by completing the square and setting the resulting expression equal to zero. The factors are then found thorough transposition, for example, $4(x + 1)^2 - 3 = 0 \Rightarrow x + 1 = \pm\sqrt{\left(\frac{3}{4}\right)} \Rightarrow x = -1 \pm \sqrt{\left(\frac{3}{4}\right)}$.

Simultaneous equations: one quadratic and one linear

- A pair of simultaneous equations may comprise one linear equation and one quadratic equation.
- The most appropriate method of solution is substitution, where the linear equation is rearranged to give one variable written in terms of the other, and this expression is substituted into the quadratic equation.

Identities and equations – what is the difference?

- An **identity** is always true for all values of the variable.
- An **equation** is true only for certain values of the variable.

Direct and inverse proportion

- Two quantities **vary directly** if they both increase or decrease at the same rate.
- Two quantities are **inversely proportional** to each other if when one decreases the other increases and vice versa.

Examination-type questions for Chapter 3

1 a Given that $a = 5$, $b = -2$ and $c = -3$, find the value of:

  i $2a + b - 3c$ ii $5abc - a^2c$

  b Write $\dfrac{5s}{4} - \dfrac{3s - 8}{3}$ as a single fraction.

2 a A jar holds 700 ml of oil.

  i What amount remained in the jar after x ml of oil was poured into a cooking pan?

  From the same jar, y ml of oil was poured into each of six small bottles.

  ii What was the final amount of oil left in the jar?

  b Five cricket bats and three balls cost $440.00. Three bats and two balls cost $280.00.

  Write a pair of simultaneous equations in c (cricket bats) and b (balls) and solve them to find the values of c and b.

3 Factorise each expression completely.

  a $10x^2 + 4xy - 2y^2 - 5xy$ b $4a^3 + 12a^2 - 10a$

  c $5x^2 - 3x - 2$ d $4y^2 - 64$

4 a Solve this equation for x.

  $\dfrac{5x - 3}{6} - \dfrac{2x + 4}{4} = \dfrac{1}{2}$

  b Solve this pair of simultaneous equations, using the most appropriate method.

  $3x - y = 3$

  $y^2 - 4 = 2xy$

5 a Given that $p \oplus q = \dfrac{pq^2 + p^2}{q}$, evaluate: i $3 \oplus 4$ ii $3 \oplus (4 \oplus 5)$.

  b Simplify: $\dfrac{8a^2b}{12ab} \div \dfrac{10b^2}{15a}$

6 a The table shows corresponding values of x and y, where x varies inversely as y. Find the values of v and w.

x	3	v	8
y	10	6	w

  b Write the expression: $4x^2 - 12x + 5$ in the form $a(x + h)^2 + k$.

7 a Make F the subject of the formula $P = \dfrac{F}{3} + \dfrac{PF}{2} - \dfrac{3}{4}$

  b Calculate the value of P, given that $F = -6$.

8 A vendor selling fruit at the market place sold 40 oranges. He sold x small oranges at $2.00 each and the remainder at $3.00 each.

   a Write an expression in x for the amount of money he took for the small oranges.

   b Write an expression in x for the total amount he took for all the oranges.

   c If he made a total amount of $100.00 from the sale of the oranges, how many were large and how many were small?

9 **a** Factorise: $4a^2 - b^2 + 2a + b$

   b Simplify: $\dfrac{2x + 1}{x^2 - 4} - \dfrac{2x - 2}{x + 2}$

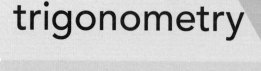

Objectives

By the end of this chapter you should be able to:

- explain concepts relating to geometry including points, lines, parallel lines, intersecting lines, perpendicular lines, line segments, rays, curves, planes, types of angle, number of faces, edges and vertices
- draw and measure angles and line segments accurately, using the appropriate instruments
- construct lines, angles, and polygons, using appropriate instruments, including the construction of
  - parallel and perpendicular lines
  - bisecting line segments and angles
  - angles: 30°, 45°, 60°, 90°, 120° and combinations of these
  - triangles, quadrilaterals, regular and irregular polygons
- identify the type(s) of symmetry possessed by a given plane figure, recognising lines of symmetry, rotational symmetry and order of rotational symmetry
- solve geometric problems by using properties of lines, angles and polygons, determining and justifying measures of lines, angles, triangles, quadrilaterals, other polygons and classes of solid
- solve geometric problems, using properties of circles and circle theorems:
  - properties: radius, diameter, chord, circumference, arc, tangent, segment, sector, semi-circle, pi
  - determine and justify angles using the circle theorems
- describe the use of transformational geometry (locating images and objects after a transformation)
- deduce the transformation, given the object and its image
- locate the image of an object under a combination of transformations
- explain Pythagoras' theorem
- describe and use trigonometrical ratios for acute angles in a right-angled triangle: sine, cosine and tangent
- solve right-angled triangles, using Pythagoras' theorem and the trigonometrical ratios
- relate objects in the physical world to geometric objects
- describe and calculate various angles of elevation and angles of depression
- explain the concept of bearings
- solve problems involving bearings and non-right angled triangles
- appreciate the importance and use of spatial geometry
- produce scale drawings
- describe the sine and cosine rules and explain when and how they are applied

Introduction

Geometry and trigonometry have been part of everyone's life: from an early age. We can observe that we are surrounded by lines, angles, shapes and curves.

In sports, participants learn about the importance of angles in launching balls, javelins and other projectiles to maximise distances of throws. Similarly, angles come into play when athletes launch their own bodies in sports involving jumps.

In weaving, sewing, art, manufacturing industry, etc. the existence of patterns – natural and manmade – the design of tools, equipment and structures. In the medical field – digital imaging. And for those of you who are crazy over your video games, the virtual realities involve a lot of geometry and trigonometry.

Trigonometry plays an extremely important role in our daily lives. On a technical and professional level, trigonometry is crucial for astronomy, aviation (one small error can bring down an entire jet), oceanography, land surveying and – and at a much lower scale, it is important in computer applications, music, modern architecture, and household redecoration.

Examples of geometry and trigonometry in real life

The geometry of tiling

Spiral staircase

Sewing a purse

Javelin – angle of throw

4.1 Geometrical concepts

Geometry is a branch of mathematics that studies, in general, the size, shape and other properties of objects in the environment: their relationships in space and, in particular, the measurement of and relationships between lines, angles, points, surfaces and solids.

You need to know and understand some basic geometrical concepts: points, lines, parallel lines, intersecting lines, perpendicular lines, line segments, rays, curves, planes, types of angles, number of faces, edges and vertices.

Key terms

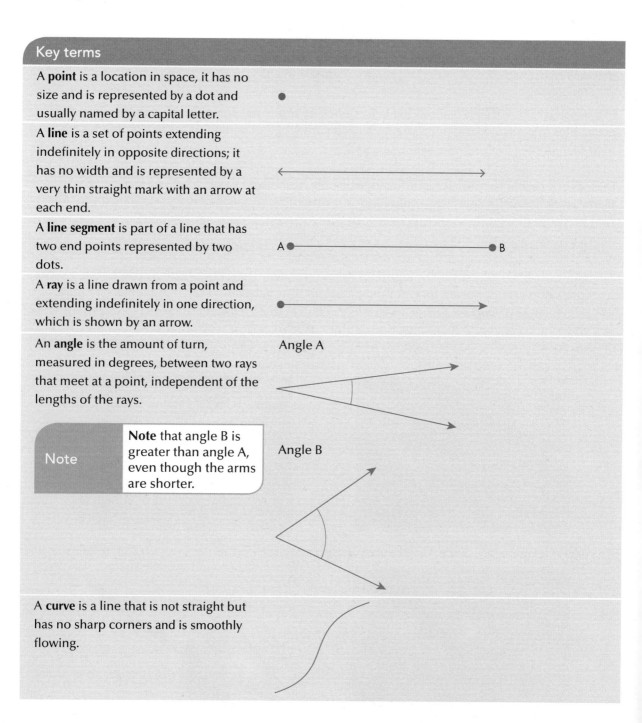

A **point** is a location in space, it has no size and is represented by a dot and usually named by a capital letter.

A **line** is a set of points extending indefinitely in opposite directions; it has no width and is represented by a very thin straight mark with an arrow at each end.

A **line segment** is part of a line that has two end points represented by two dots.

A **ray** is a line drawn from a point and extending indefinitely in one direction, which is shown by an arrow.

An **angle** is the amount of turn, measured in degrees, between two rays that meet at a point, independent of the lengths of the rays.

Note

Note that angle B is greater than angle A, even though the arms are shorter.

Angle A

Angle B

A **curve** is a line that is not straight but has no sharp corners and is smoothly flowing.

A **plane** is a flat two-dimensional surface extending indefinitely in all directions.

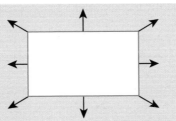

Parallel lines are lines that are a constant distance apart and never meet, in space or on a plane, no matter how far they are extended.

Intersecting lines are lines that meet in space or on a plane.

Perpendicular lines are lines that intersect at 90°.

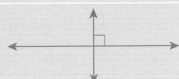

A **vertex or corner** is the point in an angle where two or more lines meet.

An **edge** is a line segment where two faces of a three-dimensional object meet or the line joining two vertices of a two-dimensional object.

A **face** is a plane surface.

A **solid** is a three-dimensional shape with several faces.

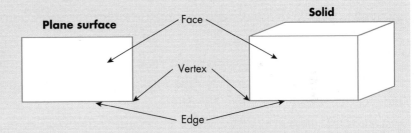

Exercise 4A

1. Draw a line of any length, placing a point at one end and an arrow at the other. How is this line described?

2. Explain, with the help of diagrams, the differences between a line, a ray and a line segment.

3. Look around your classroom and identify two objects with straight edges and two objects with curves.

4. Describe two places where you can observe line segments and two places where you can see rays.

5. Give two examples of a plane surface and two examples of solids.

6 Look around you and give three examples of perpendicular lines and three examples of parallel lines.

7 Name five objects in the classroom or in real life where you can see vertices.

8 Name five objects that contain edges and explain how these edges were formed.

9 Explain why there are no edges in a ball.

10 How many edges, vertices and faces does a shoe box have?

11 How many edges does a tin of orange juice have?

12 How many edges does an ice-cream cone have?

13 Draw a real object with six edges and four faces.

14 Draw an object with one flat surface, one curved surface and one edge. Name the object.

15 Draw an object with two triangular faces and three rectangular faces. How many vertices and edges does this object have and where will you find it in real life?

4.2 Drawing and measuring line segments and angles

You need to be able to use common geometric tools to measure and draw line segments and angles.

4.2.1 Drawing and measuring line segments

You can use a ruler to draw and measure line segments. A ruler may be a standard length (30 cm or 12 inches), or longer (a metre or 36 inches).

Most standard rulers display two scales: the metric scale shows centimetres (cm) and millimetres (mm) and the English or imperial shows inches and up to sixteenths of an inch. The scales are usually placed on opposite edges of the ruler.

A standard ruler

Using a tape to measure a tank

Measuring a shelf with a measuring tape

Using a ruler to measure an object

Place the ruler along (parallel to) the object being measured, with the zero of the ruler lining up with the edge of the object. Read the length from the scale.

You can also measure an object without necessarily beginning at the zero mark. This is the 'broken ruler' technique. Simply hold the ruler against the object, noting the marks aligning at each end, then calculate the difference between them.

Example 1

Using a ruler, draw a line segment that is about the same length as your pencil and measure it.

The ruler is used to trace the length of the pencil and the line is then measured using the ruler.

Length of segment = ? cm

Using a ruler to draw a line segment

Place the ruler where you want to draw the line and run your pencil along the straight edge. Use the scale to draw lines of specific lengths, preferably starting from zero and extending the line to your desired length.

4.2.2 Drawing and measuring angles

The mathematical instruments that you will need to draw and measure angles are:

- a protractor, semi-circular or circular, made from a transparent material, which is used to draw and measure angles

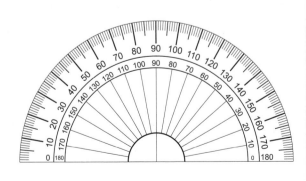

A semicircular protractor

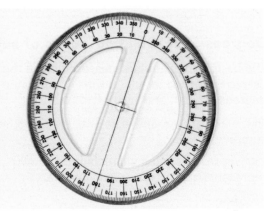

A circular protractor

- a set square, which is a right-angled triangular device, made from plastic, metal or wood, used to draw lines and measure angles.

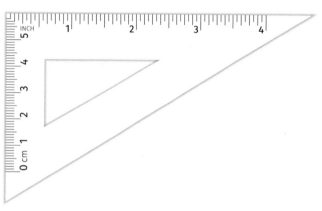

a 30°–60°–90° set square

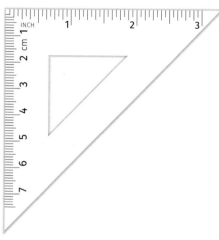

a 45°–45°–90° set square

Using a protractor to measure angles on a plan

Degrees and revolution – units of angular measurement

The basic unit of angular measurement is the **degree** (°). Protractors are normally calibrated in degrees (0°–180° or 0°–360°).

360° make one **revolution** or turn, sweeping out a complete circle.

180° represents half a full revolution and 90° represents quarter of a full revolution.

> **Key terms**
>
> Angles are measured in **degrees**. There are 360 degrees (360°) in one full turn. A **revolution** is one full turn, comprising 360°.

Using a protractor to measure and draw angles

To measure an angle

Place the protractor with its base line over one arm of the angle and the centre of the base line situated directly over the vertex of the angle (the point where the two arms of the angle meet).

Make sure that one of the zeros of the protractor is aligned with one arm of the angle, then read from this zero, round the scale, to where the other arm of the angle crosses the scale. The number on the scale represents the size of the angle being measured. The diagram shows how to measure an angle of 110°.

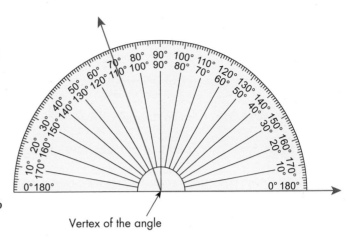

Vertex of the angle

To draw an angle

Draw a line segment, preferably horizontal, across the page.

Place the centre of the base of the protractor directly over the end of the line where you want the vertex of the angle.

Check that the zero on the scale of the protractor is also over the line you have drawn. From this zero, read around the scale to the desired angle on the protractor and place a dot on the paper to mark its position.

Draw a line from the vertex of the angle to the dot you have made, to represent the second arm of the angle. Use the protractor to check the accuracy of the drawn angle.

To draw a reflex angle, if you do not have a circular protractor, subtract the angle you are drawing from 360°, then take the angle 'outside' the angle you have drawn.

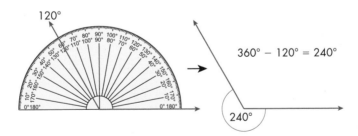

120°

360° − 120° = 240°

240°

Example 2

Use a protractor to measure the smallest angle in your 30°–60°–90° set square.

Trace the smallest angle of the set square onto paper and use your protractor to measure it.

Example 3

How many revolutions are there in 270°?

360° = 1 revolution

$270° = \dfrac{270}{360}$ revolution $= \dfrac{3}{4}$ revolution

---

Exercise 4B

1 Use your ruler to measure the length of each line segment.

A ———————————————————— B

C ———————————————— D

X ——————— Y

2 Use your ruler to measure the longest edge of your desk.

3 Using your ruler and pencil, draw a line segment equal to 14 cm.

4 Use your protractor to measure these angles.

a    **b**

5 How many degrees are there in $2\frac{1}{2}$ revolutions?

6 Using your ruler, measure the height and width of your classroom door.

7 Use your ruler and pencil to draw a line segment 0.8 cm long.

8 Use your protractor to measure this angle.

9 How many degrees are there in 0.7 of a revolution?

10 How many revolutions are there in 1260°?

11 Use your protractor to measure this angle.

12 Measure the thickness of your maths book. Hence, or otherwise, determine the thickness of one page.

13. Determine the angle the minute hand of a clock passes through when it goes from 1 to 5.

14. What fraction of a revolution does the hour hand of a clock pass through when it goes from 1:00 pm to 2:00 pm?

15. How many right angles are there in 10.75 revolutions?

4.3 Constructing lines, angles and polygons

4.3.1 Constructing parallel lines

The diagram shows two parallel lines that are a constant distance d apart. The arrows indicate that the lines are parallel.

There are two methods for drawing parallel lines.

Using a set square (45° or 60°)

Slide the set square along the edge of a ruler, with its base in contact with the ruler, as shown in the diagram. Draw lines along the perpendicular side of the set square (at 90° to the base) to produce parallel lines. Lines P, Q and R are all parallel to each other.

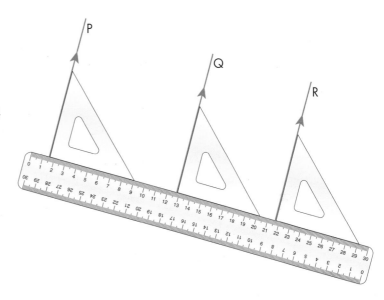

| Note | You can change the direction of the parallel lines by varying the orientation of the ruler. |

Using compasses and a ruler

You can use a pair of compasses to draw circles, to draw arcs and also to copy lengths, by setting the distance between the compass point and the pencil point to any required length.

Draw a line LM of any length and mark off a line segment AB.

Make a point R outside the line LM.

With centre R and radius AB, use your compasses to make an arc.

With centre B and radius AR make another arc to cut the first arc at point S.

Join RS. RS is now parallel to AB.

Using compasses and a pencil to draw a circle

Note

You can use this method to draw a line parallel to a given line.

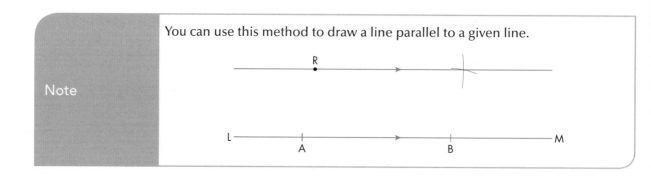

4.3.2 Constructing perpendicular lines

These diagrams show three pairs of **perpendicular lines.**

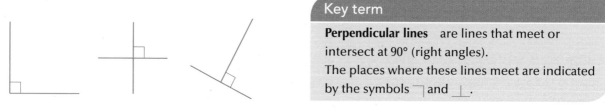

> ### Key term
>
> **Perpendicular lines** are lines that meet or intersect at 90° (right angles).
> The places where these lines meet are indicated by the symbols ⌐ and ⊥.

There are four methods for constructing perpendicular lines.

Using a set square and a ruler

Using a ruler, draw a base line in the desired direction.

Place the base of a 90° set square flush against the line and along the ruler, as in the diagram.

Draw a perpendicular line along the side of the set square, perpendicular to the base line.

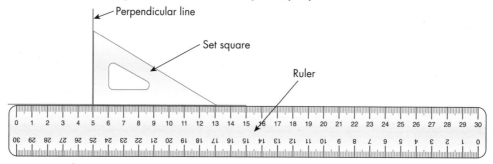

Drawing a perpendicular from a given point (A) on a straight line

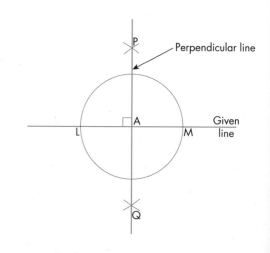

Using a ruler, draw a base line in the desired direction and mark a point, A, near the middle of the line.

Using a pair of compasses, with centre A and any radius, draw a circle to cut the line in two places: L and M. With centres L and M, and a radius greater than AL (or AM), draw arcs to cut each other at P and Q. The line PQ passes through A and is perpendicular to the straight line.

Drawing a perpendicular from a given point (K) at one end of a straight line

Draw a line of any length and label one end K.

From a point O outside the line, and with radius OK, draw a circle passing through K and cutting the line at a point S.

Draw the diameter, ST where T is the point where the line from S through O cuts the circle.

Join the points T and K to produce TK, which is the desired perpendicular at K.

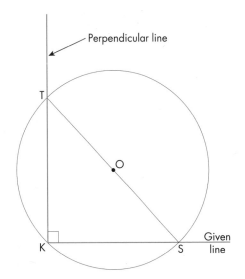

Drawing a perpendicular from a given point (P) outside a straight line

Draw a line YZ and mark a point P outside the line.

With P as centre, draw a circular arc to cut YZ at L and M.

With centres L and M, draw two more arcs to cut each other at a point W.

The line joining W to P makes an angle of 90° with the line YZ.

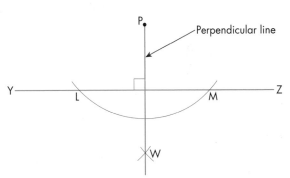

4.3.3 Bisecting line segments and angles

To bisect is to cut in half.

Bisecting line segment PQ

Draw the line segment PQ.

With P and Q as centres and radius greater than $\frac{1}{2}$PQ, draw two arcs to intersect at F and G. The straight line joining F and G bisects PQ at right angles (90°).

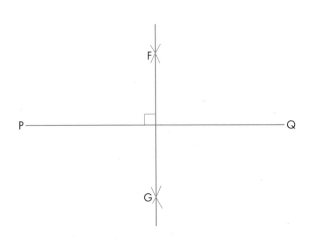

| Tip | You must always show construction lines clearly. This includes the arc lines, whenever you use compasses. |

Bisecting angle ∠PQR

To bisect angle PQR, with centre Q and any radius, draw an arc to cut PQ and RQ at D and E respectively. With centres D and E, and a radius greater than DE, draw two arcs to intersect each other at M. The line MQ is the bisector of angle PQR.

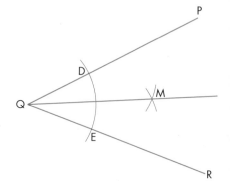

4.3.4 Constructing basic angles of 30°, 45°, 60°, 90° and combinations (105°, 120°, 135° and 150°)

Although you can measure angles, with a protractor, it is also possible to construct some angles simply by using a ruler and a pair of compasses.

Constructing 60° and 30°

Draw a line AK.

With A as centre and any radius, draw a circular arc to cut AK at P.

With P as centre and the **same** radius, draw another arc to cut the first at R.

Angle PAR is 60°.

To construct an angle of 30°, use the procedure described to bisect the angle of 60°.

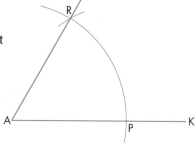

Constructing 90° and 45°

You already know how to construct an angle of 90° by constructing a perpendicular to a line.

To construct an angle of 45°, use the procedure described to bisect the angle of 90°.

In the diagram, angles NMY and LMY are both 45°.

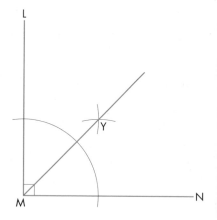

Constructing combinations: 105°, 120°, 135° and 150°

You can construct combinations of angles by constructing known angles adjacent to each other.

- 120° = 60° + 60°
- 135° = 90° + 45°

- $150° = 90° + 60°$
- $105° = 90° + 15°$ (Bisect an angle of 60° twice, to get 15°.)

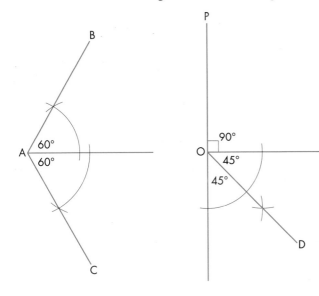

∠BAC = 120°

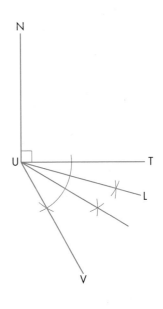

∠POD = 135°

∠NUV = 150° and ∠NUL = 105°

4.3.5 Constructing regular and irregular polygons

You need to know how to use geometrical instruments to construct **regular** and **irregular** **polygons**. In this section, you will work mainly with triangles and quadrilaterals.

Constructing a triangle, given three sides (SSS)

Example 4

Construct triangle ABC with AB = 5 cm, BC = 4 cm and AC = 8 cm.

Using a ruler, draw one side, for example, AB = 5 cm.

Set your compasses to a radius equal to the length of another side, for example, BC = 4 cm.

With centre B, draw an arc with this radius.

Set your compasses to a radius equal to the length of the third side, for example, AC = 8 cm.

With centre A, draw an arc with this radius, cutting the first at C.

Join points AC and BC to form triangle ABC.

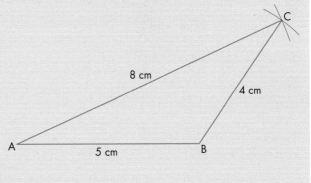

Constructing a triangle, given two sides and the angle between them (SAS)

Example 5

Construct triangle PQR with PQ = 6 cm, QR = 3 cm and angle PQR = 60°.

Using a ruler, draw one side, for example, PQ = 6 cm.

Construct an angle of 60° at Q.

Mark off the length QR = 3 cm along the 60° line.

Join P to R to complete triangle PQR.

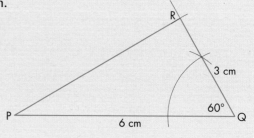

Constructing a triangle, given two angles and the side between them (ASA)

Example 6

Construct triangle LMN with LM = 10 cm, ∠LMN = 30° and ∠MLN = 30°.

Using a ruler, draw side LM = 10 cm.

Construct angle 45° at L and angle 30° at M.

Extend sides LN and MN to meet at N, hence completing triangle LMN.

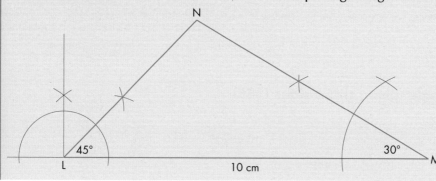

Constructing a quadrilateral

Construct the quadrilateral JKLM with side JK = 5 cm, angle JKL = 90°, side KL = 3 cm, side LM = 3 cm and angle MJK = 60°. What type of quadrilateral is it? What is the length of JM?

Using a ruler, draw JK = 5 cm and construct angle JKL = 90°.

Mark off the length KL = 3 cm.

Construct angle 60° at J.

Set your compasses to a radius equal to the length of LM = 3 cm, then with centre L draw an arc.

The point where the arc intersects the line drawn at 60° from JK is M. Join all the points, in order, to produce the quadrilateral JKLM.

JM = 4.1 cm.

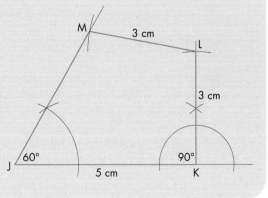

Constructing an inscribed and a circumscribed (escribed) circle

There are two types of circle associated with any triangle and with many polygons.

The inscribed circle

Look carefully at the diagram to see the method for drawing an inscribed circle of a triangle. Unlike most polygons, a triangle does not need to be regular (equilateral) to have an inscribed circle.

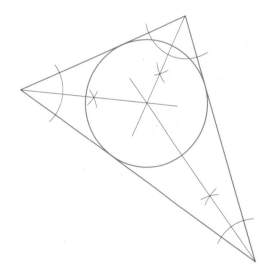

Key term

An **inscribed circle** is a circle constructed inside a regular polygon so that it touches each side of the polygon. The centre of this circle is the point of intersection of the bisectors of the interior angles of the polygon. However, note that the centre of the inscribed circle of **any** triangle is the point of intersection of the three angle bisectors.

The circumscribed or escribed circle

The centre of the **circumscribed** circle is the point of intersection of the bisectors of all the sides of the polygon. Note that, again, any triangle has a circumscribed circle.

This diagram shows the circumscribed circle of a square.

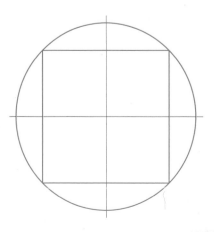

Key terms

A **circumscribed** or **escribed circle** is a circle that passes through all the vertices of a regular polygon. A polygon that has all of its vertices on a circle is a **cyclic polygon**.

Exercise 4C

1. Using set square and ruler, construct three parallel lines so that each adjacent pair is 5 cm apart.

2. Using compasses, ruler and pencil, construct two parallel lines that are 4 cm apart.

3. Using a ruler and set square, construct a line perpendicular to another line.

4. Using a ruler and pencil, draw a line 10 cm long and mark its midpoint. Label this point A. Now, using compasses, ruler and pencil only, construct a perpendicular to this line through point A.

5. Using your ruler and pencil, draw a horizontal line of length 8 cm and label it AB. Using compasses, ruler and pencil only, construct a perpendicular to the line at the point A.

6. Using your ruler and pencil, draw a straight line PQ. Mark point R anywhere outside your line. Using compasses, ruler and pencil only, construct a perpendicular from the point R to meet your drawn line at S.

7. Using protractor, ruler and pencil, draw an angle of 80°. Use your compasses and pencil to bisect the angle, showing clearly all your construction lines.

8. Draw a line segment (LM) of length 12 cm. Using ruler, compasses and pencil only, construct the perpendicular bisector of LM, showing all construction lines.

9. Using ruler, compasses and pencil only, construct angles of 60° and 90°.

10. Using ruler, compasses and pencil only, construct angles of 30° and 45°.

11. Using ruler, compasses and pencil only, construct angles of 120° and 105°.

12. Using ruler, compasses and pencil only, construct angles of 135° and 150°.

13. Using compasses, ruler and pencil only, construct triangle DEF right-angled at D, with side DE = 6 cm and DF = 8 cm. Measure the length of EF.

14. Using ruler, pencil and compasses only, construct triangle WXY with WX = 5 cm, XY = 7 cm and WY = 8 cm. Measure angle WXY.

15. Construct triangle JKL with JK = 10 cm, KL = 9 cm and angle JKL = 120°. Measure side JL.

16. Using compasses, ruler and pencil only, construct triangle ABC with side AB equal to 7 cm, ∠ABC = 135° and ∠BAC = 30°. Measure the length of side AC.

17. Construct the quadrilateral QRST with QR = 6 cm, ∠QRS = 105°, RS = 8 cm, ST = 10 cm, ∠RST = 150°. Measure the side QT and ∠QTS.

18. Construct a trapezium UVWY in which UV is parallel to WY, ∠VUY = 45°, UV = 9 cm, ∠UVW = 60° and VW = 6 cm. Find the length of UY and the size of angle VWY.

19. Construct the inscribed circle of a square of side 8 cm.

20. Construct the escribed circle of a regular pentagon.

4.4 Symmetry in plane figures

A shape has **symmetry** if moving, rotating or cutting it results in shapes that are **congruent**.

Plane shapes may have:

- reflection or line symmetry
- rotational symmetry.

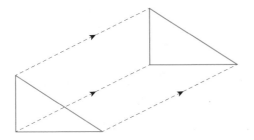

Symmetry in real life: snakes, leaves, honeycomb and tiling patterns all possess properties of symmetry.

4.4.1 Reflection symmetry

Reflection symmetry occurs when a shape can be divided into two halves that are exactly alike but one is the mirror image of the other. The two shapes are congruent.

Line symmetry is two dimensional (2D) and plane symmetry is three dimensional (3D).

Line symmetry occurs in butterflies and many other insects, as well as in leaves and reflections.

A shape can have zero, one or more **lines of symmetry**.

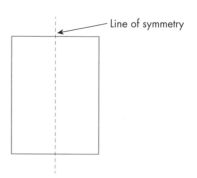
Line of symmetry

A **line of symmetry** is an imaginary line along which the shape could be folded so that one half matches the other half exactly.

Scalene triangle

Irregular quadrilateral

Camera

House

Parallelogram

Objects or shapes with no lines of symmetry

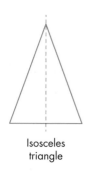
Isosceles triangle

Butterfly

Heart

Kite

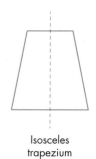
Isosceles trapezium

Objects or shapes with one line of symmetry

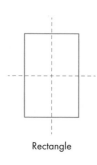
Rectangle

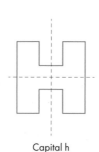
Capital h

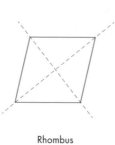
Rhombus

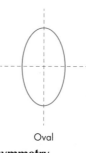
Oval

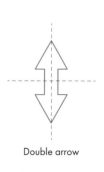
Double arrow

Objects or shapes with two lines of symmetry

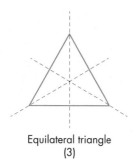
Equilateral triangle
(3)

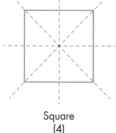
Square
(4)

Regular pentagon
(5)

Regular octagon
(8)

Objects or shapes with three or more lines of symmetry

4.4.2 Rotational symmetry

Rotational symmetry occurs when a shape can be rotated about its centre and still look the same in two or more positions. The **order of rotational symmetry** is the number of positions in which it can do this.

Note	A rotational symmetry of 1 (sometimes referred to as zero) means that the object will look exactly the same only after a complete rotation of 360°.

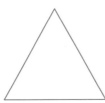 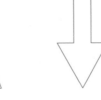  

Butterfly	Isosceles triangle	Arrow	Batman sign

Objects or shapes with order of rotational symmetry 1

Rectangle	Capital letter h	Oval	2 bladed fan

Objects or shapes with order of rotational symmetry 2

Equilateral triangle	Three-blade fan	Petals	Steering wheel

Objects or shapes with order of rotational symmetry 3

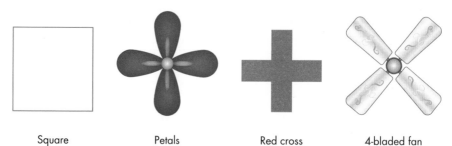

| Square | Petals | Red cross | 4-bladed fan |

Objects or shapes with order of rotational symmetry 4

| Dart board (10) | Star (5) | Circle (infinite) |

Objects or shapes with order of rotational symmetry greater than 4

Exercise 4D

1 Draw three geometrical shapes that have no lines of symmetry.

2 Name four real-life objects that have no lines of symmetry.

3 Draw three geometrical shapes that have only one line of symmetry.

4 Name four real-life objects that have only one line of symmetry.

5 Draw three geometrical shapes that have two lines of symmetry.

6 Name four real-life objects that have two lines of symmetry.

7 Draw three geometrical shapes that have three lines of symmetry.

8 Name four real-life objects that have three lines of symmetry.

9 Draw three geometrical shapes that have four or more lines of symmetry.

10 Name four real-life objects that have four or more lines of symmetry.

11 Explain why the circle has an infinite number of lines of symmetry.

12 For each order of rotational symmetry from 1 to 4, sketch or draw four objects.

13 Name four objects with order of rotational symmetry greater than four. Draw one of these objects.

14 Determine the number of lines of symmetry and the order of rotational symmetry for each of these shapes.

a   **b**   **c**

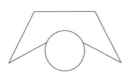 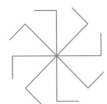 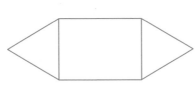 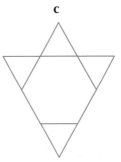

15 Think of your own shape that has six lines of symmetry and order of rotational symmetry six.

16 State the number of lines of symmetry and the order of rotational symmetry of each of these shapes.

a   **b**

4.5 Properties of angles, lines and polygons

4.5.1 Angular measurement

You have seen that angles are normally measured in degrees and that the number of degrees determines the amount of turn between the line segments or arms of the angle or the amount of turn or rotation through which one arm would have to move, to meet the other.

If one arm goes through one complete rotation (a full circle) to meet the other, this is one revolution.

A rotation of one revolution is equivalent to 360° and thus a quarter of a revolution is 90° or a right angle. Half a revolution is 180° and three-quarters of a revolution is the same as 270°.

A practical example of one revolution is the movement of the hands of a clock.

A complete revolution is equivalent to a movement of one of the hands around the clock from 12 to 12.

A quarter revolution, for example, will be from 12 to 3 and a half a revolution will be from 12 to 6.

4.5.2 Types of angle

- **Right angle** = 90°

- **Straight angle** = 180° (Angle on a straight line)

- **Acute angle:** less than 90°, for example, 30°, 60°

- **Obtuse angle:** greater than 90° but less than 180° for example, 100°, 179°

- **Reflex angle:** greater than 180° but less than 360°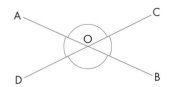

- **Complementary angles:** angles that add up to 90°, for example, 30° and 60°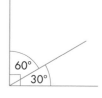

- **Supplementary angles:** angles with sum 180°, for example, 120° and 60°

4.5.3 Properties of angles and straight lines

- **Adjacent angles:** angles next to each other on a line p and q are adjacent and also supplementary, in this case $p + q = 180°$.

- **Vertically opposite** angles are formed by intersecting lines AB and CD at O.
  $\angle AOD = \angle COB$ and $\angle AOC = \angle DOB$

  They form pairs of vertically opposite angles.

  The two pairs of vertically opposite angles together can also be described as angles at the point O and they sum to 360°.

- **Alternate** and **corresponding angles** are formed when two parallel lines are cut by a **transversal**.

  - The two parallel lines WK and ST are cut by the transversal HJ.
  - The angles formed are denoted by the letters a, b, x, y, l, m, n and p.
  - The **alternate angles** are $a = m$ and $b = l$.
  - The **corresponding angles** are $l = x$, $m = y$, $a = n$ and $b = p$.
  - The **co-interior angles** are a and l, m and b (angles between the parallel lines and on the same side of the transversal). These angles are supplementary.

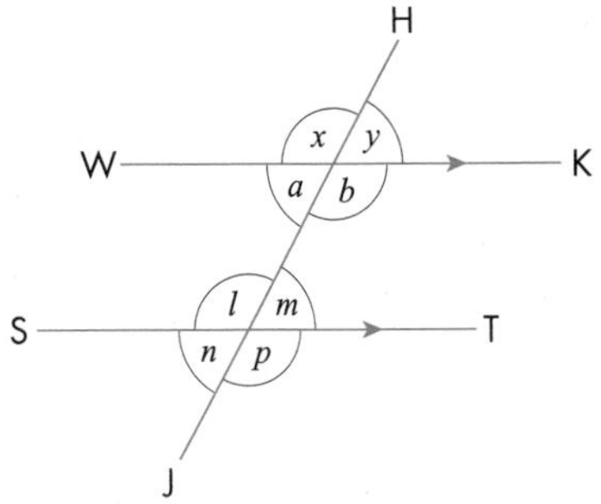

- There are other types of angle that can be identified in the diagram.

  - **Vertically opposite angles** $a = y$, $x = b$, $l = p$, $m = n$
  - **Adjacent angles** are also evident (for example, x and y, a and b, l and m, n and p, p and m). These angles are also **supplementary angles or angles on a straight line**.

> **Key term**
>
> A **transversal** is a line that passes through two or more lines.

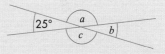

Example 8

The angle shown in the diagram, which is not drawn to scale, is formed from the intersection of the two lines. Find the sizes of the unknown angles and state the relationship of a and b and of a and c.

$a = 180° - 25°$

$\quad = 155°$

$\quad = c$

$b = 25°$ (vertically opposite to 25°)

a and b are adjacent angles or angles on a straight line.

a and c are vertically opposite angles.

4.5.4 Polygons and their properties

You have already met some **polygons**, which are plane shapes with three or more straight sides. If all of the sides are equal and all of the angles in a polygon are equal, it is a **regular polygon**. Otherwise, it is an **irregular** polygon.

> **Key term**
>
> A **regular polygon** is one in which all sides are equal and all angles are equal. Examples of regular polygons include equilateral triangles and squares. A polygon that is not regular is **irregular**.

Triangles

The simplest polygon is a triangle – a plane enclosed shape bounded by three sides.

A triangle has six important elements: three sides and three angles. The vertices are generally identified by capital letters and the sides are labelled with lower-case or common letters corresponding to the capital letter at the opposite vertex, as shown in the diagram.

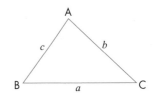

Angle properties of triangles

- The three angles in a triangle add up to 180°.

Example 9

Find the sum of the angles in triangle RST.

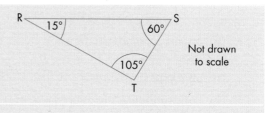

Not drawn to scale

Sum of the angles = 15° + 60° + 105° = 180°

- In any triangle the largest angle is opposite the longest side and the smallest angle is opposite the shortest side.
  In the diagram in example 9, the longest side is RS and the shortest side is ST.

- The exterior angle formed by producing one side of the triangle is equal to the sum of the two interior opposite angles.
  ∠PRS (exterior angle) = ∠RPQ + ∠RQP (sum of interior opposite angles)

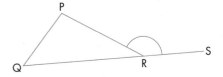

Example 10

In the diagram, which is not drawn to scale, angle PML = 40° and angle MPL = 110°.

Find the size of the exterior angle MLS.

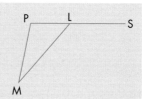

∠MLS = 40° + 110°

= 150°

Types of triangle

- **Equilateral triangle**: All the sides are equal and all the angles are equal to 60°.

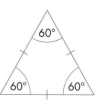

- **Isosceles triangle**: Two sides are equal and the two angles opposite the equal sides are equal.

- **Scalene triangle**: The three sides are all different lengths and the three angles are all different sizes.

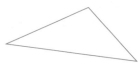

- **Right-angled triangle:** One angle is 90°. The longest side is the hypotenuse, opposite the right angle.

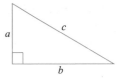
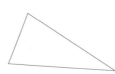

- **Acute-angled triangle:** All three angles are less than 90°.

- **Obtuse-angled triangle:** One angle is greater than 90°.

Congruent and similar triangles

Congruent triangles

In **congruent** triangles, corresponding angles are equal and corresponding sides are equal (six elements of the triangle).

> **Key term**
>
> Two triangles are **congruent** if they are identical in every respect.

Triangles ABC and XYZ are congruent. This means that:

AB = XY	BC = YZ	AC = XZ
∠A = ∠X	∠B = ∠Y = 90°	∠C = ∠Z

Additionally, area ΔABC = area ΔXYZ.

The symbol for congruency is ≡. Therefore ΔABC ≡ ΔXYZ.

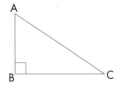
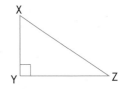

> **Note**   Non-right angled triangles can also be congruent.

Although the three corresponding angles of two congruent triangles are equal, and the three corresponding sides are equal, you do not have to prove that every pair of angles are equal and every pair of sides are equal to prove congruency. There are four conditions that you need to learn.

Proofs of congruency

- **One side and two angles in one triangle equal to the corresponding side and two angles in the other (AAS)**

  In the diagram, ∠Q = ∠M, ∠R = ∠N and PQ = LM.

- **Two sides and the included angle in one triangle equal to two sides and the included angle in the other (SAS)**

  In the diagram, ∠Y = ∠B, YZ = BC and XY = AB.

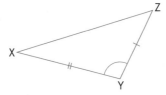
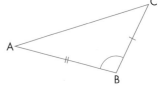

- **Three sides in one triangle equal three sides in the other (SSS)**

  In the diagram, DE = GH, DF = GI and EF = HI.

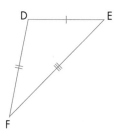
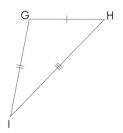

- **The hypotenuse and one other side in a right-angled triangle are equal to the hypotenuse and one side in the other (RHS)**

  In the diagram, JKL and UVW are right angles, JL = UW and KL = VW.

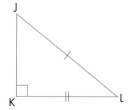
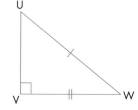

Example 11

In rectangle ABCD, prove ΔABC ≡ ΔADC.

In ΔABC and ΔADC:

AB = DC (opposite sides of a rectangle)

AD = BC (opposite sides of a rectangle)

∠ABC = ∠ADC = 90°

Therefore the triangles are congruent (SAS).

Similar triangles

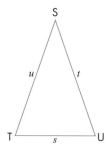
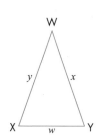

In the diagram, ∠TSU = ∠XWY, ∠STU = ∠WXY and ∠SUT = ∠WYX

Therefore the triangles are **similar**.

> **Key term**
>
> Two triangles are **similar** when their corresponding angles are equal.

Properties of similar triangles

- The ratios of the corresponding sides are equal: $\dfrac{u}{y} = \dfrac{t}{x} = \dfrac{s}{w}$ (corresponding sides are opposite equal angles)
- The ratio of the areas equals the ratio of the squares of corresponding sides:

  $$\frac{\text{area of } \Delta STU}{\text{area of } \Delta WXY} = \frac{s^2}{w^2} = \frac{t^2}{x^2} = \frac{u^2}{y^2}$$

Proofs of similarity

1   Prove that two angles in one are equal to two angles in the other.
    There are three angles in a triangle, so if you can prove two pairs of corresponding angles are
    equal the other pair must automatically be equal since all three must add up to 180°.

2   Prove that three sides of one triangle are proportional to
    the three corresponding sides of the other. Thus in the
    diagram:

$$\frac{ST}{WX} = \frac{SU}{WY} = \frac{TU}{XY} \text{ or } \frac{u}{y} = \frac{t}{x} = \frac{s}{w}$$

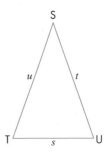
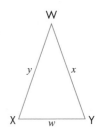

3   Prove that two sides in one triangle are proportional to
    two corresponding sides in the other triangle and the
    angles between the corresponding sides are equal.
    Referring to similar triangles in the previous diagram:

$$\frac{ST}{WX} = \frac{TU}{XY} \text{ or } \frac{u}{y} = \frac{s}{w} \text{ and } \angle T = \angle X$$

**In this steel bridge there are congruent and similar
triangles. Can you identify them?**

Example 12

Given that PQ∥SR, prove that triangle PQO is similar to
triangle SRO.

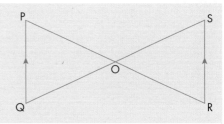

∠POQ = ∠SOR (vertically opposite angles)

∠QPO = ∠SRO (alternate angles)

Since two pairs of angles are equal, the third pair must automatically be equal.

Therefore triangle PQO is similar to triangle SRO.

Note	The symbol PQ∥SR indicates that the lines PQ and SR are parallel.

Properties of isosceles triangles

In any isosceles triangle, a perpendicular to the unequal side from the angle opposite it bisects both the side and the angle.

In the diagram, A is the 'unequal' angle of the isosceles triangle ABC. A perpendicular from A bisects the opposite side BD at the point C so that BC = DC.

The angle BAD is also bisected to produce equal angles BAC and CAD.

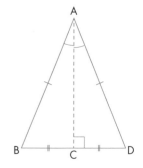

Example 13

Given an isosceles triangle YPZ, in which PY = YZ and the angle Y is bisected by a perpendicular, find the sizes of the angles at P and Z.

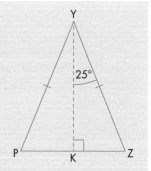

Since the perpendicular bisects the angle at Y,

$\angle PYZ = 25° \times 2 = 50°$

$\angle P = \angle Z$

$\quad = \dfrac{180° - 50°}{2}$

$\quad = 65°$

Note

Equilateral triangles also possess the properties of isosceles triangles.

Triangle LMN is an equilateral triangle.

A perpendicular from L bisects the opposite side MN, so MP = PM.

It also bisects angle MLN, so $\angle MLP = \angle PLN$.

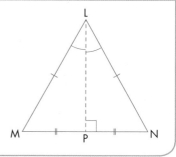

Example 14

Find the size of angle MLP in this equilateral triangle.

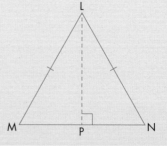

Each angle of an equilateral triangle equals 60° and since angle L is bisected, angle MLP = 60° ÷ 2 = 30°.

Properties of right-angled triangles

In the right-angled triangle ΔXYZ, the side XZ is the longest side and is opposite the right angle.

It is the **hypotenuse.**

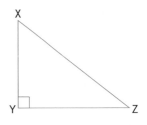

Pythagoras' theorem

Pythagoras' theorem applies only to right-angled triangles.

It can be used to find the unknown sides of a right-angled triangle.

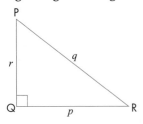

Applying Pythagoras' theorem in triangle PQR:

$PR^2 = PQ^2 + QR^2 \Rightarrow q^2 = p^2 + r^2$

Key terms

Pythagoras' theorem states that, in a right-angled triangle, the square on the **hypotenuse** is equal to the sum of the squares on the other two sides, where the hypotenuse is the side opposite the right angle.

Note

It is conventional to name the vertices of the triangle with Roman capital letters and the angles at the vertices with the corresponding italic capital letters. The sides opposite the vertices are named as the corresponding italic lower-case or common letters.

Example 15

Use Pythagoras' theorem to find the length of JL in triangle JKL.

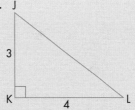

Using standard notation, JL = k.

Using Pythagoras' theorem, $k^2 = 3^2 + 4^2 \Rightarrow k^2 = 9 + 16 = 25 \Rightarrow k = \sqrt{25} = 5$

Example 16

In right-angled triangle ABC, ∠B = 90°; AC = 10 cm and AB = 8 cm. Use Pythagoras' theorem to find the length of BC.

Using Pythagoras' theorem:

$AC^2 = AB^2 + BC^2$

$\Rightarrow 10^2 = 8^2 + BC^2$

$\Rightarrow 100 = 64 + BC^2$

$BC^2 = 100 - 64 = 36$

$\Rightarrow BC = \sqrt{36}$

$BC = 6$

Properties of quadrilaterals

Angle properties of quadrilaterals

- The sum of the interior angles of **any quadrilateral** is 360°. (It can be divided into two triangles.)

> **Key term**
>
> A **quadrilateral** is a four-sided polygon or a plane enclosed figure bounded by four straight lines.

- The sum of the interior angles of a quadrilateral (or any polygon) is given by the expression $(2n - 4)90°$ where n is the number of equal sides of the polygon.
- The sum of the exterior angles of a quadrilateral or any polygon is 360°.

Example 17

Find the value of x in this quadrilateral.

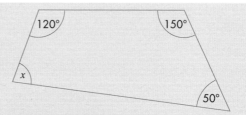

The sum of the angles in a quadrilateral is 360°.

$\Rightarrow 120° + 150° + 50° + x = 360°$

$x = 360° - (120 + 150 + 50)°$

$= 360° - 320°$

$= 40°$

Example 18

Using the expression $(2n - 4)90°$, show that the internal angles of a square add up to 360°.

A square has 4 equal sides therefore $n = 4$.

From the formula $(2n - 4)90°$, sum of internal angles $= (2 \times 4 - 4) \times 90°$

$= 4 \times 90°$

$= 360°$

Example 19

A regular polygon has three sides. What is the size of each external angle? Name the polygon.

Total of external angles $= 360°$

Value of each angle $= 360° \div 3$

$= 120°$

The polygon is a triangle.

Properties of quadrilaterals

- **Rectangles**

  A rectangle is a quadrilateral with:

  - opposite sides parallel and equal
  - all four internal angles equal to 90°
  - diagonals of the same length that bisect each other.

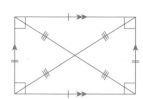

- **Squares**

  A square is a quadrilateral with:

  - all four sides equal, all four angles equal to 90°
  - opposite sides parallel
  - diagonals bisecting each other at 90°
  - diagonals bisecting the four angles.

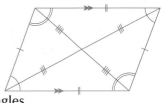

- **Parallelograms**

  A **parallelogram** is a quadrilateral with:

  - opposite sides equal and parallel
  - opposite angles equal
  - diagonals bisecting each other, forming two pairs of congruent triangles.

- **Rhombuses**

  A **rhombus** is a parallelogram with all sides equal. In addition to the properties of a parallelogram, the diagonals bisect each other at right angles and bisect the opposite angles.

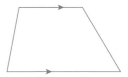

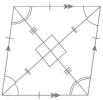

- **Trapeziums**

  A **trapezium** is a quadrilateral with **one** pair of parallel sides.

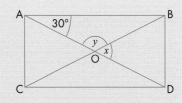

- **Kites**

  A **kite** is a quadrilateral with:

  - two pairs of equal adjacent sides
  - one pair of equal and opposite angles
  - diagonals intersecting at 90°, with one bisecting the other.

  The longer diagonal also bisects the angles at its ends and forms two congruent triangles. Together, the diagonals form two pairs of congruent triangles.

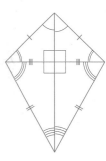

The diagram, which is not drawn to scale, represents rectangle ABCD with diagonals AD and BC. Given that angle BAO = 30°, find the values of x and y.

The diagonals AD and BC are equal and they bisect each other (property of a rectangle).

∠BAO = ∠OBA = 30° (angles of an isosceles triangle)

∠AOB = y = 180° − 60° = 120°

x = 180° − 120° = 60°

Alternatively:

x = sum of the interior opposite angles in triangle ABO

x = 30° + 30° = 60°

y = 180° − 60° = 120°

Example 21

Name the shape shown in the diagram.

Find the size of the unknown angle (θ) and the length of the unknown side (x).

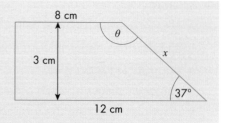

The shape is a trapezium.

$\theta = 360° - (90° + 90° + 37°)$

$\quad = 360° - 217°$

$\quad = 143°$

Using Pythagoras' theorem:

$x^2 = 3^2 + (12 - 8)^2 = 3^2 + 4^2 = 9 + 16 = 25$

$x = \sqrt{25}$

$\quad = 5 \text{ cm}$

Polygons with five or more sides

- **Pentagons**
  A **pentagon** is a polygon with five sides.
  In a regular pentagon, all sides are equal and all angles are equal.

Pentagon

- **Hexagons**
  A **hexagon** is a polygon with six sides.
  In a regular hexagon, all sides are equal and all angles are equal.

Regular hexagon

- **Other polygons**
- A **heptagon** is a polygon with seven sides.

- An **octagon** is a polygon with eight sides.

- A **nonagon** is a polygon with nine sides.

- A **decagon** is a polygon with 10 sides.

- A **hendecagon** has 11 sides.

- A **dodecagon** has 12 sides.

- A **tridecagon** has 13 sides.

- A **tetradecagon** has 14 sides.

Note	• The sum of the interior angles of **any polygon** is given by the expression $(2n - 4)90°$ where n is the number of sides of the polygon.
	• The sum of the exterior angles of **any polygon** is 360°.

Example 22

Find the sum of the interior angles and the value of each interior and exterior angle of a regular 20-sided polygon.

$$\text{Sum of interior angles} = (2n - 4)90°$$
$$= (2 \times 20 - 4)90°$$
$$= 36 \times 90°$$
$$= 3240°$$

$$\text{Value of each angle} = 3240° \div 20$$
$$= 162°$$

$$\text{Value of each exterior angle} = 360° \div 20$$
$$= 18°$$

4.6 Three-dimensional shapes

Polygons are described as two-dimensional because they have only two dimensions: length and width. They have no thickness.

Three-dimensional shapes, on the other hand, have length, width and height.

Examples of three-dimensional shapes

A solid may have a **regular** or an **irregular** shape.

Irregular solids include stones, mangoes and animals.

Regular solids include cubes and spheres.

Solids may be classified as **polyhedra** and **non-polyhedra**.

Polyhedra are solids with flat faces. Examples of polyhedra include:

- **cubes and cuboids**, each with six rectangular or square faces, eight vertices and 12 edges

- **tetrahedra** (singular tetrahedron), which are pyramids with four triangular faces, have four vertices and six edges

- **octahedra** (singular octahedron), with eight triangular faces, six vertices and 12 edges

- **dodecahedra** (singular dodecahedron), with 12 pentagonal faces, 20 vertices and 30 edges

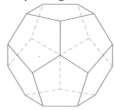

- **icosahedra** (singular icosahedron), with 20 triangular faces, 12 vertices and 30 edges.

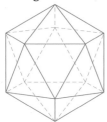

that this set of only five polyhedra in which all the faces are identical are also known as **platonic solids**, because they were described by Plato in his books? The name comes from the Greek translation of *poly* meaning 'many' and *hedron* meaning 'face'.

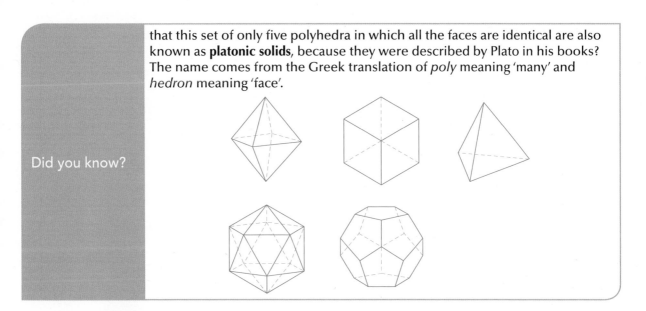

that for any polyhedron, the number of faces (*F*) + the number of vertices (*V*) – the number of edges (*E*) equals 2 (in symbols, $F + V - E = 2$)? This is known as **Euler's formula.**

Non-polyhedra are generally solids with curved surfaces. Examples of non-polyhedra include:

- **spheres** – a sphere is a perfectly symmetrical, three-dimensional figure with no edges or vertices but one curved surface upon which all points are the same distance, *r*, from the centre
- **cylinders** – a cylinder is a solid with one curved surface and two identical circular end faces; this shape is maintained throughout the length of the solid
- **cones** – a cone is a solid with one flat, circular base and one curved surface which terminates in a point called an apex
- **toruses** – a torus is a three-dimensional shape made by revolving a small circle of radius *r* along a line made by a big circle of radius *R*. The torus has no edges or vertices.

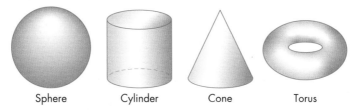

| Sphere | Cylinder | Cone | Torus |

Classification of solids

Solids can also be grouped according to their shape and the formulae for calculating their volume.

- **Prisms and cylinders** have a cross sectional area that is maintained throughout their length.

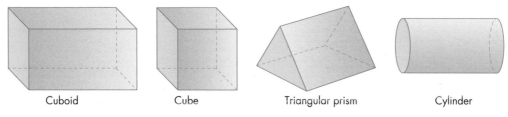

| Cuboid | Cube | Triangular prism | Cylinder |

Volume = area base × height ($V = A_b \times H$)

- **Cones and pyramids** each have a flat base (circular or polygonal) and an apex.

Cone

Pyramid

Volume of cone or pyramid = $\frac{1}{3}$ area of base × height($V = \frac{1}{3}A_b h$)

- **Spheres and toruses** have no flat faces, the distance from the centre to the surface is the radius, r.

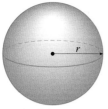

Sphere

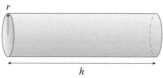

h

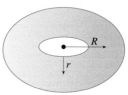

Torus

Volume of sphere = $\frac{4}{3}\pi r^3$ Volume of cylinder = $\pi r^2 h$ Volume of torus = $2\pi^2 R r^2$

Exercise 4E

1 The tip of the minute hand of a clock travels from 12 to 3. Calculate the angle through which the hand moves, in degrees, revolutions and turns.

2 When the minute hand of a clock is on 2 and the hour hand is on 8, what is the angle between the two? Give your answer in revolutions.

3 State the type of angle: 42°, 90°, 99°, 200°, 180° and 360°.

4 Give an example of:

a two complementary angles   **b** two supplementary angles.

5 Show by means of labelled diagrams and actual values:

a adjacent angles   **b** vertically opposite angles   **c** alternate angles

d corresponding angles   **e** co-interior angles.

6 In the diagram, AB is parallel to CD and XY is a transversal. Find the values of all the angles represented by letters.

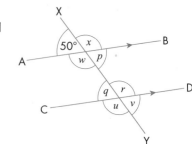

7 Look at the diagram in question 6 and write down the names for these pairs of angles:

   a *x* and *w*   **b** *w* and *q*   **c** *q* and *r*   **d** *p* and *q*   **e** *p* and *v*.

8 Look at triangle PQR.

   a Name the smallest angle.

   b Name the longest side.

   c Which two angles together sum to angle RQS?

   d What is the sum of angles PRQ, RPQ and PQR?

   e Describe angle RQS.

9 Make a sketch of these types of triangle and give two properties of each one.

   a equilateral   **b** isosceles   **c** scalene   **d** right-angled

   e acute-angled   **f** obtuse-angled

10 The diagram shows a trapezium LMNQ in which QN is parallel to LM and QL = NM.

   Prove that triangles QPL and NRM are congruent.

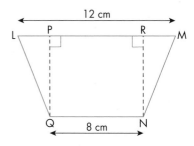

11 Triangle URV is an isosceles triangle and the line MK is parallel to RV.

   Prove that triangle URV and triangle UMK are similar.

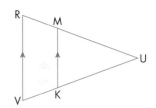

12 In the rectangle PQRS, prove that △PQR and △PSR are congruent.

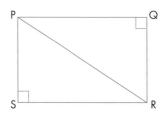

13 In the diagram, ABCD represents a square. Prove that triangles ABO and DCO are similar.

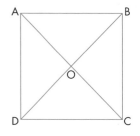

14 Sketch each quadrilateral and state two properties of each one.

  a Rectangle   **b** Square   **c** Parallelogram

  d Rhombus   **e** Trapezium   **f** Kite

15 Use Pythagoras' theorem to find the value of the unknown length in each triangle.

a

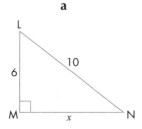

b

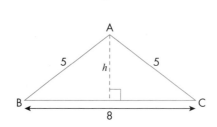

16 **a** Find the values of a, b and c in this trapezium.

  b State the perimeter of the shape.

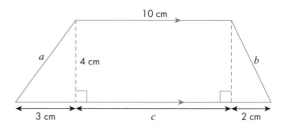

17 In the diagram, MT is parallel to YX.

  a Using the values given, prove that triangles YXZ and MTZ are similar.

  b Use the properties of similar triangles to find the values of a, l and r.

  c Find the ratio of the area of triangle YXZ to the area of triangle MTZ.

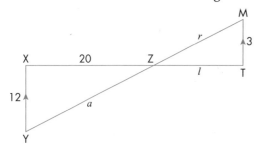

18 A boy wishes to make a kite, as shown in the diagram, with GO = 10 cm, AO = 40 cm and PO = 15 cm. GA and RP represent two pieces of light board he uses for the frame and GPARG represents the total length of thread around the frame. Find the total length of thread used, the longer piece of board (GA) and the length of the short piece RP.

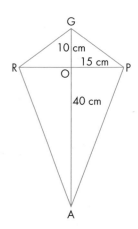

19
a Find the values of *x* and *y* in this diagram.

b Find the sizes of the exterior angles of the quadrilateral at L, M and N.

c Find the sum of all the exterior angles of quadrilateral LMNO.

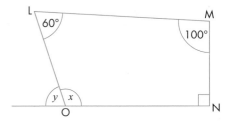

20 Find the value of each interior and exterior angle of:

a a regular pentagon   b a regular hexagon   c a regular heptagon

d a regular octagon   e a regular nonagon   f a regular decagon.

21
a Name the polygons with 11, 12, 13 and 14 sides.

b Sketch a regular polygon with 12 sides.

22 Name four polyhedra and state the numbers of faces, vertices and edges each one possesses.

23 Name and draw four non-polyhedral solids.

24 What do these solids have in common?

a Prisms and cylinders   b Cones and pyramids

4.7 Circles

4.7.1 Properties of circles

The main parts of a **circle** are:

- the **circumference** – the distance right around the circle, or the path traced by the circle

> **Key term**
>
> A **circle** is a plane shape formed by a set of points a fixed distance from a centre.

- the **radius** – the distance between the centre of the circle and a point on the circle (denoted by the letter *r*)

- the **diameter** – a line passing through the centre of the circle and connecting two points on the circle (denoted by the letter $d = 2r$); the longest chord

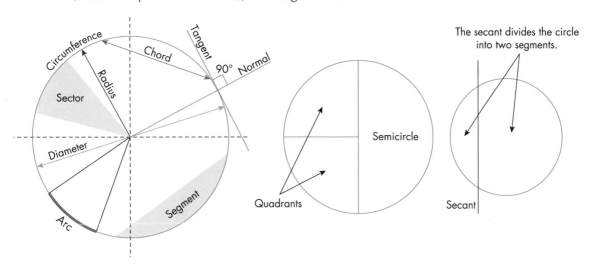

- a **chord** – a line that connects two points on a circle
- an **arc** – any curve forming part of the circumference
- a **tangent** – a line that intersects the circle at only one point
- a **secant** – a line that intersects the circle at two points and extends outside the circle
- a **sector** – the region of a circle enclosed by an arc and two radii
- a **segment** – the region of a circle bounded by a chord and the arc subtended by the chord
- a **quadrant** – quarter of a circle formed by two radii at right angles and the connecting arc
- a **semi-circle** – half of a circle consisting of an enclosed region formed by a diameter and half the circumference.

> **Note** The normal is a line that is perpendicular to the tangent.

4.7.2 Circle theorems

If you know the circle **theorems** you can use them to determine information about angles and lines associated with circles.

> **Key term**
>
> A **theorem** is a mathematical statement, formula or proposition that can be proved.

Theorem 1

The angle that an arc of a circle subtends at the centre of a circle is twice the angle it subtends at any point on the remaining part of the circumference.

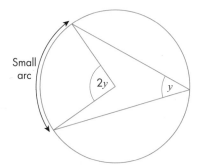
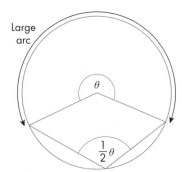

Theorem 2

Angles at the circumference in the same segment of a circle and subtended by the same arc or chord are equal.

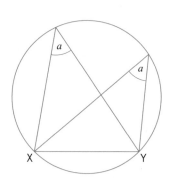

Theorem 3

The angle at the circumference subtended by the diameter is a right angle. (The angle opposite a diameter is 90°.)

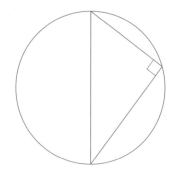

Theorem 4

The opposite angles of a **cyclic quadrilateral** are supplementary. (Angles subtended by the same arc or chord at the circumference in opposite segments add up to 180°.)

$\angle BAD + \angle BCD = 180°$

$x + y = 180°$

Note | If BD is a diameter, then the angles at A and C are right angles.

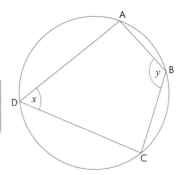

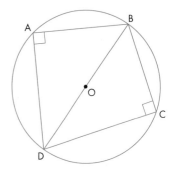

Theorem 5

The exterior angle of a cyclic quadrilateral is equal to the interior opposite angle.

In the diagram, $\theta = \angle QPS$.

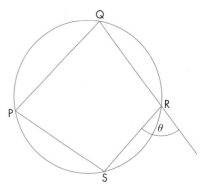

Theorem 6

The angle between a tangent to a circle and a chord through the point of contact is equal to the angle in the alternate segment.

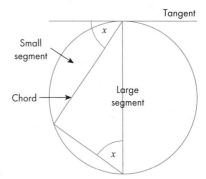

Theorem 7

A tangent to a circle is perpendicular to the diameter/radius of the circle at the point of contact.

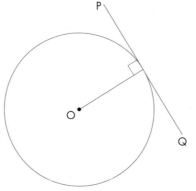

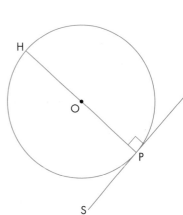

Theorem 8

The lengths of two tangents from an external point to the points of contact on the circle are equal.

Tangent TP = tangent TQ and each makes an angle of 90° with the radius.

Note also that the triangle TPQ is an isosceles triangle.

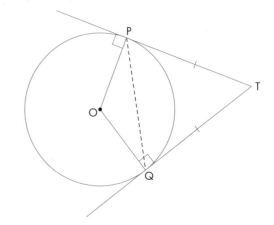

Theorem 9

The line joining the centre of a circle to the midpoint of a chord is perpendicular to the chord.

OP is perpendicular to RS and OQ is perpendicular to WV.

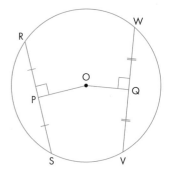

Theorem 10

If a tangent and a secant are drawn from the same point to a circle, then (length of tangent)^2 = (length of the whole secant) × (the length of the part of the secant outside the circle).

Based on the theorem: PS^2 = PR × PQ

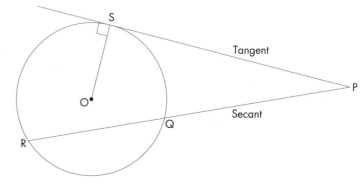

Theorem 11

Two circles touching each other internally or externally at one point will have a common tangent.

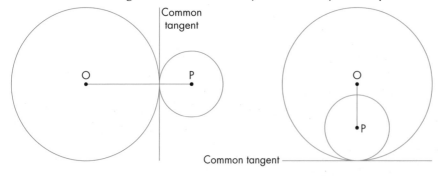

The line joining the two centres equals the sum or difference of the radii and is perpendicular to the common tangent circle.

OP = radius of large + radius of small circle   OP = radius of large − radius of small circle

Exercise 4F

1 Draw diagrams of circles and label them to show the circumference, a radius, a diameter, a chord, an arc, a tangent, a secant, a sector, a segment, a quadrant and a semicircle.

2 Find the sizes of the angles represented by letters in these circles.

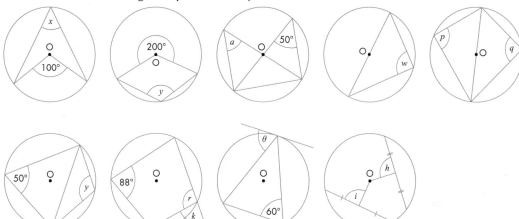

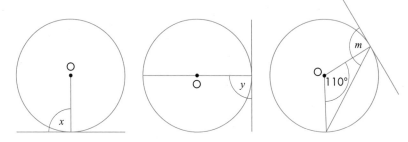

3 Find the sizes of the angles represented by letters in these circles.

4 In the diagram, the angle between the two tangents is 60°, the radius of the circle is 7.5 cm and PO = 15 cm. Find the length of the tangent and the size of angle QOP.

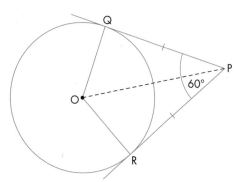

5 In the diagram, the length of the entire secant TM is 12 cm and the length of the chord MS is 4 cm. Calculate the length of the tangent.

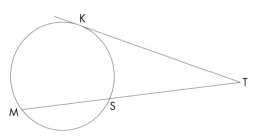

6 In each diagram, the large circle has a diameter of 12 cm and the small circle has a diameter of 4 cm. Determine the distance between the two centres in each case.

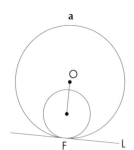

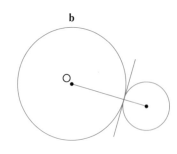

7 In the diagram, O is the centre of the circle and TP is a tangent.

Find the sizes of the angles represented by letters.

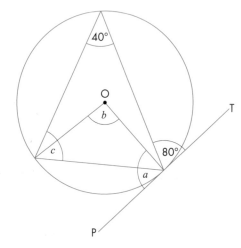

8 In the diagram, O is the centre of the circle.

Find the sizes of the angles represented by letters.

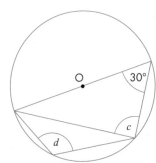

9 In the diagram, QTL is a tangent and RP is parallel to ST. Given that ∠PTL = 20°, find the sizes of ∠PRT, ∠RTS and ∠STQ.

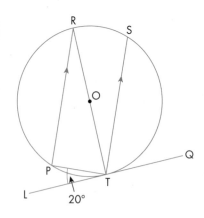

10 In the diagram, ABCD is a cyclic quadrilateral, AB is parallel to DC and ∠BCQ = 80°. Find the sizes of all the internal angles of the quadrilateral.

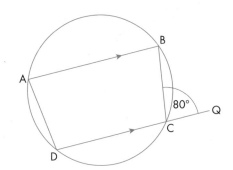

11 Use the information shown in the diagram to find the value of:

a ∠BPC

b ∠BDC.

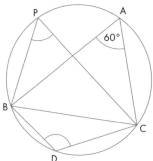

12 The diagram shows two circles with a common tangent. The line RK passes through the centres of both circles, and F is the centre of the larger circle.

Find the size of:

a ∠ABO **b** ∠AOR **c** ∠ROT

d ∠SOK **e** ∠DOC.

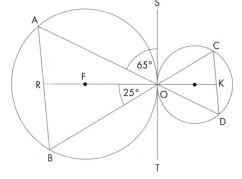

13 In the diagram, triangle OAB is isosceles, LT is parallel to AB and ∠BOT is 70°. Find the size of:

a ∠LOA **b** ∠OAB.

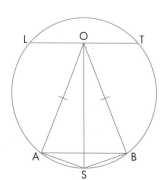

14 In the diagram, HC is a tangent making an angle of 60°
with chord DC, the angle at A is 80° and AD is parallel
to BC. Find the size of:

   a ∠BCD   **b** ∠CBD

   c ∠ADB   **d** ∠BCH.

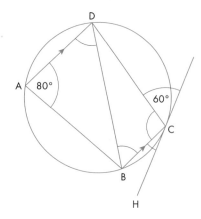

15 The diagram shows a cyclic hexagon. Determine the
sizes of the angles represented by letters.

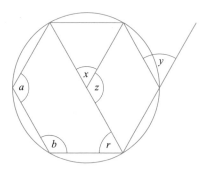

4.8 Transformation geometry

A transformation can also be defined as the
mapping of an object onto an image.

If the original position of the point (the object)
is A, then the position of the point after
transformation (the image) is A′.

4.8.1 Translations

Under a **translation**, an object moves from one
position to another, without its shape being
changed.

Look at the diagrams on the next page.

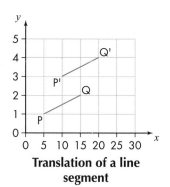

Translation of a line segment

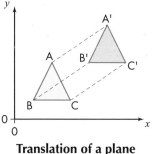

Translation of a plane shape

Properties of translations

- Corresponding line segments are equal and parallel: PQ = P'Q' and AB = A'B', BC = B'C', CA = C'A'.
- ΔABC is congruent to ΔA'B'C' (size and shape are unchanged).
- Each point in the object moves the same distance, in the same direction (parallel movement).
- The translated object (image) is obtained by joining the translated points of the object.

The translation in the first diagram above can be determined algorithmically.

Both points P and Q have undergone an identical movement parallel to the *x* and *y* axes: 5 squares horizontally (*x*-direction) and 2 squares vertically (*y*-direction). However, one square on the *x*-axis represents 5 units, so the move is actually 25 units in the *x*-direction.

You can write the mapping of P onto P' and Q onto Q' in translation notation as follows.

T(*x*, *y*) → (*x* + 25, *y* + 2)

You can also write it as a **column vector**: $\mathbf{T} = \begin{pmatrix} x \\ y \end{pmatrix} = \begin{pmatrix} 25 \\ 2 \end{pmatrix}$

When the translation is written in matrix form (as a column vector), the top number refers to a movement parallel to the *x*-axis and the bottom number refers to a movement parallel to the *y*-axis.

In a similar manner, in the translation of triangle ABC, A is mapped onto A', B is mapped onto B' and C onto C'.

> **Key term**
>
> A **column matrix** or **column vector** is a displacement vector that indicates the displacement or distance moved parallel to the *x* and *y* axes.

4.8.2 Reflections

Under a **reflection**, the object and image are mirror images of each other.

A straight line, called a **mirror line** or **line of reflection**, represents the location of the mirror. The image is located on the opposite side of the mirror line and the same distance from it as the

> **Key term**
>
> A **reflection** produces an image of the object across a line of reflection, similar to that produced in a plane mirror.

object. To find the image of each point on the object, draw a line from that point, perpendicular to the mirror line, and extend it the same distance on the other side of the line.

In order to reflect a line, LM, you need to reflect the endpoints of that line and join the images of these points to form the image L'M'.

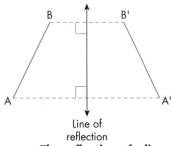

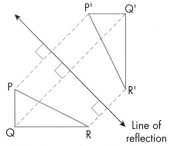

The reflection of a line segment AB.

Note that any plane shape can be reflected.

Properties of reflection

- In a reflection, the line joining a point in the image to the corresponding point in the object is always perpendicular to the line of reflection and is bisected by that line.

  object distance = image distance

- Corresponding points in the object and image are separated by a distance of twice the object distance or twice the image distance.

- In a reflection, orientation is reversed, as can be seen from the diagram above. Triangle PQR has an anti-clockwise orientation whereas the image P'Q'R' has a clockwise orientation.

- As in a translation, every point has exactly one reflection image.

- In a reflection, as in a translation, the shape and size of the object is maintained although orientation is reversed. Reflection is therefore a **congruent transformation**.

Reflection and translation

If a plane shape (such as a triangle or quadrilateral) is reflected successively in two parallel lines of reflection the result is a translation.

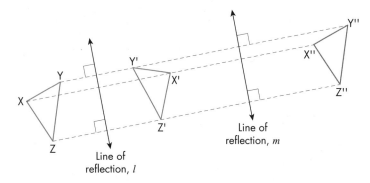

In the diagram, triangle XYZ is congruent to triangle X"Y"Z" with identical orientation. The result represents a translation.

4.8.3 Rotations

The **rotation** of an object may be clockwise or anti-clockwise.

In the diagram, the centre of rotation is the point $A(x_1, y_1)$. The angle of rotation is α and the direction of rotation is clockwise. The image of triangle ABC is the red triangle.

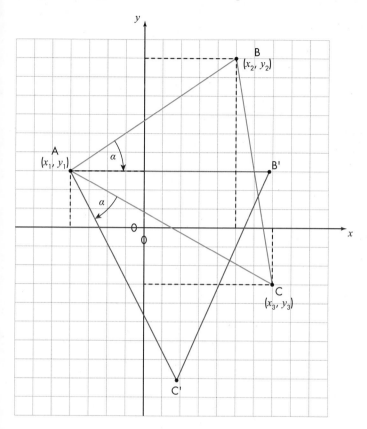

Rotation through reflection

It has already been shown that two successive reflections in parallel lines of reflection produce a translation. However, if the lines are not parallel, but meet at a point, successive reflections produce a turning of the object about the point of intersection of the lines of reflection. This is known as a rotation and the point of intersection of the lines of reflection (O) is called the **centre of rotation**.

In the diagram, $\angle LOL'' = \angle MOM'' = \angle NON'' = \theta$. The angle θ is the magnitude of the rotation.

The direction of rotation is clockwise in this case.

You can perform a rotation in the conventional manner by first marking the centre of rotation, then using a protractor and compasses to rotate each point in the appropriate direction (clockwise or anti-clockwise) through the angle of rotation.

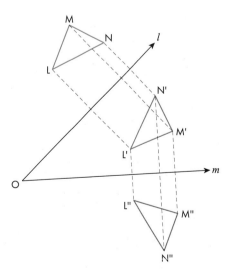

Example 23

Find the image of triangle PQR, with vertices P(1, 2), Q(4, 2) and R(3, 4), after a clockwise rotation of 90° about the origin.

Draw the triangle (PQR) on a coordinate grid. Then use a protractor to draw the line OP' at 90° (clockwise) to OP, and then the compasses to match the length OP' to OP. The vertices of the image (triangle P'Q'R') are P'(2, –1), Q'(2, –4) and R'(4, –3).

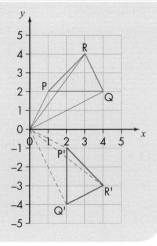

Finding the centre of rotation

Given the object and image, you need to be able to find the centre of rotation.

In the diagram the line segment AB represents the object and A'B' represents the image of AB under a rotation.

To find the centre and magnitude of rotation, join the endpoints of the lines in the image and object to produce lines AA' and BB'. Construct the perpendicular bisector of each of these lines, to meet at the point C.

C is the centre of rotation; the angle of rotation is given by the angles ACA' and BCB'.

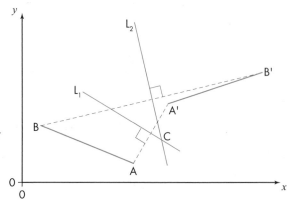

Properties of rotations

- Line segments, angles and relative orientation remain constant (congruent transformation).
- The centre of rotation remains fixed for all rotations about the centre.

4.8.4 Size transformations – enlargements and reductions

The first three transformations were congruency transformations in which the shape and size of the image remained the same as the object.

In size transformations the image may be smaller or larger than the object.

Under a size transformation, **E**, A' is the image of the point A. Given that A is at a distance OA from a fixed point O, then A' lies on OA and is at a distance kOA from O, where k is a real number.

Then k is the magnitude of the size transformation and O is the centre of the transformation. Thus **E** is a size transformation of magnitude or scale factor k and centre O.

When k is greater than 1 ($k > 1$) the transformation is an **enlargement**.

When k is positive but less than 1 ($0 < k < 1$) the transformation is a contraction or **reduction**.

Example 24

In the diagram, triangle PQR undergoes a size transformation with centre O(0, 0) and scale factor $k = 2$. Draw the image of PQR.

Since no coordinates are provided for the points P, Q and R, you need to find the image by drawing and measuring.

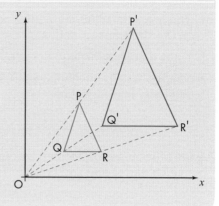

Since $k = 2$ and the centre is O, then

OP' = 2OP

OQ' = 2OQ

OR' = 2OR

Draw a line OP and extend it to P' such that OP' = 2OP.

Draw the lines OQ' such that OQ' = 2OQ and OR' such that OR' = 2OR.

Join the points to form the image P'Q'R'.

Note

When the centre is the origin and the coordinates of the object are known, you can find the coordinates of the image by simply multiplying the coordinates of the object by k (the size transformation factor), rather than by plotting and drawing as in the last example.

Example 25

Given quadrilateral ABCD with vertices A(2, 1), B(4, 3), C(5, 5) and D(3, 6), find the coordinates of the image of ABCD under a size transformation of factor 3, centre the origin O(0, 0).

The coordinates of the image A'B'C'D' are found by multiplying the coordinates of each vertex of ABCD by k.

A' = 3 × (2, 1) = (6, 3)

B' = 3 × (4, 3) = (12, 9)

C' = 3 × (5, 5) = (15, 15)

D' = 3 × (3, 6) = (9, 18)

When the centre of the size transformation is not the origin, you can find the image by scale drawing.

Example 26

Find the image of ΔLMN, with vertices L(2, 4), M(4, 6) and N(2, 6), after an enlargement, centre C(1, 1) and scale factor $k = 0.5$.

To find the image of triangle LMN, multiply the distances from the centre (C) of L, M and N by the scale factor to locate the image points of L'M'N'.

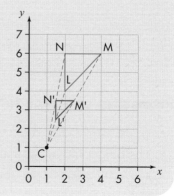

Note

This transformation is an example of a **reduction** ($k = 0.5$).

Properties of size transformations

- Angular measurements remain constant but linear measurements may increase or decrease.
- The image of a line segment is parallel to the object provided it does not pass through the centre of the size transformation.
- Lines that pass through the centre remain unchanged.
- Shapes remain unchanged although sizes may increase or decrease, so a size transformation is a **similarity transformation**.

4.8.5 Combining transformations

A **glide reflection** is performed by reflecting the object in a mirror line and then translating it along that line.

> **Key term**
>
> A **glide reflection** is a combination of two transformations: a reflection followed by a translation.

Example 27

Given triangle XYZ with vertices X(3, 5), Y(5, 8) and Z(4, 2), perform a glide reflection consisting of a reflection in the x-axis followed by a translation $\mathbf{T}\begin{pmatrix} -4 \\ 3 \end{pmatrix}$.

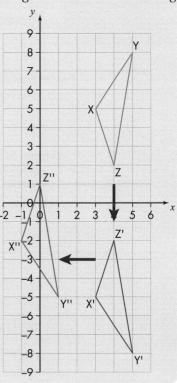

In the diagram, reflection in the x-axis produces image X'Y'Z' with vertices X'(3, –5), Y'(5, –8) and Z'(4, –2).

Translation $\begin{pmatrix} -4 \\ 3 \end{pmatrix}$ produces: X"Y"Z" where X" is the point

$(3 - 4, -5 + 3) = (-1, -2)$, Y" is the point $(5 - 4, -8 + 3) = (1, -5)$ and Z" is the point $(4 - 4, -2 + 3) = (0, 1)$.

The blue arrows in the diagram show the direction of movement of the glide reflection.

Object (blue) → reflected image (red) → translated image (green).

Properties of a glide reflection

- To perform a glide reflection you need to know the line of reflection and the column vector.
- The object and image maintain the shape and size, so it is a congruent transformation.
- The orientation of the object is reversed in the image.

Reflection followed by rotation

Example 28

Given triangle XYZ with vertices X(1, 1), Y(2, 4) and Z(3,2), perform a reflection in the *x*-axis followed by a clockwise rotation of 90° about the origin.

Reflection in the *x*-axis produces the triangle X'Y'Z' with vertices X'(1, –1), Y'(2, –4) and Z'(3, –2).

The clockwise rotation of X'Y'Z' through 90° about the origin produces the triangle with vertices X"(–1, –1), Y"(–4, –2) and Z"(–2, –3).

The blue arrows in the diagram show the sequence of transformations: object (blue) → reflected object (red) → rotated object (green).

Notice that this combination of transformations produces the same overall transformation as a reflection in the line *y* = –*x*.

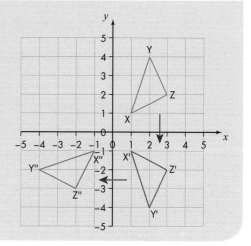

Exercise 4G

1 Perform the translation represented by the column vector $\begin{pmatrix} 3 \\ 2 \end{pmatrix}$ on a line segment between points A(1, 2) and B(5, 6).

2 Draw the line between points P(2, 1) and Q(5, 5) and reflect it in the *y*-axis.

3 Draw two lines of reflection, A and B, parallel to each other, on a sheet of plain paper, then reflect triangle LMN successively in these lines of reflection. What do you notice about the orientation of the triangle and its images?

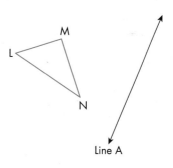

4 Reflect triangle LMN in two lines of reflection, C and D, that meet at a point, as in the diagram. What do you notice this time about the images?

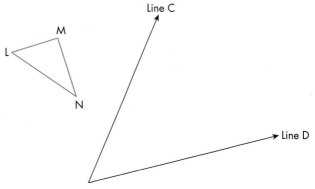

5 Plot the points W(1, 3) and K(5, 8) on a coordinate grid drawn on graph paper. Join the points to form a line segment WK. Using your compasses, centred on C(0, 0), rotate your line segment in a clockwise direction through an angle of 90°. Write the coordinates of the new points W'K'.

6 Plot the points A(2, 4), B(6, 4), C(6, 8) and D(2, 8) on a coordinate grid drawn on graph paper and join them to form rectangle ABCD. Perform an enlargement with scale factor $k = 2$ and centre of enlargement the origin. Plot the points of the image and write down their coordinates.

7 The coordinates of the vertices of ΔHKT are H(1, 8), K(6, 6) and T(4, 2). The triangle undergoes a glide reflection by first being reflected in the x-axis and then translated by $\begin{pmatrix} 3 \\ -2 \end{pmatrix}$. Find the image of HKT after the glide.

8 A triangle ABC with vertices A(0, 5), B(6, 2) and C(4, 10) undergoes a size transformation of scale factor $k = 0.8$ and centre O(0, 0). Draw the triangle and the image formed and state the coordinates of its vertices.

9 A'B'C' is the image of ABC after a clockwise rotation with unknown centre. Find, by drawing, the coordinates of the centre of rotation.

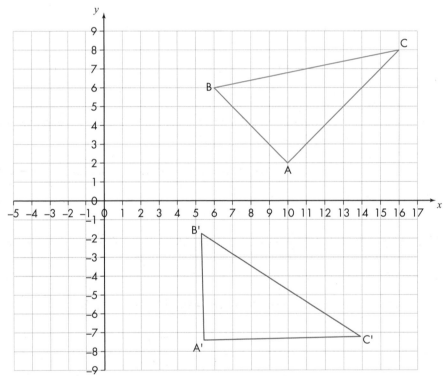

10 In the diagram, triangle A'B'C' is the image of triangle ABC after an enlargement. Find, by drawing, the centre of enlargement and the magnitude of the scale factor.

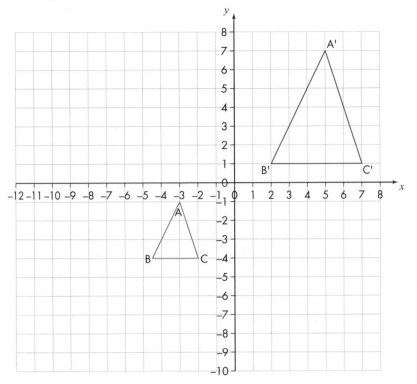

11 Trapezium JFKE has vertices J(–2, 1), F(6, 1), K(8, 5) and E(–2, 5). Find the image of the trapezium after a reduction, centre (–3, –4), scale factor 0.5.

12 Rectangle NIPK has vertices N(0, 0), I(6, 0), P(6, 4) and K(0, 4). Find the image of rectangle NIPK after an anti-clockwise rotation of 120° about centre C(–4, –5).

4.9 Trigonometry

Trigonometry is a branch of mathematics that studies the calculation of angles and sides of triangles and other polygons and the relationship between them.

4.9.1 Trigonometrical ratios of acute angles in a right-angled triangle

Sine, cosine and tangent

The sides and angles of a right-angled triangle are assigned special names and capital letters are used to denote the vertices.

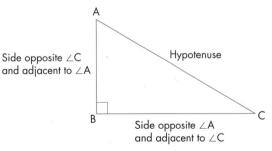

- The side opposite the right angle is the **hypotenuse.**
- The side BC facing angle A is **opposite to A**, while the side AB, next to A, is **adjacent to A.**
- The side AB facing C is **opposite to C**, while the side BC next to C is **adjacent to C.**

- Lower-case or common letters are used to denote the three sides:
  - side AB, opposite angle C, is labelled c
  - side BC, opposite angle A, is labelled a
  - side AC, the hypotenuse, opposite angle B, is labelled b.

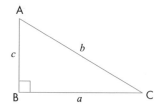

The three **trigonometrical ratios** are:

$$\textbf{sine}\ (\sin) = \frac{\text{side opposite the angle}}{\text{hypotenuse}}$$

$$\Rightarrow \sin\theta = \frac{\text{opposite}}{\text{hypotenuse}} = \frac{O}{H}$$

$$\textbf{cosine}\ (\cos) = \frac{\text{side adjacent to the angle}}{\text{hypotenuse}}$$

$$\Rightarrow \cos\theta = \frac{\text{adjacent}}{\text{hypotenuse}} = \frac{A}{H}$$

$$\textbf{tangent}\ (\tan) = \frac{\text{side opposite the angle}}{\text{side adjacent to the angle}}$$

$$\Rightarrow \tan\theta = \frac{\text{opposite}}{\text{adjacent}} = \frac{O}{A}$$

where θ is an acute angle in the triangle. You can use a calculator to find the values of the ratios.

The trigonometrical ratios can be summarised as:

$$S = \frac{O}{H} \quad C = \frac{A}{H} \quad T = \frac{O}{A}$$

> **Key term**
>
> A **trigonometrical ratio** is a ratio of the sides of a right-angled triangle, which can be used to find the value of the non-right angles as well as the sides of the triangle.

> **Key fact**
>
> In a right-angled triangle, the side opposite the right angle is the hypotenuse. For the two acute angles, the adjacent side is the side of the triangle that, together with the hypotenuse, encloses the angle.

> **Tip**
>
> If each letter is taken as the beginning of a word, the abbreviated version of the ratios shown on the right can be read out as a memorable sentence, called a **mnemonic**, which helps to remember the formulae.
>
> One commonly used mnemonic is obtained from **SOH** (**S**ine = **O**pp/**H**yp), **CAH** (**C**os = **A**dj/**H**yp) and **TOA** (**T**an = **O**pp/**A**dj), giving **SOHCAHTOA**, which you may remember as **S**ome **O**ld **H**orses **C**an **A**lways **H**ear **T**heir **O**wners **A**pproach.

Example 29

The diagram shows a right-angled triangle ABC.

(a) Determine the value of $\sin y$

(b) Determine the value of $\tan x$.

(c) Identify a trigonometric ratio that is equal to $\frac{12}{13}$.

(d) Identify a trigonometric ratio that is equal to $\frac{5}{12}$.

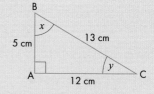

(a) $\quad \sin = \dfrac{\text{opposite}}{\text{hypotenuse}}$

$\quad\therefore \sin y = \dfrac{5}{13}$

(b) $\quad \tan = \dfrac{\text{opposite}}{\text{adjacent}}$

$\quad\therefore \tan x = \dfrac{12}{5}$

(c) Both $\sin x$ and $\cos y$ are equal to $\dfrac{12}{13}$.

(d) $\tan y \ \dfrac{5}{12}$

Example 30

Solve the right-angled triangle XYZ in which $x = 3$ cm, $y = 4$ cm and $Z = 90°$.

Sketch the triangle.

Using Pythagoras' theorem: $z^2 = x^2 + y^2$

$$z^2 = 3^2 + 4^2 = 9 + 16 = 25$$

$$z = 5 \text{ cm}$$

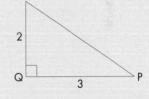

To find the unknown angles X and Y:

$$\sin Y = \frac{O}{H} = \frac{4}{5} = 0.8$$

$$\Rightarrow Y = \sin^{-1} 0.8 = 53°$$

and $X = 90° - 53° = 37°$

Note The notation $\sin^{-1} 0.8$ means 'the angle of which 0.8 is the sine'.

Example 31

In triangle PQR, angle $P = 90°$, $PQ = 3$ and $QR = 2$. Determine the unknown elements of the triangle.

Sketch the triangle.

Using Pythagoras' theorem to find the length PQ:

$$q^2 = p^2 + r^2$$

$$\Rightarrow q^2 = 2^2 + 3^2$$

$$= 4 + 9 = 13$$

$$q = \sqrt{13}$$

$$= 3.6$$

Now use the ratios to find R and Q:

$$\sin R = \frac{3}{3.6}$$

$$= 0.8333$$

$$\Rightarrow R = 56.3°$$

$$Q = 90° - 56.3°$$

$$= 33.7°$$

Note $0.5° = 30$ minutes $= 30'$ and $0.1° = 6' \Rightarrow 33.7° = 33° \ 42'$

4.9.2 Ratios for complementary angles

You know that complementary angles are angles that add up to 90°, such as 30° and 60°.

$\sin 30° = \dfrac{1}{2}$ and $\cos 60° = \dfrac{1}{2} \Rightarrow \sin 30° = \cos 60°$

- The sine of an angle is equal to the cosine of its complement.
- The cosine of an angle is equal to the sine of its complement.

Did you know? The sum of the squares of sin and cos of the same angle = 1.

$\sin^2 \theta + \cos^2 \theta = 1$

Example 32

Without using a calculator, find the sum of sin 30° + cos 60° + tan 45° and write your answer in its simplest form.

Using the relevant triangles:

$\sin 30° = \dfrac{1}{2}$, $\cos 60° = \dfrac{1}{2}$, $\tan 45° = 1$

$\dfrac{1}{2} + \dfrac{1}{2} + 1 = 2$

4.9.3 Angles of depression and elevation

Angles of elevation

When you have to look up to the top of a building or a tree, the angle through which your line of sight moves, upwards from the horizontal, is the **angle of elevation** of the object you are looking at.

In the diagram (not drawn to scale), the angle of elevation of the top of the tree is θ.

Given that the height of the man is 1.8 m and the tree has a height of 9 m and is 20 m away from the man, find the value of θ.

To find θ, use:

$\tan \theta = \dfrac{O}{A}$

$= \dfrac{9 - 1.8}{20}$ Since the angle is measured from the viewer's eye level, subtract the height of the man.

$= \dfrac{7.2}{20}$

$= 0.36$

So θ = 19.8°

Angle of depression

When you have to look down, for example, from the top of a building to a car on the street below, the angle through which your line of sight moves, downwards from the horizontal, is the **angle of depression** of the object you are looking at.

Example 34

The diagram shows the angle of depression from the top of a cliff looking out to a boat on the sea.

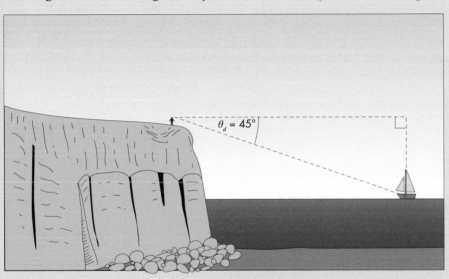

Calculate the height of the cliff, given that the horizontal distance from the boat to the base of the cliff is 120 m and the angle of depression is 45°.

To find h, use:

$$\tan \theta = \frac{O}{A}$$

$$= \frac{h}{120}$$

So $h = 120 \tan 45°$

$$= 120$$

Height of cliff = 120 m

Note

The assumption has been made that the face of the cliff is vertical, although it is unlikely to be the case.

4.9.4 Sine rule and cosine rule

The sine and cosine rules are generally used to solve **non-right angled triangles**, working with angles from 0° to 180°.

All triangles comprise six elements: three sides and three angles (3 S and 3 A). As you saw with right-angled triangles, to solve any triangle you need to know three of the elements to find the unknown three.

The sine rule

You can use the sine rule to solve triangles when you know two sides and a non-included angle or two angles and a non-included side (SSA or AAS).

The rule states that for any triangle ABC with sides a, b and c and angles A, B and C:

$$\frac{a}{\sin A} = \frac{b}{\sin B} = \frac{c}{\sin C}$$

Alternatively, it may be expressed as:

$$\frac{\sin A}{a} = \frac{\sin B}{b} = \frac{\sin C}{c}$$

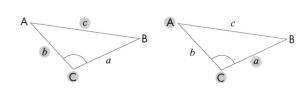

Example 35

Solve this triangle.

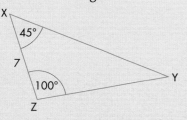

> **Tip**
>
> To solve a triangle, you need to find all the unknown angles and sides.

Since you are given two angles and the included side (ASA), you need to calculate the third angle so that you can use the sine rule.

First, find the value of Y by subtracting:

$Y = 180° - (45° + 100°)$

$Y = 35°$

> **Tip**
>
> Since the angle sum of any triangle is 180°, if you are given two angles you can always calculate the third.

Now, by the sine rule: $\dfrac{x}{\sin X} = \dfrac{y}{\sin Y} = \dfrac{z}{\sin Z} \Rightarrow \dfrac{x}{\sin 45°} = \dfrac{7}{\sin 35°} = \dfrac{z}{\sin 100°}$

$$\frac{x}{\sin 45°} = \frac{7}{\sin 35°}$$

$$\Rightarrow x = \frac{7\sin 45°}{\sin 35°}$$

$$x = \frac{7 \times 0.71}{0.57}$$

$$x = 8.7$$

Now find the value of z.

$$\frac{z}{\sin 100°} = \frac{7}{\sin 35°}$$

$$\Rightarrow z = \frac{7\sin 100°}{\sin 35°}$$

$$z = \frac{7 \times 0.98}{0.57}$$

$$z = 12.04$$

The cosine rule

The **cosine rule** is used in situations where you are given two sides and the angle between the two sides (SAS), or you are given three sides and no angles (SSS). The rule stated for triangle ABC, with sides a, b and c and angles A, B and C, is:

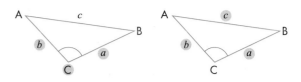

$$a^2 = b^2 + c^2 - 2bc\cos A$$

Note that the formula begins with a side and ends with the angle opposite to that side. The formula can be rewritten in terms of b or c:

$$b^2 = a^2 + c^2 - 2ac\cos B \quad \text{or} \quad c^2 = a^2 + b^2 - 2ab\cos C$$

Note	The term $2bc\cos A$ must be calculated as the product of four factors: 2, b, c and $\cos A$; they cannot be separated.

Example 36

Use the information given to find the sizes of the unknown side (c) and angles A and B in this triangle.

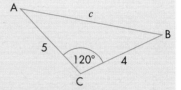

In triangle ABC (not drawn to scale) you are given two sides and the included angle (SAS).

The two known sides are $a = 4$ cm, $b = 5$ cm and the angle between them is $C = 120°$.

Applying the cosine rule:

$$c^2 = a^2 + b^2 - 2ac\cos C$$
$$\Rightarrow c^2 = 4^2 + 5^2 - 2 \times 4 \times 5 \cos 120°$$
$$c^2 = 16 + 25 - 2 \times 4 \times 5 \times (-0.5)$$
$$c^2 = 41 + 20$$
$$c^2 = 61$$
$$\Rightarrow c = 7.8$$

Applying the cosine rule again to find the value of A:

$$a^2 = b^2 + c^2 - 2bc\cos A$$
$$\Rightarrow 4^2 = 5^2 + 7.8^2 - 2 \times 5 \times 7.8 \times \cos A$$
$$16 = 25 + 61 - 2 \times 5 \times 7.8 \times \cos A$$

$$\cos A = \frac{86 - 16}{78}$$

$$\cos A = \frac{70}{78}$$

$$\cos A = 0.8974$$

$$A = 26.2°$$

Then $B = 180° - 26.2° - 120°$

$$= 33.8°$$

Example 37

Solve this triangle.

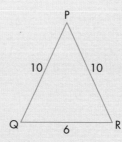

As you know three sides but no angles (SSS), use the cosine rule.

You can start with any angle, so choose P:

As you know three sides but no angles (SSS), use the cosine rule.

You can start with any angle, so choose P:

$$p^2 = q^2 + r^2 - 2qr\cos P$$

$$\Rightarrow 6^2 = 10^2 + 10^2 - 2 \times 10 \times 10\cos P$$

$$\cos P = \frac{200 - 36}{200}$$

$$\cos P = 0.82$$

$$P = 34.9°$$

To find the value of Q:

$$q^2 = r^2 + p^2 - 2rp\cos Q$$

$$\Rightarrow 10^2 = 10^2 + 6^2 - 2 \times 10 \times 6\cos Q$$

$$\cos Q = \frac{100 + 36 - 100}{120}$$

$$\cos Q = 0.3$$

$$Q = 72.5°$$

Since the triangle isosceles, $R = Q = 72.5°$

$$\Rightarrow P = 180° - 145° = 35°$$ which agrees with the result above.

Areas of non-right angled triangles (SAS and SSS)

As well as using the sine and cosine rules to solve non-right angled triangles, you can also find the areas of these triangles.

For right angled triangles or triangles where height h can be found easily, the common formula used is $A = \dfrac{1}{2} \times \text{base} \times \text{height}$.

(i) For non-right angled triangles where all three sides are known (SSS)

$$A = \sqrt{s(s-a)(s-b)(s-c)} \quad \text{and} \quad s = \frac{a+b+c}{2}$$

(ii) For non-right angled triangles where two sides and the included angle are known (SAS)

$$A = \frac{1}{2}ab\sin C$$

Example 38

(a) Find the area of triangle ABC when $a = 8$, $b = 6$ and $c = 10$.

(b) Calculate the area of triangle PQR when $p = 10$ cm, $q = 8$ cm and $R = 30°$.

(a) $A = \sqrt{s(s-a)(s-b)(s-c)}$ and $s = \dfrac{a+b+c}{2}$

$\Rightarrow s = \dfrac{8+6+10}{2}$ so $s = 12$

$A = \sqrt{12(12-8)(12-6)(12-10)}$

$= \sqrt{12 \times 4 \times 6 \times 2}$

$= 24$ square units

(b) Make a sketch of the triangle.

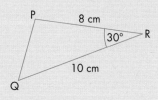

$A = \dfrac{1}{2} pq \sin R$

$= \dfrac{1}{2} \times 10 \times 8 \times \sin 30°$

$= 40 \times \dfrac{1}{2}$

$= 20$ cm^2

4.9.5 Bearings

The four cardinal directions are north, east, south and west, as shown in the diagram.

Other directions, such as north-east (NE) and south-south-east (SSE), can be identified between these major cardinal directions.

As seen in the diagram, the angle between adjacent cardinal directions is 90° and the angle between the other major directions, such as NE and E, and the adjacent cardinal directions is 45°. The angle between S and SSE is even smaller, at $22\frac{1}{2}°$.

The actual **bearing** is measured from north in a clockwise direction and is denoted by three numbers.

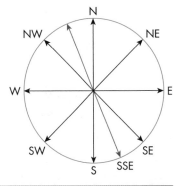

The bearing of north is given as 000°, north-east (NE) is 045°, east (E) is 090°, south-south-east (SSE) is 157.5°, north-west (NW) is 315°.

> **Key term**
>
> A **bearing** is an angle, measured clockwise from north at a fixed point, which helps determine the location of an object if you know its distance from the fixed point.

> **Tip**
>
> To find the bearing of one point from another, for example, the bearing of A from B, draw a north line at B (since the bearing is **from** B). Join points A and B with a straight line. The clockwise angle between the north line and the line BA represents the bearing of A from B.

Example 39

A is a point on the coast a distance 50 km due south of another point B. A third point C is 100 km due east of A. Find the distance between B and C and the bearing of C from A.

Make a sketch.

Use Pythagoras' theorem to find the distance BC:

$BC^2 = 50^2 + 100^2$

$\Rightarrow BC^2 = 2500 + 10\,000$

$= 12\,500$

$\Rightarrow BC = 111.8$ km

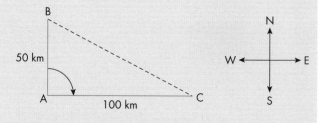

To find the bearing of C from A, first join CA then draw a north line at A, since you need the bearing 'from A'. From the north line, measure the angle between the north line and the line AC.

The distance BC is thus 111.8 km and the bearing of C from A is 090°, since you are told that C is due east of A.

Example 40

A plane takes off from an airport, A, on a bearing of 020° and flies for 400 km to point B. The bearing is then changed to 080° and the plane travels a further 500 km to C. Finally it travels to another airport D, a distance 600 km away on a bearing of 120°.

Determine the total straight distance from A to C, the distance AD and the bearing of A from C.

Make a sketch. Mark the lengths and angles you know or can derive. The angle at C, between BC and the north line, is 100°. ∠ABC is the sum of angles that are alternate to angles of 20° and 100°.

Find the distance AC by using the cosine rule:

$AC^2 = 500^2 + 400^2 - 2 \times 400 \times 500 \cos 120°$

$AC^2 = 250\,000 + 160\,000 - 400\,000 \times (-0.5)$

$= 610\,000$

$\rightarrow AC = 781$ km

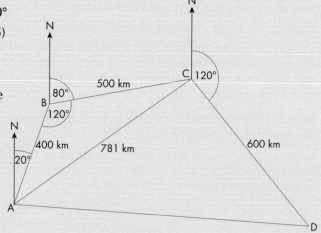

The bearing of A from C is the clockwise angle between the north line and CA.

To calculate this, start by finding ∠BCA.

Using the sine rule:

$$\frac{781}{\sin 120°} = \frac{400}{\sin BCA}$$

$$\Rightarrow \sin BCA = \frac{400 \times \sin 120°}{781}$$

$\sin BCA = 0.4435$

$\Rightarrow ∠BCA = 26°$

The bearing of A from C is $360° - (100° + 26°) = 234°$

Find the distance AD by using the cosine rule:

$AD^2 = AC^2 + CD^2 - 2 \times AC \times CD \cos ACD$

$AD^2 = (781)^2 + (600)^2 - 2 \times 781 \times 600 \times \cos(360° - 120° - 26° - 100°)$

$= 609\,961 + 360\,000 - 937\,200 \times \cos 114°$

$= 1\,351\,154$

$AD = 1162$ km

4.10 Spatial geometry: projections and scale drawing

Isometric and oblique projections

You often need to represent real, solid, three-dimensional objects, located in space, visually. Two ways of doing this are by **isometric projection** and **oblique projection**.

Isometric projection

Isometric projection is based on a set of isometric axes, consisting of three lines with equal angles (60°) between them.

The central line (OY) is vertical and the other two lines (OX and OZ) each make an angle of 30° with the horizontal and there is an angle of 120° between them, as shown in the diagram

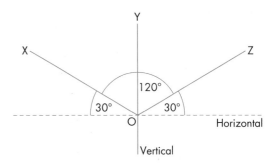

A rectangular block can be represented on these axes, as shown here.

Note	Note that each of the lines representing the edges of the solid lies on one of the three isometric axes or is parallel to it.

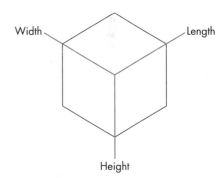

Isometric drawing of circles

Circles generally look like ellipses when represented as isometric drawings.

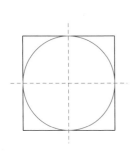
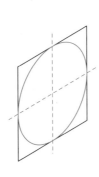
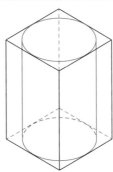

Proper circle
inscribed in
a square

Isometric circle –
must touch all
the sides

Isometric
cylinder

An example of an isometric projection

The diagram shows the front elevation and plan of an athletics podium, with its isometric projection.

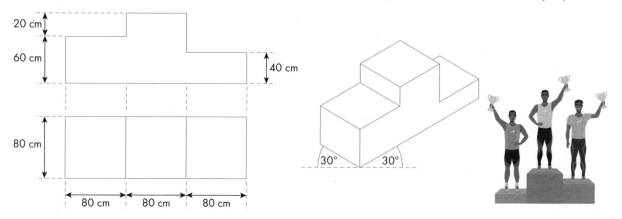

Oblique projection

Oblique projection is also based on three axes, but in this case the axis OX is horizontal, axis OY is vertical and axis OZ is drawn at 45° to the horizontal. In the projection, lengths along the axes OX and OY (and those parallel to them) are drawn full size, while lengths along OZ (or parallel to it) are drawn half size.

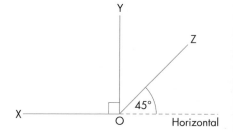

An example of an oblique drawing

This is a representation of a 10 cm × 10 cm × 20 cm cuboid drawn as an oblique projection.

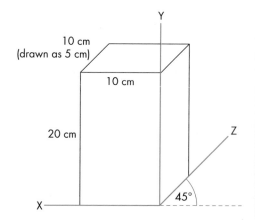

Orthographic projection

Although isometric and oblique projections can be very useful they do have a number of limitations, one of which is the limited number of 'viewpoints' – only three out of a possible six (in the case of a cube).

This limitation may be addressed by using an **orthographic projection**. The two types of orthographic projection are **first angle projection** and **third angle projection**.

First angle projection

In this type of projection the object is presented from three directions: front elevation, side elevation and plan; they should be presented in that order on your paper.

An example of first angle projection

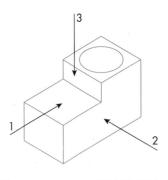

- The view from the front of the object, indicated by arrow 1, is the front elevation.
- The view from the side of the object, indicated by arrow 2, is the side elevation.
- The view from the top of the object, indicated by arrow 3, is the plan.

Conventionally, the three views are arranged in a particular way. The plan is placed at the bottom left, the front elevation appears above the plan and the side elevation is placed in the top right corner, as in this diagram. Notice that 'hidden' lines are indicated by dashed lines in the drawings.

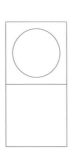

Third angle projection

For a **third angle projection**, the plan is located at the top right, with the front elevation placed directly below it. The side elevation in this case will be on the bottom left.

Scale drawing

Sometimes there is a need to draw very large or very small objects for which the actual lengths and areas cannot be represented on a sheet of paper. In this case, it is necessary to use a **scale**, to make a **scale drawing.**

> **Key term**
>
> A **scale drawing** is an accurate drawing of an object but with lengths reduced or expanded by a fixed amount called a scale.

Typical scale drawings include maps or plans of countries, houses and vehicles. Sometimes scale drawings are used for very small objects such as microbes or atoms.

Probably the most common drawings of houses are house plans, which must be produced before construction starts in order to guide the construction. A house plan shows the overall floor layout and the positions of rooms, windows and doors.

Scale drawings and plans indicate the actual dimensions of the building, which have to be calculated based on the scale. For example, a scale of 1 : 500 means that 1 cm on the drawing is equivalent to a length of 500 cm in the actual building.

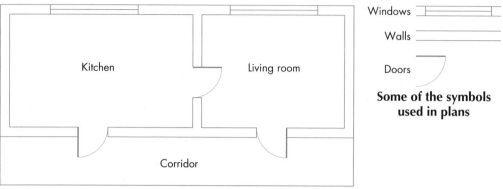

A simple plan, scale 1 : 600

To find the dimensions of the building, the measurements taken from the scale drawing must be converted to the actual measurements in real life.

Example 41

Find the actual area of a kitchen that is 5 cm × 3 cm on the plan, based on a scale of 1 : 600.

The measurement must be converted to the actual dimensions.

$5 \text{ cm} \times 3 \text{ cm} = 5 \times 600 \times 3 \times 600$

$= 5\,400\,000 \text{ cm}^2$

Converted to m²: $\dfrac{5\,400\,000}{10\,000} = 540 \text{ m}^2$

4.11 Trigonometry in three dimensions

All of the examples involving trigonometry that you have seen so far have been concerned with two dimensions only. You can apply the same methods to solving problems in three dimensions.

4.11.1 Using trigonometry to solve problems in three dimensions

The diagram shows a room in the shape of a regular cuboid. The room is 3 m long, 2 m wide and 2 m high.

Since all of the corners are right angles, you can use Pythagoras' theorem to find the lengths of the diagonals of each face, such as ED, and the angles they make with the horizontal and vertical edges. However, suppose you want to find the length of a line from one corner of the floor to the opposite corner of the ceiling, such as FD, and the angle this line makes with the floor – how will you proceed?

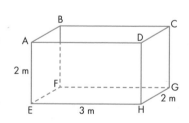

Example 42

Determine (a) the length FD (b) angle DFH in the diagram above.

(a) In order to determine the length FD you must identify a triangle in which FD is a side.

The triangle you need is FDH.

First, use Pythagoras' theorem to find the length of FH.

$FH^2 = 3^2 + 2^2$

$= 9 + 4$

$= 13$

So FH is $\sqrt{13} = 3.61$ m

Now use FH to find the length of FD.

Using Pythagoras' theorem again:

$FD^2 = FH^2 + DH^2$

$= 13 + 4$

$= 17$

So FD is $\sqrt{17} = 4.12$ m

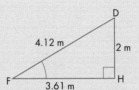

(b) Now you can use the lengths of any two sides of triangle FDH to determine angle DFH.

For example, sin DFH is $\dfrac{2}{4.12} = 0.4854$

Therefore DFH = 29°

Or tan DFH is $\dfrac{2}{3.61} = 0.5540$

Therefore DFH = 29°

Exercise 4H

1 A right-angled triangle PQR, with a right angle at Q, has angle $P = 30°$ and side $p = 6$ cm. Solve the triangle by finding the sizes of the remaining sides and angles.

2 Solve this triangle.

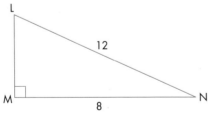

3 A boy standing 10 m away from the base of a mango tree is looking at a ripe mango at the top of the tree. If the boy is 5 feet tall and the angle of elevation from his eyes to the mango is 45° how tall is the tree?

4 A girl lying on her bed sees a cockroach on the floor at a distance of 3 m from her eyes. If the bed is 1.5 metres off the floor, find the angle of depression from the girl to the roach.

5 Write down each of these as a ratio of smaller angles.

a sin 120°   b cos 200°   c tan 300°

6 In the diagrams, all lengths are given in centimetres. Find the area of each triangle, giving your answers in square units.

a

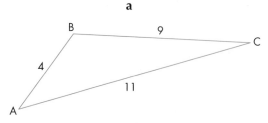

b

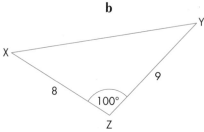

7 In triangle ABC, $a = 10$ cm, $b = 8$ cm and $c = 6$ cm. Use the appropriate rule to solve the triangle.

8 Solve triangle PQR with sides $p = 4$ cm, $q = 6$ cm and $R = 120°$.

9 Calculate the unknown sides and angle in triangle JKL.

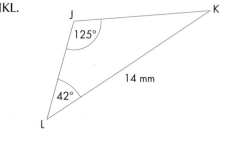

10 Determine the unknown angles and side in triangle UVW (not drawn to scale).

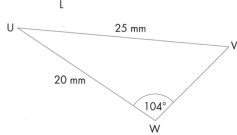

11 A ship set sail from a town A and travelled east a distance of 80 km to another town B. From B the ship travelled 110 km due south to place C and finally it travelled 50 km on a bearing of 240° from C to D. Determine the bearing of D from A and the distance AD.

12 The houses of four villagers are located in such a manner that John's house is due south of Peter's house, at a distance of 120 metres. Paul's house is 200 m away from Peter's house on a bearing 045° and Harry's house is 140 m away from John's on a bearing of 160° from John. Calculate the distance of Harry's house from Paul's house and the bearing of John's house from Paul's house.

13 Use isometric drawings to represent this object.

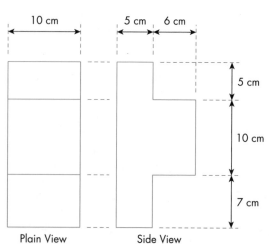

14 Use oblique projection to represent each of these shapes.

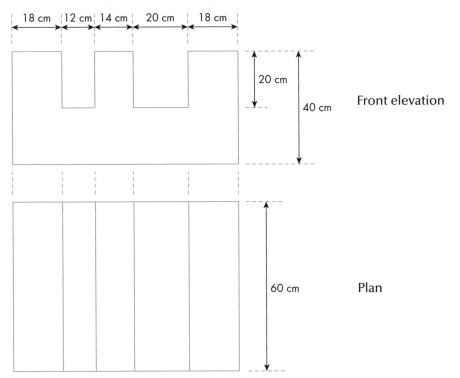

18 cm | 12 cm | 14 cm | 20 cm | 18 cm

20 cm

40 cm

Front elevation

60 cm

Plan

15 Use first angle projection to represent each of these shapes.

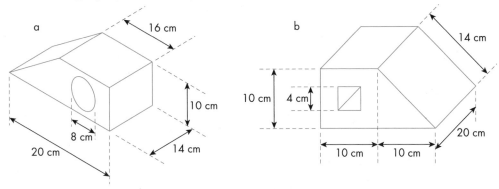

a

16 cm

10 cm

8 cm

20 cm

14 cm

b

14 cm

10 cm

4 cm

20 cm

10 cm

10 cm

16 The diagram shows a cubic aquarium tank of sides 60 cm.

Calculate the length of the line XY and angle this line makes with the base of the tank.

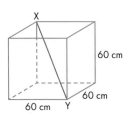

X

60 cm

60 cm

60 cm

Y

End of chapter summary

Geometrical concepts

- Important geometrical concepts include **points**, **lines**, **parallel lines**, **intersecting lines**, **perpendicular lines**, **line segments**, **rays**, **curves**, **planes**, **types of angles**, **solids** and numbers of **faces**, **edges** and **vertices**.

Drawing and measuring line segments and angles

- **Line segments:** accurate use of a ruler
- **Angles:** accurate use of a protractor
- **Geometric construction of lines, angles, and polygons**
- **Parallel and perpendicular lines**, using set square, ruler, compasses and pencil
- **Constructing a line perpendicular to another line** from points on and off the line
- **Bisecting line segments and angles**, using ruler, compasses and pencil
- **Constructing angles** of 30°, 45°, 60°, 90°, 120°, 105°, 135° and 150°
- **Constructing regular and irregular triangles and quadrilaterals** (polygons)
- **Constructing inscribed and circumscribed (escribed)** circles in polygons.

Symmetry in plane figures

- Two or more shapes are **congruent** if they are exactly the same, even if one is flipped or turned.
- An object or shape has **symmetry** if two or more parts are identical after a flip, slide or turn.
- **A given plane figure may possess:**
  - **reflection** or line symmetry (determine the number of lines of symmetry)
  - **rotational symmetry** (determine the order of rotational symmetry)
- A **line of symmetry** is an imaginary line along which the shape could be folded so that one half matches the other half exactly.
- A shape has **rotational symmetry** if it remains unchanged after undergoing a rotation or turn less than 360°.
- The **order of rotational** symmetry is the number of times the object looks exactly the same as it is rotated between 0° and 360°.

Properties of lines and angles

- Angles are normally measured in **degrees** (symbol °).
- A rotation of one **revolution** is equivalent to 360°.
- **Types of angle:** right, straight, acute, obtuse, reflex
- Pairs of angles include: complementary, supplementary.
- **Angles formed by intersecting straight line**s: adjacent angles, angles at a point, vertically opposite angles, alternate angles, corresponding angles, co-interior angles
- **Types of line:** parallel, perpendicular, intersecting, transversal

Properties of polygons

- A **polygon** is a plane enclosed figure or two-dimensional shape bounded by straight lines.
- A polygon with all its sides equal and all its angles equal is a **regular polygon**. All others are **irregular polygons**.
- **Types of triangle**: equilateral, isosceles, scalene, acute-angled, right-angled, obtuse-angled
- **Congruent triangles** are identical, corresponding sides are equal and corresponding angles are equal (four proofs).
- **Similar figures** are the same shape, with corresponding angles equal, but one is an enlargement of the other.
- **Types of quadrilateral and other polygons**: square, rectangle, rhombus, kite, parallelogram, trapezium
- A **quadrilateral** is a four-sided polygon or a plane enclosed figure bounded by four straight lines.
- **Angle properties of quadrilaterals:** sum of interior angles = 360°, sum of exterior angles = 360°
- **Angle properties of all polygons** (four or more sides):
  - Regular polygon: sum of interior angles = $(2n - 4)90°$, sum of exterior angles = 360°
  - Other named polygons: pentagon, hexagon, heptagon, octagon…
  - A polygon is **cyclic** if a circle can be drawn through all of its vertices.
- An **inscribed circle** is a circle constructed inside a regular polygon so that it touches each side of the polygon.
- A **circumscribed** or **escribed** circle is a circle that passes through all the vertices of a regular polygon.
- A **transversal** is a line that passes through two or more lines.

Three-dimensional shapes

- A **three-dimensional shape,** also called a **solid,** is any object or figure with three dimensions: length, breadth and height. All real objects, including human beings, are three-dimensional shapes.
- A **regular solid** is one in which all the faces are regular polygons and the **same** number of faces meet at each corner.
- An **irregular solid** is a solid that does not have a fixed shape like a cube or sphere but may have many edges of different lengths.

Circles

- A **circle** is a plane shape formed by a set of points a fixed distance from a centre.
- **Properties**: radius, diameter, chord, circumference, arc, tangent, segment, sector, semi-circle, pi (π)
- **Circle theorems**
  - A **theorem** is a mathematical statement, formula or proposition that can be proved
  - The angle that an arc of a circle subtends at the centre of a circle is twice the angle it subtends at any point on the remaining part of the circumference.
  - Angles at the circumference in the same segment of a circle and subtended by the same arc or chord are equal.
  - The angle at the circumference subtended by the diameter is a right angle.
  - A **cyclic quadrilateral** is one in which all four vertices touch the circumference of a circle.
  - The opposite angles of a cyclic quadrilateral are supplementary.
  - The exterior angle of a cyclic quadrilateral is equal to the interior opposite angle.

- The angle between a tangent to a circle and a chord through the point of contact is equal to the angle in the alternate segment.
- A tangent to a circle is perpendicular to the diameter/radius of the circle at the point of contact.
- The lengths of two tangents from an external point to the points of contact on the circle are equal.
- The line joining the centre of a circle to the midpoint of a chord is perpendicular to the chord.
- If a tangent and a secant are drawn from the same point to a circle, then (length of tangent)^2 = (length of the whole secant) × (the length of the part of the secant outside the circle).
- Two circles touching each other internally or externally at one point will have a common tangent.

Transformation geometry

- A **transformation** is a mathematical operation in which each point in an object or pre-image has a unique image and each point in the image corresponds to a point in the object.
- A **translation** is a sliding movement of the object (plane shape) in a straight line from one position to another.
- Using **vectors**, a **translation** in a plane may be written as a **column matrix** $\begin{pmatrix} x \\ y \end{pmatrix}$.
- A **reflection** produces an image similar to that produced in a plane mirror.
- A **rotation** is a transformation in which each point in the object turns through the same angle about a fixed point, known as the **centre of rotation**.
- In **size transformations** the image may be smaller or larger than the object.
- A **glide reflection** is a combination of two transformations: a reflection followed by a translation.
- **Deduce the transformation**, given the object and its image.
- **Locate the image** of an object under a combination of transformations.

Trigonometry

- **Trigonometry** is a branch of mathematics that studies the calculation of angles and sides of triangles and other polygons and the relationship between them.
- A **trigonometrical ratio** is a ratio of the sides of a right-angled triangle, which can be used to find the value of the non-right angles as well as the sides of the triangle.
- **sine** (sin) = $\dfrac{\text{side opposite the angle}}{\text{hypotenuse}} \Rightarrow \sin\theta = \dfrac{\text{opposite}}{\text{hypotenuse}} = \dfrac{O}{H}$
- **cosine** (cos) = $\dfrac{\text{side adjacent to the angle}}{\text{hypotenuse}} \Rightarrow \cos\theta = \dfrac{\text{adjacent}}{\text{hypotenuse}} = \dfrac{A}{H}$
- **tangent** (tan) = $\dfrac{\text{side opposite the angle}}{\text{side adjacent to the angle}} \Rightarrow \tan\theta = \dfrac{\text{opposite}}{\text{adjacent}} = \dfrac{O}{A}$
- **Pythagoras' theorem** states that, in a right-angled triangle, the square on the **hypotenuse** is equal to the sum of the squares on the other two sides.
- **Solve right-angled triangles** using Pythagoras' theorem and the trigonometrical ratios.
- The **angle of elevation** is the angle formed between the observer's line of vision and the top of an object, such as a lamppost, building or tree, when the observer needs to look up.
- The **angle of depression** is the angle formed between the observer's line of vision and the bottom of an object when the observer has to look down.
- Describe and calculate various angles of elevation and angles of depression.

- Trigonometrical ratios of angles between 0° and 360°

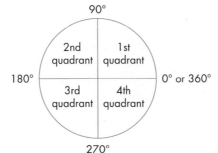

Second quadrant: 90° to 180°   First quadrant: 0° to 90°

Third quadrant: 180° to 270°   Fourth quadrant: 270° to 360°

- The sine rule states that for any triangle ABC with sides a, b and c and angles A, B and C:

$$\frac{a}{\sin A} = \frac{b}{\sin B} = \frac{c}{\sin C}$$

- The **cosine rule** states that for any triangle ABC with sides a, b and c and angles A, B and C:

$$a^2 = b^2 + c^2 - 2bc\cos A$$

- **For non-right angled triangles** where all three sides are known (SSS)

$$\text{Area} = \sqrt{s(s-a)(s-b)(s-c)} \quad \text{and} \quad s = \frac{a+b+c}{2}$$

- **For non-right angled triangles** where two sides and the included angle are known (SAS)

$$\text{Area} = \frac{1}{2}ab\sin C$$

- A **bearing** is an angle, measured clockwise from north at a fixed point, which helps determine the location of an object if you know its distance from the fixed point.

- To find the bearing of one point from another, for example, the bearing of A from B, draw a north line at B (since the bearing is **from** B). Join points A and B with a straight line. The angle between the north line and the line BA represents the bearing of A from B.

Spatial geometry: projections and scale drawing

- **Isometric projection** is based on a set of isometric axes, consisting of three lines with equal angles (60°) between them.

- **Oblique projection** is also based on three axes, but in this case the axis OX is horizontal, axis OY is vertical and axis OZ is drawn at 45° to the horizontal.

- In **first angle projection** the object is presented from three directions: front elevation, side elevation and plan, with the plan shown bottom left, the front elevation above the plan and the side elevation top right.

- For a **third angle projection**, the plan is located at the top right, with the front elevation placed directly below it and the side elevation bottom left.

- A **scale drawing** is any accurate drawing of an object but with lengths reduced or expanded by a fixed amount, called a **scale**.

Trigonometry in three dimensions

- The same methods for solving problems in two dimensions can be extended to solving problems in three dimensions.

Examination-type questions for Chapter 4

1   Using ruler, pencil and a pair of compasses only, construct triangle LMN in which LM = 8 cm, MN = 7 cm and angle LMN = 60°. Construct a line NQ parallel and equal to LM and hence construct parallelogram LMNQ.

2   In the diagram, O represents the centre of the circle. The tangent ST makes an angle of 50° with the chord KT and ∠RTP = 60°.

Find the sizes of angles KTJ, TJK, JRT, OTR and RJO.

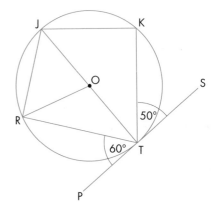

3   In the diagram (not drawn to scale), triangle FHG is a right-angled triangle and NM is parallel to FG. Angle H is 60° and GF = 10 cm.

   a   Prove that triangles HFG and HNM are similar.

   b   Find the length of HG.

   c   Find the length of FH.

   d   Given that MN = 6 cm, what is the ratio of the area of triangle FGH to that of triangle NMH?

4   Two triangles lie in a coordinate plane. Triangle QRS has vertices with coordinates Q(2, 3), R(6, 3) and S(6, 6). Triangle WXY has vertices with coordinates W(8, 3), X(14, 3) and Y(14, 6). Describe fully the transformation that maps triangle QRS onto triangle WXY. Represent both triangles on graph paper.

5 The diagram shows a map of an island. Points A, B, C and D represent four locations on that island. The scale of the map is 1 : 250 000.

   a Find the distances on the map from A to B, B to D and A to C.

   b Calculate the actual distances from A to D and from B to C.

   c The map is drawn on a grid. Use your ruler to find the dimensions of the grid. Hence determine the area of the map in cm² and the actual area of the island in km².

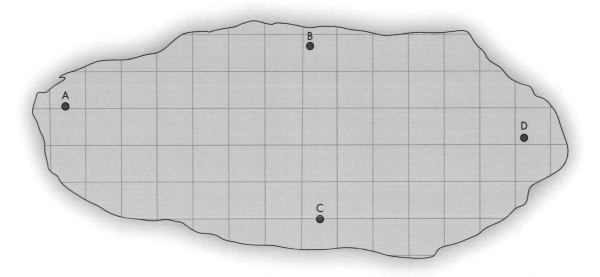

Scale 1 : 2500

6 The diagram, which is not drawn to scale, shows the positions of four churches, P, C, S and B, in a village.

C is on a bearing of 070° from P and S is on a bearing of 015° from B and due east of P. The distance of P from C is 440 m, of S from C is 300 m and PB is 250 m.

   a Copy the drawing on your own paper and show the bearings and distances given.

   b On your drawing, show the bearing of P from B.

   c Find the bearing of S from C.

   d Find the distances BS and PS.

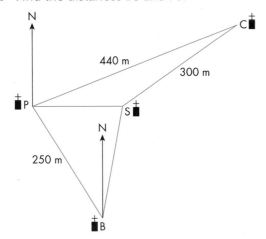

Objectives

By the end of this chapter, you should be able to:

- recognise the difference between a quantity and a unit
- convert units of length, mass, area, capacity, time and speed
- convert between SI units and imperial units of measurement
- use the appropriate SI unit of measure for length, mass, temperature and time (24-hour clock) and other quantities such as speed
- solve problems involving time, distance and speed
- identify the perimeter and area of a plane shape (two-dimensional figure), e.g. regular and irregular polygons, circles, sectors and segments
- calculate the perimeter of a polygon, a circle, an arc of a circle and a combination of polygons and circles
- calculate the area of polygons, a circle and any combination of these
- calculate the area of a sector of a circle and a segment of a circle
- calculate the area of a triangle given two sides and the included angle
- estimate the area of irregularly shaped plane figures
- estimate the margin of error for a given measurement
- use maps and scale drawings to determine distances and areas
- solve problems involving perimeter and area
- calculate the volume of prisms, cylinders, pyramids, cones, spheres, cubes and cuboids
- calculate the surface area of solids
- use the appropriate SI unit of measure for volume and surface area.

You should already know:

- how to multiply and divide by 10, 100, 1000, 10 000 etc.
- how to use indices when multiplying by powers of 10
- how to write numbers in standard form
- how to solve simple equations
- how to complete calculations involving brackets
- the names of common plane and solid shapes

- how to identify acute, obtuse and reflex angles
- Pythagoras' theorem and how to use it
- trigonometric ratios and how to use them
- how to calculate the perimeters and areas of plane shapes

Introduction

People often have to measure things in their everyday lives. In the workplace, masons, carpenters, engineers, doctors and technicians use different measuring equipment. At home too, people have to take measurements. For example, a carpenter measures the length of wood to make shelves, a parent measures the temperature of a sick baby, an athlete measures how long it took her to run a race.

In this section, you will learn about different types of measurement. This will help you to calculate with measures and to solve problems involving measurements.

Examples of measurement in real life

The pictures show everyday examples of measuring. Which of them have you seen in real life?

Length of a child's foot

Time taken for a race

Height of a house on a plan

Weighing on scales

Blood pressure

Temperature outside

5.1 Quantities and units

The International System of Units (SI) is the most widely used measurement system in the world. It is based on the metric system, which uses the decimal number system of powers of 10. The *metre* is an SI unit of the quantity *length* and the *kilogram* is an SI unit of the quantity *mass*.

Key terms

Quantity   something that can be measured, for example, the *height* of a person, the *speed* of a car

Unit   a measure of a quantity, for example, a centimetre, a gram

Did you know?   The SI metre is the distance travelled by light in a vacuum in exactly $\frac{1}{299\ 792\ 458}$ seconds.

The imperial system of measurement (also called the English system) is not based on the metric system. Examples of imperial units are the mile, the inch and the pint.

Some quantities and SI units are called **basic quantities** and **basic units** because you can derive other quantities and units from them. Here are the most frequently used basic quantities units in mathematics.

Basic quantity	Symbol	Basic unit	Symbol
Length	*l*	metre	m
Mass	*m*	kilogram	kg
Time	*t*	second	s
Temperature	*T*	degrees Celsius/kelvin	°C/K

You can combine basic quantities and units to produce **derived quantities** and **derived units**. Here are some examples of derived quantities and units.

Derived quantity	Symbol	Derived unit	Symbol
velocity/speed	*v/s*	metre per second	$m\,s^{-1}$
density	*ρ*	kilogram per cubic metre	$kg\,m^{-3}$
area	*A*	square metre	m^2
weight	*w*	kilogram metre per second squared or newton	$kg\,m\,s^{-2}$ or N
acceleration	*a*	metre per second squared	$m\,s^{-2}$
work	*W*	joule	J
force	*F*	newton	N
perimeter	*p*	metre	m
volume/capacity	*V/C*	cubic metre	m^3

5.2 Length

When you measure length, you use different units according to how long or high or far something is. For example, botanists use millimetres when they measure the length of a seed and airline pilots use kilometres when they measure distances between countries. The millimetre and the kilometre are SI units.

| Frequently used SI units of length | 1 kilometre = 1000 m = 10^3 m | 1 centimetre = 0.01 m = 10^{-2} m |
| | 1 metre = 1m = 10^0 m | 1 millimetre = 0.001 m = 10^{-3} m |

Prefixes and units

A prefix is a group of letters placed before a word. They change the meaning of the word. SI measurements use many different prefixes. The table shows the most common prefixes.

Prefix	Meaning
kilo-	one thousand
deci-	one tenth
centi-	one hundredth
milli-	one thousandth
micro-	one millionth

5.2.1 Converting between units

Sometimes, you need to convert from one unit to another. You multiply when you convert from larger to smaller units (e.g. m to mm) and divide when you convert from smaller to larger units (e.g. cm to km).

Example 1

Convert 20 km to m.

1 km = 1000 m, so the conversion is from a larger unit to a smaller unit. You will have to multiply 20 by 1000.

20 km = 20 × 1000

= 20 000 m

20 km = 20 000 m

Example 2

Convert 200 mm to cm.

In this case, since 10 mm = 1 cm and the conversion is from a smaller to a larger unit, divide by 10.

200 mm = 200 ÷ 10

= 20 cm

200 mm = 20 cm

Example 3

Convert 2 km to mm.

Convert km to m and then convert m to mm.

2 km = 2 × 1000 m

= 2000 m (since 1 km = 1000 m)

2000 m = 2000 × 1000 mm

= 2 000 000 mm (since 1 m = 1000 mm)

2 km = 2 000 000 mm

= $2 × 10^6$ mm

5.2.2 Converting between units by using indices

Another way to convert between units is to use indices (powers). Look at the following examples.

Remember	To multiply powers of 10, add the indices. $10^2 \times 10^5 = 10^{2+5}$ $\qquad\qquad = 10^7$ Dividing powers of 10 is the equivalent of multiplying by a negative index. Therefore, subtract the indices. $10^8 \div 10^3 = 10^8 \times 10^{-3}$ $\qquad\qquad = 10^{8-3}$ $\qquad\qquad = 10^5$

Example 4

Convert 2 km to mm. (As in Example 3)

Since 1 km = 1000 m = 10^3 m and 1 m = 1000 mm = 10^3 mm then:

2 km = $2 \times 10^3 \times 10^3$ mm

To multiply, add the indices.

$2 \times 10^3 \times 10^3$ mm $= 2 \times 10^{3+3}$
$\qquad\qquad\qquad = 2 \times 10^6$ mm
$\qquad\quad$ 2 km $= 2 \times 10^6$ mm

Example 5

Convert 250 cm to km.

Since 1 cm = $\frac{1}{100}$ m = 10^{-2} m and 1 m = $\frac{1}{1000}$ km = 10^{-3} km then:

$250 \times 10^{-2} \times 10^{-3}$ km $= 250 \times 10^{-2-3}$ km
$\qquad\qquad\qquad\qquad = 250 \times 10^{-5}$ km

250×10^{-5} km

You can write this answer in standard form. In this form, numbers are shown as a number between 1 and 10 multiplied by a power of 10.

250×10^{-5} km $= 2.5 \times 10^2 \times 10^{-5}$ km
$\qquad\qquad\qquad = 2.5 \times 10^{2-5}$ km
$\qquad\qquad\qquad = 2.5 \times 10^{-3}$ km
$\quad$ 250 cm $= 2.5 \times 10^{-3}$ km in standard form

There are several SI units of length greater and smaller than a metre. The table below shows these. Generally, the most commonly used prefixes from this extensive list will be:

- the multiples *kilo-* and *mega-*
- the sub-multiples *centi-, milli-, micro-, nano-*.

Prefix	Relationship to the metre	Name of unit	Symbol
exa-	$1\ 000\ 000\ 000\ 000\ 000\ 000 = 10^{18}$ m	exametre	Em
peta-	$1\ 000\ 000\ 000\ 000\ 000 = 10^{15}$ m	petametre	Pm
tera-	$1\ 000\ 000\ 000\ 000 = 10^{12}$ m	terametre	Tm
giga-	$1\ 000\ 000\ 000 = 10^{9}$ m	gigametre	Gm
mega-	$1\ 000\ 000 = 10^{6}$ m	megametre	Mm
kilo-	$1\ 000 = 10^{3}$ m	kilometre	km
hecto-	$100 = 10^{2}$ m	hectometre	hm
deca-	$10 = 10^{1}$ m	decametre	dam
deci-	$0.1 = 10^{-1}$ m	decimetre	dm
centi-	$0.01 = 10^{-2}$ m	centimetre	cm
milli-	$0.001 = 10^{-3}$ m	millimetre	mm
micro-	$0.000\ 001 = 10^{-6}$ m	micrometre	μm
nano-	$0.000\ 000\ 001 = 10^{-9}$ m	nanometre	nm
pico-	$0.000\ 000\ 000\ 001 = 10^{-12}$ m	picometre	pm
femto-	$0.000\ 000\ 000\ 000\ 001 = 10^{-15}$ m	femtometre	fm
atto-	$0.000\ 000\ 000\ 000\ 000\ 001 = 10^{-18}$ m	attometre	am

Exercise 5A

1. Convert these to metres.
   a 200 cm
   b 8 km
   c 10 mm

2. Convert these measurements.
   a 2000 mm to km
   b 40 km to cm
   c 2 m to cm
   d 5000 m to km
   e 0.5 m to mm

3. Convert these measurements. Express your answer in standard form.
   a 0.005 km to mm
   b 450 cm to mm
   c 45.8 cm to km
   d 0.0023 mm to cm

4. Using indices, convert measurements.
   a 2.5×10^{-3} km to mm
   b 0.000 008 mm to cm
   c 259 000 km to m

5. Some scientists believe that the Earth is 4.6×10^{9} years old. How many seconds will that be? Use indices to express your answer.

6. In science, the unit mass of an atom (AMU) is 1.7×10^{-27} kg. Express your answer in standard form.
   a Express 1 AMU in attograms.
   b What is the mass in picograms of Uranium 235, which contains 235 AMU?

7 One light year (the distance light travels in a year) is given as 9.46×10^{12} km. The nearest galaxy (Magellanic cloud) is 169 000 light years away. How far is this distance expressed in Em? Express your answer in standard form.

8 In science a quark is considered to be the tiniest particle of matter. Quarks combine to form protons and neutrons. The size of a quark is approximately 1 am. Express the size of a quark in Em.

5.3 Mass

The same prefixes shown above can be used to express different units of mass. For example, you use grams when you weigh flour for cooking and you use kilograms when you weigh a parcel of books at the post office.

Key term

Mass the amount of matter in an object

Frequently used SI units of mass	1 kilogram (kg) = 1000 g = 10^3 g	1 milligram = 0.001 g = 10^{-3} g
	1 gram = 1 g = 10^0 g	1 microgram = 0.000 001 g = 10^{-6} g
	1 centigram = 0.01 g = 10^{-2} g	

| Note | 1 megagram (Mg) = 1 000 000 g = 10^6 g |
| | 1 Mg = 1000 kg = 1 metric tonne (1 t) |

Example 6

Convert 0.05 mg to kg. Write your answer in standard form.

1 g = 1000 mg and 1 kg = 1000 g. You are converting from a smaller unit to a larger, and so you need to divide.

Method A: 0.05 mg = 0.05 ÷ 1000 g
$\qquad$ = 0.000 05 g
$\qquad$ = 0.000 05 ÷ 1000 kg
$\qquad$ = 0.000 000 05 kg
$\qquad$ = 5×10^{-8} kg

Method B: 0.05 mg = $0.05 \times 10^{-3} \times 10^{-3}$ kg
$\qquad$ = 0.05×10^{-6} kg
$\qquad$ = $5 \times 10^{-2} \times 10^{-6}$ kg
$\qquad$ = $5 \times 10^{-2-6}$ kg
$\qquad$ = 5×10^{-8} kg
$\qquad$ 0.05 mg = 5×10^{-8} kg

Example 7

Convert 20 000 kg to tonnes.

Since 1000 kg = 1 tonne,
$\qquad$ 20 000 kg = 20 000 ÷ 1000 tonnes
$\qquad\qquad$ = 20 tonnes
$\qquad$ 20 000 kg = 20 t

Exercise 5B

1 Convert these measurements to kg.

    a 2000 g   **b** 30 000 cg   **c** 5 Mg   **d** 100 mg

2 Convert these measurements. Use indices.

    a 2 000 000 μg to g   **b** 200 mg to Mg   **c** 70 cg to kg

3 Convert these measurements.

    a 0.000 023 kg to μg (in standard form correct to 1 decimal place)

    b 234 800 kg to mg (in standard form correct to two decimal places)

    c 2.55×10^5 μg to Mg (in standard form)

    d 0.0098×10^{-3} μg to Mg (in standard form correct to 1 significant figure)

4 Convert these measurements. Express your answers in standard form correct to 1 significant figure.

    a 2ng to pg   **b** 2000 000 Tg to cg

    c 400 fg to Gg   **d** 0.000 008 2 Mg to ng

5.4 More units of length and mass

The SI is the official measuring system in most countries, including the Caribbean countries. In the Caribbean, the imperial system is still widely used in many countries, especially by builders, electricians, plumbers, technicians, etc.

Did you know?	The change from the imperial or English system began as early as 1795 in France and was completed as late as the 1970s in most British Commonwealth countries.

Imperial measures or the English system	**Length:** inch (in), foot (ft), yard (yd), mile (mi)
	    12 in = 1 ft 5280 ft = 1 mi
	    3 ft = 1 yd 1760 yds = 1 mi
	Mass: ounces (oz), pounds (lb), ton (t)
	    16 oz = 1 lb 2240 lbs = 1 UK ton
	                   2000 lbs = 1 US ton

This table shows the relationships between units in the SI and the imperial system.

Imperial units	SI units
1 inch	2.54 cm
1 foot (12 inches)	30 cm
1 yard (3 feet)	90 cm (0.9 m)
1 mile	1.6 km (1600 m)
1 pound	453.6 grams

Comparison of actual measurements of length and mass in metric and English systems

- The door to the average house is approximately 2 metres (6.6 feet) high.
- The length of a cricket bat is about 100 cm (3.3 feet).
- The length of a typical car is 4.8 m (16 feet).
- The length and mass of the largest Carnival cruise ship are approximately 285 m (935 feet) and 130 000 tonnes (290 000 000 pounds).
- The distance between Grenada and Trinidad is approximately 165 km (103 miles).
- The average time taken by an airplane to travel between the two countries (Trinidad and Grenada) is about 45 minutes.
- The thickness of a piece of paper is approximately 0.1 mm (0.004 inches).
- A pencil dot and a small black ant are each about 1–2 mm long (0.04–0.08 inch).

Example 8

What is the metric equivalent of a $\frac{1}{2}$ inch spanner?

$$1 \text{ inch} = 25 \text{ mm}$$
$$\text{Therefore, } \frac{1}{2} \text{ inch} = \frac{1}{2} \times 25$$
$$= 25 \div 2$$
$$= 12.5 \text{ mm}$$
$$\frac{1}{2} \text{ inch} = 12.5 \text{ mm}$$

Example 9

Measure the length and width of a sheet of A4 paper ($8\frac{1}{2}$ in × 12 in) to the nearest cm.

Length: 30 cm   Width: 21 cm

Example 10

Measure the length of the line XY below. Use SI units.

X ——————————————————————— Y

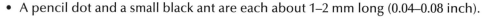

Length of line = 7.4 cm

Example 11

What is the metric equivalent of 5.8 lbs?

$$1 \text{ lb} = 453.6 \text{ g}$$
$$\text{Therefore, } 5.8 \text{ lbs} = 5.8 \times 453.6$$
$$= 2631 \text{ g}$$
$$= 2.6 \text{ kg}$$
$$5.8 \text{ lbs} = 2631 \text{ g}$$
$$= 2.6 \text{ kg}$$

Exercise 5C

1. Using a ruler, measure the length of your pencil in cm and mm.

2. Estimate the length of your thumb in cm.

3. Measure the length of the line AB in cm.   A————————————— B

4. Estimate the length and width of an exercise book in metric units.

5. An ordinary ruler measures exactly 1 foot or 12 inches or how many centimetres?

6. What is the metric equivalent of a $\frac{5}{8}$ inch spanner?

7. Estimate the width of the average front door of a house.

8. Find the mass of a leather cricket ball.

9. Estimate the mass of a soccer ball.

10. Estimate the diameter of a table-tennis ball.

11. Measure the thickness of a quarter.

12. By looking at a map, estimate the distance between Jamaica and Cuba.

13. Estimate the length, width and thickness of a king-size mattress.

14. What is the metric equivalent of 13.09 lbs?

15. What is the metric equivalent of a $5\frac{3}{8}$ inch spanner?

5.5 Time

5.5.1 Units of time

You can measure the **time** that it takes to do an activity. You can also say at what time something happens or happened or will happen.

Commonly used units of time	60 seconds (s) = 1 minute (min) 60 minutes = 1 hour (h) 24 hours = 1 day 7 days = 1 week 4 weeks = 1 month 12 months = 1 year	365 days = 1 year (366 days = 1 leap year) 52 weeks = 1 year 1 decade = 10 years 1 century = 100 years 1 millennium = 1000 years

There are also some units of time that are less than 1 second:

1 millisecond (ms) = 10^{-3} s

1 microsecond (μs) = 10^{-6} s

1 nanoseconds (ns) = 10^{-9} s

5.5.2 The 12-hour clock and 24-hour clock

A day contains 24 hours. It can be divided into two 12-hour periods:

- 12 midnight to 12 noon – a.m. (ante-meridiem or before noon)
- 12 noon to 12 midnight – p.m. (post-meridiem or after noon)

This can be seen on the 12-hour clock shown below.

On the 24-hour clock, there is only one time period. The day begins at midnight, 00:00, and the last minute of the day starts at 23:59. The day ends at midnight, 24:00, which is the same as 00:00, the start of the next day.

- Midday is shown as 12:00
- 5 a.m. is shown as 05:00
- 4 p.m. is shown as 16:00
- Here is a 24-hour clock face. The time is nearly midnight.

Example 12

Convert these times.
(a) 120 seconds to minutes   (b) 180 minutes to hours   (c) 48 hours to days
(d) 70 days to weeks   (e) 2 weeks to months   (f) 18 months to years
(g) 50 years to decades   (h) 1000 years to centuries

(a) 120 seconds = 120 ÷ 60 (60 seconds = 1 minute)
         = 2 minutes

(b) 180 minutes = 180 ÷ 60 (60 minutes = 1 hour)
         = 3 hours

(c) 48 hours = 48 ÷ 24 (24 hours = 1 day)
         = 2 days

(d) 70 days = 70 ÷ 7 (7 days = 1 week)
         = 10 weeks

(e) 2 weeks = 2 ÷ 4 (4 weeks = 1 month)
         = $\frac{1}{2}$ month

(f) 18 months = 18 ÷ 12 (12 months = 1 year)
         = $1\frac{1}{2}$ years

(g) 50 years = 50 ÷ 10 (10 years = 1 decade)
         = 5 decades

(h) 1000 years = 1000 ÷ 100 (100 years = 1 century)
         = 10 centuries

Example 13

Convert these time measurements to seconds.

(a) 4 hours   (b) 10 weeks   (c) 1 year

(a)   4 hours $= 4 \times 60$ minutes
$$= 4 \times 60 \times 60 \text{ seconds}$$
$$= 14\,400 \text{ seconds}$$

(b)   10 weeks $= 10 \times 7$ days
$$= 70 \times 24 \text{ hours}$$
$$= 70 \times 24 \times 60 \text{ minutes}$$
$$= 70 \times 24 \times 60 \times 60 \text{ seconds}$$
$$= 6\,048\,000 \text{ seconds}$$

(c)   1 year $= 365$ days
$$= 365 \times 24 \text{ hours}$$
$$= 8760 \times 60 \text{ minutes}$$
$$= 525\,600 \times 60 \text{ seconds}$$
$$= 31\,536\,000 \text{ seconds}$$

Example 14

Mary started her vacations on 7 January and returned to work after 3 weeks and 3 days. On what date did she return?

3 weeks = 21 days

$7 + 21 + 3 = 31$ days

Mary returned to work on 1 February.

Example 15

A movie started at 5:45 p.m. and ended at 7:15 p.m. How long was the movie?

5:45 – 6:00 p.m. $\rightarrow$ 15 minutes

6:00 – 7:00 $\rightarrow$ 60 minutes (1 hour = 60 minutes)

7:00 – 7:15 $\rightarrow$ 15 minutes

Total duration of the movie $= 15 + 60 + 15$
$$= 90 \text{ minutes}$$
$$= \frac{90}{60} \text{ hours}$$
$$= 1\frac{1}{2} \text{ hours}$$

Alternative solution: $07{:}15 - 05{:}45 = \begin{array}{r} {}^{6}\,{}^{7} \\ 07{:}\!\!\not{1}\!\!\not{5} \\ -\ 05{:}45 \\ \hline 1{:}30 \end{array}$

The movie lasted $1\frac{1}{2}$ hours.

Example 16

Express 2:00 a.m. in 24-hour clock time.

2:00 a.m. $= 00{:}00 + 2$ hours
$$= 02{:}00$$

Example 17

Express 5:00 p.m. in 24-hour clock time.

5:00 p.m. = 12:00 + 5 hours
         = 17:00

Example 18

The time shown on a 24-hour clock is 21:00. Express this in 12-hour clock time.

21:00 = 12 p.m. + 9 hours
      = 9 p.m.

Exercise 5D

1 Convert these times.

  a 200 seconds to minutes   **b** 600 minutes to hours   **c** 48 hours to days

  d 21 days to weeks   **e** 36 weeks to months   **f** 30 months to years

  g 80 years to decades   **h** 2000 years to centuries

2 Convert to seconds.

  a 20 minutes   **b** $2\frac{1}{2}$ hours   **c** 3 days

3 Tom went on a trip through the Caribbean. He left Barbados on 1 May and spent 2 weeks and 3 days in Jamaica. He then went to Cuba where he spent 1 week and then spent two days in Guyana. Finally, he spent 6 days in Antigua and returned to Barbados the next day. On what date did he return home?

4 Joel left home at 06:42 a.m. for school. He spent 23 minutes in the supermarket and arrived at the school an additional 48 minutes later. At what time did he arrive?

5 Express the following times in 24-hour clock time.

  a 2:50 p.m.   **b** 9:00 a.m.

6 Convert these times.

  a 2 months and 3 weeks to days, hours, minutes and seconds

  b 1000 seconds to hours

7 A gardener spent a number of days preparing his land for planting. On 15 August, he started clearing the land and spent 3 days doing so. He spent 4 days forking the land and making beds. 1 week was spent planting and another 6 days putting up a fence. On what date did he complete all his garden chores?

8 Jane left school at 2:30 p.m. and spent 15 minutes waiting for the school bus. The bus took 1 hour 13 minutes to reach its destination and Jane took a further 12 minutes to reach home. At what time did she reach home?

9 Express these 12-hour clock times, using the 24-hour clock.

  a 12:15 a.m.   **b** 11:59 p.m.

10 **a** Convert 10 seconds to years. Give your answer in standard form.

   b Convert 20 centuries to hours. Give your answer in standard form.

11 A tourist spent some time travelling to different countries. She left her home country on 5 October 2016. She spent 3 weeks and 2 days in one country, 17 days in another and 4 months and 3 weeks in a third country. After this, she returned home. What was the date of her arrival back home?

12 A student left the Caribbean on a trip to China on Monday at 14:00 and he arrived in China on Wednesday of the same week at 08:00 (he did not adjust the time on his watch). Then he took a train and arrived at his final destination on Thursday at 10:00. How long did his whole journey take?

Tip	When calculating time intervals, remember that times written in the 24-hour clock are not decimals. For example, 13:26 = 13 hours 26 minutes **not** 13.26 hours.

5.6 Weight

The **weight** of an object is often confused with its mass. However, mass is the amount of material the object contains whereas weight is a force. Weight is a derived quantity.

Key terms	
Weight	the force exerted on an object by Earth's gravity
Newton	the SI unit of force. 1 newton = 1 $kg\,m\,s^{-2}$

Weight (W)	$W = m \times g$ $(W = mg)$ g is the acceleration due to gravity.

Example 19

Find the weight of a 1 kg mass. Take gravity = 10 $m\,s^{-2}$

Weight of a 1 kg mass = 1 kg × 10 $m\,s^{-2}$
                      = 10 newtons (using $W = mg$)

5.7 Speed

Speed is another derived quantity.

Key term	
Speed	the rate at which the distance travelled by an object changes with time

Speed	speed = distance travelled ÷ time taken

The speed of an object usually varies throughout the journey. Generally, you should find the average speed.

Average speed (s)	average speed = total distance travelled ÷ total time taken $$s = \frac{d}{t}$$

Commonly used units of average speed	metre per second	$m\,s^{-1}$ (also written as m/s)
	kilometre per hour	$km\,h^{-1}$ (also written as km/h)
	To convert $km\,h^{-1}$ to $m\,s^{-1}$, multiply by $\dfrac{1000\ m}{3600\ s}$.	

Example 20

Find the average speed of an athlete who runs 100 m in 10 seconds.

Average speed = $\dfrac{\text{total distance}}{\text{total time taken}}$

$= \dfrac{100\ m}{10\ s}$

$= 10\ m\,s^{-1}$

Example 21

A car travels 80 kilometres in 10 hours. Find the average speed in $km\,h^{-1}$ and $m\,s^{-1}$.

Average speed = $\dfrac{\text{total distance}}{\text{total time taken}}$

$= \dfrac{80}{10}$

$= 8\ km\,h^{-1}$

$= 8 \times \dfrac{1000\ m}{3600\ s}$

$= 2.2\ m\,s^{-1}$

$= 8\ km\,h^{-1}$ and $2.2\ m\,s^{-1}$

Example 22

A vehicle travelled a total distance of 100 km at an average speed of $20\ km\,h^{-1}$. How long did the journey take?

Average speed = $\dfrac{\text{total distance}}{\text{total time taken}}$

total time taken = $\dfrac{\text{total distance}}{\text{average speed}}$

$= \dfrac{100\ km}{20\ km\,h^{-1}}$

$= 5\ h$

Example 23

A motorcyclist traveling at an average speed of $80\ km\,h^{-1}$, completed a journey in 45 minutes. How far did he travel?

Average speed = $\dfrac{\text{total distance}}{\text{total time taken}}$

distance = average speed × total time taken

$= 80\ km\,h^{-1} \times \dfrac{3}{4}h$

$= 60\ km$

5.8 Density

Density is a derived quantity. It describes how compact or solid the substance is. For example, a block of wood has greater density than a block of cheese.

Density (ρ)	density $= \dfrac{\text{mass}}{\text{volume}}$ $$\rho = \frac{m}{V}$$

Example 24

Find the density of an object that has a mass of 40 kg and a volume of 20 m³.

$$\rho = \frac{m}{V}$$

$$\rho = \frac{40}{20}$$

$$= 2 \text{ kg m}^{-3}$$

Exercise 5E

Use $g = 10 \text{ m s}^{-2}$.

1   Calculate the weight of a man of mass 175 kg.

2   A turtle weighs 200 N. Find its mass.

3   Which has the greater weight, a box with a mass of 90 kg or a crate with a weight of 1000 N?

4   What is the speed of an object that travels a distance of 10 m in 2 seconds?

5   A glass block is a cuboid of dimensions: 10 cm × 8 cm × 5 cm. The mass of the block is 1200 grams. What is the density of the block?

6   A cyclist travelled a distance of 10 km in 4 hours. Find his average speed in km h⁻¹ and in m s⁻¹.

7   A woman is driving to work at a speed of 30 m s⁻¹. How long will she take to arrive, given that her work place is 40 km from her home?

8   A block of wood has a mass of 200 g and a density of 0.8 g cm⁻³. Calculate its volume.

9   A horse covers a distance of 5 km with an average speed of 20 m s⁻¹. He runs the distance in three stages. He gallops the first 2 km in a time of 80 s, and the next 2 km in a time of 100 s. How long did the last part of the journey take?

10   An object weighs 180 N on Earth but 120 N in space. What is the mass of the object?

5.9 Perimeter

5.9.1 Perimeter of polygons

To find the perimeter of a polygon, you add together the lengths of all its sides. Remember the formulae below for calculating the perimeter of rectangles and regular polygons.

Formula for the perimeter of rectangles	Perimeter of a rectangle = $2(l + w)$ where l = length and w = width

Formulae for the perimeters of regular polygons	Perimeter of a regular polygon = $n \times l = nl$ where n = number of sides and l = length of one side For example: Equilateral triangle $P = 3l$ Hexagon $P = 6l$ Square $P = 4l$ Octagon $P = 8l$ Pentagon $P = 5l$

In some problems involving perimeter, the lengths of some sides are unknown. If the unknown sides are not perpendicular, you can sometimes use Pythagoras' theorem to calculate their length (see Example 27**e**).

Key term

Hypotenuse   the longest side in a right-angled triangle. It is opposite the right angle.

Pythagoras' theorem	In a right-angled triangle, the square of the length of the hypotenuse is equal to the sum of the squares of the lengths of the other two sides. $c^2 = a^2 + b^2$ $c = \sqrt{a^2 + b^2}$	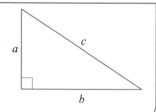

Example 25

A right-angled triangle ABC has sides AB = 3 cm and BC = 4 cm. Find the length of the hypotenuse AC.

Applying Pythagoras' theorem:

$AC^2 = AB^2 + BC^2$

$ = 3^2 + 4^2$

$ = 9 + 16$

$ = 25$

$AC = \sqrt{25}$

$ = 5 \text{ cm}$

Example 26

A plot of land in the shape of a right-angled scalene triangle has a hypotenuse of length 10 km and another side of length 6 km. Find the length of the other side of the triangle.

Applying Pythagoras' theorem:

Let the missing side be x km,

$x^2 + 6^2 = 10^2$

$ x^2 = 10^2 - 6^2$

$ = 100 - 36$

$ = 64$

$ x = \sqrt{64}$

$ = 8 \text{ km}$

Example 27

(i) Name each polygon.
(ii) Find the perimeter of each polygon. (The shapes are not drawn to scale.)

a

3 cm
9 cm

b
5 cm
5 cm

c
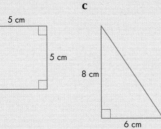
8 cm
6 cm

d
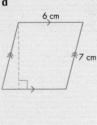
6 cm
7 cm

Not to scale

e
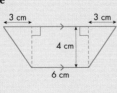
3 cm   3 cm
4 cm
6 cm

f
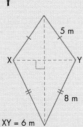
5 m
X — — — Y
8 m
XY = 6 m

g
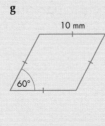
10 mm
60°

h
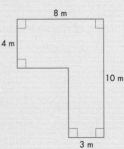
8 m
4 m
10 m
3 m

(a) (i) Rectangle

(ii) In a rectangle, opposite sides are equal.

Perimeter = 3 + 3 + 9 + 9 cm

= 24 cm

You can also use the formula for the perimeter of a rectangle:

Perimeter = 2(3 + 9) cm

= 2 × 12 cm

= 24 cm

(b) (i) Square

(ii) You can use the formula for the perimeter of a square.

Perimeter = 4 × 5 cm

= 20 cm

(c) (i) Right-angled, scalene triangle

(ii) First, you work out the length of the hypotenuse of the triangle. You can use Pythagoras' theorem.

Using Pythagoras' theorem:

$(\text{hypotenuse})^2 = 8^2 + 6^2$

$= 64 + 36$

$= 100$

$\text{hypotenuse} = \sqrt{100}$

$= 10$

Perimeter of triangle = 8 + 6 + 10 cm

= 24 cm

(d) (i) Parallelogram

> ## Key term
>
> **The symbol >** on a diagram, this shows that two or more lines or sides are parallel. On the same diagram, double or treble arrows show other sets of lines or sides that are parallel.

(ii) In a parallelogram, opposite sides are equal.

Perimeter = 7 + 6 + 7 + 6

= 2 × 7 + 2 × 6

= 14 + 12

= 26 cm

(e) (i) Trapezium

(ii) Look at the diagram of the trapezium. The trapezium can be split into a rectangle with a right-angled triangle at each end. You can use Pythagoras' theorem to find the length of the sloping sides.

$(\text{sloping side})^2 = 4^2 + 3^2$

$= 16 + 9$

$= 25$

$\text{sloping side} = \sqrt{25}$

$= 5$

Perimeter of trapezium = 3 + 6 + 3 + 5 + 6 + 5

= 28 cm

(f) (i) Kite
     (ii) In a kite, adjacent sides are equal.
          Perimeter of kite = 5 + 5 + 8 + 8
                            = 26 m

(g) (i) Rhombus
     (ii) Perimeter of rhombus = 4 × 10
                               = 40 mm

(h) (i) Irregular hexagon
     (ii) First, work out the lengths of the unmarked sides. Use subtraction.
          Horizontal unmarked side = 8 − 3
                                   = 5 m
          Vertical unmarked side = 10 − 4
                                 = 6 m
          Perimeter = 8 + 10 + 3 + 6 + 5 + 4
                    = 36 m

Key term

The symbol / on a diagram, this symbol is used to show lines or sides that are the same length. On the same diagram, double strokes indicate other sets of sides that are equal in length.

Example 28

A plot of land is in the shape of an equilateral triangle of sides 20 m. Find the perimeter of the plot of land.

$P = 3 \times l$
$ = 3 \times 20$
$ = 60$ m

Example 29

A rectangular room has dimensions: 22 m by 18 m. How many 20 cm square tiles will be needed to make a border design right around the room?

Perimeter of room = 2 (2200 + 1800) cm
$ = 8000$ cm
Length of 1 tile = 20 cm
Number of tiles = 8000 ÷ 20 cm
$ = 400$

5.9.2 Perimeter of circles, sectors and segments

Key terms

Circumference perimeter of a circle

Radius the distance from the centre of a circle to any point on the circle. The plural of *radius* is *radii*.

Chord a straight line joining two points on a circle

Diameter a chord of a circle that passes through the centre of the circle

Arc part of a curve or a circle

Sector the region enclosed by two radii and the arc they cut off the circle

Segment the region enclosed by a chord and the arc between its endpoints

You can use formulae to work out the circumference of a circle and the perimeter of a sector.

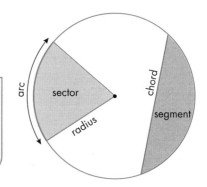

Formula for the circumference of a circle	circumference = $2\pi r$ where r = radius of the circle Since $d = 2r$, circumference = πd where d = diameter of the circle

Formula for the perimeter of a sector	perimeter of a sector = 2 × radius + length of arc The angle θ of a sector of a circle is a fraction of the whole turn, 360°. It is $\frac{\theta}{360}$. Therefore, the length of the arc is that fraction of the circumference of the circle: $2\pi r \times \frac{\theta}{360}$ perimeter of a sector = $2r + \left(2\pi r \times \frac{\theta}{360}\right)$ perimeter of a sector = $d + \left(\pi d \times \frac{\theta}{360}\right)$

Perimeter of a segment	perimeter of a segment = length of chord + length of arc

Example 30

(i) Name each of the following shapes.
(ii) Find the perimeter of each shape. (The shapes are not drawn to scale.) Use $\pi = 3.14$.

a 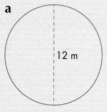   **b** 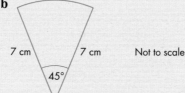   Not to scale

(a) (i) Circle
   (ii) Circumference = $2\pi r = 2 \times \pi \times 6$
                      = 37.68 m

(b) (i) Sector
        Perimeter = $2r + 2\pi r \times \frac{\theta}{360}$
                  = $(2 \times 7) + \left(2 \times 3.14 \times 7 \times \frac{45}{360}\right)$
                  = 14 + 5.5
                  = 19.5 cm

Note	Unless you are given the value of π to use, you can use the button on your calculator.

5.10 Area

The units for area are called **square units**, for example, cm² (square centimetre), m² (square metre), km² (square kilometre).

Key term	
Area	the size of a plane (flat) surface

You can use formulae to calculate the areas of plane shapes.

Rectangle $A = l \times w = lw$ where l = length and w = width

Square $A = l^2$

Parallelogram $A = bh$ where b = base and h = vertical height

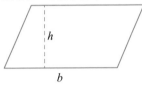

Triangle $A = \frac{1}{2} \times b \times h = \frac{1}{2} bh$ where b = base and h = vertical height

$A = \frac{1}{2} ab \times \sin C$ where a and b are the lengths of two of the sides of the triangle and C is the angle between them

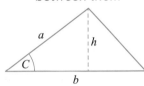

Trapezium $A = \frac{1}{2} h(a + b)$ where a and b are the lengths of the parallel sides and h = vertical height

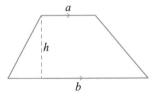

Kite $A = \frac{1}{2} \times p \times q = \frac{1}{2} pq$ where p and q are the diagonals of the kite

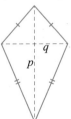

Rhombus $A = \frac{1}{2} \times p \times q = \frac{1}{2} pq$ where p and q are the diagonals of the rhombus

$A = s^2 \sin A$ where s is the length of a side and A is the angle between any two adjacent sides

Formulae for the areas of polygons

Example 31

Find the areas of these shapes (which are the same as those in Example 27).

a
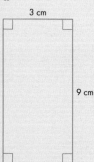
3 cm

9 cm

b
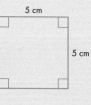
5 cm

5 cm

c

8 cm

6 cm

d
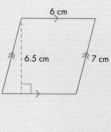
6 cm

6.5 cm   7 cm

Not to scale

e
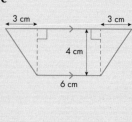
3 cm   3 cm

4 cm

6 cm

f
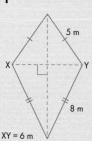
5 m

X   Y

8 m

XY = 6 m

g
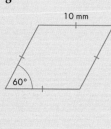
10 mm

60°

h
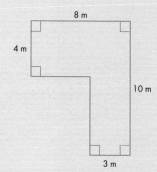
8 m

4 m

10 m

3 m

(a) Area of rectangle = $l \times w$
            = 3×9
            = 27 cm^2

(b) Area of square = $l^2 = 5 \times 5$
            = 25 cm^2

(c) Area of triangle = $\frac{1}{2} bh$

            = $\frac{1}{2} \times 6 \times 8$

            = 24 cm^2

(d) Area of parallelogram = $b \times h$
                = 6×6.5 m^2
                = 39 m^2

(e) Area of trapezium = $\frac{1}{2} h(a + b)$

            = $\frac{1}{2} \times 4 \times (12 + 6)$

            = 2×18

            = 36 cm^2

(f) Area of kite = $\frac{1}{2} (p \times q)$

     XY = horizontal diagonal = $p = 6$ m (as shown in the diagram)

     Use Pythagoras' theorem to calculate the length of q (vertical diagonal).

     Let $q = a + b$ where a is the part above XY and b is the part below XY.

     Using Pythagoras' theorem: $5^2 = a^2 + 3^2$

                    Therefore, $a^2 = 25 - 9$

                            = 16

                    Therefore, $a = 4$

Using Pythagoras' theorem: $8^2 = b^2 + 3^2$

$$\text{Therefore, } b^2 = 64 - 9$$
$$= 55$$
$$\text{Therefore, } b = 7.4$$

$$q = a + b = 4 + 7.4$$
$$= 11.4 \text{ m}$$

$$\text{Area} = \frac{1}{2} \times 6 \times 11.4$$
$$= 34.2 \text{ m}^2$$

(g) Area of rhombus = $s^2 \sin A$ (s is the length of the side, A is the angle between any two adjacent sides)

$$A = 10^2 \sin 60°$$
$$= 100 \times 0.87$$
$$= 87 \text{ mm}^2$$

(h) Area of irregular hexagon

Look at the diagram closely. You can split the hexagon into two rectangles. This is called a **compound shape**.

$$A = \text{area of horizontal rectangle + area of vertical rectangle}$$
$$= (4 \times 8) + (3 \times 6)$$
$$= 32 + 18$$
$$= 50 \text{ m}^2$$

Tip	The formulae for the area of a triangle, a parallelogram, a trapezium and a kite can all be deduced from the formula for the area of a rectangle (lw). This is useful when you can't remember these formulae.

| **Formulae for the area of circles, sectors and segments** | Circle $A = \pi r^2$
 Sector $A = \pi r^2 \times \frac{\theta}{360}$
 Segment A = area of sector OAB – area of isosceles triangle OAB | 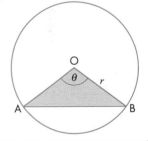 |

Example 32

Find the areas of these shapes (which are the same as in Example 30). Use $\pi = 3.14$.

a
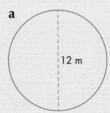
12 m

b
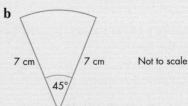
7 cm   7 cm   Not to scale
45°

(a) Area of circle = πr^2

$$A = \pi \times 6^2$$
$$= 113.04 \text{ m}^2$$

(b) Area of sector = $\pi r^2 \times \frac{\theta}{360}$

$$A = \pi \times 7^2 \times \frac{45}{360}$$
$$= 19.2 \text{ cm}^2$$

Exercise 5F

1 Calculate the perimeter and area of each of these shapes.

a

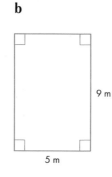

6 cm

10 cm

b

9 m

5 m

c

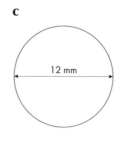

12 mm

Not to scale

d

14 cm

10 cm

8 cm

6 cm

e

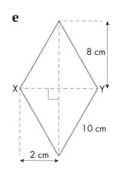

8 cm

X — Y

10 cm

2 cm

2 Calculate the perimeter and area of a rectangular field with a length of 80 m and a width of 60 m.

3 A cyclist rides around a circular track with a diameter of 200 m. Find the distance she travels.

4 Find the lengths of the sides *x* and *y*, the perimeter and the area of the shape shown below.

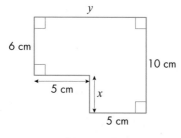

y

6 cm

10 cm

5 cm

x

5 cm

Not to scale

5 A sector of a circle has a radius of 7 cm and angle of 120°. Calculate the perimeter and area of the sector.

6 Calculate the perimeter and area of the trapezium shown below.

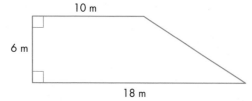

10 m

6 m

18 m

7 The irregular shape is on a grid where each square is taken as 1 cm². Estimate the area of the shape by the method of counting the squares.

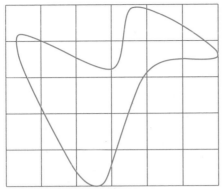

> **Tip**
>
> When estimating the area of the shape, count squares that are more than half-covered as a whole square. Ignore any squares that are less than half-covered.

8 A swimming pool has the shape of a regular hexagon in which each side is 6 m long. How much will it cost to fence the entire swimming pool if the fencing material costs $30.00 a metre?

9 A rectangular shopping mall is 60 m long by 50 m wide with a 2.5 m pavement all the way around. Find the area of the pavement and the number of 50 cm² tiles that will be needed to tile the entire pavement.

10 Find the perimeter and area of the shape shown below.

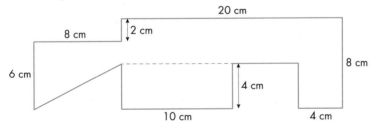

5.11 Margin of error in measurement

For an object measuring 2.8 m:

- the **greatest** possible measurement of the object = 2.8 m + 0.05 m = 2.85 m
- the **smallest** possible measurement of the object = 2.8 m – 0.05 m = 2.75 m
- the **actual** length of the object will be between 2.75 m and 2.85 m
- the **margin of error** is estimated to be: ± 0.05 m.

> **Key term**
>
> **Absolute error of length** half the smallest unit of length. For an object measuring 2.8 m, the smallest unit of length will be 0.1 m and half that unit will be 0.05 m.

5.11.1 Maximum and minimum values of measurement and most accurate results

Example 33

The lengths of three pieces of string, measured to an accuracy of 0.1 cm, are: 10.6 cm, 14.8 cm and 20.3 cm. Find the greatest possible total length, smallest possible total length and most accurate total length of the strings.

The smallest unit of length is 0.1 cm. Therefore, the margin of error is ± 0.05 cm.

The greatest possible total of the lengths:	The smallest possible total of the lengths:
10.6 cm + 0.05 cm = 10.65 cm	10.6 cm − 0.05 cm = 10.55 cm
14.8 cm + 0.5 cm = 14.85 cm	14.8 cm − 0.05 cm = 14.75 cm
20.3 cm + 0.5 cm = 20.35 cm	20.3 cm − 0.05 cm = 20.25 cm
Greatest possible total = 45.85 cm	Smallest possible total = 45.55 cm

Hence, the total of the lengths lies between 45.55 cm and 45.85 cm.

The total of the measured values = 10.6 cm + 14.8 cm + 20.3 cm
$$= 45.7 \text{ cm}$$

The number of significant figures in the actual measurement is 3.

Therefore, the **most accurate answer** = 45.7 cm (correct to 3 significant figures)

Example 34

The dimensions of a window were found to be: 1.52 m by 0.84 m correct to 2 decimal places. What is the area of the window?

Apparent area = 1.52 m × 0.84 m
$$= 1.2768 \text{ m}^2$$

Greatest possible area = 1.525 m × 0.845 m
$$= 1.288625 \text{ m}^2$$

Smallest possible area = 1.515 m × 0.835 m
$$= 1.265025 \text{ m}^2$$

The area lies between the smallest and greatest values shown above.

The most accurate area correct to 3 significant figures will be 1.28 m^2.

Exercise 5G

1 State the combined length of the following measurements correct to the appropriate number of significant figures: 2.9 cm, 8.0 cm and 1.1 cm.

2 Find the greatest possible error and the smallest possible error for each of these measurements: 12.33 kg, 13.90 kg, 50.7 kg and 33.39 kg.

3 Find the smallest and greatest possible values for the area of a rectangular plot of land measuring 20.5 m by 100.0 m each correct to 1 d.p.

4 Find the perimeter of a plot of land that is in the shape of an irregular quadrilateral of sides of lengths: 23.57 m, 30.63 m, 35.89 m, 40.55 m. Write your answer to the appropriate number of significant figures. State also the greatest and smallest possible errors.

5 A trapezium is shown with all dimensions correct to 1 decimal place.

    a Find the value of h.

    b Find the perimeter and area to the appropriate number of significant figures.

    c Determine the maximum and minimum errors for the perimeter and area.

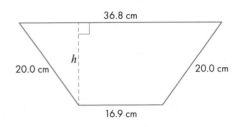

6 Find the area of a circle of radius 4.837 m to the appropriate number of significant figures.

7 A cuboid has dimensions 45.7 mm by 68.2 mm by 800.8 mm. Find the maximum and minimum volume and write the volume to a suitable degree of accuracy.

5.12 Scale drawing

Scale drawings are used to represent a large area (e.g. a country, a place, a car) on paper. A scale helps with the drawing.

The scale shows the ratio between the length on the drawing and the actual length. For example, a scale of 1 : 20 000 on a map means that every 1 unit on the map represents 20 000 units on the actual place.

Key term

Scale drawing an accurate drawing of an object in which all the lengths in the drawing are in the same ratio as the corresponding lengths in the actual object

Example 35

The scale map of St Kitts and Nevis shown below is drawn with a scale of 1 : 800 000. Find the actual distance between Basseterre and Cayon.

On the map, the distance between Basseterre and Cayon is 0.6 cm.

The scale is 1 : 800 000. This means that 1 cm on the map represents 8 km. In a scale, the units must be the same on either side of the ratio symbol :

8 km = 800 000 cm

Therefore, to find the actual distance, multiply the distance on the map by 800 000.

The actual distance is
0.6 × 800 000 cm = 480 000 cm
              = 4.8 km

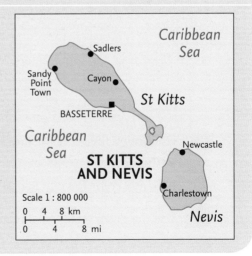

Example 36

Using the scale drawing below, find the ratio of the actual length of the horse's tail to the drawn horse. The photograph and the drawing are not to scale.

Real horse 2500 mm high　　　　　**Drawing of horse 250 mm high**

The scale of the real horse to the image is 2500 : 250 = 10 : 1.

Therefore, the ratio of the actual length of the tail to the drawn tail will also be 10 : 1.

Exercise 5H

Questions 1 to 4 relate to the map of St Kitts and Nevis (scale 1 : 800 000).

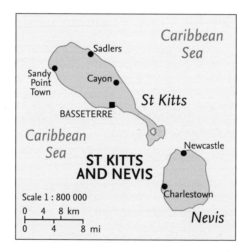

1　Determine the actual distance from:

a　Charlestown to Newcastle

b　Sandy Point Town to Saddlers.

2　Use the given scale to estimate the area of Nevis.

3　Use the given scale to estimate the area of St Kitts.

4　Use the given scale to determine the perimeter of both Nevis and St Kitts.

Questions 5 to 8 relate to the map shown, where the scale can be taken as 1 : 1 000 000.

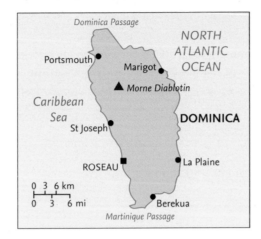

5 Measure the scale drawing distance between Roseau and St Joseph. Hence, determine the actual distance from Roseau to St Joseph.

6 Make an estimate of the perimeter and area of Dominica.

7 What is the actual distance between Portsmouth and Marigot?

8 Estimate the width (east–west) and length (north–south) of Dominica.

9 The diagram shows the plan of a lounge room. The scale used is 1 : 20. Each square on the plan is 1 cm wide.

  a Find the area of the floor, in m².

  b State the number of windows and doors.

  c What is the width of each window and door on the plan, in centimetres?

  d What is the actual width of each window and door?

  e How many 30 cm square tiles will be needed to tile the floor?

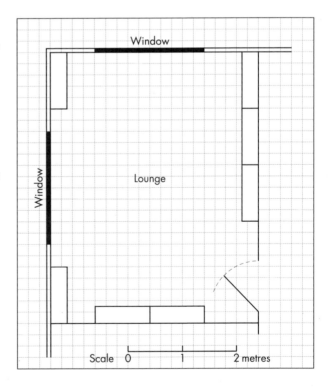

5.13 Volume

Relationship between units of capacity	1000 cm³ (or cc) = 1000 ml = 1 litre

Did you know?	A teaspoon will hold approximately 5 ml of liquid while a water tank may hold 220 000 litres (50 000 gallons).

5.13.1 Volume of prisms and cylinders

Examples of prisms are cubes, cuboids and triangular prisms.

cube

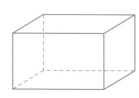
cuboid

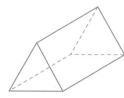
triangular prism

This is an example of a cylinder.

cylinder

You can use formulae to calculate the volumes of prisms and cylinders.

General formula for the volume of a prism or a cylinder	volume of a prism = area of base × height $V = A_b \times h$

The shape of the base depends on the type of prism or cylinder:

- cube – square
- cuboid – rectangle
- triangular prism – triangle
- cylinder – circle.

Formulae for the volumes of different prisms and cylinders			
Cube	$V = l^3$	where l is the length of an edge	
Cuboid	$V = l \times w \times d$	where l = length of cuboid, w = width of cuboid and d = depth of cuboid	
Triangular prism	$V = \frac{1}{2}b \times h \times l$	where b = base of triangular face, h = vertical height of triangular base and l = length of prism	
Cylinder	$V = \pi r^2 h$	where r = radius of circular face and h = height of cylinder	

Example 37

Calculate the volume of a cube with edges of length 5 cm.

Use the formula for the volume of a cube.

$V = l^3 = 5^3 = 125$ cm³

Example 38

(a) Name the solid in the diagram.

(b) Calculate the volume of the solid (not drawn to scale).

Angles AED and BFC are each 90°. BC = AD = 10 cm and
AE = BF = 8 cm.

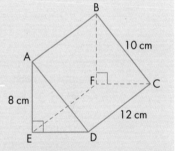

(a) Triangular prism

(b) First, use Pythagoras' theorem to find DE (b in the formula):

$AD^2 = DE^2 + AE^2$

$(10)^2 = DE^2 + (8)^2$

$100 = DE^2 + 64$

$DE^2 = 100 - 64$

$\qquad = 36$

$DE = \sqrt{36}$

$\qquad = 6$ cm

Now substitute the value of DE into the formula for the volume of a triangular prism:

$V = \frac{1}{2}b \times h \times l$

$\quad = \frac{1}{2} \times 6 \times 8 \times 12$

$\quad = 288$ cm³

Example 39

A cylindrical drum has a diameter of 2 metres and a height of 5 metres. What is the maximum amount of sand it can hold, in cm³? Use π = 3.14.

Convert 5 m and 2 m to cm.

5 m = 500 cm

2 m = 200 cm

radius = $\frac{1}{2}$ × diameter

$\qquad$ = $\frac{1}{2}$ × 200

$\qquad$ = 100 cm

Now use the formula for the volume of a cylinder.

$V = \pi r^2 h$

$\quad$ = 3.14 × 100² × 500

$\quad$ = 15 700 000 cm³

The maximum amount of sand the drum can hold is 15 700 000 cm³

Hint　All dimensions must be in the same units before you begin calculating.

Exercise 5I

Use π = 3.14 in this exercise.

1. A cardboard box has a rectangular base that measures 80 cm by 60 cm. The depth of the box is 120 cm. Find the volume of the box. Give your answer in m³.

2. A student, drinking milk from a can, wants to find out how much milk the can contains, in cm³. Given that the can is 15 cm tall and has a diameter of 10 cm, what is the capacity of the can?

3. A carpenter needs to calculate the volume of a wooden wedge that is shaped as a triangular prism. Given that the base of the prism is an equilateral triangle of sides 8 cm and the height is 20 cm, make a sketch of the wedge and find its volume.

4. Find the total volume of the compound solid shown in the diagram.

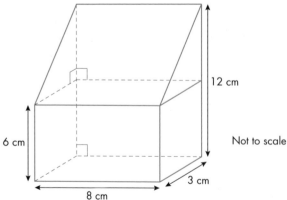

12 cm

6 cm

Not to scale

3 cm

8 cm

5 A block of modelling clay has the shape of a cuboid of base 6 cm by 14 cm and height 10 cm. (See diagram.) A hole in the shape of a wedge with a triangular base, of dimensions 3 cm by 4 cm by 5 cm, is punched through the block from the centre of the base to the top.

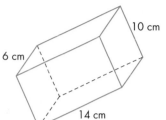

a Find the volume of modelling clay remaining.

b If this modelling clay is now reshaped to form a triangular-based prism of height 5 cm, determine the area of the base of the prism.

 6 A mason designs a concrete rectangular sink with a hole in the middle of the base to drain out the water. The external dimensions of the sink are: base 2 m by 3 m with a height of 5 m. The internal dimensions are: base 1.8 m by 2.8 m and height 4.9 m. The hole has a diameter of 6 cm.

a Make a sketch of the sink.

b What is the thickness of the walls?

c Find the volume of the concrete used to make the sink.

d Find the capacity of the sink, in m^3.

> **Tip**
> You can work out the volume of a compound shape by first finding the volumes of the component shapes.

5.13.2 Volume of pyramids and cones

Unlike prisms, pyramids and cones do not have a constant cross-section. Each has a vertex.

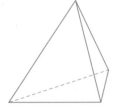
triangular-based pyramid

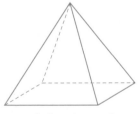
rectangular-based pyramid

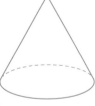
cone

The volume of a pyramid is one-third the volume of a prism with the identical base and height.

General formula for the volume of a pyramid	volume of a pyramid or cone $= \frac{1}{3}$ area of base × height $$V = \frac{1}{3} \times A_b \times h$$

The shape of the base depends on the type of pyramid:

- square-based pyramid – square
- triangular-based pyramid – triangle
- hexagonal-based pyramid – hexagon.

Cone

$V = \frac{1}{3}(\pi r^2) \times h$ where r = radius of circular base and h = vertical height of the cone

Triangular-based pyramid

$V = \frac{1}{3}(\frac{1}{2} \times b \times l) \times h$ where b = base of the triangular base and l = vertical height of the triangular base, and h = vertical height of the pyramid

$V = \frac{1}{3}(\frac{1}{2} ab \sin C) \times h$ where a and b are two sides of the triangular base and C is the angle between them, and h = vertical height of the pyramid

Rectangular-based pyramid

$V = \frac{1}{3}(l \times w) \times h$ where l = length of the rectangular base, w = width of the rectangular base and h = vertical height of the pyramid

Formulae for the volumes of a cone and different pyramids

Example 40

A cone has a base of radius 6 cm and height 10 cm and a square-based pyramid has a base with edges of length 10 cm and height 11 cm. Determine by calculation which of the two shapes has the greater volume and the difference between the volumes. Use $\pi = 3.14$.

Volume of cone $= \frac{1}{3}(\pi r^2)h$

$\qquad\qquad = \frac{1}{3} \times \pi \times 6^2 \times 10$

$\qquad\qquad = 377$ cm^3

Volume of square-based pyramid $= \frac{1}{3} \times l^2 \times h$

$\qquad\qquad\qquad\qquad\qquad\qquad = \frac{1}{3} \times 10^2 \times 11$

$\qquad\qquad\qquad\qquad\qquad\qquad = 367$ cm^3

The cone has a greater volume than the pyramid.

The difference in volume is: $377 - 367 = 10$ cm^3.

Example 41

The base of a pyramid is an isosceles triangle with two equal sides of length 30 mm and an angle of 80° between them. Its volume is 400 mm^3. Calculate the height of the pyramid.

Volume of pyramid $= \frac{1}{3}(\frac{1}{2} ab \sin C) \times h$

$\qquad\qquad 400 = \frac{1}{3}(\frac{1}{2} \times 30 \times 30 \times \sin 80°) \times h$

$\qquad\qquad\quad h = 400 \times \frac{3}{443}$

$\qquad\qquad\qquad = 2.7$ mm

Exercise 5J

Use π = 3.14 in this exercise.

1 Calculate the volumes of these solids.

  a A rectangular-based pyramid of base 30 cm by 40 cm and height 60 cm

  b A cone of base diameter 2 m and height 5 m

2 A boy wishes to make himself a party hat in the shape of a cone. Assuming his head to be roughly the shape of a sphere he measures the circumference and finds it to be 40 cm. If the required height of the hat is 20 cm, calculate the volume of the birthday hat.

3 A pyramid has a square base of sides 6 cm and a **slant** height of 5 cm. Find its volume.

4 The diagram shows a compound shape made of two pyramids. The pyramids have identical square bases but different heights. Find the total volume of the shape.

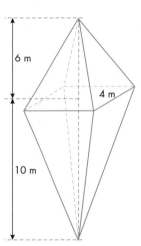

5 The diagram shows an hour glass in the shape of two identical cones.

  a What is the combined volume of the cones?

  b At the beginning of the hour, the top cone is full of sand. What will the volume of sand in the bottom cone be after 20 minutes if it flows at 17 cm³ per minute?

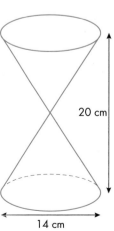

6   A cone and a square-based pyramid have the same volume but the height of the cone is three times that of the pyramid. Show that the relationship between the radius (r) of the base of the cone and the side (l) of the base of the pyramid is given by $r = \dfrac{l}{\sqrt{3\pi}}$.

7   The diagram shows a conical flask. Find the volume of liquid it contains when it is filled to the brim.

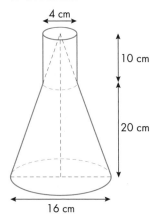

5.13.3 Volume of spheres

A sphere is a perfectly symmetrical three-dimensional shape. It has one curved surface on which every point is the same distance from a fixed point representing the centre of the sphere. A ball is considered to have a spherical shape.

Volume of a sphere	$V = \frac{4}{3}\pi r^3$ where r = radius of the sphere

Example 42

Calculate the volume of a sphere of radius 10 cm. Use $\pi = 3.14$.

$V = \frac{4}{3}\pi r^3$

$= \frac{4}{3} \times 3.14 \times 10^3$

$= 4187 \text{ cm}^3$

Example 43

A football has a circumference of 70 cm. Find its volume. Use $\pi = 3.14$.

First, find the radius of the sphere.

Circumference $= 2\pi r$

$70 = 2 \times 3.14 \times r$

Therefore, $r = \dfrac{70}{6.28}$

$ = 11.1 \text{ cm}$

Now, substitute the value for r into the volume formula:

$V = \frac{4}{3}\pi r^3$

$V = \frac{4}{3} \times 3.14 \times (11.1)^3$

$ = 5726 \text{ cm}^3$

Exercise 5K

Use π = 3.14 in this exercise.

1 Calculate the volume of a spherical cricket ball. The ball has a radius of 5 cm.

2 A hemispherical dome has a radius of 30 m.
Calculate the amount of space in the dome, in m³.

> **Key term**
>
> **Hemisphere**   half a sphere

3 Use the volume formulae to show that the volume of a cone of radius 8 cm and height 16 cm is half the volume of a sphere of diameter 16 cm.

4 A house consists of a cylinder of diameter 20 m and height 4 m and a hemisphere of the same diameter.

   a Draw a sketch of the house.

   b Find the volume occupied by the house.

5 A sphere of radius 20 cm is placed inside a cylinder of radius 40 cm and height 20 cm. What space, in cm³, remains in the cylinder?

6 A boy purchases an ice cream cone that is filled to the brim. He also receives a full scoop in the shape of a sphere of radius 3 cm. The cone shell has the same radius as the sphere and a height of 12 cm.

   a Find the total volume of ice cream received by the boy.

   b Does the cone hold more or less than a full scoop?

   c State the difference between the volume in (a) and the full scoop.

7 A container in the shape of a rectangular-based pyramid with internal dimensions of base 40 cm by 80 cm and height 100 cm is filled with tennis balls, each of diameter 14 cm. What is the maximum number of balls that can fit in the container and what is the quantity of empty space?

> **Tip**
> - Slant height is not the same as vertical height. It is usually a hypotenuse, the length of which can be found by using Pythagoras' theorem.
> - Drawing a sketch can help you to visualise solids and solve the problem.

5.14 Surface areas of solids

You use the net of a 3D shape to find the surface area of the shape. The total surface area of the shape is the total of the areas of all the faces.

Key terms

Surface area total area of the exterior surface of a solid, expressed in square units

Net a 2D shape made of polygons, which can be folded up to make a 3D shape

Surface area of a cube	A cube has 6 identical square faces. The diagrams show a cube and examples of the nets of a cube. 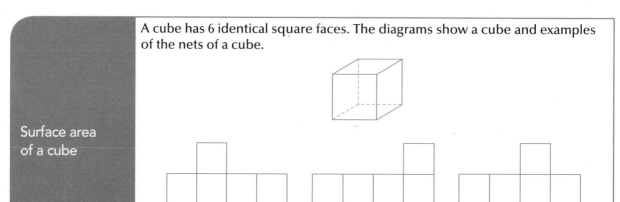 surface area of a cube = $6l^2$ where l is the length of an edge of the cube
Surface area of a cuboid	A cuboid has three pairs of identical rectangular faces. 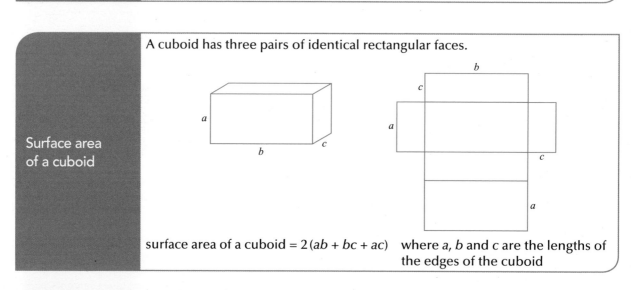 surface area of a cuboid = $2(ab + bc + ac)$ where a, b and c are the lengths of the edges of the cuboid
Surface area of a triangular-based prism	A triangular-based prism has 5 faces: 3 rectangles and 2 identical triangles. 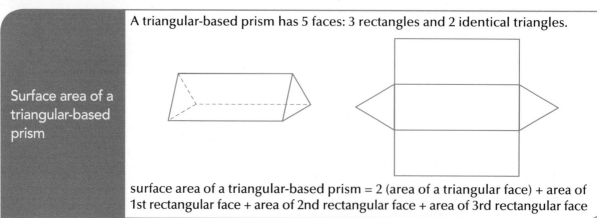 surface area of a triangular-based prism = 2 (area of a triangular face) + area of 1st rectangular face + area of 2nd rectangular face + area of 3rd rectangular face

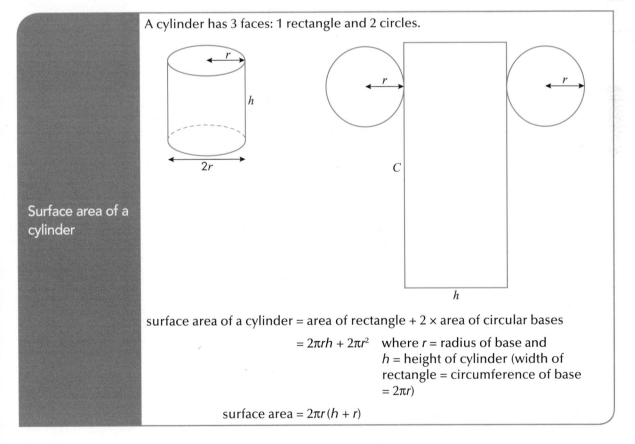

Surface area of a cone

A cone has 2 faces: a sector and a circle.

$C = 2\pi r$ = length of arc

surface area of a cone = area of sector + area of circle

$$= \pi r l + \pi r^2 \quad \text{where } l = \text{slant height of cone}$$
$$\text{and } r = \text{radius of the base}$$

Surface area of a cylinder

A cylinder has 3 faces: 1 rectangle and 2 circles.

surface area of a cylinder = area of rectangle + 2 × area of circular bases

$$= 2\pi r h + 2\pi r^2 \quad \text{where } r = \text{radius of base and}$$
$$h = \text{height of cylinder (width of}$$
$$\text{rectangle} = \text{circumference of base}$$
$$= 2\pi r)$$

surface area $= 2\pi r (h + r)$

Surface area of a square-based pyramid	A square-based pyramid has 5 faces: 1 square and 4 identical triangles 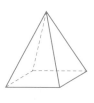 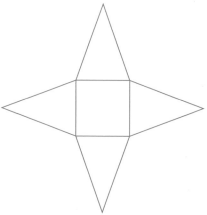

surface area = area of square base + 4 × area of triangular face

$$= l^2 + 4 \times (\tfrac{1}{2}\, lh) \quad \text{where } l = \text{length of side of square and}$$
$$h = \text{vertical height of pyramid}$$

$$= l^2 + 2lh$$

Surface area of a sphere	A sphere has 1 curved surface. It can be shown through paper folding and calculation that: surface area of a sphere = $4\pi r^2$

Example 44

Calculate the surface area of a cube of edge length 10 cm.

surface area = $6 \times l^2$
$$= 6 \times 10^2$$
$$= 6 \times 100$$
$$= 600 \text{ cm}^2$$

Example 45

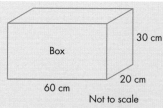

30 cm

Box

20 cm

60 cm

Not to scale

Find the surface area of the box in the diagram. The base of the box measures 20 cm by 60 cm and the depth is 30 cm.

surface area of a cuboid = sum of the area of all faces
$$= 2 \times (20 \times 60 + 20 \times 30 + 60 \times 30)$$
$$= 7200 \text{ cm}^2$$

Example 46

Calculate the area of the paper needed to make a cone-shaped hat with slant height 25 cm and radius 10 cm. Use $\pi = 3.14$.

surface area = $\pi rl + \pi r^2$
$$= 3.14 \times 10 \times 25 + 3.14 \times 10^2$$
$$= 1099 \text{ cm}^2$$

Example 47

The diagram shows a triangular-based prism. Calculate its surface area.

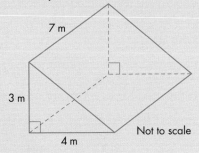

7 m

3 m

4 m

Not to scale

surface area = sum of the areas of the faces

Using Pythagoras' theorem, hypotenuse of the triangular base = $\sqrt{3^2 + 4^2} = 5$

surface area = $2 \times (\frac{1}{2} \times 4 \times 3) + (7 \times 3) + (7 \times 4) + (7 \times 5)$

$= 12 + 21 + 28 + 35$

$= 96 \text{ m}^2$

Exercise 5L

Use $\pi = 3.14$ in this exercise.

1 Find the surface area of each solid shown in the diagram. The diagrams are not to scale.

a

9 cm

5 cm

4 cm

b

4 cm   5 cm

10 cm

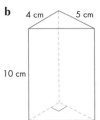

c

14 cm

12 cm

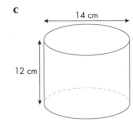

d

8 mm

20 mm

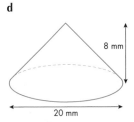

2 A sphere has a surface area of 400 cm². What is its radius?

3 A woman wants to wrap a shoe box with gift wrap. Her paper measures 40 cm by 35 cm. The box has dimensions: base 12 cm by 16 cm and height 20 cm.

   a Will the paper be large enough?

   b By how much is the area of her paper larger or smaller than the surface area of the box?

4 Find the surface area of a cylindrical drum of height 3 m and base radius 1 m.

5 A square-based pyramid has a total surface area of 400 cm². Given that the base of the pyramid is a square of side length 10 cm, find the height of the pyramid.

6 The diagram shows a trough in the shape of a triangular prism.

Using the dimensions given in the diagram, and taking ABF and CDE to be isosceles triangles of equal sides of 8 cm, find the surface area of the prism.

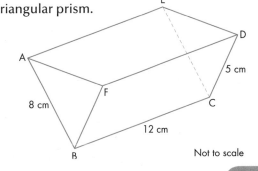

E

D

A

5 cm

F

8 cm

C

12 cm

B

Not to scale

7 An ice cream cone containing one scoop of ice cream has the combined shape of a hemisphere fitting exactly on top of a cone. The radius of the cone is 4 cm and the slant height is 10 cm.

   a Make a sketch of the shape.

   b Find the total surface area of the compound shape.

8 **a** Name the three solids that make up the compound shape shown in the diagram.

   b Find the volume and surface area.

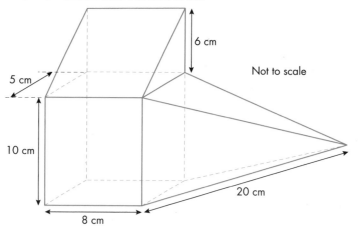

5 cm

6 cm

Not to scale

10 cm

8 cm

20 cm

Hints

- A closed container will not have the same surface area as the same open container
- The height of an isosceles triangle can be found by using Pythagoras' theorem.

End of chapter summary

Units

Prefix	Meaning
kilo-	one thousand
deci-	one tenth
centi-	one hundredth
milli-	one thousandth
micro-	one millionth

- weight = mg
- speed = distance travelled ÷ total time taken
- density = $\frac{\text{mass}}{\text{volume}}$

Perimeter formulae

- perimeter of a rectangle = $2(l + w)$
- perimeter of a regular polygon = $n \times l = nl$
- circumference of a circle = $2\pi r$
- perimeter of a sector = 2 × radius + length of arc
- perimeter of a segment = length of chord + length of arc

Area formulae

- area of a rectangle = $l \times w = lw$ where l = length and w = width
- area of a square = l^2
- area of a parallelogram = bh where b = base and h = vertical height
- area of a triangle = $\frac{1}{2} \times b \times h = \frac{1}{2} bh$ where b = base and h = vertical height
- area of a triangle = $\frac{1}{2} ab \times \sin C$ where a and b are the lengths of two of the sides of the triangle and C is the angle between them
- area of a trapezium = $\frac{1}{2} h(a + b)$
- area of a kite = $\frac{1}{2} \times p \times q = \frac{1}{2} pq$
- area of a rhombus = $\frac{1}{2} \times p \times q = \frac{1}{2} pq$
- area of a rhombus = $s^2 \sin A$
- area of a circle = πr^2
- area of a sector = $\pi r^2 \times \frac{\theta}{360}$
- area of a segment = area of sector OAB – area of isosceles triangle OAB

Margin of error

- Absolute error of length is half the smallest unit of length.

Scale drawing

- If the scale on a map is 1 : 20 000, then to find the actual distance, multiply the distance on the map by 20 000.

Volume

- $1000 \text{ cm}^3 = 1000 \text{ ml} = 1 \text{ litre}$
- volume of a prism = area of base × height
- volume of a cube = l^3
- volume of a cuboid = $l \times w \times d$
- volume of a triangular prism = $\frac{1}{2} b \times h \times l$
- volume of a cylinder = $\pi r^2 h$
- volume of a cone = $\frac{1}{3} (\pi r^2) \times h$
- volume of a triangular-based pyramid = $\frac{1}{3} (\frac{1}{2} \times b \times l) \times h$
- volume of a triangular-based pyramid = $\frac{1}{3} (\frac{1}{2} ab \sin C) \times h$
- volume of a rectangular-based pyramid = $\frac{1}{3} (l \times w) \times h$
- volume of a sphere = $\frac{4}{3} \pi r^3$

Surface area

- surface area of a cube = $6l^2$
- surface area of a cuboid = $2 (ab + bc + ac)$
- surface area of a cone = $\pi r l + \pi r^2$
- surface area of a cylinder = $2\pi r h + 2\pi r^2$
- surface area of a square-based pyramid = $l^2 + 2lh$
- surface area of a sphere = $4\pi r^2$

Examination-type questions for Chapter 5

Use $\pi = 3.14$ in this exercise.

1. A vehicle passes a town A at 05:00 with a constant speed of 40 km h^{-1}. It maintains this speed for 2 hours and 20 minutes until it stops at another town B for 40 minutes. The vehicle then travels to another town C, 50 km away, in 1 hour and 25 minutes. From C it travels to its final destination D, which is 60 km from C. It arrives in 80 minutes.

   a. Using the 24-hour clock, state the time of arrival at D.

   b. What was the total distance travelled by the vehicle?

   c. Find the time taken for the entire journey.

   d. Find the average speed of the vehicle in km h^{-1}.

   e. Express the distance travelled by the vehicle in metres. Write your answer in standard form.

2. The map below shows a scale drawing of Grenada, Carriacou and Petit Martinique on a grid of 1 cm squares. The scale of the map is 1 : 1 000 000 (1 cm : 10 km).

   a. Find, in centimetres, the distance from Saint George's to Grenville.

   b. Find by counting the squares the area occupied by the map of Grenada, in cm^2.

   c. Find the distance in centimetres from Hillsborough in Carriacou to Saint George's in Grenada.

   d. Calculate the **actual combined distance**, in kilometres, from Saint George's to Grenville and Hillsborough to Saint George's.

   e. Find the **shortest possible distance**, in kilometres, between Grenada and Carriacou.

   f. Find the **actual** distance, in kilometres, from Petit Martinique to Carriacou.

   g. Find the **actual combined area** of Grenada, Carriacou and Petit Martinique in km^2 and (miles)^2.

3. The diagram shows the base of a bathtub. It comprises a rectangle with a semi-circle at each end (not drawn to scale). The rectangle measures 3 m by 2 m as shown in the diagram. The bathtub is 1 m deep and it is filled with water to a height of 0.8 m. Calculate the volume of water in m^3. Find the internal surface area of the bathtub when empty.

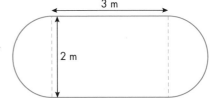

4. The diagram represents a prism with a square base PQRS of side 8 cm. The total surface area of the prism is 1280 cm^2. Calculate the length of the edge RV and the volume of the prism.

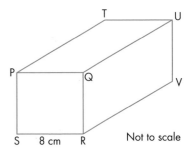

5 Three cylindrical water tanks A, B and C are filled with water. Tank A has a radius of 5 m and holds 558 m³ of water. Tank B is twice the height of tank A and has a diameter of 8 m and tank C has a height of 6 m and twice the radius of B.

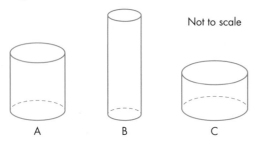

Not to scale

a Find the volume of each tank.

b Which tank has the greatest volume?

6 A piece of cheese was cut from a cylindrical block of cheese leaving the shape shown in the diagram. The original cylindrical block was 14 cm long with a radius of 6 cm.

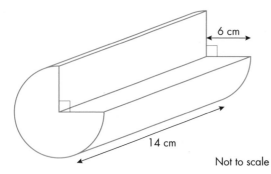

6 cm

14 cm

Not to scale

a Find the volume of the remaining cheese.

b Calculate the surface area of the remaining piece of cheese.

7

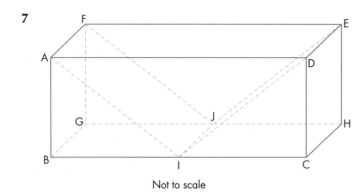

Not to scale

The diagram shows a box (cuboid) containing two identical wedges DEHCIJ and AFGBIJ. I is the midpoint of BC and J is the midpoint of GH. Calculate the volume of one wedge and the volume of empty space in the box (ADEFJI).

Note: BC = 100 cm, AF = 30 cm and AB = 40 cm

8

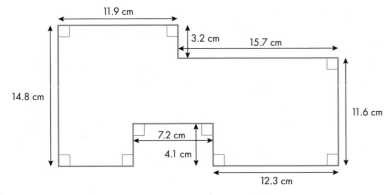

Not to scale

The diagram shows the shape of a plot of land.

a Find the greatest and smallest possible perimeters of the land.

b Find the most accurate perimeter with the correct number of significant figures.

c Find the greatest and smallest possible area.

d Find the most accurate area.

e Each tree must occupy an area of 2 m². How many trees can be planted on the land?

CHAPTER 6

Sets

Objectives

By the end of this chapter you should be able to:

- explain concepts relating to sets, give examples and non-examples of sets and provide a word description of sets
- understand set concepts: membership of a set, cardinality of a set, finite and infinite sets, universal set, empty set, complement of a set, subsets
- represent a set in various ways:
  - by description, for example, 'the set A comprises the first three natural numbers'
  - by set builder notation, for example, $A = \{x: 0 < x < 4, x \in \mathbb{N}\}$
  - by listing: $A = \{1, 2, 3, \ldots\}$
- identify and list subsets of a given set as well as determine the number of subsets of a set with n elements
- determine the number of elements in intersections, unions and complements of not more than three sets and to apply the results
- use set notation and symbols to describe relationships among sets, such as universal set, complement, subsets, equal and equivalent sets, intersection, disjoint sets and union of sets
- draw Venn diagrams to represent relationships among sets (not more than four sets, including the universal set)
- use Venn diagrams to represent the relationships among sets, including the universal set
- solve problems in number theory, algebra and geometry, using set theory concepts.

You should already know:

- number theory and computation
- algebra
- geometry and trigonometry
- measurement
- fractions and decimals.

Introduction

Since ancient times, humans have naturally categorised or grouped things according to some common criteria. For example, ancient civilisations hunted certain types of animal for food, made certain weapons and used certain types of utensil. As civilisation developed, other classifications or groups emerged, such as different classes of people in society, different types of transport, buildings, foods, weapons, clothing, political systems – the list can go on and on.

A set is a collection or grouping of objects. The concept of sets was developed in 1874 by German mathematician, Georg Cantor and is considered to be fundamental in mathematics. Set language can be used to explain almost every mathematical concept:

- in algebra, you can analyse different sets of equations
- in geometry, two-dimensional and three-dimensional shapes
- in trigonometry, types of angle and sets of ratios
- in measurement, formulae for finding areas and volumes of various shapes
- in statistics. the analysis of various sets of data.

In addition to their obvious importance in the development of mathematics, sets have wide everyday application – whether consciously or subconsciously, for example, sorting clean laundry neatly into appropriate categories, organising kitchen utensils in sets (plates, spoon, knives…), a student's school bag with different compartments for different sets of items. You can probably think of many more everyday examples.

This chapter on sets will help you to develop and use the language of sets to communicate, analyse and solve everyday problems in a logical manner.

Examples of sets in real life

A cutlery set

A set of clothing for a young man

A set of musical instruments

A set of people

6.1 Definition of a set and related concepts

6.1.1 Definition of set

A **set** is a collection of things, such as a shelf of shoes, the contents of a school locker or the even numbers up to 12. The common property might be that they are all red, or they are all items of sports kit, or that they are all even numbers.

> **Key term**
>
> A **set** is a collection or grouping of objects, called **elements** or **members** of the set, that have some common property.

> **Note**
>
> The **elements** of a set may be anything, such as colours, numbers, letters, people or items of clothing.

Non-examples of sets

It is very difficult to find a **non-example** of a set in terms of the elements since there is always likely to be some property that links them, no matter how obscure. However, a set of colours given as 'red, green, blue, colours' is a non-example since 'colours' cannot be listed as a member of the set of colours.

6.1.2 Description of a set

A set is denoted by an italic capital letter and is usually specified or represented by listing or by description.

Listing

The elements or members of the set are written as a list, inside curly brackets, not necessarily in any particular sequence or order.

> **Example 1**
>
> List the set of natural numbers from 1 to 3.
>
> By listing, $A = \{1, 2, 3\}$

Description of set

The set is described in words and sentences.

> **Example 2**
>
> Describe the set $A = \{1, 2, 3\}$.
>
> A is the set of natural numbers from 1 to 3.

6.1.3 Membership of a set

The symbol for membership of a set is $\in$ and the symbol to indicate non-membership is $\notin$.

Example 3

Given the set $P = \{2, 4, 6\}$, describe 4 and 5 in relation to P.

Since 4 is an element of P, write $4 \in P$.

However, 5 is not a member of P and this can be written as $5 \notin P$.

6.1.4 Cardinality of a set

If it is possible to count the members of a set, the number of members is the **cardinality** of the set.

The cardinality of set A in example 2, for example, is denoted as $n(A)$ and $n(A) = 3$.

Example 4

Determine the cardinality of the set $B = \{2, 4, 6, 8, 10\}$.

$n(B) = 5$

Key term

The **cardinality** of a set is the number of members there are in the set.

6.1.5 Types of set

Finite and infinite sets

Although sets such as the books on a shelf may have a countable number of elements, there are other sets, such as the counting numbers, that have an uncountable number of elements. The books on a shelf form a **finite** set, as you could count the books. The counting numbers form an **infinite** set, as it is impossible to list or count all of its members.

Example 5

Suggest a finite set.

{The set of letters in the alphabet} = {a, b, c , d, …, x, y, z} is a finite set since its members are countable.

Key terms

A **finite** set is a set in which all the elements can be listed. An **infinite** set is a set in which the elements cannot all be listed or counted.

Example 6

Suggest an infinite set.

{The set of natural numbers} = {1, 2, 3, 4, …} is an infinite set.

Note

The three dots indicate that there are missing elements. If they appear in the middle of a set, for example, in {1, 2, 3, …, 99, 100} they indicate the numbers 4–98. If they appear at the end, they generally indicate that the sequence goes on infinitely.

Null or empty set

So far you have considered sets that have at least one member, but it is possible for sets to have no members, for example, the set of even prime numbers greater than 3. Such a set is a **null** or **empty** set.

> **Key term**
>
> The **null** or **empty** set is a set with no elements. It is represented by { } or ∅.

> **Note**
>
> The set {0} is not an empty set, since it has the member zero.

Example 7

Suggest an empty set.

{A set of weeks with 8 days} = { } = ∅

Well-defined sets

When you describe a set, it is important that you make clear exactly what characterises the members of the set. Your description must be sufficiently specific that there is no doubt, in which case your set will be a **well-defined** set.

> **Key term**
>
> A set is **well defined** if it is perfectly clear what elements belong to the set.

Example 8

Suggest a well-defined set.

{the first four letters of the alphabet} = {a, b, c, d} is a well-defined set.

However, {a set of letters} is not well defined since it does not specify which letters are included.

Sets of numbers

The sets of numbers are defined as:

- **natural numbers** (counting numbers): {1, 2, 3, …}
- **whole numbers**: {0, 1, 2, 3, …}
- **integers**: negative and positive whole numbers and zero: {…, –3, –2, –1, 0, 1, 2, 3, …}
- **rational numbers**: integers together with numbers that may be expressed as $\frac{p}{q}$, where p and q are both integers, including fractions, terminating and repeating decimals, for example, $\left\{\frac{3}{4}, \frac{1}{3}, 5, …\right\}$
- **irrational numbers**: decimals that cannot be expressed as $\frac{p}{q}$, where p and q are both integers, and neither terminate nor include a group of repeating digits and non-repeating decimals, for example, $\left\{\pi, e, \sqrt{2}\right\}$
- **real numbers**: all rational and irrational numbers, for example, $\left\{…, -3, 0, \frac{1}{2}, \pi, 8, …\right\}$

Equal and equivalent sets

If two finite sets can be matched, element for element, so that they are identical, then they have the same cardinality and they are **equal** sets.

> **Key term**
>
> Two or more sets are **equal** if they contain exactly the same elements.

When comparing sets, the order in which the elements are listed and the repetition of elements are inconsequential.

Give examples of equal and unequal sets.

$P = \{1, 2, 3, 4\}$, $Q = \{3, 4, 1, 2\}$, $R = \{a, b, c, d\}$

$P = Q$ (P and Q are equal sets.)

$P \neq R$ (P and R are unequal sets.)

If two finite sets have the same cardinality but their elements do not match, one for one, they are **equivalent** sets.

Key term

Two or more sets are **equivalent** if they have the same cardinality (same number of elements). The individual elements may be different but the number of elements must be the same.

Give examples of equivalent sets.

$J = \{1, 2, 3, 4\}$, $K = \{2, 4, 6, 8, 10\}$, $M = \{a, b, c, d\}$

J is equivalent to M. (They have the same number of elements.)

Subsets

The members of a set may be split up or divided, arbitrarily or according to some rule. The smaller sets thus formed are **subsets** of the initial set.

Key term

A **subset** is any set that is contained in, or can be obtained from, another set. Every set has itself and the null (empty) set as subsets.

Find all the subsets that can be obtained from the set $K = \{1, 2, 3\}$.

The subsets are: $\{1, 2, 3\}$, $\{1, 2\}$, $\{1, 3\}$, $\{2, 3\}$, $\{1\}$, $\{2\}$, $\{3\}$, $\{\ \ \}$.

You can see that there are 8 subsets that can be formed from set K, which contains 3 elements.

Key fact

The **number of subsets** (N) in a set can be determined from the formula: $N = 2^n$ where n is the number of elements in the set.

How many subsets can be obtained from the set $W = \{a, b, c, d, e\}$?

The number of subsets is given by the equation: $N = 2^n$.

Since $n = 5$, $N = 2^5 = 32$.

The number sets can be written as a sequence of nested subsets (one inside another):

natural numbers $\subset$ whole numbers $\subset$ integers $\subset$ rational numbers $\subset$ real numbers

where the symbol $\subset$ means 'is contained in' or 'is a subset of'.

Using symbols: $\mathbb{N} \subset \mathbb{W} \subset \mathbb{Z} \subset \mathbb{Q} \subset \mathbb{R}$

Also, irrational numbers ($\mathbb{I}$) $\subset$ real numbers ($\mathbb{R}$)

Universal sets

Any set may be considered to be part of a larger set, called its **universal set**. Although every set is therefore a subset of its universal set, it is not conventional to describe the sets you are considering as such.

Examples of universal sets include:

- A = {all fruits}
- O = {all shapes}
- M = {all animals}

> **Key term**
>
> A **universal set** is one that contains all the members of a particular group or class of sets. Generally, it is a large set from which smaller sets and subsets can be obtained.

Complement of a set

Every set is part of a universal set. It follows therefore that some of the members of the universal set will not be members of that set. These members of the universal set form the **complement** of that set.

> **Key term**
>
> The **complement** of a set comprises all the elements not included in that set. The complement of set A is written as A'.

Example 13

Given the universal set $\mathbb{N}$ = {natural numbers from 1 to 10} and A = {even numbers between 1 and 10} find A', the complement of A.

$A = \{2, 4, 6, 8\}$

The complement of A comprises the elements that are in the universal set, $\mathbb{N}$, that are not included in A. These elements are 1, 3, 5, 7, 9, 10.

$A' = \{1, 3, 5, 7, 9, 10\}$

Set builder notation

Set builder notation is a way of describing the members of a set in a concise way.

> **Key term**
>
> In **set builder notation**, a set is identified by stating the conditions satisfied by the elements or members of the set.

Example 14

State the set that is described as $P = \{x: x^2 = 25\}$ in set builder notation.

Read $P = \{x: x^2 = 25\}$ as 'P is the set x such that $x^2 = 25$.'

The members of P are the solutions to the equation: $x^2 = 25 \Rightarrow x = +5, -5$.

$P = \{5, -5\}$

> **Note**
>
> Note that in this context, the symbol : is read as 'such that'.

Exercise 6A

1 Identify the non-examples of sets and give reasons why they are non-examples.

    a $A = \{a, b, c, d\}$ **b** $\mathbb{N} = \{1, 2, 3, \text{natural numbers}\}$

    c $K = \{\text{whole numbers}\}$ **d** $M = \{\text{letters of the alphabet from g to z}\}$

2 Give three non-examples of sets.

3 List these sets.

    a {five countries in the Caribbean}   **b** {five different fruits}

    c {six models of vehicles}   **d** {five wild animals}

    e {5 popular careers}

4 Describe each set.

    a {car, bus, boat, aeroplane}   **b** {yam, potato, dasheen}

    c {plate, cup, saucer, spoon}   **d** {pencil, exercise book, pen, eraser, ruler}

    e {sheep, cow, donkey, goat}

5 **a** Given the set $P = \{\text{nail, hammer, saw, ...}\}$, use appropriate set notation to state two more members of P.

    b $K = \{1, 2, 3, c, 6, 7\}$. Using set notation, state which element of K is not a member of $\mathbb{N} = \{\text{natural numbers}\}$.

    c Write in set notation: 'Mango is a member of the set of fruits.'

    d Use set notation to write: 'Cabbage is not a member of the set of fruits.'

    e Write down the meaning of 'house $\notin$ {vehicles}'.

    f Write down the meaning of 'radio $\in$ {electronic equipment}'.

6 **a** Given that $W = \{\text{apple, plum, orange, banana}\}$, what is the value of $n(W)$?

    b State the number of elements in the set $V = \{\text{vowels}\}$.

    c State the number of elements in the set $P = \{\text{letters of the alphabet}\}$.

    d Given that $G = \{\text{letters between h and k}\}$, state $n(G)$.

    e Given that $R = \{\text{cat, dog, rabbit}\}$, state $n(R)$.

    f Given that M is the set of letters between k and l, write down $n(M)$.

7 **a** Give three examples of a finite set and three examples of an infinite set.

    b Give three examples of a null set.

    c Give three examples of a well-defined set and three examples of a poorly defined set.

    d Give one example of a set of:

        i natural numbers   **ii** whole numbers   **iii** integers

        iv rational numbers   **v** irrational numbers   **vi** real numbers.

8 Which of these sets are equal and which are equivalent?

    $K = \{1, 2, 3, 4\}$ $P = \{a, b, c, d\}$ $J = \{5, 3, 2, 1\}$

    $L = \{c, a, b, d, e\}$ $H = \{3, 4, 1, 2\}$

9 List all the subsets of each of these sets.

    a {1, 2}   **b** {a, b, c}   **c** {dog, cat, pig, goat}

10 Describe each of these sets.

    a $R = \{1, 3, 5, 7, 9\}$ **b** $W = \{2, 4, 6, 8, 10\}$ **c** $L = \{2, 3, 5, 7, 11, 13\}$

    d $P = \{1, 10, 10^2, 10^3, 10^4\}$ **e** $T = \{-4, -16, -64, -256\}$

11 State the number of subsets that can be made from each set.

    a $S = \{6\}$ **b** $T = \{w, x, y, z\}$ **c** $F = \{grape, berry, cashew, pear, avocado\}$

12 Determine the number of subsets that can
be made from $H = \{cricket, football, netball,$
$volleyball, table\ tennis, basketball\}$.

13 State the number of subsets in $Z = \{seven$
$different\ makes\ of\ vehicle\}$.

14 Give five examples of an infinite
universal set.

15 Given the universal set $U = \{all\ the\ letters\ in\ the\ alphabet\}$ and set $D = \{letters\ from\ b$
to $y\}$, list the members of the set D'.

16 Given $A = \{whole\ numbers\ up\ to\ 30\}$ and set $Q = \{multiples\ of\ 3\}$, list the elements of
Q' in A.

17 List the elements of each set.

    a $Y = \{x: x^2 = 81\}$

    b $M = \{x: x \in last\ ten\ letters\ of\ the$
       $alphabet\}$

    c $E = \{x: x\ is\ a\ primary\ colour\}$

18 Use set builder notation to describe these
sets.

    a the set of windward islands

    b the set of OECS states

    c {reptiles}

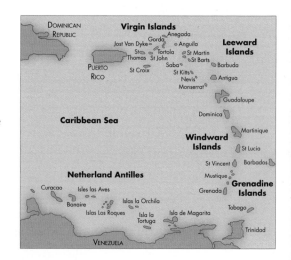

19 Name each type of set.

    a a set of girls

    b a set of stars in the heavens

    c a set of continents on Earth

    d $G = \{garden\ tools\}$

    e a set of capital letters from A to Z

    f $J = \{factors\ of\ 7\ between\ 1\ and\ 6\}$

    g $N = \{a, b, c, d\}$ and $L = \{c, b, a, 4\}$

6.2 Intersection, union and complements of not more than 3 sets

6.2.1 Intersection of sets and disjoint sets

Sometimes, two or more sets may include the same elements. The set of elements that belong to two or more sets comprise the **intersection** of those sets.

Example 15

Find the intersection of the sets:
$A = \{1, 2, 3, 4\}$, $B = \{3, 4, 5, 6\}$ and $C = \{3, 4, 7, 8\}$.

$A \cap B \cap C = \{3, 4\}$ since these elements appear in all three sets.

Key term

The **intersection** of two or more sets is the set of elements common to both or all sets. It is denoted by the symbol $\cap$.

It is not always the case that two or more sets have elements in common. If two sets have no common elements they are **disjoint** and their intersection is the empty set.

Example 16

$P = \{a, b, c\}$, $Q = \{x, y, z\}$. What is the intersection, $P \cap Q$?

P and Q have no members in common, they are disjoint.

For two disjoint sets, the intersection is the null set: $P \cap Q = \{\ \}$.

Key term

Two sets are **disjoint** if they have no common elements.

6.2.2 Union of sets

If two or more sets are combined, the resulting set is the **union** of the sets.

Example 17

Determine the union of sets $A = \{1, 2, 3, 4\}$, $B = \{3, 4, 5, 6\}$ and $C = \{3, 4, 7, 8\}$.

$A \cup B \cup C = \{1, 2, 3, 4, 5, 6, 7, 8\}$. The three sets are combined without repeating any element.

Key term

The **union** of two or more sets is the set of all the elements that are in a combination of all the sets. The union is denoted by the symbol $\cup$.

6.2.3 Problems involving intersections, unions and complements

Now that you have learnt the basic information about sets, you can start to solve problems involving them.

Example 18

Given a universal set $H = \{\text{whole numbers up to } 15\}$ with $K = \{\text{odd numbers up to } 15\}$, $E = \{\text{even numbers less than } 15\}$ and $M = \{\text{multiples of 5 up to } 15\}$, find: (a) $E \cap M$ (b) $M \cup K$ (c) K' (d) $(E \cup K)'$.

First list all the individual sets.

$H = \{0, 1, 2, 3, 4, 5, 6, 7, 8, 9, 10, 11, 12, 13, 14, 15\}$

$E = \{2, 4, 6, 8, 10, 12, 14\}$ $K = \{1, 3, 5, 7, 9, 11, 13, 15\}$

$M = \{5, 10, 15\}$

(a) $E \cap M = \{10\}$ (b) $M \cup K = \{1, 3, 5, 7, 9, 10, 11, 13, 15\}$

(c) $K' = \{0, 2, 4, 6, 8, 10, 12, 14\}$ (d) $(E \cup K)' = \{0\}$

Exercise 6B

1 Find the intersection of each pair of sets, using appropriate set notation to write your answers.

   a $I = \{1, 2, 3\}$, $W = \{3, 4, 5\}$ **b** $P = \{a, b, c, d, e\}$, $R = \{e, f, g, h, j\}$

   c $A = \{\text{numbers from 1 to 18}\}$, $B = \{\text{numbers from 14 to 22}\}$

   d $D = \{\text{apples, mangoes, plums, grapes}\}$, $W = \{\text{plums, oranges, cherries}\}$

2 Using appropriate set notation, write down the union of each pair of sets given in question 1.

3 Given $S = \{10, 20, 30, 40, 50\}$ and $T = \{20, 30, 60\}$, find $S \cap T$ and $S \cup T$.

4 Using set notation, write down the intersection of each pair of sets.

   a $F = \{\text{factors of 12}\}$, $G = \{\text{factors of 9}\}$

   b $M = \{\text{multiples of 4 from 4 to 16}\}$, $K = \{\text{multiples of 3 between 3 and 15}\}$

5 Using appropriate set notation, write down the union of each pair of sets given in question 4.

6 Given $R = \{x: -4 < x < 10\}$ and $Z = \{x: 0 \leqslant x < 15\}$, find $R \cap Z$ and $R \cup Z$.

7 Given a universal set $U = \{\text{whole numbers from 0 to 30}\}$, set $Y = \{\text{even numbers}\}$, $H = \{\text{odd numbers}\}$ and $F = \{\text{multiples of 5}\}$, find:

   a H' **b** $Y \cup F$ **c** $H \cap F$.

8 A universal set of integers $U = \{x: -20 \leqslant x \leqslant 20\}$ includes $K = \{x: 0 \leqslant x \leqslant 10\}$, $M = \{x: 15 > x \geqslant 5\}$ and $Z = \{x: -15 \leqslant x < -5\}$.

   a List the elements of K, M and Z.

   b Which two subsets are disjoint sets?

   c Find $K \cap M$.

   d Find $K \cap Z$.

   e Find $(K \cup M \cup Z)'$.

9 $P = \{\text{prime numbers between 1 and 10}\}$, $E = \{\text{even numbers between 1 and 10}\}$ and $O = \{\text{odd numbers from 1 to 10}\}$.

   a Suggest the smallest universal set that will contain all the elements of P, E and O.

   b List the elements of $P \cap E$.

   c List the elements of $O \cup P$.

   d List the elements of $O \cap P \cap E$.

10 $P = \{\text{prime factors of 20}\}$, $M = \{\text{multiples of 2 up to 20}\}$ and $F = \{\text{factors of 20}\}$.

   a List the elements of P, F and M.   **b** List the elements of $P \cap M \cap F$.

   c List the elements of $P \cup M \cup F$.   **d** List the elements of $(P \cup F)'$.

11 Using your own examples, show the relationship between natural numbers, whole numbers, integers, rational numbers, irrational numbers and real numbers.

12 Given the words 'encyclopedia', 'mathematics' and 'figurative', form three sets (E, M and F respectively), using the letters of each word as elements and the alphabet as the universal set, U.

a List the elements of each set.

b Write down, using set notation, the elements of $E \cup M \cup F$.

c List the elements of $(E \cup M \cup F)'$.

d List the elements of $E \cap M \cap F$.

e State the value of $n(E \cup M \cup F)$.

13 **a** Write down a universal set of each of these sets.

    i the days of the week

    ii the months of the year

    iii the weeks of the year

b List the elements of part **a(i)**, using proper set notation.

c List the elements of part **a(ii)**, using appropriate set notation.

d Describe part **a(iii)**, using set builder notation.

e List the elements of the intersection of the sets.

f State the numbers of elements in all three sets, expressing your answer in set notation.

14 The universal set $U = \{$natural numbers up to 100$\}$ includes $Q = \{$factors of 100$\}$, $M = \{$multiples of 5$\}$, $S = \{$square numbers$\}$ and $P = \{$prime numbers$\}$.

List all the elements of:

a Q, M, S and P

b $(Q \cap M \cap S) \cup P$

c $(Q \cup M \cup P \cup S)'$

d $(Q \cup M \cup P) \cap S$

e $(Q \cup M)' \cup (P \cup S)'$.

15 In a school of 500 students, all students study English, 400 study Mathematics, 300 study History. All the students who study Mathematics also study English. Half of the students learning History also study English and Mathematics.

a Use set notation to show these relationships.

b Write down:

    i $n(E)$ **ii** $n(M)$ **iii** $n(E \cap M)$

    iv $n(H \cap E \cap M)$ **v** $n(H)$.

c What would you say is the universal set?

6.3 Venn diagrams

Sometimes it can be helpful to visualise the sets you are working with. You can do this by drawing a **Venn diagram**, in which a rectangle represents the universal set, and the sets you are considering are drawn inside this rectangle, as loops that may or may not overlap.

Key term

A **Venn diagram** is a diagram, consisting of a rectangle and circles or ellipses, used to represent a universal set and the sets within them, and to show relationships among sets.

6.3.1 Drawing and interpreting Venn diagrams

Using Venn diagrams is an important skill as they have many applications in other areas and offer a concise and clear way to express problems visually.

Did you know? Venn diagrams were introduced in the 19th century by mathematician Joseph Venn.

Example 19

(a) Draw a Venn diagram to represent the universal set of natural numbers from 1 to 10 and subsets E = {even numbers} and O = {odd numbers}.

(b) A fruit vendor has two baskets of fruit. Basket A contains apples, mangoes and plums. Basket B contains mangoes, pears and oranges. Represent this information in a Venn diagram with fruits as the universal set and subsets A and B.

(a)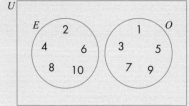

(b)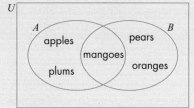

Example 20

(a) Represent in a Venn diagram the universal set U = {letters of the alphabet} and the sets P = {a, b, c, d, e}, Q = {d, e, f, g, h, k, l, m, n} and R = {s, t, u, v, w}.

(b) List the elements of: **i** $P \cap Q$ **ii** $Q \cap R$ **iii** $P \cup R$ **iv** $(P \cup Q \cup R)'$

(c) Write down $n(U)$.

(a)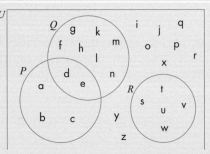

(b) **i** $P \cap Q$ = {d, e} (These two elements belong to both sets.)

   ii $Q \cap R$ = { } (There are no common elements.)

   iii $P \cup R$ = {a, b, c, d, e, s, t, u, v, w} (Elements are a combination of both sets.)

   iv $(P \cup Q \cup R)'$ = {i, j, o, p, q, r, x, y, z} (The complement comprises all elements outside of $P \cup Q \cup R$.)

(c) $n(U)$ = 26 (Total number of elements in the universal set.)

6.3.2 Problems involving Venn diagrams and the use of the formula for the number of elements in the intersection of two sets

For two intersecting sets A and B, $n(A \cup B) = n(A) + n(B) - n(A \cap B)$

This formula is used in the solution of set-related problems such as those in examples 21 and 22.

Example 21

In a class of 40 students, 13 like Mathematics, 15 like English and 20 like neither Mathematics nor English. How many like both Mathematics and English? Represent your answer in a Venn diagram.

Let M = {students who like Mathematics} $\Rightarrow n(M) = 13$

Let E = {students who like English} $\Rightarrow n(E) = 15$

$n(M \cup E)' = 20 \Rightarrow n(M \cup E) = 20$

Applying the formula:

$n(M \cup E) = n(M) + n(E) - n(M \cap E)$

$20 = 13 + 15 - n(M \cap E)$

$\Rightarrow n(M \cap E) = 28 - 20$

$\qquad\qquad = 8$

$\quad n(M \cap E) = 8$

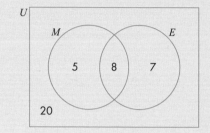

Tip

In set notation the word 'only' is quite significant as it refers to the elements that are not part of any intersection. For example, if there is an intersection of two sets A and B, then 'A only' refers to the elements that belong only to A and are not in B.

Tip

When completing a Venn diagram it is advisable to begin with the intersection and work outwards.

Example 22

In a survey to discover children's favourite ice-cream flavour, it was found that among a group of children, 30 liked vanilla, 20 liked both vanilla and chocolate, 10 liked chocolate only and 5 liked none of them. How many children were surveyed? Draw a Venn diagram to illustrate your answer.

Let V represent students liking vanilla and C represent students liking chocolate.

In the Venn diagram, begin with the intersection.

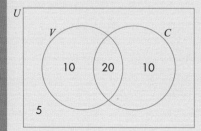

Note

$n(V_{\text{only}}) = 30 - 20$

$\qquad\qquad = 10$

$n(U) = 10 + 20 + 10 + 5$

$\qquad\quad = 45$

6.3.3 Problems involving Venn diagrams and the use of the formula for the number of elements in the intersection of three sets

For three intersecting sets A and B and C,

$n(A \cup B \cup C) = n(A) + n(B) + n(C) - n(A \cap B) - n(A \cap C) - n(B \cap C) + n(A \cap B \cap C)$

This formula is used in the solution of set-related problems such as those in examples 23 and 24.

Example 23

The Venn diagram represents the membership of a sports club and their participation in three different sports: cricket, football and basketball.

Determine and write in set notation:

(a) the number who play football only

(b) the number who play all three sports

(c) the number who play football and cricket

(d) the number who play basketball and football

(e) the number who play basketball and cricket

(f) the number who play football and cricket only

(g) the number who play none of these sports

(h) the total membership of the sports club.

Notice the set notation in this solution.

(a) $n((B \cup C)' \cap F) = 9$ (b) $n(B \cap C \cap F) = 2$ (c) $n(F \cap C) = 27$

(d) $n(B \cap F) = 15$ (e) $n(B \cap C) = 22$ (f) $n(B' \cap (F \cap C)) = 25$

(g) $n(F \cup B \cup C)' = 8$ (h) $n(U) = 92$

Example 24

A number of students were asked about their preferences between three fruits: mango, apple and plum. 10 students liked mango only, 12 students liked apples, 14 students liked plums, 5 students liked none of the three, 4 students liked all three, 8 students liked mangoes and plums, 6 students liked plums and apples, 5 students liked mango and apple only.

(a) How many students like apples only?

(b) How many students like plums only?

(c) How many students were asked?

The information is displayed in the Venn diagram, working from the middle intersection of the three sets outwards.

(a) $n((M \cup P)' \cap A) = 12 - (5 + 4 + 2) = 1$

(b) $n((M \cup A)' \cap P) = 14 - (4 + 4 + 2) = 4$

(c) $n(U) = 10 + 5 + 1 + 4 + 4 + 2 + 4 + 5 = 35$

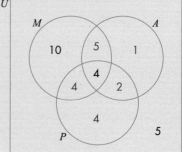

Exercise 6C

 1　**a** Draw a Venn diagram to represent the universal set U = {numbers 1 to 20} with sets
A = {1 to 15} and B = {10 to 20}.

　b From your Venn diagram find:

　　i $A \cap B$　　　　　　　　**ii** $(A \cup B)'$.

2　**a** Draw a Venn diagram to represent the universal set S = {letters a to m} and sets
P = {a, b, c, d, e, f} and Q = {f, g, h, I, j, k}.

　b From your Venn diagram find:

　　i $P \cup Q$　　　**ii** $(P \cap Q)'$　　　**iii** P'　　　**iv** Q'.

3　Draw a Venn diagram to represent universal set $\mathbb{N}$ = {natural numbers up to 50} and sets
M = {multiples of 2}, G = {multiples of 5} and F = {factors of 36}.

4　In the Venn diagram, the universal set (F) consists of different fruits and sets H and G
represent the favourite fruits of two groups of
people.

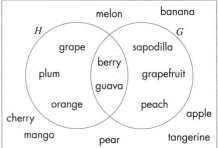

　a List the elements of F, H and G.

　b List $H \cap G$.　　　　　**c** List $H \cup G$.

　d List $(H \cup G)'$.　　　　**e** List $(H \cap G)'$.

5　The numbers of students studying Mathematics
and English are shown in this Venn diagram.

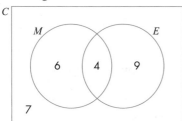

The number of students in the class is represented by the universal set C. M represents
the students studying Mathematics and E represents the students studying English.

　a How many students are there in the class?

　b How many students study Mathematics?

　c How many students study English?

　d How many students study both Mathematics and English?

　e How many students study neither Mathematics nor English?

　f How many students study Mathematics only?

6 The Venn diagram represents the membership of a sports club. The universal set *S* represents the total number of members of the club. *C* represents the number playing cricket, *F* represents those playing football, *T* represents those playing tennis. Some are involved in more than one sport and some do not play any sport.

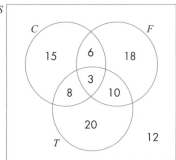

a How many members take part in three sports?

b How many take part in two sports?

c How many are involved in only one sport?

d How many are involved in none of the three sports?

e What is the total membership of the club?

7 In a survey carried out with a class to determine students' favourite ice-cream flavour, 20 students like vanilla, 15 like chocolate and 12 like both vanilla and chocolate.

a Given that 8 students like neither of these flavours, how many students are there in the class?

b How many like vanilla only?

c How many like chocolate only?

8 A total of 50 students were asked about their favourite fruit. 30 like oranges, 26 like pears and 2 like neither.

a How many liked both oranges and pears?

b How many students like only one fruit?

9 There are 40 foreign language students in a class. 10 students study Spanish only, 8 study French only and 20 study another language. How many study both Spanish and French?

10 A cultural group has 60 members. 20 like dancing and singing, and 10 like neither dancing nor singing. If 25 like only dancing, how many like dancing and how many like singing?

11 In a survey carried out among a number of women it was discovered that 20 of them enjoyed baking bread, 25 enjoyed baking cakes and 30 enjoyed making pastries. 8 of them did not enjoy making any of these, 10 enjoyed making all three, 15 enjoyed making both pastries and cakes, 12 enjoyed making bread and cakes and 14 enjoyed making bread and pastries.

   a How many women were surveyed?

   b How many enjoyed baking:

      i pastries only   **ii** bread only   **iii** cakes only?

12 120 men were interviewed for a job requiring a carpenter, mason or painter. 40 men could only paint, 30 could only do carpentry and 20 could only do masonry. Given that 10 men could do masonry and carpentry, 14 men could do painting and carpentry and 12 men could do masonry and painting but 6 men could do none, how many could do all three jobs?

13 **a** A school's male sport section has a number of boys who are involved in at least one sport. Some students play football, some play cricket, some play basketball. Short students, who are those less than 6 feet tall, play either cricket or football and some play both. Only students 6 feet tall or more play basketball, but not cricket or football. Draw a Venn diagram to represent this information.

   b John is 6 feet 2 inches tall. Mark is 5 feet 8 inches, he does not like cricket but is fond of basketball and football. Harry, who is 5 feet 11 inches in height, plays two sports. Locate each boy in your Venn diagram.

14 A total of 80 athletes normally take part in four events: 100 m, 200 m, 400 m and a field event.

   5 athletes take part in 100 m, 200 m and 400 m; 8 athletes do the 100 m and 200 m; 16 athletes do the 200 m and 400 m and 11 athletes do the 100 m and 400 m. If 20 athletes do the 100 m only, 18 do the 200 m only and 33 do the 400 m, how many athletes only take part in field events?

15 Three sets A, B and C intersect such that $n(A \cap B \cap C) = 1$, $n(A \cap B) = 5$, $n(A \cap C) = 8$, $n(B \cap C) = 7$; $n(A) = 30$, $n(B) = 40$, $n(C) = 50$ and $n(A \cup B \cup C)' = 3$. Evaluate $n(A \cup B \cup C)$.

16 Study the Venn diagram and list the members of:

   a $P \cap Q \cap R$ **b** $P \cap Q$ **c** $Q \cap R$ **d** $P \cap R$ **e** P

   f Q **g** R **h** $P \cup Q \cup R$ **i** $(P \cup Q \cup R)'$

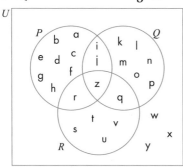

6.4 Using set theory to solve problems in number theory, algebra and geometry

Sets can be used to solve many different problems in number theory, algebra and geometry.

6.4.1 Sets and number theory

Example 25

Use a Venn diagram to represent the different types of number.

natural numbers $\subset$ whole numbers $\subset$ integers $\subset$ rational numbers $\subset$ real numbers

Using symbols: $\mathbb{N} \subset \mathbb{W} \subset \mathbb{Z} \subset \mathbb{Q} \subset \mathbb{R}$

Also: irrational numbers ($\mathbb{I}$) $\subset$ real numbers ($\mathbb{R}$)

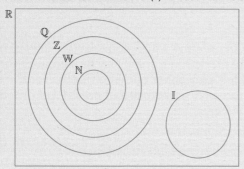

Example 26

Represent in a Venn diagram:

universal set U = {natural numbers up to 30}

A = {even numbers}, B = {odd numbers}, C = {prime numbers}.

Begin by listing elements of all sets (universal and subsets).

Then add them to a Venn diagram.

U = {1, 2, 3, 4, 5, 6, 7, 8, 9, 10, 11, 12, 13, 14, 15, 16, 17, 18, 19, 20, 21, 22, 23, 24, 25, 26, 27, 28, 29, 30}

A = {2, 4, 6, 8, 10, 12, 14, 16, 18, 20, 22, 24, 26, 28, 30}

B = {1, 3, 5, 7, 9, 11, 13, 15, 17, 19, 21, 23, 25, 27, 29}

C = {2, 3, 5, 7, 11, 13, 17, 19, 23, 29}

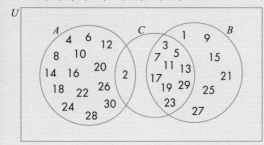

6.4.2 Sets and algebra

Example 27

The Venn diagram represents the preferences for cats or dogs as pets among 100 home owners. Given that C represents the set of homeowners who love cats and D the set of homeowners who love dogs, calculate, from the algebraic expressions in the Venn diagram:

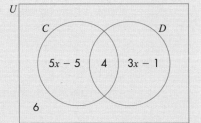

(a) the number that like cats only

(b) the number that like dogs only

(c) the total number who like cats or dogs.

For the two intersecting sets C and D, apply the formula:

$n(C \cup D) = n(C) + n(D) - n(C \cap D)$

Substituting:

$(100 - 6) = 5x - 5 + 4 + 3x - 1 + 4 - 4$

$94 + 5 - 4 + 1 - 4 + 4 = 8x$

$\Rightarrow x = \dfrac{96}{8}$

$= 12$, or use:

$n(U) = 100$

$\quad = 5x - 5 + 4 + 3x - 1 + 6$

$100 - 4 + 1 - 6 + 5 = 8x$

$\Rightarrow 96 = 8x$

$\quad x = 12$

(a) $n(D' \cap C) = 5x - 5$
$\qquad\qquad = 5 \times 12 - 5$
$\qquad\qquad = 55$

(b) $n(C' \cap D) = 3x - 1$
$\qquad\qquad = 3 \times 12 - 1$
$\qquad\qquad = 36 - 1$
$\qquad\qquad = 35$

(c) $n(C \cup D) = 5x - 5 + 4 + 3x - 1$
$\qquad\qquad = 8x - 2$
$\qquad\qquad = 8 \times 12 - 2$
$\qquad\qquad = 96 - 2$
$\qquad\qquad = 94$

6.4.3 Sets and geometry

Set builder notation can be used to define geometrical sets.

Example 28

List the elements of the set K where $K = \{x: x = \text{polygon with 3 sides}\}$.

$K = \{\text{triangle}\}$

Example 29

List two members of the set of quadrilaterals.

$\{\text{parallelogram, square}\}$

Exercise 6D

1 Draw a fully labelled Venn diagram showing the relationships between natural numbers, whole numbers and integers.

2 Represent rational numbers, irrational numbers and real numbers in a Venn diagram similar to the one for Question 1. What is the relationship between the sets of rational and irrational numbers?

3 Draw a Venn diagram to represent these sets.

U = {rational numbers from –10 to 10} I = {integers from –6 to 6}

W = {whole numbers up to 6} N = {natural numbers between 4 and 10}

4 In a large family of 15 people, 8 like fruit, 10 like vegetables, x like both fruit and vegetables and three like neither fruit nor vegetables. Determine the value of x and state the number liking fruit only or vegetables only.

5 Draw a Venn diagram to represent the different types of three-sided polygon.

6 List and sketch five members of the set of quadrilaterals.

7 **a** Draw a Venn diagram in which the universal set is $U = \{-5, -4, -3 -2, -1, 0, 1\frac{1}{2}, 2.5, 3\frac{1}{4}, 4,$ $5\frac{3}{4}, 6, 7, 8, 9.75, 10\frac{1}{3}, \sqrt{2}, \sqrt{4}\}$.

   Include in the diagram the sets A = {integers}, B = {whole numbers}, C = {rational numbers}, D = {irrational numbers}.

   b From your Venn diagram determine:

   i $A \cap B \cap C$ **ii** $(A \cap B) \cup D$ **iii** $(B \cap C) \cup A$ **iv** D'

   v $(A \cup B)'$ **vi** $n(U)$ **vii** $C \cap D$

8 **a** Draw a Venn diagram in which the universal set is the natural numbers up to 20. Show on your diagram the sets M = {multiples of 2}, F = {factors of 20}, O = {odd numbers} and P = {prime numbers}.

   b List the elements of:

   i F **ii** $M \cap F$ **iii** $P \cap M$ **iv** $P \cup M \cup F$ **v** $(P \cup M)'$

9 Draw a Venn diagram to represent a universal set of plane shapes consisting of 5 three-sided polygons (set A) and 4 four-sided polygons (set B). List the elements of $A \cup B$ and $A \cap B$.

10 The Venn diagram shows students' usage of social media, with S as the universal set. A total of 100 students were interviewed on their preference of social media between: WhatsApp (W), Instagram (I) and Facebook (F). The results, shown as algebraic expressions, indicate also that $(2x + 5)$ students did not use any of the aforementioned social media.

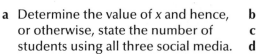

   a Determine the value of x and hence, or otherwise, state the number of students using all three social media.

   b How many students use WhatsApp only?
   c How many use Facebook only?
   d How many use Instagram only?

   e How many use only two forms of social media?

11 The Venn diagram below represents a universal set U representing members of a girls' sports club.

    a Given that 30 girls like netball (N), ($40 - x$) like basketball (B), 20 like basketball only, 5 like both netball and basketball and 6 liked neither of these sports, determine the value of x and state the total number of girls in the club.

    b How many girls like basketball or netball only?

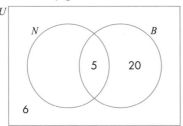

12 A total of 60 students were asked about their preference of drinks. 22 preferred soda, 16 preferred water and 30 preferred orange juice. The number who liked all three was given as $2x - 3$ and the number who liked none of the three was given as $x + 10$. Given that 8 liked soda and orange juice, 10 liked water and soda and 12 liked water and orange juice:

    a determine the value of x

    b hence calculate:

      i the number liking all three drinks

      ii the number that preferred only one type of drink

      iii the number that liked none of these drinks

      iv the number liking two types of drink.

13 **a** Draw a Venn diagram showing the relationship between the major types of number systems: natural numbers ($\mathbb{N}$), whole numbers ($\mathbb{W}$), integers ($\mathbb{Z}$), rational numbers ($\mathbb{Q}$), irrational numbers ($\mathbb{I}$) and real numbers ($\mathbb{R}$).

    b From your diagram, determine:

      i $\mathbb{N} \cap \mathbb{W}$ **ii** $\mathbb{W} \cap \mathbb{Q}$ **iii** $\mathbb{Q} \cap \mathbb{I}$.

14 Draw a Venn diagram to represent a universal set of solid shapes that also contains sets A = {shapes with same base and top}, B = {shapes with a polygonal base and apex} and C = {shapes with no flat surface}.

15 Represent these solids in a Venn diagram, based on the number of flat faces they have: A = {shapes with four faces}, B = {shapes with six faces}, C = {shapes with eight faces} with the universal set being solids. Include set D consisting of five named irregular solids.

End of chapter summary

Definition of sets and related concepts

- A **set** is a collection or grouping of objects, called **elements** or **members** of the set.
- A set of colours given as 'red, green, blue, colours' is a **non-example** since 'colours' cannot be listed as a member of the set of colours.
- A set is denoted by an italic capital letter and is usually specified or represented by listing or by description.
- The elements or members of the set may be **listed**, inside curly brackets, not necessarily in any particular sequence or order.
- The set may be **described** in words and sentences.
- The symbol for membership of a set is $\in$ and the symbol to indicate non-membership is $\notin$.
- The **cardinality** of a set is the number of members there are in the set.
- A **finite** set is a set in which all the elements can be listed.
- An in**finite** set is a set in which the elements cannot all be listed or counted.
- The **null** or **empty set** is a set with no elements. It is represented by $\{\ \}$ or $\varnothing$.
- A set is **well defined** if it is perfectly clear what elements belong to the set.
- The sets of numbers are defined as:
    - **natural numbers** (counting numbers): $\{1, 2, 3 \dots\}$
    - **whole numbers**: $\{0, 1, 2, 3, \dots\}$
    - **integers**: Negative and positive whole numbers and zero $\{\dots, -3, -2, -1, 0, 1, 2, 3, \dots\}$
    - **rational numbers**: Integers together with fractions, terminating and repeating decimals $\left\{\dfrac{3}{4}, \dfrac{1}{3}, 5\dots\right\}$
    - **irrational numbers**: non-terminating and non-repeating decimals $\left\{\pi, e, \sqrt{2}\right\}$
    - **real numbers**: All rational and irrational numbers, for example, $\left\{\dots, -3, 0, \dfrac{1}{2}, \pi, 8, \dots\right\}$
- Two or more sets are **equal** if they have exactly the same elements.
- Two or more sets are **equivalent** if they have the same cardinality (same number of elements).
- A subset is any set that is contained in, or can be obtained from, another set. Every set has itself and the null (empty) set as subsets.
- The **number of subsets (N)** in a set can be determined from the formula: $N = 2^n$ where n is the number of elements in the set.
- A **universal set** is one that contains all the members of a particular group or class of sets.
- The **complement** of a set comprises all the elements not included in that set. The complement of set A is written as A'.
- In **set builder notation**, a set is identified by stating the conditions satisfied by the elements or members of the set, for example, $K = \{x: x = N\}$.

Intersection, union and complement of not more than three sets

- The **intersection** of two or more sets is the set of elements common to both sets. It is denoted by the symbol $\cap$.
- **Two sets are disjoint** if they have no common elements.
- The **union** of two or more sets is the set of all the elements that are in a combination of all the sets. The union is denoted by the symbol $\cup$.

Venn diagrams

- A **Venn diagram** is a diagram, consisting of a rectangle and circles or ellipses, used to represent a universal set and the sets within them, and to show relationships among sets.
- For two intersecting sets A and B, $n(A \cup B) = n(A) + n(B) - n(A \cap B)$
- For three intersecting sets A and B and C,

$$n(A \cup B \cup C) = n(A) + n(B) + n(C) - n(A \cap B) - n(A \cap C) - n(B \cap C) + n(A \cap B \cap C)$$

Using set theory to solve problems in number theory, algebra and geometry

- Sets can be used to solve many different problems in number theory, algebra and geometry.

Examination-type questions for Chapter 6

1   The Venn diagram shows the numbers of students playing cricket and football from a group of 40 boys.

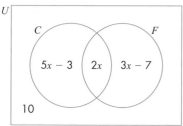

The universal set (U) is represented by the group of 40 boys. The set of boys playing cricket is C and the set of boys playing football is F.

a   Write an expression in x for the total number of boys in the group.

b   Determine the value of x.

c   Determine the number of boys playing both cricket and football.

d   Evaluate $n(C \cap F)'$.

e   How many boys play cricket only?

2   A universal set consists of natural numbers between 1 and 29. A = {odd numbers between 10 and 20} and B = {factors of 28}.

a   Determine $n(U)$ **b**   List the elements of A.   **c**   List the elements of B.

d   Draw a Venn diagram to represent the information.

e   List the elements of $A \cup B$ and $A \cap B$.

3   In the Venn diagram, the universal set U represents the students in a school's sport team. The diagram shows that there are 18 students on the athletics team, A, and also gives an algebraic expression for the number of students that swim only, x. Three students are on both the athletics and the swim teams and x + 5 students are on neither team.

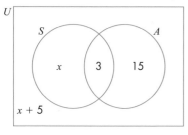

a   Solve for x if the number of students on the swim team is 20.

b   Determine the total number of students on the sports team.

c   Determine the number of students not involved in either sport.

4 A universal set is given by $U = \{x : x \in W, 0 \leqslant x \leqslant 20\}$

It contains sets $K = \{$multiples of 5$\}$ and $H = \{$factors of 20$\}$.

  a Draw a Venn diagram to show the information.

  b List the elements of:

    i $K \cap H$ **ii** $K \cup H$ **iii** $(K \cup H)'$.

5 **a** From the Venn diagram, write down:

    i U **ii** P **iii** Q **iv** R **v** $P \cap Q$

    vi $Q \cap R$ **vii** $P \cap R$ **viii** $P \cup Q \cup R$ **ix** $(P \cup Q \cup R)'$ **x** $n(U)$

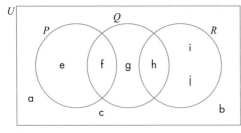

  b On a separate diagram, shade $(P \cap Q) \cup (Q \cap R)$.

Objectives

By the end of this chapter you should be able to:

- explain basic concepts associated with relations
- represent a relation in various ways
- state the characteristics that define a function
- distinguish between a relation and a function graphically
- use functional notation, for example, $f : x \rightarrow x^2$; or $f(x) = x^2$
- draw graphs of linear functions, understand the concept of linear functions, and recognise types of linear function
- determine the intercepts of the graph of linear functions, x-intercepts and y-intercepts, graphically and algebraically
- determine the gradient of a straight line, and understand the definition of gradient or slope
- determine the equation of a straight line using the graph of the line; the coordinates of two points on the line; the gradient and one point on the line; and one point on the line or its gradient, and its relationship to another line
- solve problems involving the gradients of parallel and perpendicular lines
- understand the concepts of magnitude or length and midpoint, determine from coordinates on a line segment the length of the line and the coordinates of the midpoint
- solve a pair of simultaneous linear equations in two unknowns graphically, interpret the intersection of graphs
- represent the solution of linear inequalities in one variable, using set notation, the number line, and a graph
- draw a graph to represent a linear inequality in two variables
- use linear programming techniques to solve problems involving two variables graphically
- derive composite functions of no more than two functions, for example f, g, f^2, given f and g
- understand the non-commutativity of composite functions (fg ≠ gf)
- state the relationship between a function and its inverse
- evaluate a function $f(x)$ at a given value a of x ($f(a)$, $fg(a)$), where $a \in \mathbb{R}$
- draw and use the graph of a quadratic function to identify its features

- interpret the graph of a quadratic function to determine concepts of the gradient of a curve at a point, tangent; turning point, roots of the function; and an estimate of the value of the gradient at a given point; intercepts of the function
- determine the equation of the axis of symmetry and the maximum or minimum value of a quadratic function expressed in the form $a(x + h)^2 + k$
- sketch the graph of a quadratic function expressed in the form $y = (x + h)^2 + k$ and determine the number of roots
- interpret graphs of functions including distance–time graphs and speed–time graphs
- solve problems involving graphs of linear and non-linear functions.

You should already know:

- number theory and computation
- algebra
- sets
- measurement

Introduction

Relationships are very common in today's world. Apart from natural relationships between human beings, there can be relationships between other things. Living things, such as species of plants and animals, are related. Non-living things such as numbers, letters, symbols and objects in general also relate in one way or another.

The nature of the relationship can be different, of course, but many of these relations can be expressed mathematically. A mathematical equation, for example, can be used to express many physical relationships present in electronics, physics and the sciences as a whole. In physics, a relationship exists between force, mass and acceleration: the force on a body = mass × acceleration ($F = ma$), which represents Newton's second law.

Working through this chapter, you should appreciate the importance of relations and functions in mathematics and that many mathematical relations may be represented in symbolic (use of letters and formulae), tabular (use of tables) or pictorial (use of drawings, graphs and pictures) forms. Additionally, you should be able to appreciate the usefulness of concepts in relations, functions and graphs and the application of these in the solution of real-world problems.

Examples of relations, functions and graphs in real life

Parabolas can often be seen in real life

Graph showing temperature change during a year

Linear programming can be used to optimise relationships between different variables

Sales and purchases of produce may be controlled by relationships agreed by buyers and sellers.

7.1 Relations and functions

7.1.1 Basic concepts associated with relations

A mathematical **relation** links one set of numbers or objects to another. Examples of mathematical relations include {(1, 2), (3, 4), (5, 6)} and {(John, Mary), (Paul, Ann)}.

You should be able to present a relation as a set of ordered pairs.

Key term

A **relation** may be represented by a set of input and output values or items expressed as ordered pairs.

Examples of relations include: {(1, 2), (3, 4), (5, 6), (7, 8)}

Non-examples of relations include: {1, 2, 3, 4, 5}, {(1), (1, 2, 3) and (1, 2, 3, 4)}

When a relation is represented as a set of ordered pairs, the first elements in the ordered pairs belong to the **domain** of the relation and the second elements belong to the **range**.

> **Example 1**
>
> State the domain and range of the relation $P = \{(2, 4), (6, 8) (10, 12)\}$.
>
> Domain = {2, 6, 10}
>
> Range = {4, 8, 12}

Domain values are also known as **input values** (what goes into the relation).

Range values are also known as **output values** (what comes out of the relation).

The **codomain** is the set of values that contains the range of a relation. It can be the same as the range or larger than the range.

The **image** of a relation is also considered to be the range or output values of the relation.

For example, suppose a relation is defined by:

- domain: {first four odd numbers} = {1, 3, 5, 7} (This also represents the input.)
- range = image: {first four even numbers} = {2, 4, 6, 8} (This also represents the output.)

Then the codomain = {natural numbers}.

Key terms

In a set of ordered pairs representing a relation, the first elements belong to the **domain** of the relation and are **input values**. The second elements belong to the **range**, also known as the **image**, and are **output values**. The **codomain** contains all the values in the range.

7.1.2 Ways of representing relations

As well as being listed as a set of ordered pairs, relations may be represented in five different ways.

- **Set notation**
  For example, $R = \{(1, 2), (3, 4), (5, 6), (7, 8)\}$

  The first elements in each pair belong to the domain and the second elements belong to the range.

  The domain is {1, 3, 5, 7} and the range is {2, 4, 6, 8}.

- **An arrow diagram**

For example, 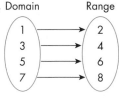 Domain   Range

In an arrow diagram, the set on the left lists the elements of the domain while the set on the right lists the elements of the range. Arrows are drawn from domain to range to connect each pair of elements.

- **A table**

For example,

x = domain	y = range
1	2
3	4
5	6
7	8

The values in the domain (input) and the values in the range (output) are tabulated, with the domain values (normally referred to as x-values) in the left-hand column and the range values (the y-values) in the right-hand column

- **A graph**

The values in a table can be used to draw a graph on a set of coordinate axes. The horizontal axis (normally called the x-axis) represents the values of the domain and the vertical or y-axis represents the values of the range.

The numbers in each pair of values are plotted against each other, to obtain points or coordinates. The points are joined by a straight line or curve to obtain a graph. Note that each point is marked by a neat dot or cross.

The information in the set, arrow diagram and table above are represented in this graph. The values of the domain are placed on the horizontal or x-axis while the values of the range are placed on the vertical or y-axis. Note that the scales used on the axes are different.

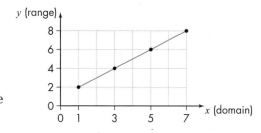

- **An equation** describing the relationship between the domain and the range

To continue the example, the relationship is: The domain $+ 1 =$ range $\Rightarrow x + 1 = y$. This relation must be true for each ordered pair.

You can produce a table of values from an equation by assigning three values to the domain and calculating three corresponding values for the range; for example, if the relation is given by $y = x + 4$, selecting three values of the domain (1, 2 and 3) and calculating the values of the range gives $y_1 = 1 + 4 = 5$, $y_2 = 2 + 4 = 6$ and $y_3 = 3 + 4 = 7$. The table will be like this.

x	1	2	3
y	5	6	7

7.1.3 Characteristics of a function

A **function** is a special type of relation.

All functions are therefore relations but not all relations are functions.

A function, being a relation, can therefore be represented in five different ways, as already listed. A function however has some unique characteristics. When a set is used to represent a function, based on the definition, the elements of the domain cannot be repeated.

> **Key term**
>
> A **function** is a relation in which each element in the domain is paired with one and only one element in the range; the domain and range are considered to be two different sets.

Examples of a function include: $A = \{(1, 5), (2, 10), (3, 15), (4, 20), (5, 25)\}$

Non-examples of functions include: $P = \{(1, 5), (2, 10), (2, 15), (3, 20), (4, 20)\}$. This is not a function because 2 is repeated in the domain.

You can use an arrow diagram to represent a function that is a **one-to-one** function, where each element in the domain is paired with one element in the range, or a **many-to-one** function, where more than one element in the domain may be paired with or mapped onto one element in the range.

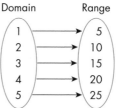
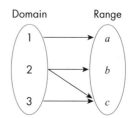

One-to-one function **Many-to-one function**

Both the one-to-one and many-to-one arrow diagrams are consistent with the definition of a function as being a relation in which one element in the domain is paired with only one element in the range.

The next arrow diagram shows a **one-to-many** relation in which one element in the domain is paired with more than one element in the range.

This is not consistent with the definition of a function; therefore the one-to-many arrow diagrams depicts a relation that is not a function.

One-to-many relation (not a function)

Graphical method – the vertical line test

When representing a relation in the form of a graph, to determine whether it is a function or not, draw a vertical line parallel to the y-axis. If this line cuts the graph at one point only, then this is consistent with the definition of a function (one element in the domain mapped onto one element in the range).

This diagram provides a graphical example of a function. The vertical lines cut the graph at only one point.

On the right is a non-example of a function, the vertical line cuts the graph in more than one point.

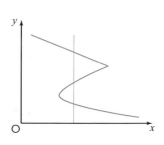

Functional notation

A relation can be written as a mapping, for example, $x \rightarrow 3x - 7$. This relation is a function and is written as:

- $f(x) = y$
- $f: x \rightarrow 3x - 7$

- $f(x) = 3x - 7$
- $y = 3x - 7$

These all mean the same thing: the function of x is $3x - 7$.

Other letters can be used to denote a function, for example:

- $f(g) = h \Rightarrow$ domain g and range h
- $s(h) = m \Rightarrow$ domain h and range m

The image of a function

To find the image of a function you need to evaluate the function.

Given f(x) = 3x – 3, evaluate f(2) and f(–5).

Substituting for x:

f(2) = 3(2) – 3

$\quad$ = 6 – 3

$\quad$ = 3

f(–5) = 3(–5) – 3

$\quad$ = –15 – 3

$\quad$ = –18

Note In this example, 2 and –5 are members of the domain while 3 and –18 are members of the range or image of the function.

Inverse functions

To derive the **inverse** of a function you interchange the domain and range values (x and y values or input and output values).

Key term

An **inverse function** is a function that reverses the operation of another function.

Notes Only bijective functions have inverses.

The next example shows how to find the inverse of a function that is in the form of an equation.

Find the inverse of the function f(x) = 2x – 1.

f(x) = 2x – 1, rewritten as: $\quad y = 2x - 1$

interchange the x and y: $\quad x = 2y - 1$

Make y the subject: $\quad \Rightarrow 2y = x + 1$

$$\Rightarrow y = \frac{x + 1}{2}$$

$f^{-1}(x) = y = \dfrac{x + 1}{2}$ (Inverse function)

This function is presented in the form of a set. Represent the inverse in five different ways.

F = {(1, 2), (3, 4), (5, 6), (7, 8), (9, 10)}

To find the inverse, interchange domain and range values (x and y) for each format.

Represented as a set

f^{-1} = {(2, 1), (4, 3), (6, 5), (8, 7), (10, 9)}

Represented as a table

x	y
2	1
4	3
6	5
8	7
10	9

Represented as an arrow diagram

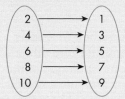

Example 4 continues on next page

Represented as an equation

The original function F can be written as: $f(x) = y = x + 1$
(deduced from the set).

To find $f^{-1}(x)$ (inverse function):

$y = x + 1$

$x = y + 1$ (interchanging x and y)

$y = x - 1$ (inverse function)

$f^{-1}(x) = x - 1$

Represented as a graph

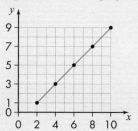

Example 5

Represent the function $f(x) = -2x - 3$ as an arrow diagram and a set.

Represented as an arrow diagram

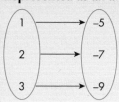

Represented as a set

$x = 1: f(1) = -2(1) - 3 = -2 - 3 = -5$

$x = 2: f(2) = -2(2) - 3 = -4 - 3 = -7$

$x = 3: f(3) = -2(3) - 3 = -6 - 3 = -9$

$f(x) = \{(1, -5)\ (2, -7)\ (3, -9)\}$

Example 6

Given the function $g(x)$ in the form of this graph, represent the function in a table and as an equation.

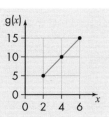

Represented as a table

x	2	4	6
$g(x)$	5	10	15

Represented as an equation

$x = 2 \Rightarrow g(2) = 2.5 \times 2$

$= 5$

$x = 4 \Rightarrow g(4) = 2.5 \times 4$

$= 10$

$g(x) = 2.5x$

Composite functions

A **composite function** is formed when functions are combined or operated, one after the other. The range of the first function is included in the domain of the next, and so on. The next example shows how to form a composite function from two functions.

> **Key term**
>
> A **composite function** is a function obtained from combining two functions so that the range of one function is contained in the domain of the other.

Example 7

Given two functions: $f(x) = 5x + 2$; $g(x) = 3x - 4$ find the composite function $fg(x)$.

Function $f(x)$ appears first so write it first but with an open set of brackets in place of the variable x.

$f(x) = 5 (\quad) + 2$

Now place function $g(x)$ inside the empty bracket to complete the operation.

$fg(x) = 5(3x - 4) + 2$

Simplifying the answer: $fg(x) = 15x - 20 + 2$

$\qquad\qquad\qquad = 15x - 18$

Generally, finding the composite function is **not** commutative: $fg(x) \neq gf(x)$.

Therefore, using the same data: $gf(x) = 3 (\quad) - 4$

$\qquad\qquad\qquad\qquad = 3(5x + 2) - 4$

$\qquad\qquad\qquad\qquad = 15x + 2$

The next example shows how to form a composite function from three functions.

Example 8

Given $f(x) = 3x + 4$, $g(x) = 9x$ and $h(x) = x + 1$, write $fgh(x)$ in terms of x and simplify your answer.

$fg(x) = 3(9x) + 4$

$\qquad = 27x + 4$

$fgh(x) = 27(x + 1) + 4$

$\qquad = 27x + 31$

Tip	Irrespective of the number of functions, the procedure for finding a composite function is progressively from left to right.

Inverse of a composite function

After determining the composite function, you need to be able to calculate the inverse.

Example 9

Given $f(x) = 2x + 2$ and $g(x) = 3x - 1$,

(a) determine:

    (i) $(gf)^{-1}(5)$ (ii) $ff^{-1}(x)$

(b) show that $f^{-1}g^{-1}(x) = (gf)^{-1}(x)$.

(a) (i) $gf(x) = 3(2x + 2) - 1$

$\qquad = 6x + 6 - 1$

$\qquad = 6x + 5$

To find $(gf)^{-1}(x)$,

let $gf(x) = y = 6x + 5$

$\Rightarrow x = 6y + 5$ (interchanging domain and range)

$y = (gf)^{-1}(x)$

$= \dfrac{x - 5}{6}$ (making y the subject to find the inverse function)

Evaluating the function: $(gf)^{-1}(5) = \dfrac{5 - 5}{6}$

$= \dfrac{0}{6}$

$= 0$

(ii) To determine $ff^{-1}(x)$,

$y = 2x + 2 \Rightarrow x = 2y + 2$

$\Rightarrow y = f^{-1}(x)$

$= \dfrac{x - 2}{2}$

$ff^{-1}(x) = 2\left(\dfrac{x - 2}{2}\right) + 2$

$= x - 2 + 2$

$= x$

(b) To determine: $f^{-1}g^{-1}$:

$f^{-1}(x) = \dfrac{x - 2}{2}$ and $g^{-1}(x) = \dfrac{x + 1}{3}$

$f^{-1}g^{-1} = \dfrac{\dfrac{(x + 1)}{3} - 2}{2}$

$= \dfrac{x + 1 - 6}{6} = \dfrac{x - 5}{6}$

$gf(x) = 6x + 5$

$(gf)^{-1}(x) = \dfrac{x - 5}{6}$

$= f^{-1}g^{-1}(x)$

Tip	When determining the inverse of composite functions, the bracket makes a big difference:

When determining the inverse of composite functions, the bracket makes a big difference:

$gf^{-1} \neq (gf)^{-1}$ since gf^{-1} represents the composition of the function $g(x)$ and $f^{-1}(x)$ whereas $(gf)^{-1}(x)$ refers to the inverse of the composite function: $gf(x)$.

$f^{-1}g^{-1}(x) = (gf)^{-1}(x)$

conversely:

$g^{-1}f^{-1}(x) = (fg)^{-1}(x)$

Exercise 7A

1   Give one example and one non-example of a relation.

2   Given the relation $R = \{(2, 5), (4, 10), (6, 15), (8, 20)\}$, write down the domain and range.

3   Given the domain $\{1, 2, 3, 4, 5\}$ and the range $\{a, b, c, d, e\}$, write down one possible codomain.

4   Represent the relation $M = \{(2, 5), (3, 6), (4, 7), (5, 8), (6, 9)\}$ as:
    a   an arrow diagram   b   a table   c   a graph   d   an equation.

5   Represent the function $f(x) = 2x - 3$ in four different ways.

6   Give one example and one non-example of a function.

7   Give one example and one non-example of a function in the form of a set.

8   Draw arrow diagrams to represent a one-to-one, a many-to-one and a one-to-many relation and indicate, giving reasons, which one of these represents a function.

9   Sketch two graphs, one that is a function and one that is not a function.

10 Given the function $f(x) = 3x - 5$, evaluate: a   f(2)   b   f(–3).

11 This table represents the function $g(x)$.

x	1	2	3	4	5	6	7	8	9	10
y	1	3	5	7	9	11	13	15	17	19

Given that x represents the elements of the domain and y the elements of the range, represent the inverse function in the form of:
    a   a set   b   a table   c   an arrow diagram   d   a graph.

12 Given the function g: $x \rightarrow \dfrac{2 - 6x}{5}$, evaluate: a   g(–3)   b $g\left(\dfrac{1}{2}\right)$.

13 Given the relation $\{(x, y): y = $ square numbers between 2 and 20; $x \in \mathbb{W}$, $\mathbb{W} = \{$whole numbers$\}\}$, state the elements of the domain and the range.

14 List the domain and range of each relation and state the codomain in each case.
    a $(p, q): 2p = 3q - 1, p \in H, H = \{$integers between 0 and 5$\}$
    b $(r, s): r = s^2 + 1, r, s \in \mathbb{W}, \mathbb{W} = \{$whole numbers up to 50$\}$.

15 a   Represent this arrow diagram as a graph.
    b   Does this arrow diagram represent a relation only or a function? Explain.

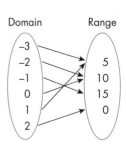

16 Given the function: $f(x) = 3x - 5$, write the inverse function.

17 Given $f(x) = 2x - 3$ and $g(x) = 3x - 7$, write down and simplify the composite function fg(x). Hence, evaluate fg(–1).

18 Given the relation $\{(a, b): b = \dfrac{0.5a - 3}{4}$, a is an even number, $a \in \mathbb{N}$,
$\{\mathbb{N}$ = natural numbers $< 10\}\}$, write down the ordered pairs of the relation.

19 Represent the function $\{(2, 2), (3, 5), (4, 8), (5, 11), (6, 14)\}$ as an equation.

20 Determine the inverse function of $g(x) = \dfrac{2x - 7}{3} - \dfrac{3}{4}$.

21 Given the functions $f(x) = \dfrac{2x - 3}{5}$ and $g(x) = \dfrac{15x - 7}{2}$, determine the composite function $gf(x)$ and hence, evaluate $gf(-4)$.

22 Given the functions $f(x) = 2x - 3$ and $g(x) = 5x$, determine $ff^{-1}(x)$, $fg^{-1}(2)$ and $f^{-1}g^{-1}(3)$.

23 Given functions $f(x) = x + 4$, $g(x) = 5x - 2$ and $h(x) = \dfrac{2x - 3}{4}$, determine:

a $fgh(x)$ **b** $gfh^{-1}(2)$ **c** $hfg^{-1}(-5)$.

7.2 Coordinate geometry and graphs of linear functions

7.2.1 Coordinate geometry

Cartesian coordinates and plotting points

You already know how to plot a graph. First, you need a set of coordinate axes, the x and y axes, that are perpendicular to each other and intersect at a point called the origin (O). At the origin the values of x and y are both 0. A set of values, in sequence, is evenly placed along each axis, and marked on the heavier grid lines on the graph paper. The scale must be uniform on both axes, although each axis may use a different scale.

The axes are drawn in a horizontal plane, called the **Cartesian plane**, in which you can plot **Cartesian** or **rectangular coordinates**, which are ordered pairs of values, to obtain points that may be connected by straight lines or curves to produce graphs.

The x and y-axes extend in both directions. Values on the x-axis are positive to the right of the origin and negative to the left of it. Values on the y-axis are positive above the origin and negative below it.

> **Key term**
>
> **Cartesian** or **rectangular coordinates** are the ordered pairs of values representing the domain and range of a relation (x and y values) that are plotted in a **Cartesian plane** to obtain points. These points are then joined by lines or curves to produce graphs.

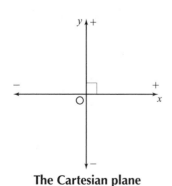

The Cartesian plane

> **Remember**
>
> When choosing the scale for plotting points, consider the greatest and smallest values in the domain and range. Unless you can use a scale of 0, 1, 2, 3… on an axis, you should choose scales that count in 'easy' numbers, such as in multiples of 2, 4, 5 or 10.

Drawing shapes from connecting the points

Plot the points: A(1, 1), B(5, 1) and C(5, 4) on a Cartesian plane, then join the points to make a closed shape. Name the figure formed and calculate its perimeter and area.

The shape is a triangle.

Its area is $\dfrac{1}{2} \times 4 \times 3 = 6$ square units

By Pythagoras' theorem, $AC = \sqrt{4^2 + 3^2}$

$$= \sqrt{25}$$

$$= 5$$

The perimeter is $4 + 3 + 5 = 12$ units

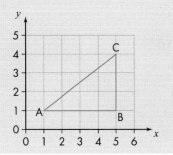

The length of a straight line segment

If you know the coordinates of the endpoints of a straight line segment, you can find the **length** of the line segment.

The diagram shows line segment PQ, with endpoints $P(x_1, y_1)$ and $Q(x_2, y_2)$.

Use Pythagoras' theorem to find the length of PQ.

Since PQ represents the hypotenuse of a right-angled triangle ($\triangle PQR$):

$$PQ^2 = PR^2 + QR^2$$

$$= (x_2 - x_1)^2 + (y_2 - y_1)^2$$

$$PQ = \sqrt{(x_2 - x_1)^2 + (y_2 - y_1)^2}$$

> **Key term**
>
> The **length** of a straight line segment is the distance between the points that mark the endpoints of the line segment.

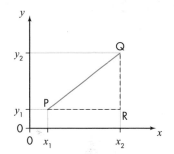

You can use this formula to find the length of any straight segment, without having to draw the graph, if you know the coordinates of the endpoints. Of course, you can also use the graphical method to find the length of the line segment when the points are plotted.

The midpoint of a line segment

The point M is the halfway point between O and P. It is the **midpoint** of OP.

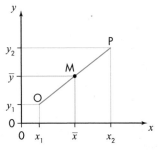

> **Key term**
>
> The **midpoint** of a straight line segment is the halfway point between the endpoints of the line segment.

You can determine the midpoint of a line segment graphically (by observation) or by using a formula.

You can find the coordinates, $(\bar{x}, \bar{y})$, of the midpoint of the lines OP and LM graphically by determining half the distance covered by the line along the x-axis and half the distance covered by the line along the y-axis.

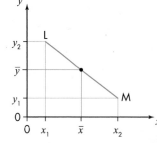

Mathematically, $\bar{x} = \dfrac{x_1 + x_2}{2}, \bar{y} = \dfrac{y_1 + y_2}{2}$.

Example 11

Given the straight line with endpoints U(–1, 3) and V(5, 5), determine the length of UV and the coordinates of the midpoint of UV.

$x_1 = -1, x_2 = 5, y_1 = 3, y_2 = 5$

$UV = \sqrt{(x_2 - x_1)^2 + (y_2 - y_1)^2}$

$\quad = \sqrt{\{5 - (-1)^2\} + (5 - 3)^2}$

$\quad = \sqrt{6^2 + 2^2}$

$\quad = \sqrt{40}$

$\quad = 6.3$

For the midpoint,

$\bar{x} = \dfrac{x_1 + x_2}{2}, \bar{y} = \dfrac{y_1 + y_2}{2}$

$\bar{x} = \dfrac{-1 + 5}{2} \qquad \bar{y} = \dfrac{3 + 5}{2}$

$\quad = 2 \qquad\qquad = 4$

$(\bar{x}, \bar{y}) = (2, 4)$

Gradient, intercept and equation of a straight line

The general equation of a straight line is $y = mx + c$, where m is the **gradient** of the line and c is the **y-intercept**.

When the line slopes upwards from left to right the gradient is positive and when it slopes down from left to right, the gradient is negative.

To find the equation of any straight line, you need to determine the values of m and c and substitute them into the general equation.

Key terms

The **gradient** or slope (m) of a line is really the steepness of the line or the angle of inclination of the line with the horizontal.

The **y-intercept** is where the line cuts the y-axis. (The **x-intercept** is where the line cuts the x-axis.)

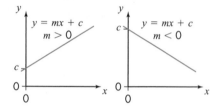

Example 12

Find the equation of the line passing through the points A(2, 1) and B(4, 2).

The gradient m is the steepness of the line.

$\Rightarrow \dfrac{\text{the change in } y}{\text{the change in } x} = \dfrac{y_2 - y_1}{x_2 - x_1}$

$\qquad\qquad\qquad = \dfrac{2 - 1}{4 - 2}$

$\qquad\qquad\qquad = \dfrac{1}{2}$

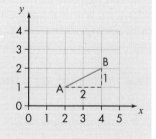

For the y-intercept c, substitute for x, y and m in the general equation $y = mx + c$.

The values for x and y can be taken from A or B since the line passes through both.

$x = 2, y = 1, m = \frac{1}{2}$

$\Rightarrow 1 = \frac{1}{2}(2) + c$

$\Rightarrow c = 1 - 1$

$\quad\; = 0$

$\quad y = \frac{1}{2}x + 0$

The equation is $y = \frac{1}{2}x$.

Sometimes there is no need to calculate c because the y-intercept is quite clear. The coordinates can also indicate the value of c, for example, if a line passes through the point $(0, a)$, then the y-intercept must be a since the first coordinate is the x-value and the second is the y-value.

Example 13

Determine the equation of the straight line passing through the point K(0, 5), with a gradient of $1\frac{1}{2}$.

$y = mx + c$

Since the line passes through K(0, 5), $c = 5$. Also, $m = 1\frac{1}{2}$ (given).

The equation is $y = 1\frac{1}{2}x + 5$.

Parallel and perpendicular lines

If two lines are parallel to each other in the Cartesian plane, their gradients are equal.

Therefore, for parallel lines L_1 and L_2 with gradients m_1 and m_2 respectively, $m_1 = m_2$.

Key fact

The gradients of two lines in a Cartesian plane are equal if the lines are parallel.

If two lines are perpendicular to each other, the product of their gradients is –1.

Key fact

The product of the gradients of two lines in a Cartesian plane is –1 if the lines are perpendicular to each other.

For perpendicular lines D_1 and D_2 with gradients m_1 and m_2 respectively, $m_1 \times m_2 = -1$ or $m_1 = -\dfrac{1}{m_2}$.

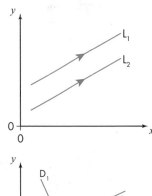

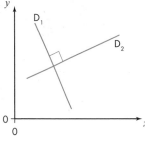

Example 14

Determine the equation of a straight line parallel to the line $y = 3x + 4$ and passing through the point P(0, 8).

Parallel lines have the same gradient. The gradient of $y = 3x + 4 \Rightarrow m = 3$.

For the line through P: $m = 3$ and $c = 8$

The equation is $y = 3x + 8$.

Example 15

Determine the equation of a line passing through the point R(–3, 2) and perpendicular to the line between S(6, 4) and T(8, 8).

The gradient of ST is $\dfrac{y_2 - y_1}{x_2 - x_1} = \dfrac{8 - 4}{8 - 6}$

$$= \dfrac{4}{2}$$

$$= 2$$

The gradient of the line through R $= -\dfrac{1}{2}$ (minus the reciprocal of the gradient of ST).

The equation of the line through R(–3, 2) is $m = -\dfrac{1}{2}$ and $x = -3$ when $y = 2$.

Find c by substituting in $y = mx + c$.

$$\Rightarrow 2 = -\dfrac{1}{2}(-3) + c$$

$$\Rightarrow 2 = \dfrac{3}{2} + c$$

$$\Rightarrow c = 2 - 1\dfrac{1}{2}$$

$$= \dfrac{1}{2}$$

Substituting for m and c, the equation is $y = -\dfrac{1}{2}x + \dfrac{1}{2}$.

Example 16

Study these linear equations and determine whether the lines in each pair are parallel, perpendicular or neither.

(a) $2x - 3y = 6$ (b) $3x + 12 = 6y$

     $5y = 10x + 15$ $2y + 4x = 1$

Rewrite the equations in the format $y = mx + c$.

(a) $-3y = -2x + 6 \Rightarrow y = \dfrac{2}{3}x - 2$

     $5y = 10x + 15 \Rightarrow y = 2x + 3$

     $m_1 = \dfrac{2}{3}$ and $m_2 = 2$

     Since $m_1 \neq m_2$ and $m_1 \times m_2 \neq -1$, the lines are neither parallel nor perpendicular.

(b) $3x + 12 = 6y \Rightarrow y = \dfrac{3x}{6} + \dfrac{12}{6}$

               $\Rightarrow y = \dfrac{1}{2}x + 2$ (dividing by 6)

               $\Rightarrow m_1 = \dfrac{1}{2}$

     $2y + 4x = 1 \Rightarrow 2y = -4x + 1$

               $\Rightarrow y = -2x + \dfrac{1}{2}$ (dividing by 2)

               $\Rightarrow m_2 = -2$

     $m_1 \neq m_2$ but $m_1 \times m_2 = \dfrac{1}{2} \times -2 = -1$

Therefore the lines are perpendicular to each other.

Equation of the perpendicular bisector

You know from your work in geometry that the perpendicular bisector of a line segment AB is the line perpendicular to AB and passing through its centre or midpoint. To find the equation of the perpendicular bisector of a line algebraically, first find the gradient of the line and the coordinates of its midpoint. Then determine the equation of the line passing through the midpoint of the line and perpendicular to it.

Example 17

Determine the coordinates of the midpoint of the line through P(–3, –5) and Q(7, 3). Hence determine the equation of the perpendicular bisector of PQ.

The midpoint of PQ is M $(\bar{x}, \bar{y})$, where:

$$\bar{x} = \frac{x_1 + x_2}{2} \qquad\qquad \bar{y} = \frac{y_1 + y_2}{2}$$

$$= \frac{-3 + 7}{2} \qquad\qquad = \frac{-5 + 3}{2}$$

$$= \frac{4}{2} \qquad\qquad\qquad = \frac{-2}{2}$$

$$= 2 \qquad\qquad\qquad\quad = -1$$

$$\Rightarrow M(2, -1)$$

The gradient of PQ is $\dfrac{y_2 - y_1}{x_2 - x_1} = \dfrac{3 - (-5)}{7 - (-3)}$

$$= \frac{8}{10}$$

$$= \frac{4}{5}$$

Thus the gradient of the perpendicular bisector of PQ $= -\frac{5}{4}$

Since the perpendicular bisector passes through M(2, –1):

$$y = mx + c \Rightarrow -1 = -\tfrac{5}{4}(2) + c$$

$$\Rightarrow c = 1\tfrac{1}{2}$$

The equation of the perpendicular bisector is $y = -\frac{5x}{4} + 1\frac{1}{2}$.

Exercise 7B

1 State, giving reasons, whether each function is a linear function.
   a $f(x) = 2x - 5$ **b** $g(x) = x^2$ **c** $y = 9$ **d** $h(x) = x$

2 Draw a graph of the function $f(x) = 5x + 1$.

3 Draw a graph of the function $g(x) = 2 - x$.

4 Draw a graph of the function: **a** $y = 10$ **b** $x = 2$.

5 Draw a graph of the function **a** $y = 7 - 3x$ **b** $2y + 4 = 8x$.

6 Plot the points P(1, 7) and Q(5, 2), join the points with a straight line and determine the length of PQ.

7.2 Coordinate geometry and graphs of linear functions 345

7 Plot the points A(1, 2), B(9, 2) and C(9, 8).

    a Join the points to make a shape.

    b Name the shape formed.

    c Determine the perimeter and area of the shape.

    d Determine the length of AC and the coordinates of the midpoint of AC.

8 Rewrite the equation of the straight line $2x - 3y = 6$ in the form $y = mx + c$.

    Write down the values of m and c.

9 Write down the equation of the straight line with gradient $3\frac{1}{2}$ that makes an intercept of $\frac{3}{4}$ on the y-axis.

10 Draw a graph of the function $5x - 10y = 20$.

11 Plot the points P(3, 5), Q(7, 5), R(7, 11) and S(3, 11).

    a Join the points with straight lines to make a shape.

    b Name the shape formed.

    c Calculate the perimeter and area of the shape.

    d Determine the length of PR and the coordinates of the midpoint of QS.

12 Determine whether the lines in each pair are parallel to each other, perpendicular to each other or neither parallel nor perpendicular. Justify your answers.

    a $2x - 3y = 4$ **b** $7x = 21y - 42$ **c** $10y - 20x + 40 = 0$

       $5y + 1 = 15x$ $2 + 4y = -12x$ $6x + 24 = 3y$

13 Determine the equation of a straight line that passes through the points L(0, 1) and M(2, 8).

14 Determine the equation of the straight line passing through K(2, 5) and J(7, 9).

15 Determine the equation of a line parallel to the line $2y = x + 1$ and passing through Q(0, 7).

16 Determine the equation of the line perpendicular to the line $2y - 3x = 6$ and passing through A(2, 6) and B(5, 8).

17 Determine the equation of the line passing through (0, 0) and (5, 5).

18 Draw the graph of the function $\frac{3}{2}x - \frac{1}{4} = \frac{3}{8}y + \frac{1}{2}$.

19 Plot the points T(–5, –6), U(7, –6), V(4, 8) and W(–2, 8).

    a Join the points with straight lines to make a shape.

    b Name the shape formed.

    c Calculate the perimeter and area of the shape.

    d Determine the length of TV, using the equation for the length of a straight line.

    e Determine the coordinates of the midpoint of UW.

20 Determine the equation of a straight line that passes through the two points A(0, 8) and B(16, 0).

7 Relations, functions and graphs

21 **a** Determine the equation of a line passing through L(–3, 6) and M(6, –8).

      b Determine the equation of a line perpendicular to LM and passing through P(9, 0).

22 Determine the equation of the perpendicular bisector of the line HK through the points H(2, 10) and K(8, –4).

23 Determine the equation of the line passing through C(0, 4) and D(0, 7).

24 Determine the equation of the line passing through E(5, 0) and F(10, 0).

25 Determine the equation of the perpendicular bisector of the line joining R(–14, 0) and T(0, 14).

26 Plot the points W(–4, 0), X(3, –7), Y(7, 2) and Z(–1, 5).

      a Join the points and name the shape formed.

      b Calculate the perimeter and area of the shape.

      c Determine the coordinates of the midpoint of XZ.

27 Draw the graph of the function $\dfrac{2.3}{1.7}x - \dfrac{2.8}{0.4} = \dfrac{3.2}{6.5}y + \dfrac{5.1}{3.5}$.

7.2.2 Drawing graphs of linear functions

Drawing graphs

You have seen that the graph of a **linear function** is a straight line.

These examples show some of the forms a linear function can take.

- $f(x) = 8x - 7$
- $g(x) = 9 - x$
- $h(x) = x$
- $2x + y = 9$
- $k(x) = 2$
- $f: x \rightarrow 1 - x$
- $y = 4$
- $y = x$

> ### Key term
>
> A **linear function** is a function in which the highest power of the variable is not greater than one (1). A graph of such a function results in a straight line.

For each of these forms, you could draw a graph. The graph of a linear function is always a straight line and the minimum number of points you need, to be able to draw a straight line, is two. However, to cater for any error in computation it is sensible to use a minimum of three points to draw the graph.

Example 18

Draw the graph of the function $f(x) = 8x - 7$.

Evaluate the function for three values of x: 0, 1, 2 (preferably), although any three values can be used.

$f(0) = 8 (0) - 7 = -7$

$f(1) = 8 (1) - 7 = 8 - 7 = 1$

$f(2) = 8 (2) - 7 = 16 - 7 = 9$

Use these to complete a table of values.

x	0	1	2
y	–7	1	9

Choose suitable scales for the axes and plot the points, joining them to give a line that extends to the full range of the domain and range.

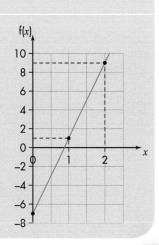

Example 19

Draw the graph of h(x) = x.

The values of the domain and the range are equal.

Here is a typical table of values for this function.

x	1	2	3
h(x)	1	2	3

Now draw the graph.

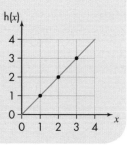

The graph bisects the angle between the axes, if the scale on both axes is the same.

Example 20

Draw the graph of the function $2x + y = 9$.

Make y the subject: $y = -2x + 9$.

Calculate some values for the function $y = -2x + 9$.

$x = 0 \Rightarrow y = -2(0) + 9 = 9$

$x = 1 \Rightarrow y = -2(1) + 9 = 7$

$x = 2 \Rightarrow y = -2(2) + 9 = 5$

Draw up a table of values.

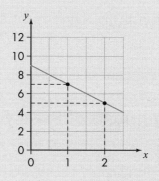

x	0	1	2
y = -2x + 9	9	7	5

Draw the graph.

Example 21

Draw the graphs of these functions.

(a) $x = 4$ (b) $y = 3$

(a) At any point on this graph, x will have a value of 4.

The graph is a straight line passing through 4 on the x-axis and parallel to the y-axis.

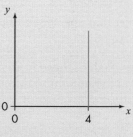

(b) At any point on this graph, y will have a value of 3.

The graph is a straight line passing through 3 and parallel to the x-axis.

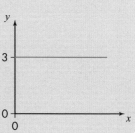

7.2.3 Graphical solution of simultaneous equations

In algebra, **simultaneous equations** are solved in a number of ways: using matrices, by elimination and by substitution. It is also possible to solve them by drawing graphs of linear equations.

For the graphical method, you plot and draw the graph of each of the simultaneous equations on the same set of coordinate axes, so that they intersect. The coordinates of the point of intersection represent the solution to the simultaneous equations.

Example 22

Solve the simultaneous equations $2x - y = 2$ and $x - y = -1$ graphically.

Make y the subject in each equation.

$y = 2x - 2$

$y = x + 1$

Draw up a table of values, ready for plotting both equations on the same axes.

x	1	2	3
y_1	0	2	4
y_2	2	3	4

Plot the graphs on the same axes.

Scale (x-axis) 1 : 1

Scale (y-axis) 1 : 1

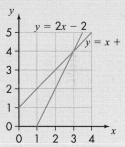

The graphs intersect at (3, 4) so the solution is $x = 3$, $y = 4$.

Example 23

A student wanting to buy exercise books and pens found that two exercise books and one pen cost $4:00 and that one exercise book and 1 pen cost $3.00. Use a graphical method to determine the individual cost of one pen and one exercise book.

Let the cost of an exercise book be $$x$ and the cost of a pen be $$y$.

Write two simultaneous equations based on the data given.

$2x + y = 4$

$x + y = 3$

Rewrite each equation with y as the subject.

$y = 4 - 2x$

$y = 3 - x$

Plot two graphs on the same axes, noting the point of intersection of the graphs as the values of x and y.

x		0	1	2
$y_1 = 4 - 2x$		4	2	0
$y_2 = 3 - x$		3	2	1

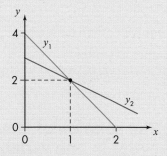

The graphs intersect at the point (1, 2) where $x = 1$; $y = 2$.

This means that the cost of a book is $1.00 and the cost of a pen is $2.00.

Tip

The intercept of the y-axis is found when $x = 0$ and the intercept of the x-axis is found when $y = 0$.

Example 24

Solve the simultaneous equations $2y - 4x = 6$ and $3y - 9x = 12$ graphically.

Rewrite and simplify both equations.

$y - 2x = 3$ (dividing by 2)

$y - 3x = 4$ (dividing by 3)

Make y the subject.

$y = 2x + 3$

$y = 3x + 4$

Draw up a table of values, ready for plotting both equations on the same axes.

x	0	1	2
$y_1 = 2x + 3$	3	5	7
$y_2 = 3x + 4$	4	7	10

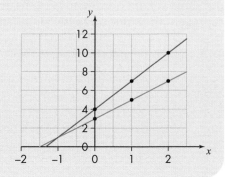

The graphs intersect at the point $(-1, 1)$ so the solution is $x = -1$, $y = 1$.

Note

If the graphs do not intersect within the range of values you have plotted, extend the lines in the appropriate direction until they do. This is the case in the example above.

7.3 Linear inequalities

7.3.1 Definition of linear inequalities

You know that an inequality is a statement that two quantities are unequal. A **linear inequality** is a mathematical way of expressing this.

Here are some everyday examples of the use of symbols to express inequalities.

John is older than Harry. ($J > H$)

You have to be at least 18 years old to vote. ($V \geq 18$)

The minibus cannot hold more than 14 passengers. ($M \leq 14$)

Trinidad is bigger than Grenada. ($T > G$)

Mary runs faster than Jane. ($S_M > S_J$)

The number of students in class 5 is less than the number in class 4. ($C_5 < C_4$)

Earth is larger than Pluto. ($E > P$)

7.3.2 The solution of linear inequalities in one variable

Inequalities are solved in a similar way to equations, but you must take care if you multiply through by a negative number. This results in a change of sign and a reversal of the inequality symbol.

Key term

A **linear inequality** is a mathematical statement that shows one quantity or number to be greater than or less than another and is represented by the symbols:

- $<$ (less than)
- $>$ (greater than)
- $\geq$ (greater than or equal to)
- $\leq$ (less than or equal to).

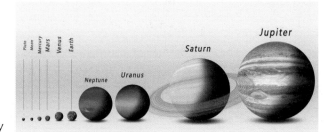

Example 25

Solve $5x - 10 > 20$.

$5x > 20 + 10 \Rightarrow 5x > 30$

$\qquad\qquad \Rightarrow x > 6$ (dividing each side by 5)

Example 26

Solve $-2x > 4$.

$-2x > 4 \Rightarrow 0 > 4 + 2x$

$\qquad\qquad \Rightarrow -4 > 2x$

$\qquad\qquad \Rightarrow -2 > x$

$\qquad\qquad \Rightarrow x < -2 \qquad$ (reversing the inequality to get x on the LHS)

Alternatively:

$-2x > 4 \Rightarrow 2x < -4 \qquad$ (multiplying through by -1)

$\qquad\qquad \Rightarrow x < -2 \qquad$ (dividing through by 2)

In either case, the inequality sign is changed from $>$ to $<$ to get from $-2x$ to x.

Representing the solution of linear inequalities in one variable

The solutions of a linear inequality may be represented as set, on the number line or in a graph.

Representation as a set

Example 27

Solve the inequality $4x - 6 \geq 18$, $x \in \mathbb{N}$, $\mathbb{N}$ = {natural numbers}, representing the solution as a set.

$4x - 6 \geq 18 \Rightarrow 4x \geq 18 + 6$

$\qquad\qquad \Rightarrow 4x \geq 24$

$\qquad\qquad \Rightarrow x \geq 6$ (dividing through by 4)

Represented as a set, the solution is $x = \{6, 7, 8 \ldots\}$.

Example 28

Solve the inequality $9 - 7x < 3x - 5$, $x \in \mathbb{W}$, $\mathbb{W}$ = {whole numbers}, representing the solution as a set.

$9 - 7x < 3x - 5 \Rightarrow -7x - 3x < -9 - 5$

$\qquad\qquad \Rightarrow -10x < -14$

$\qquad\qquad \Rightarrow \dfrac{-10}{-10}x < \dfrac{-14}{-10}$

$\qquad\qquad \Rightarrow x > 1.4 \qquad$ (**Note** that the inequality sign is reversed to make x positive.)

The solution set is $x = \{x: x \geq 2\}$, $x \in \mathbb{W}$, $\mathbb{W}$ = {whole numbers}. (The minimum integer value is 2.)

Representation on a number line

The solution of inequalities may be represented on a number line, like this one.

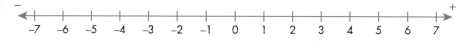

Arrows are drawn over the number line, with the pointer indicating the direction of the possible solutions and a circle at the other end situated over the minimum or maximum value, a. An open circle indicates an inequality $<$ or $>$, while a solid or closed circle indicates $\leqslant$ or $\geqslant$.

If the solution of the inequality lies within a specific range of values, the arrow is replaced with a line having a circle at each end, again being open or closed depending on the symbol $\geqslant$ or $>$, $\leqslant$ or $<$.

Example 29

Solve the inequality $5 > 3x - 4$, $x \in \mathbb{Z}$, $\mathbb{Z} = \{\text{integers}\}$. Represent the solution on an integer number line.

$5 > 3x - 4 \Rightarrow 5 + 4 > 3x$

$\Rightarrow 9 > 3x$

$\Rightarrow 3 > x$

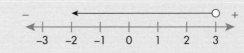

Example 30

Solve these inequalities: and represent the solutions on a number line for real numbers.

(a) $3x + 1 > 4x - 8$ (b) $3x + 1 \leqslant -5x + 12$

(a) $3x + 1 > 4x - 8 \Rightarrow 8 + 1 > 4x - 3x$

$\Rightarrow x < 9$

(b) $3x + 1 \leqslant -5x + 12 \Rightarrow 8x \leqslant 11$

$\Rightarrow x \leqslant \dfrac{11}{8}$

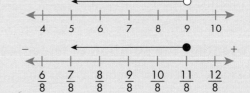

Example 31

Solve the inequality $-8 < 3x - 5 \leqslant 16$, representing your solution on a number line.

$-8 + 5 < 3x - 5 + 5 \leqslant 16 + 5$ (adding 5 to every expression to get rid of the -5)

$-3 < 3x \leqslant 21$

$-1 < x \leqslant 7$ (dividing through by 3)

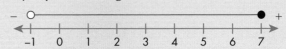

Representation of linear inequalities in one variable on a graph

To represent the solution of an inequality as a graph, start by drawing the graph of the corresponding equation, replacing the inequality sign with an equals sign, for example, for the inequality $x > 2$, draw the graph of $x = 2$. However, if the inequality is $<$ or $>$, draw the graph as a **broken line** and if the inequality is $\leqslant$ or $\geqslant$, draw the graph as a **solid** line.

To indicate the solution, shade the region that satisfies the inequality:

- for an inequality $x < a$, shade the region to the left of the line $x = a$, for an inequality $x > a$, shade the region to the right of the line $x = a$
- for an inequality $y < a$, shade the region below the line $y = a$, for an inequality $y > a$, shade the region above the line $y = a$.

Example 32

Solve each inequality, representing your solutions on separate graphs.

(a) $2x - 3 > 5$, $x \in \mathbb{W}$, $\mathbb{W} = \{$whole numbers$\}$

(b) $3y - 6 \leqslant y + 2$, $y \in \mathbb{W}$, $\mathbb{W} = \{$whole numbers$\}$

(a) $2x - 3 > 5 \Rightarrow 2x > 3 + 5$

$\Rightarrow 2x > 8$

$\Rightarrow x > 4$

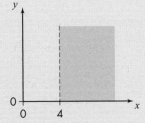

(b) $3y - 6 \leqslant y + 2 \Rightarrow 2y \leqslant 8$

$\Rightarrow y \leqslant 4$

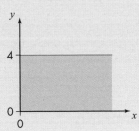

7.3.3 Graphs of linear inequalities in two variables

Just as there are linear equations in two variables, there are also linear inequalities in two variables. To represent a linear inequality in two variables on a graph:

- write the inequality with y as the subject
- replace the inequality symbol by an equals sign and draw the graph
- use a solid line if the sign is $\leqslant$ or $\geqslant$ and a broken line for $<$ or $>$
- shade the appropriate region, based on the inequality.

Example 33

Represent the solution of the inequality $2y - 6x \geqslant 4$ on a graph.

$2y \geqslant 6x + 4 \Rightarrow y \geqslant 3x + 2$ (dividing by 2)

Draw up a table of values for $y = 3x + 2$.

x	0	1	2
$y = 3x + 2$	2	5	8

Draw the graph, using a solid line since the inequality is $\geqslant$.

Shade the region above the line, since the inequality is $\geqslant$.

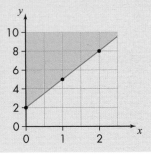

Tip

If you are unsure which region to shade, choose a point on the graph and substitute the values into the inequality. If your point satisfies the inequality, shade that region. If it does not, shade the other region.

The region enclosed by inequalities

Sometimes, you may need to solve a problem involving three (or more) inequalities. This may result in a shaded region in the shape of a polygon that contains all the points that satisfy all inequalities.

Example 34

Shade the region satisfied by the inequalities $y \geqslant 4x$, $y \leqslant 10$ and $10x + 5y - 50 \geqslant 0$.

$10x + 5y - 50 \geqslant 0 \Rightarrow 5y \geqslant -10x + 50$

$$\Rightarrow y \geqslant -2x + 10, \ x, y \in \mathbb{W}, \ \mathbb{W} = \{\text{whole numbers}\}$$

Draw up a table of values for the three equations. Then draw the graphs.

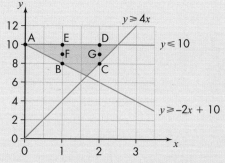

x	0	1	2
y = 10	10	10	10
y = 4x	0	4	8
y = −2x + 10	10	8	6

Shade the region according to the inequalities. It is actually the region representing the intersection of all shaded areas. Since x and y are both whole numbers, only certain points in the region meet the condition of all inequalities. These are A(0, 10); B(1, 8); C(2, 8), D(2, 10), E(1, 10), F(1, 9) and G(2, 9).

7.3.4 Linear programming

Linear programming is very useful, especially in business where it is used to maximise profits and minimise loss through the creation of linear functions and inequalities to represent and establish relationships between cost, profit and other financial quantities. The solution and graphing of these inequalities result in a feasible region, the vertices of which are used to determine the maximum and minimum values of the functions.

> **Key term**
>
> **Linear programming** is a mathematical method used to find the most useful solution to a set of conditions.

Example 35

Draw the graph of the solution set and hence determine the maximum and minimum values of $10x + 5y - 1$ that satisfy the solution set for $x \geqslant 0$, $x \leqslant 1$, $x + y \leqslant 5$ and $y - 2x \geqslant 0$.

Draw the graphs by changing the inequalities to equalities and then determine the feasible region by shading.

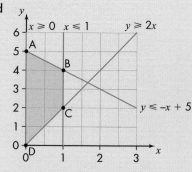

x	0	1	2
x = 0			
x = 1			
y = −x + 5	5	4	3
y = 2x	0	2	4

The four vertices of the region are: A(0, 5), B(1, 4), C(1, 2) and D(0, 0).

Find the maximum and minimum value of the function by evaluating $10x + 5y - 1$ at each vertex.

At A(0, 5): $10x + 5y - 1 = 10(0) + 5(5) - 1 = 24$

At B(1, 4): $10x + 5y - 1 = 10(1) + 5(4) - 1 = 29$

At C(1, 2): $10x + 5y - 1 = 10(1) + 5(2) - 1 = 19$

At D(0, 0): $10(0) + 5(0) - 1 = -1$

The maximum value of the function is at B, where it is 29.

The minimum value is at D, where it is −1.

Exercise 7C

1 Using symbols, express each statement as an inequality.

    a Spain is larger than Jamaica.

    b The snail is slower than the rabbit.

    c John is at least 12 years old.

    d The height of the tree was between 8 m and 9 m.

    e Class B had at most 35 students.

2 Solve these inequalities.

    a $2x - 3 > 7$ **b** $5x + 4 \leqslant 7$

3 Solve these inequalities, representing each answer as a solution set.

    a $3x - 9 \geqslant 11x + 7$ **b** $14x + 8 - 10x - 12 < 0, x \in \mathbb{R}$ ($\mathbb{R}$ = {real numbers})

4 Solve each inequality, representing your answer on a number line.

    a $5(2x - 4) \leqslant 3x + 4$ **b** $17(x - 1) > 6(2x - 8), x \in \mathbb{Z}$ ($\mathbb{Z}$ = {integers})

5 Solve each inequality, representing your solution on a graph.

    a $1\frac{1}{4}x - 3 \leqslant 4\frac{3}{4}x + 11$ **b** $\dfrac{5x - 3}{2} > \dfrac{3}{8}$

6 Represent the solution to each inequality on a graph.

    a $y > -3$ **b** $x \leqslant 7$ **c** $y > x$ **d** $y \geqslant 2x + 1$ **e** $2y - 4x > 12$

7 Solve the inequality $3(4x - 9) \geqslant 2x + 3, x \in \mathbb{W}$, representing your answer in a set.

8 Solve the inequality: $8\left(\dfrac{1}{2}x - \dfrac{3}{4}\right) \leqslant 5\left(2x - \dfrac{2}{5}\right), x \in \mathbb{Z}$ ($\mathbb{Z}$ = {integers}), representing your answer on a number line.

9 Represent on a graph the solution to the inequality $\dfrac{12x - 2}{4} < \dfrac{4x - 5}{2}$.

10 Represent on a graph the solution to the inequality $\dfrac{3x - 4}{5} + \dfrac{3}{10} \leqslant \dfrac{x}{10} + \dfrac{1}{2}$.

11 Solve $\dfrac{2x - 3y}{4} > \dfrac{1}{2}$, representing your solution on a graph.

12 Shade the region enclosed by the inequalities $x \geqslant 0$, $y \leqslant -x + 8$ and $2y \geqslant x + 2$.

13 Draw the graph of the solution set for $x \geqslant 2$, $x \leqslant 8$, $y \leqslant 10$ and $y \geqslant 3$. Use this solution set to find the maximum and minimum values of $3x - 7y + 2$.

14 Solve the simultaneous equations $y = 2x - 3$ and $y = x + 1$ by a graphical method.

15 The sum of two numbers a and b is 13. When twice the first number is added to three times the second number the result is 36. Use a graphical method to determine a and b.

16 Solve, representing your answer in a set, $\dfrac{2x - 3}{5x - 4} \geqslant \dfrac{3}{4} + \dfrac{1}{8}$ ($x \in \mathbb{W}$, $\mathbb{W} = \{\text{whole numbers}\}$).

17 Solve the inequality $2x - 7 < 15x + 1 \leqslant 11x + 8$, $x \in \mathbb{Z}$, $\mathbb{Z} = \{\text{integers}\}$. Represent your solution on a number line.

18 Represent on a graph the solution to the inequality $\dfrac{\sqrt{12x - 5}}{2} > \dfrac{1}{2}$.

19 Solve $\dfrac{4x - 3}{5} \leqslant \dfrac{5y + 1}{2} + \dfrac{3}{4}$, representing your answer on a graph.

20 Shade the region enclosed by the inequalities $\dfrac{2x - y}{5} > 2$, $y \geqslant 2$, $y \leqslant 8$ and $y \leqslant \dfrac{-5x - 10}{3}$.

21 Draw the graph of the solution set for $y \leqslant 5x$, $x + y \leqslant 10$, $x < 4$ and $y + x \geqslant 0$. Use this solution set to find the maximum and minimum integer values of $12x - 14y + 22$.

22 The midpoint of a line with gradient 2 is M(3, 4). Determine the equation of the line and the equation of the perpendicular bisector of the line.

23 Use intersecting graphs to solve these simultaneous equations.

$2x + 3y = 8$

$5x + 4y = 7$

7.4 Quadratic functions

A quadratic function is a function of the form $f(x) = ax^2 + bx + c$, where the variable is x, a and b are coefficients and c is a constant.

> **Key term**
>
> A **quadratic function** is one in which the highest power of the variable is 2.

The quadratic function can also be written as: $y = ax^2 + bx + c$, with a, b and c defined as before.

Note that in the equation $y = ax^2 + bx + c$:

a is the coefficient of x^2

b is the coefficient of x

c is a constant term (the point at which the curve cuts the y-axis)

x is the independent variable

y is the dependent variable.

The graph of a quadratic function is a smooth curve, called a **parabola**.

> **Key term**
>
> A parabola is a curve produced by plotting the graph of a **quadratic function**.

Depending on the value of the constants a, b and c the quadratic function can have different formats, for example:

- $f(x) = 2x^2 + 5x + 3$ (a, b, $c \neq 0$)
- $f(x) = 10x^2 - 6x$ ($c = 0$)
- $f(x) = 12x^2$ ($b = c = 0$)
- $f(x) = -2x^2 + 3$ ($b = 0$)

These graphs of quadratic functions illustrate functions based on the values of a, b and c.

$f(x) = ax^2 + bx + c$ ($a > 0$; $b, c \neq 0$), turning point is a minimum, two roots (values of x) exist.

If $b^2 - 4ac > 0$, then two real roots exist. Note that if $a \leqslant 0$, the graph is inverted, see below.

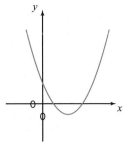

$f(x) = ax^2$ ($a > 0$; $b = c = 0$), turning point is a minimum, only one root exists.

If $b^2 - 4ac = 0$, then tonly one real root exists.

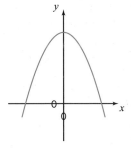

$f(x) = -ax^2 + c$; ($a < 0$; $b = 0$), turning point is a maximum, two roots exist.

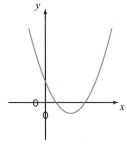

$f(x) = ax^2 + bx + c$ ($a > 0$, $b = 0$, $c > 0$), turning point is a minimum, no real roots.

Note that when a and c are both negative the graph has a maximum turning point and no real roots.

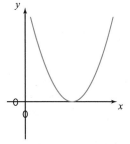

$f(x) = ax^2 + c$ ($a > 0$, $b = 0$), turning point is a minimum, no real roots exist.

If $b^2 - 4ac < 0$, then no real roots exist.

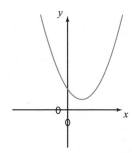

7.4.1 Plotting the graph of a quadratic function

Generally, to plot the graph of a linear function you need three points and the graph is a straight line.

For a quadratic function, which results in a parabola, you generally need seven points. For each member of the domain you obtain an image.

Example 36

Plot the graph of the function $y = 2x^2 - 4x + 2$ for the domain $x: -2 \leqslant x \leqslant 4$.

Draw up a table of values by evaluating the function for different values of the domain.

$f(-2) = 2(-2)^2 - 4(-2) + 2$
$\qquad = 8 + 8 + 2$
$\qquad = 18$

$f(-1) = 2(-1)^2 - 4(-1) + 2$
$\qquad = 2 + 4 + 2$
$\qquad = 8$

$f(0) = 2(0)^2 - 4(0) + 2$
$\qquad = 0 + 0 + 2$
$\qquad = 2$

$f(1) = 2(1)^2 - 4(1) + 2$
$\qquad = 2 - 4 + 2$
$\qquad = 0$

$f(2) = 2(2)^2 - 4(2) + 2$
$\qquad = 8 - 8 + 2$
$\qquad = 2$

$f(3) = 2(3)^2 - 4(3) + 2$
$\qquad = 18 - 12 + 2$
$\qquad = 8$

$f(4) = 2(4)^2 - 4(4) + 2$
$\qquad = 32 - 16 + 2$
$\qquad = 18$

x	-2	-1	0	1	2	3	4
$f(x)$	18	8	2	0	2	8	18

Then plot the graph from values in the table.

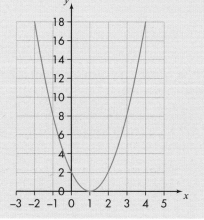

Tip

The graph of a quadratic function in which the coefficient of x^2 is negative ($a < 0$) takes the shape ∩ (inverted ∪).

7.4.2 The axis of symmetry

Looking carefully at the table in the preceding example, you can see that for values of the domain on either side of the central value ($x = 1$), the value of the range is the same. This central value is known as the **axis of symmetry**.

Key term

The **axis of symmetry** is a mirror line that divides the curve into two identical halves.

For the quadratic function, the axis of symmetry is given by the formula $x = \dfrac{-b}{2a}$.

Once you have determined the axis of symmetry, you need only one set of values in the domain (to the left or right of the axis of symmetry) to find the range, since the other set will produce the same values, as you saw in example 36.

7 Relations, functions and graphs

Example 37

Use the axis of symmetry to plot the graph of the function $f(x) = -2x^2$.

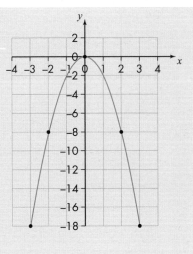

The axis of symmetry is $x = \dfrac{-b}{2a} \Rightarrow x = \dfrac{-0}{-4} = 0$ (from the function, $a = -2$, $b = 0$).

Since $x = 0$, the domain will consist of three values to the right of the axis of symmetry and three values to the left. You need only calculate the four corresponding range values (including the axis of symmetry).

$f(0) = -2(0)^2 = 0$ $f(2) = -2(2)^2 = -8$

$f(1) = -2(1)^2 = -2$ $f(3) = -2(3)^2 = -18$

Only four values of the range have been calculated. The other values may be deduced.

Produce the table of values and use it to plot the graph.

x	–3	–2	–1	0	1	2	3
$f(x)$	–18	–8	–2	0	–2	–8	–18

7.4.3 The maximum or minimum of the quadratic function

A function of the form $f(x) = ax^2 + bx + c$ has a **minimum value** if $a > 0$, since the graph is in the form of a $\cup$-shaped parabola. In example 36 the minimum point of the curve corresponds to a point 0 on the y-axis, therefore the minimum of that function is $f(x) = 0$.

A function of the form $f(x) = ax^2 + bx + c$ has a **maximum value** if $a < 0$, since the graph is in the shape of an inverted $\cup$, as shown in example 37. The maximum point is the highest point on the graph, which again turns out to be 0.

7.4.4 Interpreting the graph of a quadratic function

The gradient of a curve

The graph of a linear function is a straight line that has a constant gradient. The graph of a quadratic function, however, is a parabola or curve and the gradient is constantly changing. To find a gradient of a curve you generally find the gradient at a point on the curve. To do this, draw

> **Key term**
>
> A **tangent** is a line that touches the curve at only one point.

a **tangent** at the point and find the gradient of the tangent from the formula $m = \frac{y_2 - y_1}{x_2 - x_1}$, where (x_1, y_1) and (x_2, y_2) are two suitable points on the tangent.

Example 38

Study these sketches of quadratic functions and write down the coordinates of the points A, B, C and D. Draw a tangent at the given point in each curve.

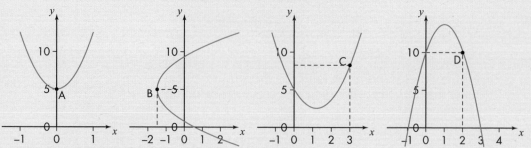

The points are A(0, 5), B(–1.5, 5), C(3, 8) and D(2, 10).

Here are the tangents at the points.

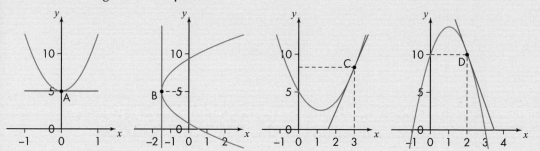

Example 39

Determine the gradient at each point, A, B, C and D, in example 38.

At point A, the tangent is horizontal which results in a gradient = 0.

At B, the tangent is vertical and hence the gradient is infinite or undefined.

At C the tangent has a positive gradient determined from the formula $\dfrac{y_2 - y_1}{x_2 - x_1}$

The gradient at C is therefore $\dfrac{8 - 0}{3 - 1.6} = \dfrac{8}{1.4} = 5.7$

At D the tangent is sloping down from left to right and as a result it has a negative value. This can also be determined from the formula $\dfrac{y_2 - y_1}{x_2 - x_1}$.

The gradient at D is therefore $\dfrac{10 - 0}{2 - 3.5} = -\dfrac{10}{1.5} = -6.7$

You can make an estimation of the gradient of a tangent to a curve at a specific point, based on the slope of the tangent.

Example 40

Find the gradient of this curve at the point P(5, 0).

The gradient at P = the gradient of the tangent at P. You can find this by determining the vertical displacement of any segment of that tangent divided by a corresponding horizontal displacement (rise/run).

From the diagram, this works out to be $\frac{5}{1} = 5$.

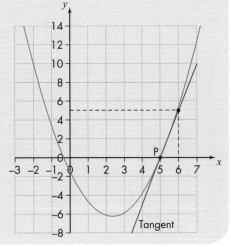

Turning points

At a **turning point** of a curve, the tangent to the curve is parallel to the x-axis, resulting in a gradient of 0.

From one side of the turning point to the other, the gradient changes sign (positive to negative or negative to positive).

> **Key term**
>
> A **turning point** of a curve is the point on the curve at which it changes direction.

A **maximum point** is one in which the curve moves from a positive gradient to a negative gradient, see point A(4, 10) in the diagram.

A **minimum point** is one in which the curve moves from a negative gradient to a positive gradient, see point B(5, –10) in the diagram.

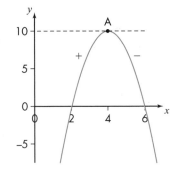

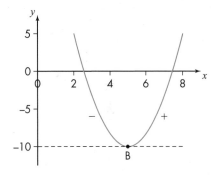

> **Tip**
>
> The maximum or minimum of a curve is not necessarily the lowest or highest point on the curve but maxima and minima are associated with the turning points.

Roots of a quadratic function or equation

The **roots** of a quadratic equation or function $f(x) = ax^2 + bx + c$ are the solutions to the equation $f(x) = 0$.

You can determine the roots of the quadratic equation by:

> **Key term**
>
> The **roots** of a quadratic equation or function ($f(x) = ax^2 + bx + c$) are the solutions of the equation when $f(x) = 0$, or the x-intercepts of the graph representing the function, and are real numbers.

- an algebraic method, by factorisation, completing the square or the quadratic formula, as described in Chapter 3

- a graphical method, plotting the graph of the function (parabola) and noting the x-intercepts, which constitute the roots of the function.

In the graphical method, it may be necessary to rearrange the equation so that the graph cuts a line $y = a$ rather than $y = 0$, which is the x-axis.

Example 41

Use a graphical method to solve the quadratic equation: $2x^2 - 4x - 2 = 0$.

Axis of symmetry $x = \dfrac{-b}{2a} = \dfrac{-(-4)}{2 \times 2} = \dfrac{4}{4} = 1$

x	-2	-1	0	1	2	3	4
$f(x) = 2x^2 - 4x - 2$	14	4	-2	-4	-2	4	14

The intercepts on the x-axis are 2.4 and –0.4.

Therefore the roots of the equation are 2.4 and –0.4.

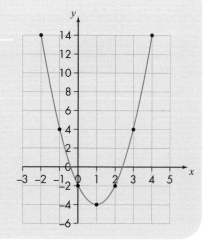

7.4.5 Rearranging a quadratic function to find the roots graphically

It is sometimes possible to use an existing function or graph to solve another function by rearranging the equation. For example, the function $f(x) = 2x^2 - 4x - 2$ can be used to find the roots of the function $f(x) = 2x^2 - 4x - 10$.

You can find the roots of the function $f(x) = 2x^2 - 4x - 10$ by first rearranging the equation to bring it into line with $2x^2 - 4x - 2 = 0$. Adding 8 to both sides of the function gives:

$2x^2 - 4x - 10 \Rightarrow 2x^2 - 4x - 10 + 8 = 8$

$\Rightarrow 2x^2 - 4x - 2 = 8$

Now you can determine the roots by locating the points of intersection of the graphs of $f(x) = 2x^2 - 4x - 2$ and $f(x) = 8$. The graph of $f(x) = 8$ is a line through $y = 8$, parallel to the x-axis.

This straight line, $f(x) = 8$, intersects the curve $f(x) = 2x^2 - 4x - 2$ at $x = 3.4$ and $x = -1.4$.

The roots of the function $f(x) = 2x^2 - 4x - 10$ are therefore: $x = 3.4$ and $x = -1.4$.

Generally, this method of rearranging the terms in a known function to solve a similar is effective, provided that the rearrangement produces a linear as well as a quadratic function.

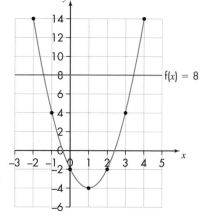

Example 42

Use the function $f(x) = 5x^2 - 3x + 4$ to find two simultaneous functions that will produce the roots of the function $f(x) = 5x^2 - 5x$.

Starting with the function $5x^2 - 3x + 4$ and adding $-2x - 4$ leads to:

$5x^2 - 3x + 4 - 2x - 4 = 5x^2 - 5x$

Therefore the two simultaneous functions required to derive the roots of $5x^2 - 5x$ are $5x^2 - 3x + 4$ and $2x + 4$.

7.4.6 Finding the roots of a quadratic expression by completing the square

A quadratic expression ($ax^2 + bx + c$) may be written in the form $a(x + h)^2 + k$ by the method called 'completing the square'.

- First, equate the quadratic equation to 0, for example:

    $3x^2 - 6x = 4 \Rightarrow 3x^2 - 6x - 4 = 0$

- Factorise the quadratic expression to make the coefficient of x^2 equal to 1 and place the constant term outside the brackets:

    $3x^2 - 6x - 4 \Rightarrow 3(x^2 - 2x\ \ ) - 4$

- Make the expression inside the brackets into a perfect square by adding the relevant term, which is ($\frac{1}{2}$ coefficient of x)^2, in this case is

    $\left(\frac{1}{2}\times-2\right)^2 = 1$

    $3(x^2 - 2x + 1) - 4 = 3(x - 1)^2 - 4$

- Add or subtract the extra term to complete the square:

    $3(x - 1)^2 - 4 - 3 \times 1$

- Write the quadratic expression in the form $a(x + h)^2 + k$:

    $3(x - 1)^2 - 7$.

> **Tip**
>
> In this example, the original expression to be written in the form $a(x + h)^2 + k$ is $3x^2 - 6x - 4$. Adding half the coefficient of x^2 into the expression in brackets to complete the square introduces an additional term to the original expression. The additional term is 3×1 (remember the 3 outside the brackets), which must be subtracted; hence the expression:
>
> $3(x - 1)^2 - 4 - 3 \times 1 = 3(x - 1)^2 - 7$
>
> which gives the original expression, $3x^2 - 6x - 4$.

Calculating the roots

- Set the expression you have found equal to 0:

    $3(x - 1)^2 - 7 = 0$

- Transpose and solve for x:

    $(x - 1)^2 = \frac{7}{3} \Rightarrow x - 1 = \pm\sqrt{\frac{7}{3}}$

    $\Rightarrow x = 1 \pm 1.5$

- State the roots:

    $x = 1 \pm 1.5 \Rightarrow x = 2.5$ and $x = -0.5$

> **Note**
>
> If the coefficient of x^2 in the original expression is negative, the extra term will be negative and must be eliminated by addition.

Example 43

Determine the roots of the quadratic equation $4x^2 = 8x - 1$ by completing the square.

Equate to zero:	$4x^2 - 8x + 1 = 0$
Factorise the first two terms:	$4(x^2 - 2x) + 1 = 0$
Complete the square:	$4(x^2 - 2x + (-1)^2) + 1 - 4(1) = 0$ (Subtract $- 4(1)$ to remove the excess)
Write in the form $a(x + h)^2 + k = 0$:	$4(x - 1)^2 - 3 = 0$
Determine the roots through transposition:	$(x - 1)^2 = \frac{3}{4} \Rightarrow x = 1 \pm\sqrt{\frac{3}{4}}$

Example 44

Determine the roots of the quadratic equation $-12x^2 = 6x - 9$ by completing the square.

Equate to zero and change all the signs:	$12x^2 + 6x - 9 = 0$
Factorise the first two terms:	$-12\left(x^2 + \dfrac{1}{2}x\right) + 9 = 0$
Divide by common factor 3:	$4x^2 + 2x - 3 = 0$
Complete the square and eliminate the excess:	$4\left(x^2 + \dfrac{1}{2}x + \left(\dfrac{1}{4}\right)^2\right) - 3 + \left(4 \times \left(\dfrac{1}{4}\right)^2\right) = 0$
Simplify:	$4\left(x + \dfrac{1}{4}\right)^2 = \dfrac{13}{4}$
	$\left(x + \dfrac{1}{4}\right)^2 = \dfrac{13}{16}$
	$x = -\dfrac{1}{4} \pm \dfrac{\sqrt{13}}{4}$

7.4.7 Finding the minimum and maximum values by completing the square

Minimum value

When the coefficient of x^2 is positive ($a > 0$) the quadratic function has a minimum value because of the $\cup$-shape of the curve. You can determine this value from the expression obtained from completing the square by setting the contents of the brackets equal to 0.

Example 45

Find the minimum value and roots of the function $f(x) = 3x^2 - 6x - 4$.

In a previous example the quadratic function $f(x) = 3x^2 - 6x - 4$ was rewritten as $3(x - 1)^2 - 7$.

Since a is positive the function will have a minimum value, which you can find by setting $x - 1 = 0$. This means that the minimum value of $f(x)$ is $0 - 7 = -7$.

You can calculate the value of x that produces that minimum value from the equation $x - 1 = 0$ $\Rightarrow x = 1$.

Maximum value

A quadratic function has a maximum value if the coefficient of x^2 is negative ($a < 0$).

Example 46

Given the function $f(x) = -2x^2 + 8x - 1$, find the maximum value, the value of x that produces that maximum and the roots of the function.

Factorising: $-2(x^2 - 4x) - 1$

Completing the square: $-2(x^2 - 4x + 4) - 1 + 8$

> **Note**
>
> **Note** that the amount added to complete the square (4) is multiplied by -2 outside the brackets, producing -8. To eliminate the extra value of -8, a value of $+8$ must be added to the overall expression.

Final expression: $-2(x - 2)^2 + 7$

The maximum value of the function is for $x - 2 = 0 \Rightarrow$ the maximum value of $f(x)$ is 7.

The value of x producing that maximum is when $x - 2 = 0 \Rightarrow x = 2$.

To find the roots:

$$-2(x - 2)^2 + 7 = 0 \Rightarrow (x - 2)^2 = \frac{-7}{-2}$$
$$= 3.5$$
$$\Rightarrow x - 2 = \pm\sqrt{3.5}$$
$$= 1.9$$

The roots are $x = 2 + 1.9 = 3.9$ and $x = 2 - 1.9 = 0.1$.

7.4.8 Sketching the graph of a quadratic function

Once you have obtained the roots of a quadratic function and the minimum or maximum value, you can make a simple sketch of the function.

Example 47

In example 46, the roots of the function $f(x) = -2x^2 + 8x - 1$ are 3.9 and 0.1. This means that the curve intersects the x-axis at these points. The maximum value of the function is 7 at $x = 2$. This represents the highest point of the curve ($y = 7$). Use these values to sketch the curve.

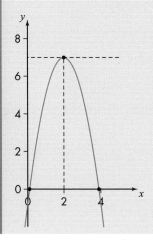

7.4.9 Quadratic inequalities

To solve a quadratic inequality, write it as an equation and solve it. If you then make a sketch of the graph of the function you can use this to determine the limits of the function.

Example 48

Solve the quadratic inequality $f(x) = x^2 - 5x - 6 \leqslant 0$.

$x^2 - 5x - 6 = 0 \Rightarrow (x - 6)(x + 1) = 0$

$\Rightarrow x = -1$ or $x = 6$ (the intercepts of the x-axis)

The minimum value of $f(x)$ occurs when $x = \dfrac{-b}{2a}$ (axis of symmetry) $\Rightarrow x = \dfrac{5}{2} = 2.5$.

The minimum value of $f(x)$ is $(2.5)^2 - 5(2.5) - 6 = -12.25$.

Make a sketch of the curve:

Examining the points of intersection of the curve with the x-axis and the part of the curve below and above the x-axis, you can see that the solution set for the inequality $f(x) \leqslant 0$ is given by the domain $x: \{-1 \leqslant x \leqslant 6\}$.

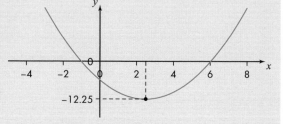

You can also see from the graph that the function is less than zero (under the x-axis) for those values of x stated in the solution set.

Tip

You can also determine the maximum or minimum value of a function by completing the square or by solving the equation for the axis of symmetry ($x = \frac{-b}{2a}$) and substituting for x in the equation $y = ax^2 + bx + c$. The roots of the function are found by solving for x when $y = 0$.

7.5 Graphs of other functions

7.5.1 Graphs of other non-linear functions

As well as linear functions and quadratic functions, there are many other non-linear functions given by the general formula $f(x) = ax^n$ (a is a real number; $n = 3, -1, -2$).

Examples of non-linear functions include:

- $f(x) = x^3$ (cubic function)
- $f(x) = \dfrac{1}{x} = x^{-1}$ (also called the reciprocal function)
- $f(x) = x^{-2}$.

To plot the graph of any of these functions, assign values to x and determine corresponding values for y, formulating a table of values.

Here are some typical graphs for these functions.

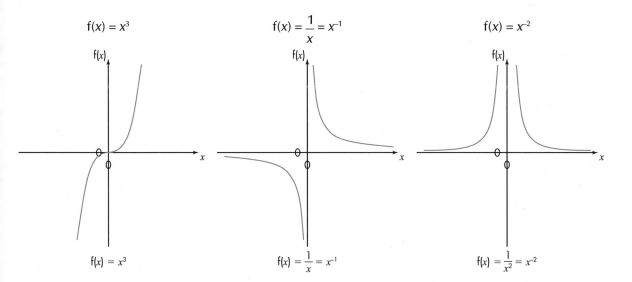

$f(x) = x^3$

$f(x) = \frac{1}{x} = x^{-1}$

$f(x) = x^{-2}$

$f(x) = x^3$

$f(x) = \frac{1}{x} = x^{-1}$

$f(x) = \frac{1}{x^2} = x^{-2}$

7.5.2 Distance–time graphs and velocity–time graphs

Distance–time graphs

The velocity or speed of a moving object is normally calculated from the formula

$v = \dfrac{\text{distance}}{\text{time}}$.

You can plot a **distance–time** graph showing distance travelled (vertical axis) against time taken (horizontal axis). If the distance travelled is proportional to the time taken the graph results in a straight line. The gradient of this graph will have a constant value, which will represent the velocity.

Example 49

Draw the distance–time graph for a vehicle that travelled a distance of 1000 m in 10 seconds at constant speed. Use your graph to find the velocity.

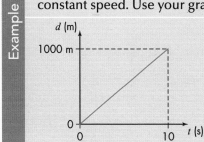

The velocity is given by the gradient, which is $\dfrac{1000}{10} = 100$ m s^{-1}.

If the speed is not constant and is constantly changing, the graph may look more like this.

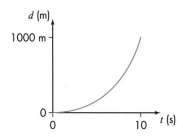

Example 50

This distance–time graph represents the motion of a moving vehicle. Determine the velocity of the vehicle after 10 seconds and the distance travelled after 20 seconds.

The velocity is variable and can be found for a specific time, for example, after 10 seconds.

The velocity in this case is the gradient of the tangent to the curve at $t = 10$ s.

From the graph the gradient of the tangent at $t = 10$ s is

$$\frac{d}{t} = \frac{140}{20} = 7 \text{ m s}^{-1}.$$

Notes

When the gradient is determined from the graph, its precision depends on how accurately the graph is drawn.

After 20 s the distance travelled from the graph = 190 m.

In general, you should be able to determine distance, time and velocity from a distance–time graph.

Velocity–time graphs

The change in velocity of an object divided by the time taken normally gives the **acceleration**, provided the acceleration is constant.

The gradient of a **velocity–time** graph will be equal to the acceleration. The graph will be a straight line for constant acceleration or a curve for variable acceleration.

From a velocity–time graph, you can obtain:

Key terms

The **acceleration** of an object is the rate of change of the velocity with time $\left(a = \dfrac{v}{t} \right)$.

A **velocity–time** graph shows how velocity varies with time for a moving object.

- the velocity and time (read from the axes)
- the acceleration (the gradient of the graph)
- the distance travelled by the object (the area under the graph).

Example 51

This velocity–time graph shows motion of a car. Calculate the initial acceleration, the velocity after 20 seconds, the total distance travelled and the acceleration after 60 seconds.

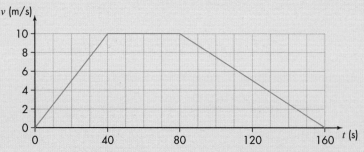

The acceleration, a is given by the gradient which is $\frac{5}{20} = 0.25 \text{ m s}^{-2}$.

The velocity, v (after 20 s) is 5 m s^{-1} (reading directly from the graph).

The total distance travelled, d, is given by the area of the trapezium, which is $\frac{1}{2}(40 + 160) \times 10 = 1000$ m

The acceleration after 60 s = 0 m s^{-2} (reading directly from the graph).

7 Relations, functions and graphs

Exercise 7D

1 Write down the values of a, b and c for each quadratic function.

   a $f(x) = x^2$ **b** $f(x) = 2x^2 - 1$ **c** $f(x) = 9x^2 - 20x$ **d** $f(x) = -3x^2 - 20x - 2$

2 Sketch five types of quadratic function.

3 Draw the graph for the function $f(x) = 3x^2 - 12x + 8$ for $5 \geqslant x \geqslant -1$.

4 Given the function $f(x) = x^2 - 6x + 2$, find the axis of symmetry and plot the graph of the function.

5 Identify the axis of symmetry and then plot the graph of the function $y = -x^2$.

6 Plot the graph of the function $y = 4x^2 + 2$ and find the value of the gradient of the curve at the point where $x = 2$.

7 Plot the graph of each function and hence find the maximum or minimum value of the function and the x-value corresponding to that maximum or minimum.

   a $y = 10x^2 + 20x - 5$ **b** $y = -2x^2 + 12x$

8 Identify and label the turning points of this graph and state whether each is a maximum or minimum, giving reasons.

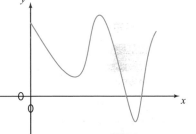

9 Use a graphical method to determine the roots of the function $f(x) = 3x^2 - 12x + 4$.

10 Use the function in question 9 to find the roots of the function $f(x) = 3x^2 - 12x + 10$.

11 Use the method of completing the square to determine the maximum or minimum of each function and its roots.

   a $4x^2 + 8x + 4$ **b** $-5x^2 + 20x$

12 Determine the axis of symmetry of the function $f(x) = x^2 - 20x + 100$ and hence find its maximum or minimum value and its roots.

13 **a** Use the method of completing the square to find the maximum or minimum value of the function $f(x) = 5x^2 + 10x + 9$.

   b Find the value of x that corresponds to the maximum or minimum and the roots of the function, then sketch the curve.

14 Write down the solution set for the inequality $f(x) = 9x^2 - 9 \geqslant 0$.

15 Identify the axis of symmetry and then plot the graph of the function $f(x) = 3x^2 - 24x + 1$.

16 Draw the graph of the function $f(x) = 4x^2 - 2x + 10$ and find the gradient of the curve at the point corresponding to a domain value of $x = 1$.

17 Write down the solution set for the inequality $f(x) = 2x^2 - x + 6 \geqslant 0$.

18 Using your own values for the domain, draw the graph for the function $y = x^3$.

19 Using your own table of values, draw the graph of the function $f(x) = 5x^2 - 7x + 13$.

20 Using the axis of symmetry, draw the graph of the function $-10x^2 + 10x - 8$.

21 **a** Draw the graph of the function $f(x) = 5x^3 + 2x^2 - 1$, using an appropriate domain of seven values.

    b Calculate a positive and negative gradient at any two points and state the coordinates of two turning points.

22 Using the method of completing the square, sketch the curve for the function $f(x) = 8x^2 - 48x + 12$.

23 Determine the axis of symmetry for the function $100x^2 - 20$ and hence sketch the function.

24 Find the roots of the function $f(x) = 6x^2 - 8x + 10$ and use them to find the roots of the function $6x^2 + 12x - 3$ by rearranging the first function.

25 Sketch the curve and write down the solution set for the inequality $f(x) = 6x^2 - 2x - 4 \leqslant 0$.

26 Using appropriate values for the domain, plot the graph of the function $f(x) = \dfrac{1}{x}$.

27 Using an appropriate domain, plot the graph of the function $f(x) = -\dfrac{4}{x^2} + \dfrac{2}{x}$.

End of chapter summary

Relations and functions

- A **relation** may be represented by a set of input and output values or items expressed as ordered pairs.

- In a set of ordered pairs representing a relation, the first elements belong to the **domain** of the relation and are **input values**. The second elements belong to the **range**, also known as the **image**, and are **output values**. The **codomain** contains all the values in the range.

- A **relation** may be represented:

  - in set notation

  - by an arrow diagram

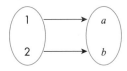

  - in a table

x	1	3	5	7
y	2	4	6	8

  - as a graph drawn from a table

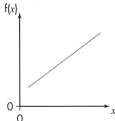

    or by an equation, such as $f(x) = y = x + 1$.

- A **function** is a relation in which each element in the domain is paired with one and only one element in the range; the domain and range are considered to be two different sets.

- An **inverse function** is a function that reverses the operation of another function.

- A **composite function** is a function obtained from combining two or more functions so that the range of one function is contained in the domain of the next.

Coordinate geometry: graphs of linear functions

- **Cartesian** or **rectangular coordinates** are the ordered pairs of values representing the domain and range of a relation (x and y values) that are plotted in a Cartesian plane to obtain points.

- The **length** of a straight line segment is the distance between the points that mark the endpoints of the line segment

- The **midpoint** of a straight line segment is the halfway point between the endpoints of the line segment

- If two lines are parallel to each other in the Cartesian plane, their gradients are equal.
- If two lines are perpendicular to each other, the product of their gradients is –1.
- A **linear function** is a function in which the highest power of the variable is not greater than one (1). A graph of such a function results in a straight line. Various graphs of a linear function:

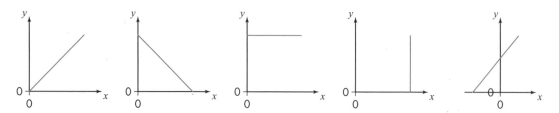

- **Simultaneous equations** are two or more equations in two unknowns that share a solution that satisfies both equations.
- Simultaneous equations may be solved graphically, by drawing their graphs and noting where the lies intersect.

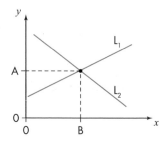

Linear inequalities

- A **linear inequality** is a mathematical statement that shows one quantity or number to be greater than or less than another and is represented by the symbols:
  - < (less than)
  - > (greater than)
  - ⩾ (greater than or equal to)
  - ⩽ (less than or equal to).
- The solutions of a linear inequality may be represented as set, on the number line or in a graph.
- If the inequality is < or >, draw the graph as a **broken line** and if the inequality is ⩽ or ⩾, draw the graph as a **solid** line.
- To indicate the solution, shade the region that satisfies the inequality:
  - for an inequality $x < a$, shade the region to the left of the line $x = a$, for an inequality $x > a$, shade the region to the right of the line $x = a$
  - for an inequality $y < a$, shade the region below the line $y = a$, for an inequality $y > a$, shade the region above the line $y = a$.
- **Linear programming** is a mathematical method used to find the most useful solution to a set of conditions.

Quadratic functions

- A **quadratic function** is one in which the highest power of the variable is 2.
- The graph of a quadratic function is a smooth curve, called a **parabola**.

- The **axis of symmetry** is a mirror line that divides the curve into two identical halves.
- A function of the form $f(x) = ax^2 + bx + c$ has a **minimum point** when $a > 0$ and a **maximum point** when $a < 0$.
- A **tangent** is a line that touches the curve at only one point.
- To find the gradient at a point on a curve, draw a tangent at the point and find the gradient of the tangent from the formula $\frac{y_2 - y_1}{x_2 - x_1}$, where (x_1, y_1) and (x_2, y_2) are two suitable points on the tangent.
- A **turning point** of a curve is a point on the curve at which it changes direction and the gradient also changes sign (positive to negative or negative to positive).
  - At a **maximum point** the curve moves from a positive gradient to a negative gradient.
  - At a **minimum point** the curve moves from a negative gradient to a positive gradient.
- The **roots of a quadratic equation** or function $(f(x) = ax^2 + bx + c)$ are the solutions of the equation when $f(x) = 0$, or the x-intercepts of the graph representing the function, and are real numbers.
- It is sometimes possible to use an existing function or graph to solve another function by rearranging the equation.
- A quadratic expression $(ax^2 + bx + c)$ may be written in the form $a(x + h)^2 + k$ by the method called 'completing the square'.
- Once you have completed the square, you can find the roots by setting the whole expression for the function, $a(x + h)^2 + k$, equal to 0.
- You can find the minimum or maximum value by setting the expression $x + h$ equal to 0.
- Once you have obtained the roots of a quadratic function and the minimum or maximum value, you can make a simple sketch of the function.

Graphs of other functions

- As well as linear functions and quadratic functions, there are many other non-linear functions given by the general formula $f(x) = ax^n$ (a is a real number; $n = 3, -1, -2$).
- The velocity or speed of a moving object is normally calculated from the formula $v = \dfrac{\text{distance}}{\text{time}}$.
- A **distance–time** graph shows how distance travelled varies with time for a moving object.
- You can determine distance, time and velocity from a distance–time graph.
- The **acceleration** of an object is the rate of change of the velocity with time $\left(a = \dfrac{v}{t} \right)$.
- A **velocity–time** graph shows how velocity varies with time for a moving object.
- From a velocity–time graph, you can obtain:
  - the velocity and time (read from the axes)
  - the acceleration (the gradient of the graph)
  - the distance travelled by the object (the area under the graph).

Examination-type questions for Chapter 7

1 The function $f(x) = \frac{3}{4}x - 1\frac{1}{2}$. The domain of this function is $x: -3 \leqslant x < 0$.

   a **i** Calculate $f(-2)$. **ii** Find the value of x for which $f(x) = -3$.

   iii Determine the range or image of the function, expressing your answer as a solution set.

   b Solve the inequality $10x - 3 \leqslant 2$ and represent your answer on a number line.

   c Solve the inequality $3x - 5 > x + 1$ and represent your answer on a graph.

2 **a** Given $f(x) = 12x - 9$ and $g(x) = \dfrac{2x + 5}{2}$, evaluate $f(-1)$ and $g(\frac{1}{2})$.

   b Evaluate $gf(-2)$.

   c Find $f^{-1}(10)$.

3 Given the points $H(-1, 5)$ and $K(9, 1)$ determine:

   a the gradient of HK

   b the coordinates of the midpoint of HK

   c the equation of the perpendicular bisector of HK.

4 The table shows values for the two variables x and y, which are inversely proportional to each other.

x	2	1.5	p	12
y	3	4	6	q

   a Write an equation to show the relationship between x, y and the constant k, showing inverse proportionality.

   b Calculate the values of p and q.

   c Sketch the graph of the function $f(x) = 2x^2 - 12x - 5$ by completing the square to obtain the intercepts on the x-axis, the maximum or minimum value of the function and the corresponding value of x.

   d Use your graph to solve the equation $2x^2 - 12x - 5 = x - 1$.

5 The diagram shows a sketch of a quadratic function.

   a State the roots of the quadratic function.

   b Given that the general equation of the function is $f(x) = x^2 + bx + c$ find c from the graph.

   c Determine the value of b.

   d Determine the equation of the axis of symmetry and hence state the coordinates of the minimum point.

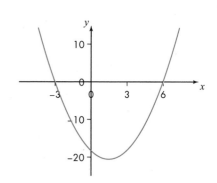

6 The equation of a straight line is given by: $9 = -6y - 3x$.

   a Write the equation in the form $y = mx + c$ and hence find the gradient of the line and the intercept on the *y*-axis.

   b Determine the equation of a line perpendicular to the line in part **a** and passing through the point Q(3, 5).

   c Draw a coordinate grid and sketch the inequalities $x \geqslant 0$, $y \geqslant 0$, $y \leqslant -x + 5$ and $x < 2$, then shade the region enclosed.

   d Given the function $f(x) = 3x^2 - 7x + 12$, determine which vertex in that shaded region will produce the greatest value for the function and which will produce the least value. State these values.

7 John started off on his long island tour from his home H on his bicycle. The journey is shown in this distance–time graph.

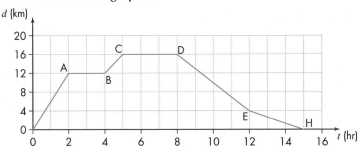

   a How many stops did John make on the journey and what duration was his longest stop? Why do you think it was the longest?

   b How much time did the entire journey take and what was his average speed?

   c What stage of the journey was his slowest and what stage was his fastest? What was his speed at both stages?

   d When was his average speed greater, towards his destination or when he was returning?

   e What was the total stoppage time for the entire journey?

Objectives

By the end of this chapter you should be able to:

- appreciate the meaning, importance and application of Statistics in real life
- differentiate between sample and population attributes
- construct a frequency table for a given set of data
- determine class features for a given set of data, including class interval, class boundaries, class limits, class width
- construct statistical diagrams, including pie charts, bar charts, line graphs, histograms with bars of equal width and frequency polygons
- determine measures of central tendency for raw, ungrouped and grouped data
- determine when it is most appropriate to use the mean, median and mode as the average for a set of data
- determine the measures of dispersion (spread) for raw, ungrouped and grouped data – range, interquartile range and semi-interquartile range – estimating these measures for grouped data
- use standard deviation to compare sets of data
- draw a cumulative frequency curve (ogive)
- analyse statistical diagrams by finding the mean, mode, median, range, quartiles, interquartile range, semi-interquartile range and identify trends and patterns
- determine the proportion or percentage of the sample above or below a given value from raw data, a frequency table or a cumulative frequency curve
- identify the sample space for a simple experiment, including the use of coins, dice and playing cards and the use of contingency tables
- determine experimental and theoretical probabilities of simple events, explore the use of contingency tables, implementing the addition rule for exclusive events and multiplication for independent events
- make inference(s) from statistics.

You should already know:

- Number theory and computation
- Algebra
- Geometry and trigonometry
- Measurement
- Fractions and decimals
- Relations and functions
- Sets

Introduction

The study of statistics involves the observation, collection, analysis and representation of data or information obtained from any source, so that it can be used for a wide variety of purposes. The sources from which data is obtained include weather forecasts, medical research, political activities, sports, business, education, financial institutions and reports on the consumption of products and services. The collected data is normally converted into visual media such as tables, charts and graphs and analysed to produce important results.

The results obtained from the use of statistics usually have very important applications. For example, the level of consumption of products and services is analysed and this helps to determine where emphasis should be placed in production processes. In the field of medicine, the effectiveness of certain medications is studied. Financial analyses are carried out in order to discover the best ways to maximise profits and minimise loss. The weather pattern can be studied to help make useful weather predictions. In education, statistical information is of great importance, from the classroom to the Ministry of Education and beyond. The wide use and importance of statistics becomes quite evident when you look at newspapers, government reports and financial statements.

There are many other everyday uses of statistics that help make life easier. At the same time, though, there is a great need to look out for erroneous and deliberately misleading analyses and results.

This chapter presents and compares the various statistical methods to be carried out in a safe and responsible manner and seeks to explain the use of statistics and probability in everyday life.

Examples of statistics in real life

Weather reports and records, such as rainfall

Agricultural statistics

Sports reports and reviews

Comprehensive information about government provision of housing, transport, education

8.1 Sample statistics and population parameters

8.1.1 Populations and sampling

Populations

In statistics, a **population** is a set of items or objects that have at least one common property or characteristic that enables it to be the subject of a study from which facts, conclusions and inferences can be obtained.

The **data** that is collected is then sorted and analysed, to produce information that can be used in many useful ways.

> **Key term**
>
> A **population** is a set of items or objects (such as people, material things, concepts) with at least one common property or characteristic that can be studied.

> **Key term**
>
> **Data** consists of information or facts collected and interpreted in order for statistical analyses to be made. Data may take the form of words, numbers, measurements or descriptions of objects.

> **Note**
>
> Populations can be any size. Sometimes they are very large but they can also be small.

Examples of populations include:

- the population of a country
- all the females or males in a country
- the dogs in a neighbourhood
- a type of vehicle in a country

- a particular interest among a class of people
- fruit or vegetables
- species of plant or animal.

Sampling

In a survey, it is not possible to observe all of the individuals or objects in a given population, especially large populations. Therefore, **samples** of the population being studied are selected. For example, if the population is all secondary students in a particular school, the sample could be the students in one class.

> **Key term**
>
> A **sample** is a small group taken from a population and used to represent the entire population.

8.1.2 Parameters

A **parameter** of a population is a selected feature or characteristic of the population, such as belonging to a particular age range of a number of children, or some **statistic** that is a typical number such as an average or percentage, representing some measurable aspect of the population.

> **Key terms**
>
> A **parameter** of a population is a fixed characteristic of that population.
>
> A **statistic** is a characteristic of a sample – a summary number (average or percentage) that describes the sample.

Example 1

Give one example of a population in statistics and a possible sample of that population.

Population: A primary school with ten class rooms and a student population of 400

Sample: 40 students chosen randomly (two boys and two girls from each class)

8.2 Classifying and representing data

You know that data is information that has to be collected, before being analysed to produce useful statistical information. Data can be classified according to various criteria.

8.2.1 Classification of data

These are the categories of data that you need to recognise.

- **Qualitative data** is descriptive data, for example, colours, types of music, design of carnival costumes.
- **Quantitative data** is numerical information, such as numbers of people in a crowd, or measurements.

  Quantitative data can be further categorised.

  - **Continuous data** is data that is measured and can take any value, often within two given values (a range), for example, the height of a person.
  - **Discrete data** can take only specific values, such as the number of children in a family.

8.2.2 Classification of data as four levels of measurement

Nominal level of measurement

The nominal level of measurement is the simplest or lowest way to characterise data. It means 'in name only'; therefore nominal data refers, for example, to names, categories, labels, signs, featured athlete numbers and eye colour. This is a qualitative type of data requiring 'yes' or 'no' responses in surveys. Shoe size and favourite foods are other examples of nominal data.

Nominal data cannot be organised statistically; this form of analysis has little meaning.

Ordinal level of measurement

This is the second level of measurement. Unlike data at the nominal level, this type of data can be ordered. However, not much significance or meaning can be assigned to the arrangement of the data. Examples include the top 100 beaches, rankings with no specific measurable criteria. This type of data is not normally used in calculations and statistical analysis.

Interval level of measurement

In this, the third level, the data again can be ordered and the differences between data are useful and have analytical application. Examples of this level of measurement would be degrees Celsius or degrees Fahrenheit. The data can be used in calculation and analysis but the extreme values, such as 0, are not as significant as the differences in values.

Ratio level of measurement

This level of measurement represents the highest level. It is superior to the interval level because not only are the differences significant but the zero and other extreme values are used in calculation and comparisons can be made between measurements. Statements such as '2 times', '3 times', 'n times' are all valid. Operations such as sums, differences, divisions, multiplication, ratios, etc. can all be used with the given data.

8.2.3 Raw data

Raw data is data that is still in the form in which it was collected, before it is presented in a usable form.

Example 2

Give an example of raw data.

The ages of 20 men, in years:

30, 35, 40, 50, 60, 64, 44, 38, 32, 28, 36, 66, 48, 62, 55, 52, 70, 42, 34, 46

Key term

Raw data is data that has not been formatted or organised in any way, for example, the marks of 20 students in an examination, presented in random order.

8.2.4 Tally charts

A **tally chart** is a table used to record data as it is collected. It consists of three columns: name of the **variable**, the count or **tally** and the **frequency**. It is used to collect raw quantitative data, with each data item being recorded by means of a tally stroke.

Tip

In order to ease the process of tallying and improve accuracy, every fifth tally is used to cross the four preceding tallies, thus making bundles of five (卌). Additionally, if the total tally is found and compared with the total data analysed the result should be the same.

Key terms

A **tally chart** is a table used to collect data.

A **variable** is any characteristic of the individuals in a data set.

A **tally** is a way of counting the individuals in a data set by means of tally strokes.

The **frequency** is the number of times a particular data value occurs.

Information from the tally chart is used to construct statistical tables and diagrams.

Example 3

Here are the marks of 25 students in a mathematics test, marked out of 10.

Marks obtained by students:

1, 3, 9, 8, 7, 4, 6, 7, 5, 5, 10, 3, 7, 9, 8, 10, 6, 2, 1, 7, 4, 9, 10, 3, 7

Draw up a tally chart to represent the data.

Draw up the table and use tallies to record the information in the middle column.

Then count the tallies for each mark and fill in the frequencies.

Marks	Tally	Frequency
1	II	2
2	I	1
3	III	3
4	II	2
5	II	2
6	II	2
7	卌	5
8	II	2
9	III	3
10	III	3

8.2.5 Frequency tables

Once the data has been collected, it has to be analysed and organised into useful information. The first step is to draw up a **frequency table** or **frequency distribution**.

8.2.6 Grouped distributions

For large quantities of data, it may not be feasible to record individual values in a frequency distribution. In this case, the data can be arranged in groups or classes. The tally chart is modified to indicate all the members of a particular group or class. This is called a **grouped distribution** and the frequency table is called a **grouped frequency distribution**.

Having too many groups, however, can disguise or even lose the pattern of the distribution and having too few can compromise the details in the data. The recommended number of groups is therefore from 5 to 20.

When data is arranged in groups or classes, it is sorted into ranges of values. Each range is a class, defined by **class intervals** such as 40–45, and the values that fall into this class are 40, 41, 42, 43, 44, 45. To determine the class intervals for a set of data, decide how many groups or classes are required and then divide the full range of values by the number of groups, bearing in mind that the recommended number of groups is between 5 and 20.

Alternatively, you can determine the class interval first, with a particular class size in mind, which will define the number of classes or groups that make up the distribution.

The class intervals may also be considered to be numerically equal to 'upper limit – lower limit' of a class. This is also called the **class width**.

Example 4

Here are the marks of 20 students in a mathematics examination, graded out of 20.

19, 7, 2, 18, 16, 8, 3, 2, 1, 12, 15, 9, 10, 14, 11, 18, 2, 11, 15, 20

Represent the information in the form of a tally chart and a grouped frequency distribution.

The range of the distribution is 20 − 1 = 19.

Assuming the number of groups to be 5 then $\frac{19}{5} \approx 4$. The class intervals can therefore be:

0–4; 5–9; 10–14; 15–19; 20–24.

Draw up the tally chart, using the class intervals.

Class (marks)	Tally	Frequency
0–4	ℍℍ	5
5–9	|||	3
10–14	ℍℍ	5
15–19	ℍℍ	6
20–24			1

Note

The class interval could have been smaller, for example, 0–3, 4–6; …, resulting in more groups.

Class limits

The **class limits** are the upper and lower values of the class interval. For example, for the class interval 5–9 in the previous example, the lower limit is 5 and the upper limit is 9. Subtracting 5 from 9 (upper limit – lower limit) gives the class interval, in this case 4.

Example 5

Given the class: 23–56, what are the upper and lower limits?

The upper limit is 56 and the lower limit is 23.

Key term

Class limits   are the highest and lowest values of any class or group in a grouped distribution.

Class boundaries

Class limits are inadequate to represent groups of continuous data, since there is a 'gap' between the upper limit of one group and the lower limit of the next. Therefore, it is more appropriate to use **class boundaries** to represent continuous data.

The upper class boundary of a distribution is equal to the upper limit of one class added to the lower limit of the following class divided by 2. The lower class boundary of one class is the upper boundary of the preceding class. The lower boundary of the first class is normally deduced from the lower boundaries of the other classes.

Key terms

Class boundaries   are used to eliminate gaps between successive classes in a grouped distribution. Each of these values is calculated as the midpoint between the upper limit of one class and the lower limit of the next class.

Example 6

Using the data and corresponding tally chart in Example 4, use class boundaries to formulate a frequency table.

The upper boundary of the first class is $\frac{4+5}{2} = 4.5 \Rightarrow$ the lower boundary of the second class is 4.5.

Following the same pattern, determine the upper boundary of each class by adding 0.5 to the upper limit and the lower boundary by subtracting 0.5 from the lower limit. The frequency table will therefore be as shown.

Class boundaries	Frequency
–0.5–4.5	5
4.5–9.5	3
9.5–14.5	5
14.5–19.5	6
19.5–24.5	1

Example 7

Determine the class width in the grouped frequency distribution shown in Example 6.

The class width can be determined from any class in the distribution.

The class width is 9.5 – 4.5 = 5.0

Tip

The class interval is found by subtracting the lower limit from the upper limit and class width = upper boundary – lower boundary. Both remain constant throughout the grouped distribution but the width is always greater than the interval.

8.3 Statistical diagrams

8.3.1 Bar charts

Proportional bar chart

A **proportional bar chart** is a single bar or column that conveys proportions through the heights or areas of its component parts.

A fruit basket contains 80 different fruits.

Type of fruit	mango	apple	orange	pear	grape
Number of fruits	10	5	20	15	30

Represent this data on a proportional bar chart.

Draw a vertical rectangle with a scale down the side. The highest point on the scale represents the sum of all the values and the lowest point is 0.

Use the scale to position the sections, and colour them appropriately.

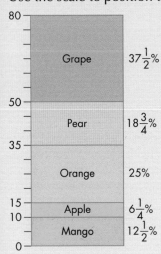

The information can also be represented as percentages. The percentages are added on the right of the chart.

Mango: $\dfrac{10}{80} \times 100 = 12\dfrac{1}{2}\%$

Apple: $\dfrac{5}{80} \times 100 = 6\dfrac{1}{4}\%$

Orange: $\dfrac{20}{80} \times 100 = 25\%$

Pear: $\dfrac{15}{80} \times 100 = 18\dfrac{3}{4}\%$

Grape: $\dfrac{30}{80} \times 100 = 37\dfrac{1}{2}\%$

Checking the percentages:

$$12\dfrac{1}{2} + 6\dfrac{1}{4} + 25 + 18\dfrac{3}{4} + 37\dfrac{1}{2} = 100\%$$

> **Note**
>
> To write any quantity as a percentage of a total, write the quantity over the total and multiply the result by 100.

> **Tip**
>
> The sum of all the percentages representing the various quantities that make up the total must be 100%. This represents a way to check your answer.

Simple vertical bar charts

In a **simple vertical bar chart**, vertical bars represent the data categories. The height of the bar represents the frequency for the category it represents. The bars are drawn along the horizontal axis, they are all the same width and are separated by equal spaces. There is a scale on the vertical axis. The highest point on this scale will correspond to the **frequency** of the data category that occurs most often and the lowest point will be 0.

Draw a simple vertical bar chart to represent the information from Example 8.

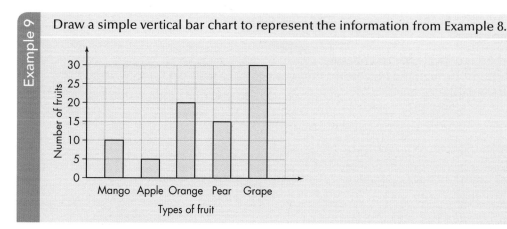

Simple horizontal bar charts

The **simple horizontal bar chart** is similar to the vertical bar chart in all respects, except that the scale is written along the horizontal axis and the bars are drawn down the vertical axis, parallel to the horizontal axis.

> ### Key term
>
> In a **simple horizontal bar chart** each data category is represented by a horizontal bar, the length of which represents the frequency of the data category.

Represent the information given in Example 8 on a simple horizontal bar chart.

The information on the axes is interchanged, with 'Types of fruit' on the vertical axis and 'Number of fruits' on the horizontal axis.

Chronological bar charts

The word 'chronological' refers to time, therefore a **chronological bar chart** is a vertical bar chart that compares some quantity or occurrence with time, which is always placed on the horizontal axis.

> ### Key term
>
> A **chronological bar chart** displays data in the time order in which it occurs.

The table shows the number of accidents taking place in a certain country over five years. Represent the information in a chronological bar chart.

Year	2013	2014	2015	2016	2017	2018
Number of accidents	8	12	10	15	9	20

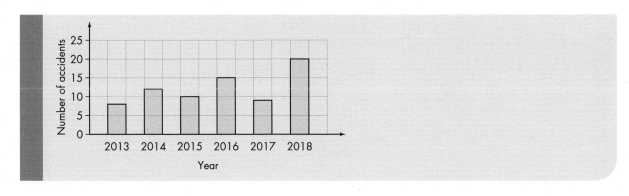

8.3.2 Line graphs

A **line graph** is drawn to show a relationship between two variables. Generally, one of the variables is time and this is always represented on the horizontal axis. A line graph is thus another way of representing chronological data and is also useful in indicating upward or downward trends.

In both types of graph time may be measured in seconds, minutes, days, weeks, months, years or decades, for example.

This line graph is drawn for the data in Example 11.

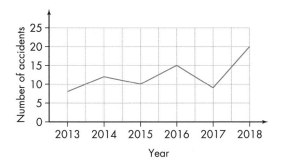

8.3.3 Pie charts

In a **pie chart** each data category is represented as a sector of a circle. The angle of each sector is calculated by dividing the frequency of the individual data category by the total frequency and multiplying by 360° ($\frac{D}{T} \times 360°$). Each data category is represented by the appropriate fraction of the whole circle or 'pie'.

Just as for the proportional bar chart (see Example 8), you can represent the data in a pie chart as percentages. First, express each data category as a percentage of the whole data set, by dividing the category frequency by the total frequency and multiplying by 100. Then calculate the angle for the data category by finding the appropriate percentage of 360°.

Example 12

At the end of a hard day a fruit vendor has sold 10 mangoes, 20 apples, 25 oranges, 5 pears and 40 grapes. Represent this information in a pie chart.

Calculate the sector angle for each individual data category.

10 mangoes: $\frac{10}{100} \times 360 = 36°$ 20 apples: $\frac{20}{100} \times 360 = 72°$

25 oranges: $\frac{25}{100} \times 360 = 90°$ 5 pears: $\frac{5}{100} \times 360 = 18°$

40 grapes: $\frac{40}{100} \times 360 = 144°$

Use these angles to create the sectors of the circle.

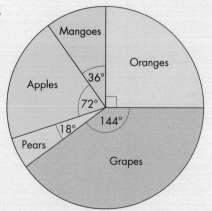

Note	In this case, because the vendor sold 100 pieces of fruit in total, expressing the data categories as percentages would have required exactly the same working.

Tip	The calculations could be simplified by identifying a relationship between quantities, for example, number of apples = 2 × number of mangoes, so sector angle for apples = 2 × sector angle for mangoes. The sum of the sector angles must be 360°.

8.3.4 Histograms

At first glance, **histograms** may resemble bar charts, but there are important differences. A histogram represents a frequency distribution and consists of bars or rectangles drawn without any space between them. The heights of these bars represent the frequencies (which are always plotted on the vertical axis). The area of each bar is also directly proportional to the frequency since the width is kept constant.

Note	A histogram may only be used to plot the frequency of score occurrences in a continuous data set that has been divided into classes.

Key term

A **histogram** is a diagram used to represent a frequency distribution, consisting of vertical rectangles drawn without any space between them; the height of the rectangle represents the frequency of the data.

Tip	The starting point of the width of a histogram does not have to be zero. It could be a negative or a larger number. The vertical side of the first rectangle may or may not coincide with the y-axis, as shown in this diagram.

  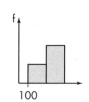

Example 13

Construct a histogram to represent the data in this tally chart.

Marks	1	2	3	4	5	6	7	8	9	10
Frequency	II	I	III	II	II	II	IIII	II	III	III

Draw the histogram, based on the tally chart.

The bars are of equal width and represent the frequency.

There are no spaces between them.

In this case, the area of each bar (rectangle) equals the frequency (the width of each rectangle = 1 and the height = frequency).

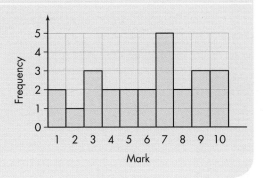

Histograms of discrete distributions

A discrete distribution contains specific values, generally values that can be counted. There are two ways to represent this type of distribution, as shown in the next example.

Example 14

A batsman made a century in a cricket match. A breakdown of the runs he scored is represented in this frequency distribution table. Represent the information in a histogram.

Runs	Frequency
singles (1)	30
twos (2)	11
threes (3)	5
fours (4)	4
fives (5)	1
sixes (6)	2

This is an example of a discrete distribution – the numbers of runs can only be whole numbers (no fractional values). The information can be displayed like this.

(a)

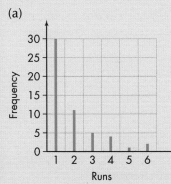

(b)

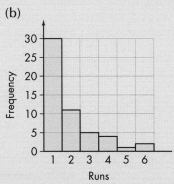

In diagram (a) the data is shown as a vertical bar chart with bars of almost zero width. In diagram (b) the same data is given as a histogram in which the area of each rectangle is the same as the frequency. In this case the data is treated as continuous data.

Histograms of continuous distributions

In a continuous distribution data can take any numerical values. The next example illustrates this.

Example 15

These are the heights (in feet) of 15 students.

4.0, 4.4, 5.0, 5.4, 5.2, 4.0, 4.4, 5.0, 5.4, 4.8, 4.0, 5.0, 4.6, 4.2, 4.4

Draw up a tally chart and frequency table, then draw a histogram for continuous data.

Height (feet)	Tally	Frequency
4.0	|||	3
4.2			1
4.4	|||	3
4.6			1
4.8			1
5.0	|||	3
5.2			1
5.4	||	2

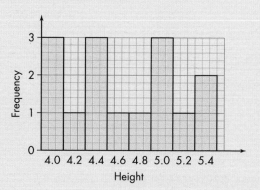

The histogram above shows that bars of equal width but different heights can be used to represent continuous data since any height can be obtained between 4.0 and 5.4 ft in this particular example.

Histograms of grouped distributions

The histogram of a grouped distribution is drawn using class boundaries, which are arranged on the horizontal axis. The frequencies of the distribution are represented by the heights of the bars and the class width is reflected in the width of the bars, calculated from the difference between the upper boundary and the lower boundary.

Example 16

Draw a histogram to represent the grouped frequency table shown in Example 6.

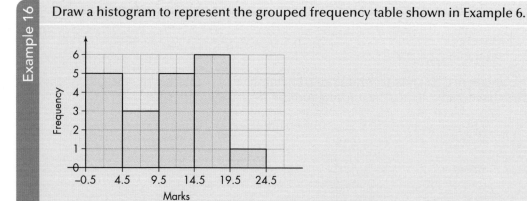

8.3.5 Frequency polygons and frequency curves

> **Key term**
>
> A **frequency polygon** is a line graph used to represent a frequency or grouped distribution and to indicate the spread of the distribution. It is formed by joining the midpoints of the top of the bars in a histogram by straight lines.

Example 17

This histogram represents the masses of 88 packages. Draw a frequency polygon to represent the data.

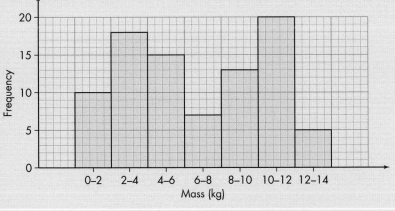

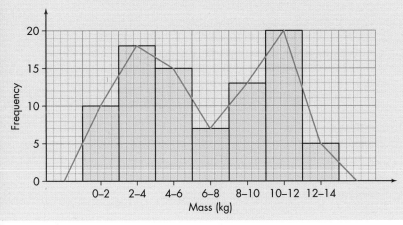

The frequency polygon is drawn by connecting the midpoints of the tops of adjacent bars by straight lines. The lines are extended to the midpoint of the interval before and the one after the histogram on the horizontal axis, extending right down to the horizontal axis to complete the polygon. This ensures that the area of the frequency polygon coincides with the area of the histogram and by extension the total frequency.

You can draw the frequency polygon for the grouped distribution in example 16 by joining the midpoints of the tops of adjacent bars in the histogram. You must use straight lines, as shown in the next example.

Example 18

Draw a frequency polygon to represent the grouped frequency distribution and histogram in Example 16.

Join the midpoints of the bars, using straight lines, and then join the midpoints of the first and last bars to the horizontal axis.

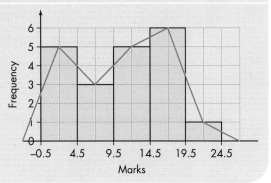

Observe that the polygon is a closed shape bounded by straight lines; hence the requirement to begin at the mid-interval of the class before the first class, and to end at the mid-interval of the class after the last on the horizontal axis.

1. Give two examples of population, as used in Statistics, and suggest a sample for each.

2. Create your own example of raw data to suggest the marks of 20 students in a mathematics examination marked out of 10.

3. Give two practical examples of each of these types of data:

   a qualitative   **b** continuous   **c** discrete.

4. This raw data represents the runs made by 22 players in a cricket match:

   40, 25, 10, 1, 2, 6, 17, 100, 50, 66, 0, 19, 10, 17, 25, 45, 50, 9, 25, 15, 70, 30.

   Represent the information in a tally chart.

5. Represent the data in question 4 in a frequency table.

6. This data represents the types of 300 fruits sold by a fruit vendor.

Fruit	Pears	Apples	Mangoes	Grapes	Bananas	Plums
Number sold	20	30	50	100	60	40

   a Show the information on:

      i a proportionate bar chart

      ii a vertical bar chart.

   b Calculate the percentage of each type of fruit sold.

7. This data represents 100 members of a sports club.

Sport	football	cricket	netball	basketball	table tennis	athletics
Number who play	15	18	20	30	7	10

   a Represent the information in:

      i a horizontal bar chart

      ii a pie chart.

   b Calculate the percentage of the membership that play each sport.

8. The data represents the numbers of hurricanes affecting a particular country in different years.

Year	2012	2013	2014	2015	2016	2017
Number of hurricanes	5	4	2	1	3	9

   Construct a chronological bar chart and a line graph to represent the data.

9. Construct a histogram from this frequency table.

Marks	1	2	3	4	5	6	7	8	9
Number of students	5	8	12	20	14	6	3	2	1

10 This data represents the masses of forty men, in kilograms.

60, 85, 60, 80, 67, 90, 87, 82, 68, 89, 73, 72, 86, 88, 84, 83, 70, 72, 66, 75, 77, 70, 61, 62, 69, 74, 78, 85, 81, 64, 71, 61, 90, 74, 82, 87, 89, 90, 65, 76

  a Formulate a grouped distribution table and tally chart for the data.

  b State the upper and lower limits of the first class.

  c Write down the class interval of the last class.

11 This proportionate bar chart represents the monetary value of goods in a housewife's shopping cart. The percentages shown represent the percentages of her total expenditure.

  a Which of the five items takes most of the money?

  b Calculate the total value of the goods in the shopping cart.

Flour 25%
Rice 10%
Potato 20%
Vegetables $105.00
Sugar 10%

12 A vertical bar chart shows the number of goals scored by six footballers in a season.

A scored 5 goals, B scored twice the number scored by A, C scored three times the number A scored, D scored 4 goals, E scored half as many goals as scored by D, and F also scored a number of goals.

  a Given that 40 goals were scored altogether, how many goals did F score?

  b Draw the vertical bar chart.

13 The data shows the lengths of different objects. Represent this information on a horizontal bar chart.

Object	match	nail	phone	pencil	fly	pen	finger
Length (cm)	4.5	10.0	14.0	18.5	0.5	15.0	8.5

14 These are the best times for a 100 m sprint scored by an athlete over a number of years. Represent the information in a chronological bar chart.

Year	2012	2013	2014	2015	2016	2017	2018	2019
Time (s)	14.0	13.5	12.1	11.8	11.3	10.6	10.5	10.0

15 Represent the information in question 14 in a line graph.

16 Draw a frequency polygon to represent the data in question 9.

17 Using the grouped distribution table in question 10:

  a formulate class boundaries for each class

  b calculate the class width

  c construct a histogram and a frequency polygon.

18 This grouped frequency distribution table shows the heights of 100 men, in centimetres. Determine the class boundaries and represent the information in a histogram and frequency polygon.

Height (cm)	Number of men (frequency)
150–155	5
157–162	8
164–169	23
171–176	28
178–183	20
185–190	12
192–197	3
199–204	1

19 Draw up a frequency polygon to represent the diameters of different sized ball bearings, as represented in this table.

Diameter (cm)	0.5–1.0	1.0–1.5	1.5–2.0	2.0–2.5	2.5–3.0	3.0–3.5
Number of ball bearings	1	3	14	23	18	6

20 This discrete data represents the numbers of children in several families in a village. Construct two types of histogram to represent the data.

Number of children	1	2	3	4	5	6	7	8	9	10
Number of families	8	6	5	9	7	4	3	2	2	1

21 The table lists the areas, in km², of several different countries. Represent the information on a horizontal bar chart.

Country	Russia	Mexico	Indonesia	Nigeria	Congo	Taiwan
Area (km²)	17 098 246	1 964 375	1 910 931	923 768	342 000	35 193

22 The table represents the fractions of students in a class studying different subjects. Using percentages as well as degrees, represent the data on a pie chart.

Fraction	$\frac{1}{8}$	$\frac{1}{4}$	$\frac{3}{16}$	$\frac{1}{16}$	$\frac{3}{32}$	$\frac{1}{32}$	$\frac{5}{32}$	$\frac{3}{32}$
Subject	history	mathematics	physics	biology	French	art	English	chemistry

23 The pie chart represents the students in a class and their preferences for various fruits.

a What is the sector angle for students who prefer melons?

b Given that 8 students prefer cherries, how many students prefer melons?

c How many students are there actually in the class?

d Represent the preferences as percentages.

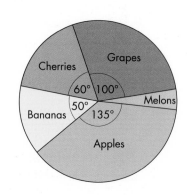

24 The line graph indicates the numbers of vehicles imported into a country over a seven-year period.

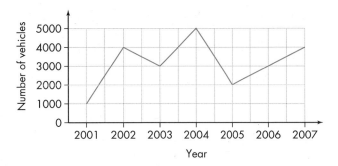

a How many vehicles were imported in the seven-year period?

b In which year was the number imported the highest? How many vehicles were brought in that year?

c In which year was the least number of vehicles brought in? How many?

d State the periods (years) in which there was **i** an increase and **ii** a decline in importation and state by how much.

25 Represent the data given in question 23 in the form of a vertical bar chart and a horizontal bar chart.

26 The table shows the shoe sizes of students between the ages of 9 to 12 years in a school.

Shoe sizes	3	$3\frac{1}{2}$	4	$4\frac{1}{2}$	5	$5\frac{1}{2}$	6	$6\frac{1}{2}$	7
Number of children	20	30	45	50	60	40	24	16	8

a Construct a frequency polygon to represent the information.

b What is the type of data presented?

27 The diagram represents a histogram of a grouped frequency distribution for the ages of a number of boys and men.

Draw up a frequency table, with class limits, class boundaries and frequencies to represent this data. Use the ranges used in the diagram.

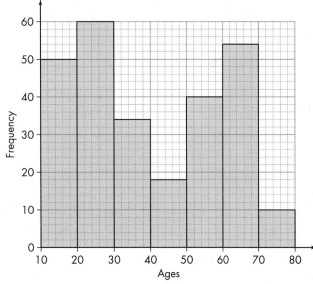

28 The frequency polygon analyses the performances of a number of bowlers over a period of time, in terms of wickets obtained in cricket series. Construct a grouped frequency table and histogram to represent this data.

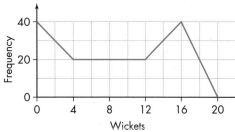

8.4 Measures of central tendency

A measure of central tendency is a central value, measurement or data value, sometimes called an **average**, which represents all the values in the distribution.

There are three commonly used measures of central tendency.

- The **mean**, sometimes called the arithmetic mean or average = $\dfrac{\text{sum of all the values}}{\text{number of values}}$

- The **median** is the value that is halfway along the distribution when the data is arranged in ascending order. If there is an even number of values the median is calculated as the mean of the two middle values.

- The **mode** is the value that occurs most frequently in a distribution.

These statistical averages can be determined for each type of distribution.

8.4.1 Mean, median and mode of a simple distribution

Example 19

The ages of five children are 2, 4, 7, 6 and 11. Calculate the mean age.

The mean is $\bar{x} = \dfrac{2 + 4 + 7 + 6 + 11}{5}$

$= \dfrac{30}{5}$

$= 6$

Example 20

The marks of 11 students in an examination are:

2, 3, 3, 5, 8, 4, 9, 2, 3, 7, 1

Determine the median mark.

First, arrange the data in ascending order: 1, 2, 2, 3, 3, ③, 4, 5, 7, 8, 9

Identify the 'middle' mark, in this case, 3.

The median is 3.

Example 21

These are the lengths of eight pieces of string (in cm):

9, 7, 2, 4, 3, 4, 8, 6

Determine the median length.

First arrange the lengths in ascending order: 2, 3, 4, 4, 6, 7, 8, 9

As the number of lengths is even, identify the two middle lengths: 4 and 6.

The median length is $\dfrac{4 + 6}{2} = \dfrac{10}{2}$

$= 5$

The median is 5.

Example 22

This distribution represents the marks of 12 students in a test marked out of 10.

1, 2, 6, 4, 8, 5, 9, 3, 2, 2, 7, 6

State the mode.

The mode is the value with the highest frequency.

The mode is 2 since it occurs 3 times.

The mode is 2.

8.4.2 Mean, median, mode of a frequency distribution

Mean

To determine the mean of a frequency distribution you need to add an additional column to the right in the frequency table.

The first two columns represent the variable (x) and the frequency (f). The third column represents the product of the value of the variable and its frequency, fx. Then the mean is calculated from the total of all the fx values and the total of the frequencies:

$\bar{x} = \dfrac{\Sigma fx}{\Sigma f}$ where Σ means 'sum of'.

Example 23

Calculate the mean of the frequency distribution represented by this frequency table, where x represents the marks of 60 students in an assignment graded out of 10.

x	1	2	3	4	5	6	7	8	9
f	5	9	2	10	6	15	8	4	1

Turn the frequency table into a vertical array and include an additional column, fx.

Calculate the values of fx.

Calculate the sum of fx and the sum of f.

$\bar{x} = \dfrac{\Sigma fx}{\Sigma f}$

The mean $= \dfrac{286}{60}$

$= 4.8$

Marks (x)	Frequency (f)	fx
1	5	5
2	9	18
3	2	6
4	10	40
5	6	30
6	15	90
7	8	56
8	4	32
9	1	9
	$\Sigma f = 60$	$\Sigma fx = 286$

Median

To deduce the median for a frequency distribution from the frequency table you need to identify the value that corresponds to the middle frequency. If there are two middle frequencies (in the case of an even number of distributions) the median is the mean of the two middle values.

Another method for determining the median is to formulate a cumulative frequency table with two columns: one for the marks and the other for the cumulative frequency, which you find by adding successive frequencies.

Mark	Cumulative frequency
1	5
2	14
3	16
4	26
5	32
6	47
7	55
8	59
9	60

As you can see from the cumulative frequency table, the middle cumulative frequencies (30–31) correspond to a mark of 5.

Example 24

Determine the median of the frequency distribution given in Example 23.

There are 60 values, so the median lies between the 30th and 31st values, which are 5 and 5.

The median is $\dfrac{5+5}{2} = \dfrac{10}{2}$

$= 5$

Mode

The mode of a frequency distribution is actually the simplest measure of central tendency that can be found from a frequency table. Based on the definition of mode as the value that occurs most often in a distribution, the mode of the distribution is naturally the value with the greatest frequency.

Example 25

Find the mode of the frequency distribution in Example 23.

From the frequency table in example 23 it is clear that the mark with the greatest frequency is 6 (frequency 15). Therefore, the mode is 6.

8.4.3 Mean, median, mode of a grouped distribution

Mean

There are two methods for calculating the mean of a grouped frequency distribution.

The class midpoint method

This method uses the formula mean $\bar{x} = \dfrac{\sum f\bar{x}_m}{\sum f}$ where $\bar{x}_m$ is the class midpoint.

Example 26

Use the class midpoints to calculate an estimate of the mean mark for this grouped frequency distribution.

Class	0–4	5–9	10–14	15–19	20–24
Frequency	5	3	5	6	1

Turn the original frequency table and add two extra columns for $\bar{x}_m$ (class midpoint) and $f\bar{x}_m$ (frequency × class midpoint).

Class	Frequency (f)	Class midpoint ($\bar{x}_m$)	$f(\bar{x}_m)$
0–4	5	2	10
5–9	3	7	21
10–14	5	12	60
15–19	6	17	102
20–24	1	22	22
	$\Sigma f = 20$		$\Sigma f\bar{x}_m = 215$

The mean $\bar{x} = \dfrac{\Sigma f\bar{x}_m}{\Sigma f} = \dfrac{215}{20} = 10.75$

The coded method

This is an alternative to the class midpoint method, although it is somewhat more complicated. It consists of:

- choosing an **assumed mean** – any value (x) from the class but preferably the middle value

  (In example 26 a good assumed mean could be 12, the middle value of the class midpoint.)
- determining the **unit size**, which is the difference between successive values of the class midpoint
  (Again, in example 26, the unit size will be $7 - 2 = 12 - 7 = 17 - 12 = 22 - 17 = 5$.)
- calculating the coded value (x_c) corresponding to each value of the class midpoint. This coded value is the difference between the assumed mean and the class midpoint values (which will be positive or negative) expressed in terms of the of unit size.

  (Examining the previous example, the coded value for the first class midpoint will be:

  assumed mean – first class midpoint = $12 - 2 = 10$ ($2 \times$ unit size); coded value for second class midpoint = $12 - 7 = 5$ ($1 \times$ unit size); coded value for assumed mean = $12 - 12 = 0$, etc.)

The table of values used with this method contains the columns: Class, Class midpoint ($\bar{x}_m$), Frequency (f), Coded value (x_c), Product of frequency and coded value (fx_c). From the table the sums Σfx_c and Σf are determined.

The coded value of the mean is calculated from the formula $\bar{x}_c = \dfrac{\Sigma fx_c}{\Sigma f}$.

The mean itself is calculated from the formula mean = assumed mean + $\bar{x}_c \times$ unit size.

Use the coded method to calculate an estimate of the mean of the grouped distribution in Example 26.

Class	Class midpoint ($\bar{x}_m$)	Frequency (f)	Coded value (x_c)	fx_c
0–4	2	5	–2	–10
5–9	7	3	–1	–3
10–14	12	5	0	0
15–19	17	6	1	6
20–24	22	1	2	2
		$\Sigma f = 20$		$\Sigma fx_c = -5$

$$\bar{x}_c = \frac{\Sigma fx_c}{\Sigma f}$$

$$= \frac{-5}{20}$$

$$= -0.25$$

In this example, assumed mean = 12, unit size = 5 and $\bar{x}_c = -0.25$.

The mean = assumed mean + $\bar{x}_c \times$ unit size = 12 – 0.25 × 5 = 12 – 1.25 = 10.75

Note Both methods yield the same result; however, the second method, although more complicated, involves smaller values and does not really require the use of a calculator.

Median

Unlike the medians for simple and frequency distributions, the median of a grouped distribution is determined by a graphical method.

First, you need to draw up a **cumulative frequency table**, which includes a column in which you record a 'running total' of the frequencies of the data values or classes, in order.

Here are cumulative frequency tables for the data used in examples 23 and 26.

Cumulative frequency table for ungrouped data

Marks (x)	Frequency (f)	Cumulative frequency
1	5	5
2	9	14
3	2	16
4	10	26
5	6	32
6	15	47
7	8	55
8	4	59
9	1	60

Cumulative frequency table for grouped data

Class	Frequency (f)	Cumulative frequency
0–4	5	5
5–9	3	8
10–14	5	13
15–19	6	19
20–24	1	20

The cumulative frequencies are plotted against class boundaries on a set of axes with the data values or class boundaries on the horizontal axis and the cumulative frequency on the vertical axis. The cumulative frequency for each data value or class is plotted against the data value, or upper class boundary for that class.

The resulting graph is called a **cumulative frequency curve** or **ogive** and it possesses a unique shape. The ogive can be used to determine the value in the distribution corresponding to the middle frequency, which is the median.

Note	When plotting an ogive from a cumulative frequency table, the first lower boundary is plotted against a cumulative frequency of 0 and each successive upper boundary is plotted against the cumulative frequency value for that value in the distribution. The diagram shows the typical shape of the cumulative frequency curve (ogive).	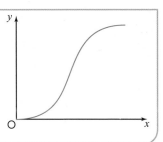

Example 28

Draw up a cumulative frequency table and use it to draw the cumulative frequency curve for this grouped frequency distribution.

Hence evaluate the median.

Class	0–4	5–9	10–14	15–19	20–24
Frequency	5	3	5	6	1

First, determine the class boundaries.

Then draw up the cumulative frequency table. The final cumulative frequency must be equal to the total frequency.

Class boundary	Frequency (*f*)	⇒	Class boundary	Cumulative frequency
–0.5 – 4.5	5		–0.5	0
4.5 – 9.5	3		4.5	0 + 5 = 5
9.5 – 14.5	5		9.5	5 + 3 = 8
14.5 – 19.5	6		14.5	8 + 5 = 13
19.5 – 24.5	1		19.5	13 + 6 = 19
	$\Sigma(f) = 20$		24.5	19 + 1 = 20

Now draw the cumulative frequency curve (ogive).

To find the median of this grouped frequency distribution, locate the middle frequency (10) and draw a straight horizontal line from here to intersect the curve. Then draw a vertical line from this point to the horizontal axis. That value represents the median.

From the ogive, the median is 12 marks.

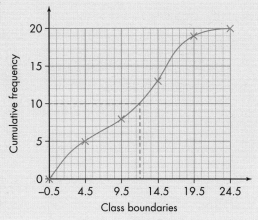

Modal class

Rather than trying to find the mode for a grouped frequency distribution, you determine the **modal class**. This is the class corresponding to the greatest frequency. This can be identified easily from the grouped frequency table.

Key term

For a grouped frequency distribution, the mode is taken as the class with the highest frequency, which is the **modal** class.

Example 29

Find the modal class for this grouped frequency distribution.

Class	0–4	5–9	10–14	15–19	20–24
Frequency	5	3	5	6	1

The class with the greatest frequency is clearly 15–19 so the modal class is 15–19.

8.4.4 Measures of dispersion

Measures of central tendency or statistical averages are important in that they give an idea of the position or location of the data in a distribution, focusing more on an average for the data. They are even sometimes referred to as measures of location.

However, it is often necessary to analyse the spread or dispersion of a distribution relative to its central tendency. This is achieved through the four measures of dispersion.

The range

The **range** is the difference between the greatest and smallest values in a distribution.

For a simple or discrete distribution the range = largest data value – smallest data value

Key term

The **range** is the difference between the greatest and smallest values.

For an ungrouped frequency distribution,

range = upper boundary limit of last class – lower boundary limit of first class.

Example 30

(a) The numbers of goals scored by seven footballers in a competition are 2, 4, 5, 8, 2, 4, 7. Find the range of the distribution.

(b) The table shows a frequency distribution indicating the ages of 50 people in a survey. Determine the range of the distribution.

Age (years)	10	11	12	13	14	15	16	17	18	19
Number of people (frequency)	5	8	7	3	2	10	1	11	2	1

(a) The range is largest data value (by observation) – smallest data value (by observation), which is $8 - 2 = 6$.

(b) The range is upper boundary limit – lower boundary limit, which is $19.5 - 9.5 = 10$

Quartiles

You have seen that you can find the median of a grouped frequency distribution by identifying the middle value of the cumulative frequency and finding the value of the variable that corresponds to this cumulative frequency. The cumulative frequency can be further divided, into four equal parts. Each part is called a **quartile**.

The **lower quartile** (Q_1) is the data value that relates to the 25th percentile (25%) of the cumulative frequency.

The **middle quartile** (Q_2) is the median and corresponds to the 50th percentile (50%) of the cumulative frequency.

The **upper quartile** (Q_3) corresponds to the 75th percentile (75%) of the cumulative frequency.

The interquartile range

The interquartile range = upper quartile – lower quartile = $Q_3 - Q_1$

The semi-interquartile range

The semi-interquartile range is half the interquartile range, $\dfrac{Q_3 - Q_1}{2}$. It is also known as the **quartile deviation**.

The advantage of using the semi-interquartile range is that it does not take into account any extreme values. It gives a measure of how the scores are spread about the median or middle values. It is used extensively in business and education.

Example 31

The heights of 9 trees, in metres, are: 4, 6, 10, 3, 2, 8, 7, 5, 9. Calculate the interquartile range and semi-interquartile range of the heights.

First, organise the distribution in ascending order: 2, 3, 4, 5, 6, 7, 8, 9, 10

The median Q_2 is 6, Q_1 is $\dfrac{3+4}{2} = 3.5$, Q_3 is $\dfrac{8+9}{2} = \dfrac{17}{2} = 8.5$.

To determine Q_1 and Q_3, consider the middle of the first half of the distribution and the middle of the second half respectively.

The interquartile range is $Q_3 - Q_1 = 8.5 - 3.5 = 5$

The semi-interquartile range is $\dfrac{1}{2}(Q_3 - Q_1) = \dfrac{1}{2} \times 5 = 2.5$

Standard deviation

The standard deviation is another way of describing the spread of the data in a set. It is a measure of how much the members of the set differ from their mean value.

The following graphs show the results of a mathematics examination for Class A and Class B. The mean mark for both classes was 55%, which might suggest that the ability in mathematics is the same for both classes, but this is not supported by the distribution of marks.

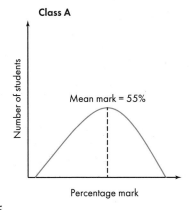

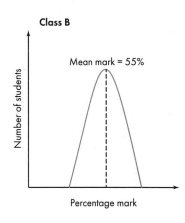

In Class A, the marks are widely spread out about the mean. The students show a broad range of ability ranging from very weak to very strong. The data for Class A has a high standard deviation.

In Class B, the marks are distributed much more narrowly about the mean. The ability of the students is less varied with no very weak students or very strong students. The data for Class B has a low standard deviation.

The standard deviation is expressed as a number and takes its units from the data is describes. The greater the value, the more spread out the data. As a general rule, the value of the standard deviation is about one quarter the value of the range.

How data is distributed about a given value

It is often useful to determine the proportion, or the percentage, of a set of data which falls above or below a particular value.

For example, a teacher might evaluate how well, or how badly, the class has understood a particular topic by setting them a test in which the pass mark is 7 out of 10. If a high proportion of the class score 7 or more, the teacher can be confident that the topic has been well understood. Conversely, if a high proportion of the class score less than 7, this indicates the topic has not been understood and some remedial teaching is needed.

You might wish to determine the proportion of a particular value from raw data. Here are the suits of the first 20 cards dealt from a pack of playing cards.

♣ ♦ ♣ ♥ ♦ ♥ ♥ ♠ ♦ ♣ ♥ ♠ ♦ ♦ ♣ ♥ ♠ ♥ ♥ ♣ ♦

What proportion of the cards are hearts (♥)? To find out simply count the number of hearts in the raw data.

There are 6 hearts so the proportion of hearts is 6 out of 20. We can also express this as a percentage: $\frac{6}{20} \times 100 = 30\%$.

In order to determine the proportion of a particular value in a frequency table, simply add the frequencies of the values above or below, depending on what you wish to find.

The following frequency table shows the results of a survey carried out by scientists who investigated the number of eggs laid by Caribbean doves. During the breeding season they visited 50 nests and recorded the number of eggs in each.

To find out what proportion of birds laid at least 3 eggs add the frequencies of 3, 4, 5 and 6 eggs.

Frequency of 3 or more eggs = 11 + 9 + 5 + 2 = 27

The proportion of birds that laid 3 or more eggs is therefore 27 out of 50. Expressed as a percentage this is $\frac{27}{50} \times 100 = 54\%$.

To find out what proportion of birds laid less than 2 eggs add the frequencies of 1 and 0 eggs.

Frequency of less than 2 eggs = 6 + 1 = 7

The proportion of birds that laid less than 2 eggs is therefore 7 out of 50, or 14%.

Number of eggs in the nest	Frequency
0	1
1	6
2	16
3	11
4	9
5	5
6	2

A cumulative frequency curve provides a running total of the frequency up to and including a particular value.

Example 33

The following cumulative frequency curve shows the results of a questionnaire given out by a restaurant in order to find the age groups of its customers. They used this information when creating a new menu.

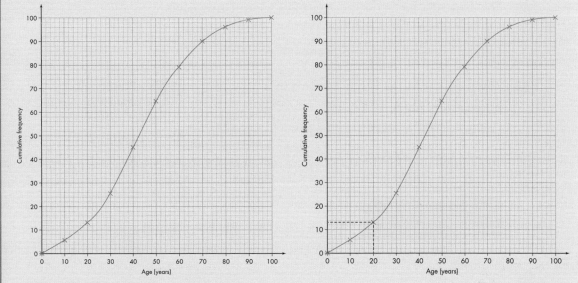

To find the proportion of diners who are 20 years or younger we draw a vertical line from 20 years to the curve, and then a horizontal line to the cumulative frequency axis. Please note the lines in red on the right side graph on above.

The cumulative frequency curve shows that 13% of the diners were 20 years or younger.

To find the proportion of diners within the age group 31–60 years we need to subtract the frequency of those who are 30 years or younger from the frequency of those who are 60 years or younger. See the red lines in the graph on right side.

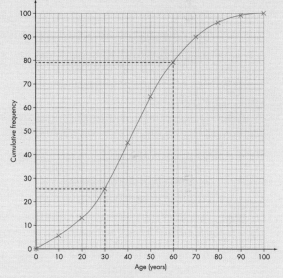

79% of the diners were 60 years old or less while 25% of the diners were 30 years or less.

The percentage of diners in the age range 31–60 years is therefore $79 - 25 = 54\%$.

If 250 customers completed the questionnaire, we can say the $250 \times \dfrac{54}{100} = 135$ customers were between 31 and 60 years old.

Exercise 8B

1 Calculate the mean of this set of numbers.

1, 5, 9, 2, 13

2 Determine the median of this set of numbers.

12, 4, 9, 1, 3, 8, 11

3 State the mode of this set of numbers.

3, 6, 8, 2, 5, 7, 3, 9

4 This data set represents the ages of 9 children in a group.

1, 9, 4, 3, 6, 7, 3, 5, 2

Determine the median, mode and mean.

5 These are the marks of ten students in an examination marked out of 10.

8, 9, 4, 4, 6, 3, 7, 8, 5, 1

Determine the median, mode and mean of the distribution.

6 This frequency distribution gives the numbers of subject passes achieved by students of a particular school in an external examination.

Number of subjects	1	2	3	4	5	6	7	8	9	10
Number of students	12	18	22	30	35	40	28	15	12	8

Use your frequency table to calculate the mean, median and mode of the distribution.

7 Use the class midpoint method to determine the mean of this grouped frequency distribution.

Class	1–3	4–6	7–9	10–12	13–15	16–18	19–21
Frequency	10	6	18	1	16	7	2

8 Use the coded method to calculate an estimate of the mean of the data in question 7.

9 a Using class boundaries, construct a cumulative frequency curve from the data in question 7.

b Hence derive an estimate of the median of the distribution.

c What is the modal class?

10 a Determine the lower quartile, second quartile (median) and upper quartile from your cumulative frequency curve in question 9.

b Determine values for the range, interquartile range and semi-interquartile range of the distribution.

11 Calculate the range, interquartile range and semi-interquartile range for each of these distributions.

a The number of fruits consumed by seven students: 2, 3, 5, 6, 4, 1, 9.

b
Variable	1	2	3	4	5	6	7	8	9	10
Frequency	3	8	12	20	15	10	14	4	2	1

12 Two classes, A and B, took an examination in science. Here is some information about their results.

Class	Mean mark	Standard deviation
A	65	10
B	65	15

a What does standard deviation measure?

b In which class was the ability shown by students in the examination more varied? Explain your answer.

c Estimate the range of marks obtained by Class A.

13 The following frequency chart shows the number of bedrooms in all of the homes in one street.

Number of bedrooms	Frequency
1	14
2	22
3	20
4	3
5	1

a How many homes are in the street?

b What proportion of homes have more than 3 bedrooms?

c What percentage of homes have 2 bedrooms or less?

14 At the end of their school year, 70 students sat a mathematics examination in which the maximum possible score was 80 marks. Their results are shown on the following cumulative frequency curve.

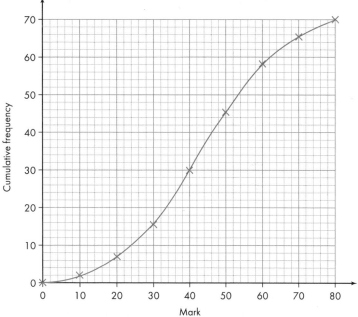

a How many students scored half marks of less?

b In order to pass the examination a student needed to obtain at least 75% of the marks. How many students passed?

c The teacher decided to award a prize to any student scoring 75 marks or more. How many prizes will be needed?

8.5 Probability

Many activities in daily life include an element of chance, for example, when you decide to go out, your decision as to whether to take a coat may depend on the **probability** that it is going to rain. The actions of tossing a coin to obtain a head or tail, drawing a specific card form a normal pack, rolling a dice are all called **events** and the possible results are called **outcomes**.

Outcomes that are equally likely to occur are **equiprobable**.

Probability can therefore be calculated as: $\dfrac{\text{the number of ways of producing a favourable outcome}}{\text{the total number of equiprobable outcomes}}$

Probability normally has a numerical value on a scale from 0 to 1. An event that is impossible has a probability of zero while an event that is 100% certain to happen has a probability of one (1).

All other probabilities lie between these two extremes and can therefore be expressed as fractions, decimals or percentages.

> **Key terms**
>
> **Probability** is a measure of the likelihood or possibility of an event occurring.
>
> Activities such as tossing a coin to obtain a head or tail are **events** and the possible results are **outcomes**.
>
> Equally likely outcomes are **equiprobable**.

Example 34

What is the probability of obtaining a head on tossing a coin?

A coin has 1 head and 1 tail. The probability is therefore

$$\frac{\text{number of favourable outcomes}}{\text{number of equiprobable outcomes}} = \frac{1}{2}$$

8.5.1 Total probability

The total probability of an event covers all possible outcomes.

Total probability = Pr(favourable outcomes) + Pr(unfavourable outcomes)

$$= 1$$

So the total probability = 1

For example, when tossing a coin, Pr(heads) $= \dfrac{1}{2}$, Pr(tails) $= \dfrac{1}{2}$ $\Rightarrow$ total probability is $\dfrac{1}{2} + \dfrac{1}{2} = 1$

Example 35

A bag contains two blue balls and three red balls.

(a) What is the probability of selecting a blue ball from the bag?

(b) Apply the principle of total probability to find the probability of choosing a red ball.

(a) Pr(blue ball) = $\dfrac{\text{number of favourable outcomes}}{\text{number of equiprobable outcomes}} = \dfrac{2}{5}$

(b) Pr(red ball) = $1 - \dfrac{2}{5} = \dfrac{3}{5}$

8.5.2 Experimental probability

Probability may be investigated by repeatedly experimenting, to establish how frequently different outcomes occur.

Key term

Experimental probability is the probability determined by experiment.

$$\text{Experimental probability} = \frac{\text{number of successful experiments (trials)}}{\text{total number of experiments (trials)}}$$

Example 36

One hundred bolts are examined and tested for flaws. Ten of these are found to have some defects. What is the probability that a bolt chosen at random from this batch is defective?

$$\text{Pr(defective bolt)} = \frac{\text{number found to be defective}}{\text{total number tested}} = \frac{10}{100} = \frac{1}{10}$$

8.5.3 Theoretical probability

Experimental probability is not always feasible, due to the impractical nature of some events. In these situations, you need to apply theoretical probability, based on mathematical rules and formulae.

Key term

Theoretical probability is the likelihood that an outcome will occur, based on all the known conditions.

Example 37

250 coins, consisting of 100 ten cents and 150 five cents are in a bag. What is the probability of drawing a ten cents coin from the bag?

$$\text{Pr(10c)} = \frac{100}{250} = \frac{2}{5}$$

8.5.4 Sample spaces

The concept of **sample spaces** links probability with set theory. A sample space includes all the possible outcomes, whatever their probability.

In the sample space, therefore the probability of an outcome A can be written as:

$\text{Pr}(A) = \dfrac{n(A)}{n(U)}$ where $n(A)$ is the number of ways A can occur.

Key terms

The **sample space** or **universal set (U),** is the set of all possible outcomes – favourable and unfavourable – of a random experiment. Every outcome in the sample space is a member of the sample space and is called a **sample point.**

Applying this notation to total probability $\text{Pr}(A) + \text{Pr}(A') = 1$.

Example 38

A boy writes the numbers 1 to 15 on 15 separate pieces of paper, puts them in a bag and shakes them up. He now attempts, without looking into the bag, to draw an even number. What is the probability that he will draw an even number?

The subset of even numbers, $E = \{2, 4, 6, 8, 10, 12, 14\} \Rightarrow n(E) = 7$

The universal set $U = \{1, 2, 3, 4, 5, 6, 7, 8, 9, 10, 11, 12, 13, 14, 15\} \Rightarrow n(U) = 15$

$$Pr(E) = \frac{n(E)}{n(U)} = \frac{7}{15}$$

The sample space (U) can be represented in a Venn diagram.

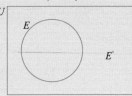

8.5.5 Mutually exclusive outcomes

Mutually exclusive outcomes are events that cannot happen simultaneously (together). For example, you cannot obtain both a head and a tail in a single toss of a coin. So you can find the probability of obtaining either a head or a tail but not both.

Key term

Mutually exclusive outcomes are events that cannot happen at the same time, such as obtaining both a head and a tail in a single toss of a coin.

In a sample space, mutually exclusive outcomes are shown as disjoint subsets.

$Pr(A \text{ or } B) = Pr(A \cup B) = Pr(A) + Pr(B)$ (addition law of probability)

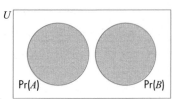

Example 39

A card is pulled from a standard pack of 52. What is the probability of pulling either an ace of spades or an ace of clubs?

$$Pr(\text{ace of spades}) = \frac{1}{52}$$

$$Pr(\text{ace of clubs}) = \frac{1}{52}$$

$$\Rightarrow Pr(\text{ace of spades or ace of clubs}) = \frac{1}{52} + \frac{1}{52} = \frac{2}{52} = \frac{1}{26}$$

Non-mutually exclusive outcomes can occur at the same time, for example, you can draw a king and a diamond card by drawing the king of diamonds. In this case, the probability becomes:

$$Pr(P \text{ or } Q \text{ or both } P \text{ and } Q) = Pr(P) + Pr(Q) - Pr(P \cap Q)$$
$$= Pr(P) + Pr(Q) - Pr(P) \times Pr(Q)$$

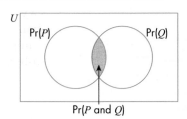

8.5.6 Independent outcomes

Some outcomes have no effect on the outcomes of other events, for example, pulling an ace from a pack of cards has absolutely no effect on rolling a dice to obtain a three or one. These are called **independent outcomes**. Since both can take place simultaneously:

$Pr(P \text{ and } Q) = Pr(P \cap Q) = Pr(P) \times Pr(Q)$

Example 40

A dice with 6 faces numbered 1 to 6 is rolled and a card is pulled from a pack of 52 cards at the same time. Determine the probability that the card will be an ace and the dice will show a 3.

$Pr(\text{rolling } 3) = \dfrac{1}{6}$ $Pr(\text{pulling ace}) = \dfrac{4}{52} = \dfrac{1}{13}$

Since the outcomes are independent, $Pr(\text{rolling a dice and pulling an ace}) = \dfrac{1}{6} \times \dfrac{1}{13} = \dfrac{1}{78}$

8.5.7 Dependent outcomes

Sometimes one outcome can affect the next and subsequent outcomes. For example, if you choose one item from a collection, the probabilities of all the remaining objects in the collection are affected. The probability of choosing any one item will be affected if another item is removed and not replaced. These are **dependent outcomes**.

Example 41

A bag contains two mangoes, three oranges and five grapes.

(a) What is the probability of drawing a grape from the bag?

(b) If the grape is eaten, what is the probability of drawing another grape?

(a) $Pr(\text{first grape}) = \dfrac{5}{10} = \dfrac{1}{2}$

(b) $Pr(\text{second grape}) = \dfrac{4}{9}$

Note

The number of grapes in the bag has decreased from 5 to 4 and the number of fruits from 10 to 9.

8.5.8 Contingency tables

The **contingency table** is a very useful tool in solving conditional probability problems. The table is used to display sample values of two dependent variables, contingent (dependent) on each other.

Example 42

This is an incomplete contingency table displaying data about Form 1 and Form 2 students who drink or do not drink cola.

Students	Form 1	Form 2	Totals
Drinks cola			
Does not drink cola	10	20	30
Totals	40	45	85

(a) Complete the contingency table.

(b) What is the probability that a Form 1 student does not drink cola?

(c) What is the probability that a Form 2 student drinks cola?

(d) Determine whether being a Form 1 student and drinking cola are independent.

(a) Complete the table by subtracting from the totals to fill in the missing row of data.

$40 - 10 = 30$ $45 - 20 = 25$ $85 - 30 = 55$

Students	Form 1	Form 2	Totals
Drinks cola	30	25	55
Does not drink cola	10	20	30
Totals	40	45	85

(b) $\text{Pr(Form 1 and does not drink cola)} = \dfrac{\text{number not drinking}}{\text{total}} = \dfrac{10}{40} = \dfrac{1}{4}$

(c) $\text{Pr(Form 2 and drinks cola)} = \dfrac{\text{number of Form 2 drinking}}{\text{total}} = \dfrac{25}{45} = \dfrac{5}{9}$

(d) Being a Form I student and drinking cola are considered independent if:

$\text{Pr(Form 1 and drinks cola)} = \text{Pr(Form 1)} \times \text{Pr(drinks cola)}$

$\text{Pr(Form 1 and drinks cola)} = \dfrac{30}{85} = 0.35$

$\text{Pr(Form 1)} = \dfrac{40}{85}$

$\text{Pr(drinks cola)} = \dfrac{55}{85}$ (Takes into consideration both forms.)

$\text{Pr(Form 1)} \times \text{Pr(drinks cola)} = \dfrac{40}{85} \times \dfrac{55}{85} = 0.30$

$\text{Pr(Form 1 and drinks cola)} \neq \text{Pr(Form 1)} \times \text{Pr(drinks cola)} \Rightarrow \text{dependency}$

Being a Form 1 student and drinking cola are thus considered to be dependent and not independent events.

Exercise 8C

1 What is the probability of obtaining tails on tossing a coin?

2 What is the probability of drawing an ace of spades from a pack of 52 cards?

3 Calculate the probability of scoring a 5 when a dice with faces marked from 1 to 6 is rolled.

4 A bag contains three red marbles and six blue marbles. Calculate the probability of picking a blue marble.

5 A bag contains 10 cards with the numbers 1 to 10 written on them. Calculate the probability of picking a card with the number 15 written on it.

6 Calculate the probability of drawing a king from a pack of cards.

7 Calculate the probability of picking a yellow ball from a bag containing 20 yellow balls.

8 The letters of the word 'independent' are written on separate cards and placed in an envelope. What is the probability of arbitrarily taking an e from that envelope?

9 If a dice is cast, what is the probability of scoring an even number?

10 A bag contains a number of red and yellow balls. The probability of picking a red ball is $\frac{5}{17}$. What is the probability of picking a yellow ball?

11 A physics student carried out the same experiment on seven occasions to determine the acceleration due to gravity. He was successful on three occasions but unsuccessful on four occasions. What is the probability that any one of his experiments was successful?

12 A girl picks out a letter of the alphabet. What is the probability that she chooses a vowel? Show your answer in a sample space in the form of a Venn diagram.

13 A universal set is given as whole numbers from 1 to 10. What is the probability of selecting a square number from that set? Use a sample space showing all favourable and unfavourable results to record your answer.

14 A dice is rolled and a coin is tossed at the same time. What is the probability of scoring a 5 and a head? How do you describe these two events?

15 Robin drew a card from a pack of 52 cards. He then replaced it and drew another card. Calculate the probability that he drew a jack and a spade in any order.

16 A bag contains two red balls, four green balls, six yellow balls and five white balls.

a What is the probability of choosing one white ball from the bag?

b If the white ball is not replaced what is the probability of choosing a yellow ball next?

c If the yellow ball is also not replaced, what is the probability of pulling another yellow ball from the bag next?

17 The students in Class V obtained the following marks out of 10 in an examination.

Marks	1	2	3	4	5	6	7	8	9	10
Number of students	3	5	9	8	9	5	4	3	2	1

a What is the probability that a student got more than 7 marks?

b What is the probability that a student got fewer than 3 marks?

c Calculate the probability that a student got 75% or more in the exam.

18 A survey was carried out among a group of boys and girls to determine their social media preference: Facebook or Instagram. The result of the survey is displayed in this contingency table.

Youths	Boys	Girls	Totals
Facebook	40	60	
Instagram			
Totals	70	70	

   a Complete the contingency table.

   b What is the probability that boys preferred Instagram?

   c What is the probability that girls preferred Facebook?

   d Determine whether or not the students' preference Facebook is independent of whether they are a boy or girl.

End of chapter summary

Sample statistics and population parameters

- A **population** in statistics is a set of items or objects (such as people, material things, concepts) with at least one common property or characteristic that can be studied.
- A **sample** is a small group taken from a population and used to represent the entire population.
- **Data** consists of information or facts collected in order for statistical analyses to be made. It may take the form of words, numbers, measurements or descriptions of objects.
- A **parameter** of a population is a fixed characteristic of that population.
- A **statistic** is a characteristic of a sample – a summary number (average or percentage) that describes the sample.

Statistical diagrams

- A **proportional bar chart** is a single bar or column divided into portions that represent the components of the whole.
- In a **simple vertical bar chart** each data category is represented by a vertical bar, the height of which represents the frequency of the data category.
- The **frequency** of a data value is the number of times it occurs.
- In a **simple horizontal bar chart** each data category is represented by a horizontal bar, the length of which represents the frequency of the data category.
- A **chronological bar chart** displays data in the time order in which it occurs.
- A **line graph** shows the relationship between two variables.
- In a **pie chart** each data category is represented as a sector of a circle that represents all the data. The sectors may represent the data as proportions or percentages.

Classifying and representing data

- **Qualitative data** is descriptive data, for example, colours, types of music, design of carnival costumes.
- **Quantitative data** is numerical information, such as numbers of people in a crowd, or measurements.
- **Continuous data** is data that is measured and can take any value, often within two given values (a range), for example, the height of a person.
- **Discrete data** can take only specific values, such as the number of children in a family.
- **Raw data** is data that has not been formatted or organised in any way, for example, the marks of 20 students in an examination, presented in random order.
- A **tally chart** is a table used to collect data.
- A **variable** is any characteristic of the individuals in a data set.
- A **tally** is a way of counting the individuals in a data set by means of tally strokes.
- A **frequency table or frequency distribution** is a list, table or graph that displays the frequencies of the data.
- A **histogram** is a diagram used to represent a frequency distribution, consisting of vertical bars drawn without any space between them; the height of the rectangle represents the frequency of the data.

- A **grouped distribution** is a large amount of data arranged in 5 to 20 groups or classes.
- A **grouped frequency distribution** is a frequency distribution for grouped data.
- The **class interval** is the range of values that make up a particular class or group or the difference between the upper and lower limit of a class or group.
- **Class limits** are the highest and lowest values of any class or group in a grouped distribution.
- The **class width** is defined as the difference between the lower limit or boundary and the upper limit or boundary of any class in a grouped distribution.
- **Class boundaries** are used to eliminate gaps between successive classes in a grouped distribution. Each of these values is calculated as the midpoint between the upper limit of one class and the lower limit of the next class.
- A **frequency polygon** is a line graph used to represent a frequency or grouped distribution and to indicate the spread of the distribution. It is formed by joining the mid-points of the top of the bars in a histogram by straight lines.
- A **frequency curve** is a frequency diagram based on a histogram for large quantities of data.

Measures of central tendency

- The **mean**, sometimes called the arithmetic mean or average $= \dfrac{\text{sum of all the values}}{\text{number of values}}$
- The **median** is the value that is halfway along the distribution when the data is arranged in ascending order. If there is an even number of values the median is calculated as the mean of the two middle values.
- The **mode** is the value that occurs most frequently in a distribution.
- It is possible to have two modes (two different values repeated the same number of times). This is known as a **bimodal** distribution.
- If there are more than two modes the distribution is said to be **multi-modal.**
- To determine the **mean of a frequency distribution** you need to add an additional column for fx the frequency table. Then the mean is calculated from the total of all the fx values and the total of the frequencies as $\bar{x} = \frac{\Sigma fx}{\Sigma f}$.
- To deduce the **median for a frequency distribution** from the frequency table you need to identify the value that corresponds to the middle frequency. If there are two middle frequencies (in the case of an even number of values) the median is the mean of the two middle values.
- **The mode of a frequency distribution** is the value with the greatest frequency.
- To deduce the **median for a grouped frequency distribution** you need to draw up a cumulative frequency chart and draw the cumulative frequency diagram, called the **ogive**, and use it to determine the middle value.
- For a grouped frequency distribution, you determine the **modal class** rather than the mode.
- The mean may be affected by extreme values so it can be misleading.
- The median (as the middle value) of a distribution, presents a clearer picture of the distribution, especially when there are extremely low or high values.
- The mode or modal class is used when the most repeated value is required, for example, when a shopkeeper is replenishing stock.
- The cumulative frequency can be divided into four equal parts and each part is called a **quartile**.
  - The **lower quartile** (Q_1) is the data value that relates to the 25th percentile (25%) of the cumulative frequency.
  - The **middle quartile** (Q_2) is the median and corresponds to the 50th percentile (50%) of the cumulative frequency.
  - The **upper quartile** (Q_3) corresponds to the 75th percentile (75%) of the cumulative frequency.

- To analyse the spread or dispersion of a distribution relative to its central tendency you use the measures of dispersion.
- The **range** is the difference between the greatest and smallest values in a distribution.
- The **interquartile range** = upper quartile – lower quartile = $Q_3 - Q_1$
- **The semi-interquartile range** is half the interquartile range = $\frac{Q_3 - Q_1}{2}$. It is also known as the **quartile deviation**.

Probability

- **Probability** is a measure of the likelihood or possibility of an event occurring.
- Activities such as tossing a coin to obtain a head or tail are **events** and the possible results are **outcomes**.
- Outcomes that are equally likely to occur are **equiprobable**.
- Probability can be calculated as $\dfrac{\text{the number of ways of producing a favourable outcome}}{\text{the total number of equiprobable outcomes}}$
- Total probability = Pr(favourable outcomes) + Pr(unfavourable outcomes)
- **Experimental probability** is the probability determined by experiment.
- **Experimental probability** $= \dfrac{\text{number of successful experiments (trials)}}{\text{total number of experiments (trials)}}$
- **Theoretical probability** is the likelihood that an outcome will occur, based on all the known conditions.
- The **sample space or universal set (U),** is the set of all possible outcomes – favourable and unfavourable - of a random experiment.
- Every outcome in the sample space is a member of the sample space and is called a **sample point**.
- **Mutually exclusive outcomes** are outcomes that cannot happen at the same time, such as obtaining both a head and a tail in a single toss of a coin.
- An **independent outcome** has no effect on other outcomes.
- **Dependent outcomes** can affect each other's probabilities.
- A **contingency table** is a special table in which the data is organised in such a way as to facilitate the calculation of probabilities.

Examination-type questions for Chapter 8

1   The table represents the distances a group of students were able to throw the cricket ball during their physical education lesson.

Distance thrown (m)	Frequency	Cumulative frequency
0–8	1	1
9–17	3	4
18–26	6	
27–35	12	
36–44	8	
45–53	4	

   a   Write down the upper and lower class limits for the third class.
   b   Determine the class width for the third class.
   c   Complete the table by including the missing values for cumulative frequency.
   d   How many students were not able to throw more than 26 metres?
   e   Draw a cumulative frequency curve to represent the information in the table.
   f   On your graph draw a line to indicate the number of students that were able to throw more than 44 m.
   g   What is the probability that a student chosen at random would have been able to throw more than 26 m?

2   a   The masses, in kilograms, of 8 men were recorded as 60, 65, 72, 70, 90, 86, 82, 76. Calculate:

   i   the range   ii   the interquartile range   iii   the median value.

   b   The heights of 50 women were tallied and recorded in this grouped frequency table.

Height of women (cm)	Number of women
150–154	2
155–159	18
160–164	16
165–169	8
170–174	6

   i   Draw a histogram and frequency polygon to represent the data.
   ii   Calculate an estimate of the mean height of the women.
   iii   What is the modal class?
   iv   What is the probability of a woman randomly selected having a height more than 159 cm but less than 165 cm?

3 The table shows the intake of children in a pre-primary school over the last seven years.

Year	2012	2013	2014	2015	2016	2017	2018
Intake	100	120	150	80	50	200	240

   a Draw a bar chart to represent the information given.

   b In which year was there: **i** the greatest intake   **ii** the least intake?

   c In which two consecutive years was there:

     i the greatest increase in intake

     ii the greatest drop in intake?

   d How many students were taken in over the 7-year period?

4 The pie chart shows the preferences of a number of children for different flavours of ice cream.

   a What is the value of x?

   b Which flavour was chosen by most children?

   c What percentage of the children liked chocolate?

   d Given that 20 children like mango, how many children were actually surveyed?

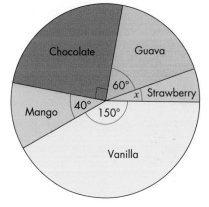

5 The diagram represents a frequency polygon showing the numbers of fruits consumed by some students on a daily basis, for example, 10 students consume not a single fruit daily.

   a Draw up a frequency table to represent the information.

   b Find the median and mode of the distribution.

   c Find the mean of the distribution.

   d How many students were interviewed?

   e What is the probability that a student chosen at random eats 8 fruits a day?

CHAPTER 9

Vectors, matrices and transformations

Objectives

By the end of the chapter you should be able to:

- explain concepts associated with vectors: concept of a vector, magnitude, unit vector, direction, parallel vectors, equal vectors, inverse vector as well as scalars: scalar multiples, magnitude, etc

- simplify expressions involving vectors; work with vector algebra: addition, subtraction, scalar multiplication and vector geometry, triangle law, parallelogram law

- write down the position vector of a point P(a, b) as $\overrightarrow{OP} = \begin{pmatrix} a \\ b \end{pmatrix}$ where O is the origin (0, 0)

- describe displacement and position vectors, including the use of coordinates in the x–y plane, to identify and determine displacement and position vectors

- determine the magnitude of a vector, including unit vectors (magnitude 1)

- determine the direction of a vector

- use vectors to solve problems in geometry and to explain the significance of points in a straight line, parallel lines, displacement velocity, weight

- explain basic concepts associated with matrices: concept of a matrix, types of matrix, row, column, square, rectangular, order of a matrix

- solve problems involving matrix operations such as addition and subtraction of matrices of the same order, scalar multiples, multiplication of conformable matrices, equality and non-commutativity of matrix multiplication

- find the determinant of a 2 × 2 matrix, define the multiplicative inverse of a non-singular square matrix, work with identity square matrices

- find the inverse of a non-singular 2 × 2 matrix, the determinant and adjoint of a matrix

- determine a 2 × 2 matrix associated with a specified transformation, for example:
    - reflection in the x-axis, y-axis, the lines $y = x$ and $y = -x$
    - rotation in a clockwise or anticlockwise direction about the origin (the general rotation matrix)
    - enlargement with centre at the origin
    - other important transformations

- use matrices to solve simple problems in arithmetic, algebra and geometry

- discuss the use of matrices to present data, the equality of matrices, the use of matrices to solve linear simultaneous equations with two unknowns. (Problems involving determinants are restricted to 2 × 2 matrices.)

- Number theory and computation
- Algebra
- Geometry and trigonometry
- Measurement
- Fractions and decimals
- Relations and functions

Introduction

These instructions were given to someone trying to locate a building:

'Walk straight ahead for two blocks, turn right and walk for another block then walk three more blocks in a southern direction.' The instructions include direction as well as distance or magnitude. This is what vectors are all about. There are many other everyday examples of vector applications in real life, ranging from the simple to the complex.

Vectors are used, as you have just seen, to locate objects and individuals. They are also used to describe the behaviour of people and objects subjected to forces. In sports, the force and direction with which the javelin, football, other missiles, even including the human body (in high jump and long jump) are launched, are crucial in acquiring the desired results or success, for example, the greatest distance a javelin can be thrown is achieved when it is launched at 45° to the horizontal.

In the more professional and technical domain, vectors are essential in engineering, whether in building a bridge sturdily and safely to withstand forces in various directions, designing a vehicle or other machines, navigation – ensuring that wind speed and direction are adequate for aeroplanes, boats and other crafts. One of the expressions used by air traffic control personnel on the approach of a plane is, 'Expect vectors for the visual approach.'

In this chapter you will learn how to use vectors and matrices to model and solve real-world problems. Additionally, you will learn to define and describe various linear transformations, their application in everyday problem-solving situations and how they can be combined with vectors and matrices to simplify the solution of these everyday problems further.

Examples of vectors and matrices in real life

Aircraft are given instructions by the control tower involving direction and distance as they land.

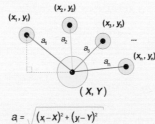

$$a_i = \sqrt{(x_i - X)^2 + (y_i - Y)^2}$$

Distances between points on a graph have direction and distance.

Communications between countries involve messages moving various distances in many directions.

The instrument panels of a helicopter give the pilots vital information, especially when they are flying over airfields at night.

9.1 Concepts associated with vectors

In previous work the quantities that you have worked with have, in the main, been **scalar** quantities, which have **magnitude** but no direction. However, you will be aware that some quantities, such as forces, act in particular directions. Weight, for example, acts downwards while velocity acts in the direction of motion. Quantities that have direction as well as magnitude are **vectors**.

9.1.1 Vector representation

A vector may be represented by a line segment of given magnitude or length, with an arrow at one end indicating the direction of action of the vector. The line segment may be drawn to scale.

Example 1

Represent the vector 20 N due west.

The vector is represented by a horizontal line, drawn to a scale of 1 cm : 2 N, with an arrow pointing west.

$\longleftarrow$ _____

20 N

A vector starting at P and ending at Q can be represented with capital letters and an arrow as $\overrightarrow{PQ}$.

Vectors are sometimes represented by a lower-case or common bold letter, for example, $\overrightarrow{PQ} = \mathbf{a}$. The magnitude of vector $\overrightarrow{PQ}$ may be expressed as $|\overrightarrow{PQ}|$ and the magnitude of $\mathbf{a}$ may be expresse as $|\mathbf{a}|$ or a.

9.1.2 Scalar multiplication

You know that a scalar quantity has size, or magnitude, but no direction. Scalar multiplication increases or decreases the magnitude of a vector by a factor.

In general, if k is a real number, the vector $k\overrightarrow{LM}$ is also a vector in the direction of $\overrightarrow{LM}$ and of magnitude k times the magnitude of $\overrightarrow{LM}$.

Example 2

A vector is given as $\overrightarrow{PQ} = 5$ m/s^2 due south. Evaluate $5\overrightarrow{PQ}$.

$5\overrightarrow{PQ} = 5 \times 5 = 25$ m/s^2 due south

Example 3

M is the midpoint of PR and $\overrightarrow{MR} = \mathbf{b}$. Write the vector $2\overrightarrow{PR}$ in terms of **b**.

$2\overrightarrow{PR} = 4 \times 2\overrightarrow{MR}$

$\qquad = 4 \times \mathbf{b}$

$\qquad = 4\mathbf{b}$

9.1.3 Parallel, equal and inverse vectors

- Two vectors are **parallel** if they act in the same direction. In this case, one can be written as a positive scalar multiple of the other. Parallel vectors are represented by parallel lines with arrows pointing in the same direction, as shown in the diagram, where $\overrightarrow{AB} \parallel \overrightarrow{LM}$.

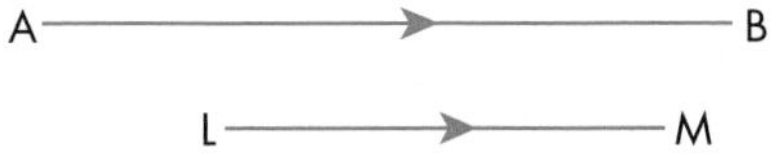

  A ————————→———— B

      L ———→———— M

- Two vectors are **equal** if they are parallel and have the same magnitude, for example, in the diagram, $\overrightarrow{PQ} = \overrightarrow{RS} = 10$ N.

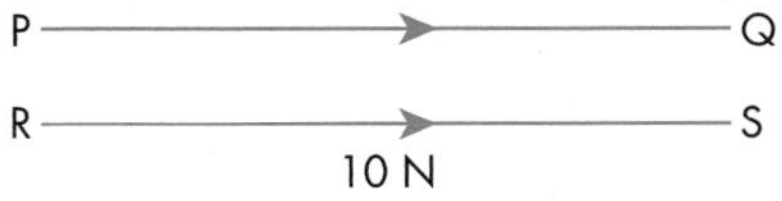

  P ————————→———— Q

  R ————————→———— S
               10 N

- Two vectors are **inverse** if they have the same magnitude but opposite directions, for example, $\overrightarrow{HK}$ and $\overrightarrow{SR}$ are inverse vectors.

  H ————————→———— K

  R ————————←———— S

  Generally the inverse of any vector $\overrightarrow{AB}$ is $-\overrightarrow{AB} = \overrightarrow{BA}$.

9.1.4 Vector algebra

Vector algebra involves addition and subtraction of vectors and multiplication of a vector by a scalar.

Vectors are sometimes represented by lower-case or common letters in bold, which makes their addition, subtraction and multiplication clearer.

Example 4

Determine the sum of the three parallel vectors **a**, **b** = 3**a** and **c** = $\frac{1}{2}$**a**, as in the diagram.

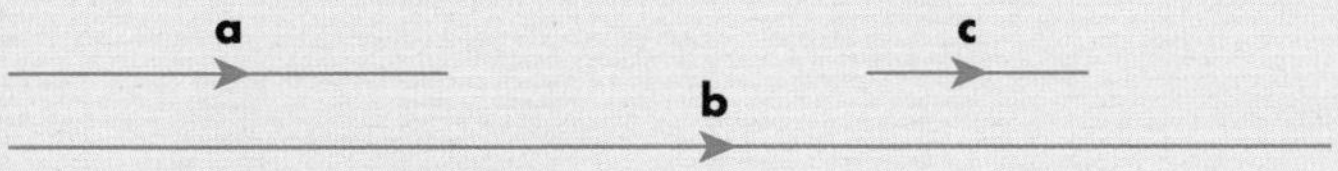

a ——→—— c —→——

——————————→——————————
                b

Since **b** = 3**a** and **c** = $\frac{1}{2}$**a**:

$$a + b + c = a + 3a + \tfrac{1}{2}a$$

$$= 4\tfrac{1}{2}a$$

Example 5

Given the vectors in example 4, simplify **a** + **c** − **b**.

$$a + c - b = a + \tfrac{1}{2}a - 3a$$

$$= 1\tfrac{1}{2}a - 3a$$

$$= -1\tfrac{1}{2}a$$

9.1.5 Addition and subtraction of non-parallel vectors

Addition

The general rule for adding two or more vectors is that the vectors must be aligned head to tail while maintaining their original direction. The combined vector is the **resultant**, often represented by **r**.

Note

Note that the head of the vector is considered to be the end with the arrow and the tail is the opposite end.

Example 6

A van travels 3 km due north from A to B and then 4 km due east from B to C. If the van had taken a shortcut directly from A to C, what would the distance AC be?

The displacement vector $\overrightarrow{AB}$ = 3 km north and $\overrightarrow{BC}$ = 4 km east. These are represented head to tail. The vector $\overrightarrow{AC}$ is the resultant vector obtained from adding $\overrightarrow{AB}$ and $\overrightarrow{BC}$. These vectors have a common point at B. The resultant always goes from A, the start of $\overrightarrow{AB}$, to C, the end of $\overrightarrow{BC}$.

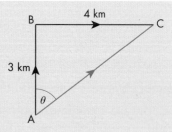

The direction of the resultant $\overrightarrow{AC}$ is given by the angle θ between the resultant and the northerly direction, AB. Use trigonometry to find the angle θ:

$$\tan\theta = \frac{4}{3}$$

$$= 1.33$$

$$\Rightarrow \theta = 53.1°$$

Now use Pythagoras' theorem in the right-angled triangle ABC to find the magnitude of $\overrightarrow{AC}$:

$$AC^2 = AB^2 + BC^2 \Rightarrow AC^2 = 3^2 + 4^2$$

$$= 9 + 16$$

$$= 25$$

$$\Rightarrow AC = \sqrt{25} \text{ which is 5.}$$

The magnitude of the resultant vector $\overrightarrow{AC}$ is 5 km.

Note

Note that since the vectors $\overrightarrow{AB}$ and $\overrightarrow{BC}$ have a common point (B), you can write $\overrightarrow{AB} + \overrightarrow{BC} = \overrightarrow{AC}$. This is true for any vectors that have a common point.

Example 7

Determine the resultant of three vectors of magnitude: 1 N, 2 N and 3 N acting in the directions shown in the diagram (not drawn to scale).

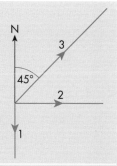

Make a scale drawing, aligning the vectors head to tail and maintaining their original directions and magnitudes. The resultant is **r** in the diagram.

Based on the scale drawing $\mathbf{r} = 4\frac{1}{2}$ N and its direction is $45° + 30° = 75°$ from north.

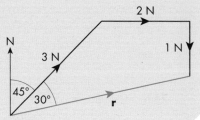

Example 8

Four children are pulling on a cart. Tom pulls with a force of $\mathbf{t} = 20$ N east, Mary pulls with a force of $\mathbf{m} = 30$ N west, Harry pulls with a force of $\mathbf{h} = 60$ N north-west and Ana pulls with a force of $\mathbf{a} = 10$ N south. Determine the magnitude and direction of the resultant force, **r**, on the cart.

Make a scale drawing of the forces (vectors) aligned head to tail and maintaining their original directions and magnitudes.

From the scale drawing it can be determined that the magnitude, r, of the resultant is approximately 60 N and **r** acts in a direction 300° from north.

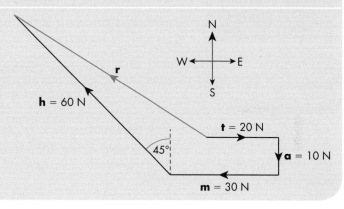

Note

Note: A precise measurement of magnitude and direction can be obtained using ruler and protractor.

Subtraction

Subtracting one vector from another is equivalent to adding the inverse of the subtracted vector to the first vector.

Example 9

In the vector diagram, **a** = 3 N north and **b** = 4 N east. Determine the resultant of **a** – **b**.

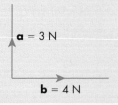

Draw a line to represent **a** with the same magnitude and direction as in the original diagram. Then draw a line for –**b** (the reverse of **b**), head to tail with **a**, and draw the resultant, as shown.

Apply Pythagoras' theorem to find the magnitude of the resultant:

$r^2 = 4^2 + 3^2$

$\quad = 25$

$r = 5$ N

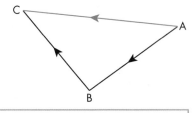

The direction is:

$\theta = \tan^{-1}\dfrac{3}{4}$

$\quad = 37°\quad$ from the direction of –**b**

Tip	Whereas the vectors to be added or subtracted are represented by arrows drawn head to tail, the resultant is always represented by an arrow drawn from the head of the first vector to the tail of the final vector.

9.1.6 The triangle and parallelogram laws of vector addition

The triangle law

If two adjacent sides of a triangle, drawn in order, represent the magnitudes and directions of two vectors acting from the same point, then the third side of the triangle represents the resultant of the two vectors in magnitude and direction.

From the diagram: $\overrightarrow{AB} + \overrightarrow{BC} = \overrightarrow{AC}$

Remember	Remember, the two vectors have a common point so can be combined.

The parallelogram law

If two vectors acting at a point are represented in magnitude and direction by two adjacent sides of a parallelogram, drawn from the same point, then the resultant of the two vectors is represented in magnitude and direction by the diagonal of the parallelogram, drawn from the same point.

In the diagram: $\overrightarrow{AC} + \overrightarrow{AB} = \overrightarrow{AD}$ and $\overrightarrow{AD} = $ **r**.

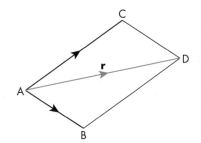

Example 10

In the triangle CAB, CA and AB represent vectors **a** and **b** in magnitude and direction. Given that **a** = 4 N, **b** = 5 N due east and the angle between CA and AB is 150°, use the triangle law to determine the magnitude and direction of the resultant $\overrightarrow{CB}$ = **r**.

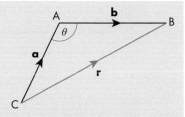

Use the cosine rule (non-right angled triangle) to find the magnitude of the resultant:

$r^2 = a^2 + b^2 - 2ab \cos\theta$

The direction of **r** can be determined since it is the angle **r** makes with the northerly direction, known as the bearing of **r**.

If **a** = 4 N, **b** = 5 N and the angle between CA and AB is 150°,

$r^2 = 4^2 + 5^2 - 2 \times 4 \times 5 \cos 150°$

$\quad = 16 + 25 - 40 \cos 150°$

$r^2 = 41 + 34.6$

$\quad = 75.6$

$r = 8.7$

Now use the sine rule to find the angle at C.

$\dfrac{\sin C}{5} = \dfrac{\sin 150°}{8.7}$

$\sin C = \dfrac{5 \sin 150°}{8.7}$

$\quad = 0.2873\ldots$

$C = 16.7°$

r = 8.7 N on a bearing of 076.7° from CA.

> The angle at B will therefore be 180° − (150° + 16.7°) = 13.3°.
>
> Then the direction of **r** will be 90° − 13.3° = 76.7°.
>
> The bearing of **r** is 076.7°.

Tip

In order to determine a resultant vector from the addition or subtraction of two or more vectors, it may be necessary to reverse the direction of vectors so that they may be placed head to tail. When this happens, the order of the letters is reversed and the sign of the vector also changes (+ to − or − to +).

Exercise 9A

1. Using an appropriate scale, draw diagrams to represent these vectors.

   a 5 N north   b 15 m west   c 40 m s⁻¹ south   d 100 m s⁻² east

2. Given that the magnitude of $\overrightarrow{PQ}$ is 18 N, evaluate the magnitude of:

   a $2\overrightarrow{PQ}$ b $24\overrightarrow{PQ}$ c $9.5\overrightarrow{PQ}$ d $24\frac{1}{2}\overrightarrow{PQ}$.

3. Given vector $\overrightarrow{LM}$ = 60**m**, with K the midpoint of LM, express in terms of **m**:

   a $3\frac{1}{4}\overrightarrow{LM}$ b $1\frac{1}{2}\overrightarrow{KM}$ c $5\overrightarrow{LK}$ d $12\frac{1}{4}\overrightarrow{LK}$.

4   **a** Using an appropriate scale, represent the vector $\overrightarrow{ST}$ = 20 N north-east by an arrow.

     b Draw a vector with twice the magnitude of $\overrightarrow{ST}$.

     c Draw the vector $2\overrightarrow{TS}$.

     d Draw a vector $\overrightarrow{HK}$ that is parallel to $\overrightarrow{ST}$ and equal to $\dfrac{3}{4}\overrightarrow{ST}$.

5   Given the vectors:

     B = **a**   **D** = 2**a**

               C = 3**a**

     express in terms of **a**:

     a   **B** + **C** + **D**   **b**   **B** + **C** – **D**   **c**   **C** – 2**B** – **D**   **d** $\dfrac{1}{2}$**D** – $\dfrac{2}{3}$**C** + **B**

6   A vehicle travels 60 km east from town T to town V and 40 km south from V to W. If the driver had travelled directly from T to W along a straight line how much shorter, in kilometres, would his journey have been?

7   A bird sees a ripe mango at the very top of the tree, flies directly to it and pecks it. If the mango is at a vertical height of 20 m from the ground and the bird flies from a spot on the ground 15 m from the base of the tree, how far does the bird have to travel and in what direction with respect to the horizontal?

8   Determine the resultant of the two vectors **p** and **q** in the directions given.

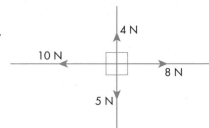

9   The vector **p** has magnitude of 7 m in a direction 60° from north and **r** has magnitude 13 m and makes an angle of 40° with **p**. Determine the resultant of the two vectors.

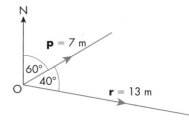

10   The diagram (not drawn to scale) shows four vectors acting at a point O. Using scale drawing or otherwise, determine the resultant vector. State its magnitude and direction.

11   These vectors (not drawn to scale) are shown acting from the same point O. Using the magnitudes and directions given, determine the magnitude and direction of the resultant.

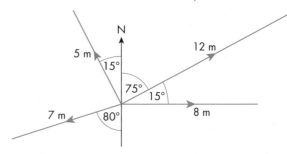

12 From a position P on the playing field a girl runs 20 m in a north-easterly direction, then 15 m east, 24 m south, 12 m west and then returns to her starting point.

    a What was the total distance covered?

    b In what direction was her final run back to her starting point?

13 A vector $\overrightarrow{LM}$ has a magnitude of 2a units. R is the midpoint of LM so $\overrightarrow{LR} = \overrightarrow{RM} = $ **a**. Another vector $\overrightarrow{LP}$ is twice $\overrightarrow{LM}$. Determine the resultant of this combination of the vectors, in terms of **a**.

$$2\overrightarrow{LM} - \tfrac{1}{2}\overrightarrow{RM} - 3\overrightarrow{RL} - 1\tfrac{3}{4}\overrightarrow{PL} + 2\tfrac{1}{2}\overrightarrow{RP}$$

14 Determine the magnitude of resultant of these three vectors, in terms of **a**, **b** and **c**.

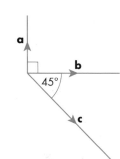

15 In the diagram, AD represents vector **a**, BC represents vector **b** and DC represents vector **c**. P is the midpoint of AD and Q is the midpoint of BC.

Write, in terms of **a**, **b** and **c**, the vectors:

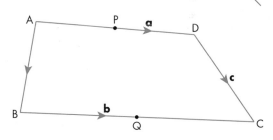

    a $\overrightarrow{AC}$ **b** $\overrightarrow{BD}$

    c $\overrightarrow{AB}$ **d** $\overrightarrow{DQ}$

    e $\overrightarrow{BP}$ **f** $\overrightarrow{PC}$

    g $\overrightarrow{PQ}$.

9.2 Position vectors

9.2.1 Concept and description of position vectors and unit vectors

The position of any point may be described in terms of any other fixed point. Graphically, the position vector of point P from the origin O is drawn as an arrow from O to P.

Mathematically it is expressed in terms of **unit vectors**. In diagrams, lower-case or common letters such as **a**, **b**, **p**, **q** may also be used to denote the magnitude and direction of a position vector.

A position vector can be two-dimensional $\begin{pmatrix} x \\ y \end{pmatrix}$ or three-dimensional $\begin{pmatrix} x \\ y \\ z \end{pmatrix}$.

> **Key term**
>
> A **unit vector** is a vector with a magnitude of 1. It is sometimes called a **direction vector**.

> **Key term**
>
> A **position vector**, sometimes called a **location** vector, is a vector that represents the position of a point P in space in relation to a fixed reference point such as the origin (O).

> **Note**
>
> Using column vectors to describe vectors in this way is explained in Section 9.2.4.

Two dimensional	Three dimensional
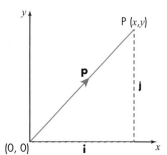	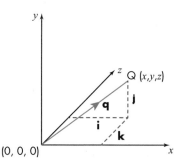

For the two-dimensional representation, the unit vectors are **i** along the *x*-axis and **j** along the *y*-axis, so the position vector of point P is given by:

$$\overrightarrow{OP} = x\mathbf{i} + y\mathbf{j}$$

$$= \begin{pmatrix} x \\ y \end{pmatrix}$$

$$= \mathbf{p}$$

For the three-dimensional representation the unit vectors are: are **i** along the *x*-axis, **j** along the *y*-axis and **k** along the *z*-axis, so the position vector of point Q is given by:

$$\overrightarrow{OQ} = x\mathbf{i} + y\mathbf{j} + z\mathbf{k}$$

$$= \begin{pmatrix} x \\ y \\ z \end{pmatrix}$$

$$= \mathbf{q}$$

9.2.2 Scalar magnitude of a position vector and unit vector

Like any other vector, a position vector has **magnitude** and **direction**. You need to be able to calculate these.

$$OP^2 = (x - 0)^2 + (y - 0)^2 = x^2 + y^2 \Rightarrow \overrightarrow{OP} = \sqrt{x^2 + y^2}$$

For a position vector $\overrightarrow{OQ}$, represented three dimensionally, the magnitude is given by:

$$OQ^2 = x^2 + y^2 + z^2 \Rightarrow \overrightarrow{OQ} = \sqrt{x^2 + y^2 + z^2}$$

> **Key term**
>
> The **magnitude** of a position vector is a scalar quantity represented by the distance from the origin to the point and can be found by using Pythagoras' theorem.

Example 11

The position vector $\overrightarrow{OA}$ is shown in the diagram.

(a) Express the vector in terms of the unit vectors **i** in the *x*-direction and **j** in the *y*-direction.

(b) Evaluate the magnitude of $\overrightarrow{OA}$ and determine its direction with respect to the *x*-axis.

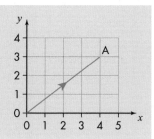

(a) $\overrightarrow{OA} = 4\mathbf{i} + 3\mathbf{j}$

(b) The magnitude of $\overrightarrow{OA} = |OA|$

$$= \sqrt{4^2 + 3^2}$$
$$= \sqrt{25}$$
$$= 5$$

The magnitude of $\overrightarrow{OA} = 5$ units

The direction of $\overrightarrow{OA}$ is:

$$\theta = \tan^{-1}\frac{3}{4}$$
$$= \tan^{-1}0.75$$
$$= 37°$$

Note

In the Cartesian plane, the angle of a vector is usually measured anticlockwise from the positive x-direction.

It is possible to convert any vector into a direction or unit vector by dividing the vector by its magnitude.

In general, the **unit vector** $\overline{\mathbf{a}} = \frac{\mathbf{a}}{|\mathbf{a}|}$, where $\overline{\mathbf{a}}$ is the unit vector associated with vector **a** and $|\mathbf{a}|$ is the modulus or magnitude of **a**. Note that the magnitude of **a** may also be represented by a.

Key term

The **unit vector** $\overline{\mathbf{a}}$ associated with vector **a** is calculated as $\overline{\mathbf{a}} = \frac{\mathbf{a}}{|\mathbf{a}|}$ where $|\mathbf{a}|$ is the magnitude of **a**.

Example 12

Find the unit vector that acts in the same direction as vector $6\mathbf{i} + 8\mathbf{j}$.

The length or magnitude of vector $6\mathbf{i} + 8\mathbf{j}$ is given by:

$$\sqrt{6^2 + 8^2} = \sqrt{100}$$
$$= 10$$

The unit vector parallel to $6\mathbf{i} + 8\mathbf{j}$ is given by:

$\frac{1}{10}(6\mathbf{i} + 8\mathbf{j})$

Example 13

Given two position vectors; $\overrightarrow{OA} = \mathbf{a}$ and $\overrightarrow{OB} = \mathbf{b}$, write in terms of **a** and **b** the vectors $\overrightarrow{AB}$ and $\overrightarrow{BA}$.

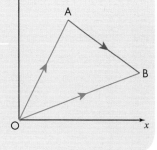

Using the rule of vector addition, $\overrightarrow{AB}$ is the resultant arising from the addition $\overrightarrow{AO} + \overrightarrow{OB}$.

$\overrightarrow{AB} = -\mathbf{a} + \mathbf{b}$

$\overrightarrow{BA} = -(-\mathbf{a} + \mathbf{b})$ $\quad(\overrightarrow{BA} = -\overrightarrow{AB})$

$\quad\quad = \mathbf{a} - \mathbf{b}$

Tip

Remember that you sometimes need to reverse the direction of vectors so that they are placed head to tail, as in the previous example. You must reverse the order of the letters and change the sign of the vector (+ to − or − to +).

9.2.3 Collinearity

When a vector is multiplied by a scalar the result is another vector of possibly different magnitude but with the same direction. If these vectors act along the same line, then they are **collinear vectors** (co means 'same' and linear means 'line').

Mathematically, collinearity can be expressed as follows:

Vectors **a** and **b** are parallel if: $\mathbf{a} = k\mathbf{b}$ where k is a constant. For collinearity there must be an added condition that the vectors are in the same straight line.

Graphically:

$\overrightarrow{AB}$ and $\overrightarrow{BC}$ are collinear.

Consider the graphical representation of three vectors: **a**, **b** and **c** as shown in the diagram.

a, **b** and **c** are parallel to each other but not necessarily acting along the same line, so therefore are not collinear.

9.2.4 Representing a position vector as a column vector

In addition to the various ways of representing position vectors already discussed, another very common form of representation is as **column vectors**.

Example 14

Represent the position vectors $\overrightarrow{OP}$ and $\overrightarrow{OQ}$ as column vectors and find their sum.

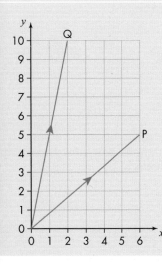

$$\overrightarrow{OP} = \begin{pmatrix} 6 \\ 5 \end{pmatrix}, \overrightarrow{OQ} = \begin{pmatrix} 2 \\ 10 \end{pmatrix} \text{ and}$$

$$\overrightarrow{OP} + \overrightarrow{OQ} = \begin{pmatrix} 6 \\ 5 \end{pmatrix} + \begin{pmatrix} 2 \\ 10 \end{pmatrix}$$

$$= \begin{pmatrix} 8 \\ 15 \end{pmatrix}$$

The same result can be obtained graphically by aligning the two vectors head to tail and drawing the resultant.

Exercise 9B

1 Identify the position vectors in the diagram, representing them as:

   a unit vectors   **b** column vectors.

2 Represent the position vectors $\mathbf{a} = 4\mathbf{i} + 3\mathbf{j}$ and $\mathbf{b} = \begin{pmatrix} 2 \\ 6 \end{pmatrix}$ on one diagram.

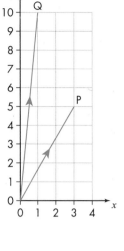

3 Draw a diagram to represent the position vectors $\overrightarrow{OA} = \begin{pmatrix} -3 \\ 8 \end{pmatrix}$ and

$\overrightarrow{OB} = \begin{pmatrix} 2 \\ -9 \end{pmatrix}$.

4 Determine the magnitude and direction of these position vectors.

   a $\overrightarrow{OR} = 3\mathbf{i} + 4\mathbf{j}$ **b** $\overrightarrow{OM} = \begin{pmatrix} 6 \\ 8 \end{pmatrix}$

5 Using vector notation:

   a name the position vector shown in the diagram in two ways and determine its magnitude and direction

   b write two equations representing the position vector.

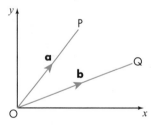

6 Write the unit vector that acts in the same direction as position vector $8\mathbf{i} - 12\mathbf{j}$.

7 The diagram shows two position vectors, $\overrightarrow{OP} = \mathbf{a}$ and $\overrightarrow{OQ} = \mathbf{b}$.

Express $2\overrightarrow{PQ} + \overrightarrow{QP}$ in terms of $\mathbf{a}$ and $\mathbf{b}$.

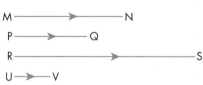

8 The vectors in the diagram (not drawn to scale) are parallel to each other.

$\overrightarrow{MN} = 3\mathbf{a}$, $\overrightarrow{PQ} = 2\mathbf{a}$, $\overrightarrow{RS} = 5\mathbf{a}$, $\overrightarrow{UV} = \mathbf{a}$

   a Express their sum in terms of $\mathbf{a}$.

   b Express $2\overrightarrow{MN} - \overrightarrow{PQ} + \overrightarrow{RS} - 8\overrightarrow{UV}$ in terms of $\mathbf{a}$.

9 The diagram shows two position vectors, $\overrightarrow{OW}$ and $\overrightarrow{OS}$.

   a Represent each position vector as a column vector.

   b Determine the sum of the position vectors, expressing your answer as a column vector.

   c Express $\overrightarrow{OS} - \overrightarrow{OW}$ as a column vector.

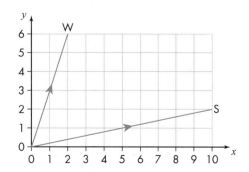

10 The position vectors of $\overrightarrow{OB}$, $\overrightarrow{OC}$ and $\overrightarrow{OD}$ are **a**, **b** and **c** respectively. AB = BC = CD = DE.

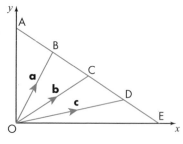

   a Write the vector $\overrightarrow{BC}$ in terms of **a** and **b**.

   b Express $\overrightarrow{CD}$ in terms of **b** and **c**.

   c Write $\overrightarrow{AB}$ and $\overrightarrow{DE}$ in terms of **a**, **b** or **c**.

11 The diagram represents a three-dimensional drawing of the position vector $\overrightarrow{OP}$.

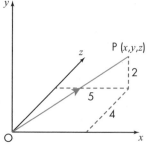

   a Write the position vector $\overrightarrow{OP}$ in terms of its unit vectors.

   b Evaluate the scalar magnitude of $\overrightarrow{OP}$.

   c Calculate the angle that $\overrightarrow{OP}$ makes with the x-axis.

12 A position vector $\overrightarrow{OJ}$ is represented by the coordinates: J(–5, 3, 10). Draw the position vector on a diagram and evaluate its scalar magnitude.

13 Determine the resultant of these three vectors and draw it on the same diagram.

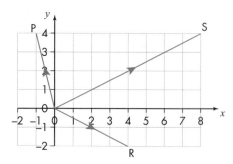

14 State the unit vector that acts in the same direction as position vector $-2\mathbf{i} + 3\mathbf{j} - 8\mathbf{k}$.

15 The points L, M and N have position vectors $\mathbf{p} + 2\mathbf{r}$, $3\mathbf{p} + 6\mathbf{r}$ and $-2\mathbf{p} - 4\mathbf{r}$ relative to the origin.

   a Prove that L, M and N are collinear.

   b State, in terms of **p** and **r**, $\overrightarrow{LM} + \overrightarrow{MN}$.

   c Represent all three position vectors in the same diagram.

9.3 Matrices

9.3.1 Concept of matrices and their real-world applications

Matrices can be used to store large amounts of data. The information is arranged in rows and columns. The data may be numerical, although matrices can also display networks, such as roads, communications and electrical circuits.

This matrix has two rows and three columns.

$$\begin{pmatrix} 1 & 4 & 8 \\ 3 & 9 & 6 \end{pmatrix} \!\!\!\!\triangleright \text{Rows} \qquad \begin{pmatrix} 1 & 4 & 8 \\ 3 & 9 & 6 \end{pmatrix}$$

Columns

> **Key term**
>
> A **matrix** is a square or rectangular array of numbers, symbols or expressions arranged in rows and columns.

Real-world applications of matrices

The diagram shows a network of roads between three towns, A, B and C.

The connections between the three towns can be represented in this matrix.

The zeros in the leading diagonal, from top left to bottom right, show that there is no route from a town to itself, as you would expect.

For each of the towns, there are two routes to any other town.

$$\begin{pmatrix} - & A & B & C \\ A & 0 & 2 & 2 \\ B & 2 & 0 & 2 \\ C & 2 & 2 & 0 \end{pmatrix}$$

Another example of matrices in real life is in the categorisation of objects.

Example 15

Use a matrix to classify these groups of people.

Type Nationality	Men	Women	Children
American	100	80	60
British	80	120	50
Grenadian	140	30	90
Barbadian	60	90	20

The information can be displayed in this matrix.

$$\begin{pmatrix} 100 & 80 & 60 \\ 80 & 120 & 50 \\ 140 & 30 & 90 \\ 60 & 90 & 20 \end{pmatrix}$$

Note

Matrices are always enclosed in large brackets. The more you use matrices, the more you will appreciate their usefulness and understand how to manipulate them.

9.3.2 Orders and types of matrices

The order of a matrix

A matrix may be defined by the numbers of rows and columns it has. This can be written as $r \times c$, where r is the number of rows and c is the number of columns, and is the **order** of the matrix.

Key term

The **order** of a matrix gives the number of rows and the number of columns. In the matrix
$$\begin{pmatrix} 1 & 3 & 4 & 8 \\ 5 & 0 & 9 & 1 \\ 3 & 6 & 8 & 7 \end{pmatrix},$$ the order is three by four, written as 3×4.

Types of matrix

- A **row matrix** has only one row, for example, $(2 \quad 3 \quad 1)$.

- A **column matrix** has only one column, for example, $\begin{pmatrix} 7 \\ 1 \end{pmatrix}$.

- A **null matrix** has zero as every element, for example, $(0 \quad 0 \quad 0)$.
- A **square matrix** has equal numbers of rows and columns, for example, $\begin{pmatrix} 9 & 3 \\ 5 & 2 \end{pmatrix}$.
- A **diagonal matrix** is a square matrix with at least one element of its leading or principal diagonal being non-zero but all other elements being zero, for example, $\begin{pmatrix} 3 & 0 \\ 0 & 4 \end{pmatrix}$.
- A **unit** or **identity matrix** is a square matrix in which all the leading diagonal elements equal 1 and all others are 0, for example, $\begin{pmatrix} 1 & 0 \\ 0 & 1 \end{pmatrix}$.
- A **rectangular matrix** is one in which the numbers of rows and columns are different and the elements are arranged in a rectangular shape, for example, $\begin{pmatrix} 3 & 8 \\ 4 & 2 \\ 0 & 1 \end{pmatrix}$.

9.3.3 Solving problems involving matrix operations

Addition and subtraction of matrices

Matrices can be added and subtracted provided that they have the same order. Corresponding elements of the matrices are added or subtracted.

Example 16

Add these matrices.

$$\begin{pmatrix} 4 & 3 \\ 2 & 1 \end{pmatrix} + \begin{pmatrix} 5 & -6 \\ 0 & 8 \end{pmatrix}$$

$$\begin{pmatrix} 4 & 3 \\ 2 & 1 \end{pmatrix} + \begin{pmatrix} 5 & -6 \\ 0 & 8 \end{pmatrix} = \begin{pmatrix} 4+5 & 3-6 \\ 2+0 & 1+8 \end{pmatrix}$$

$$= \begin{pmatrix} 9 & -3 \\ 2 & 9 \end{pmatrix}$$

Example 17

Subtract these matrices.

$$\begin{pmatrix} 7 & -5 \\ 20 & 9 \end{pmatrix} - \begin{pmatrix} 12 & 3 \\ -2 & 6 \end{pmatrix}$$

$$\begin{pmatrix} 7 & -5 \\ 20 & 9 \end{pmatrix} - \begin{pmatrix} 12 & 3 \\ -2 & 6 \end{pmatrix} = \begin{pmatrix} 7-12 & -5-3 \\ 20--2 & 9-6 \end{pmatrix}$$

$$= \begin{pmatrix} -5 & -8 \\ 22 & 3 \end{pmatrix}$$

Multiplication of matrices

Any matrix may be multiplied by a scalar but there are conditions that apply when you multiply two matrices together.

Multiplying a matrix by a scalar

This is quite straightforward, you simply multiply each element of the matrix by the scalar quantity outside the brackets.

Example 18

Multiply $5 \times \begin{pmatrix} 3 & 1 \\ -2 & 0 \end{pmatrix}$.

$$5 \times \begin{pmatrix} 3 & 1 \\ -2 & 0 \end{pmatrix} = \begin{pmatrix} 5 \times 3 & 5 \times 1 \\ 5 \times -2 & 5 \times 0 \end{pmatrix}$$

$$= \begin{pmatrix} 15 & 5 \\ -10 & 0 \end{pmatrix}$$

Multiplying a matrix by a matrix

Two matrices can be multiplied provided that the number of columns in the first matrix is equal to the number of rows in the second (the two inner values in the multiplication of the orders). To multiply the matrices, you successively multiply the rows of the first matrix by the columns of the second.

$$\begin{pmatrix} a & b \\ c & d \end{pmatrix} \times \begin{pmatrix} p & q \\ r & s \end{pmatrix} = \begin{pmatrix} ap + br & aq + bs \\ cp + dr & cq + ds \end{pmatrix}$$

Example 19

Multiply these matrices.

$$\begin{pmatrix} 2 & 3 \\ 1 & 5 \end{pmatrix} \times \begin{pmatrix} 6 & -1 \\ 4 & 8 \end{pmatrix}$$

$$\begin{pmatrix} 2 & 3 \\ 1 & 5 \end{pmatrix} \times \begin{pmatrix} 6 & -1 \\ 4 & 8 \end{pmatrix}$$

$2 \times 2 \quad 2 \times 2$

inner values

Looking at the orders of the matrices, the number of columns in the first matrix is equal to the number of rows in the second matrix, and this satisfies the condition for matrix multiplication.

$$\begin{pmatrix} 24 & 22 \\ 26 & 39 \end{pmatrix}$$

The order is 2×2.

Multiply row 1 of matrix 1 (2   3) by column 1 of matrix 2 $\begin{pmatrix} 6 \\ 4 \end{pmatrix}$ to get $2 \times 6 + 3 \times 4 = 24$.

Repeat for column 2 of matrix 2: $2 \times -1 + 3 \times 8 = 22$

Move on to row 2 of matrix 1 to repeat the process.

$1 \times 6 + 5 \times 4 = 26$ and $1 \times -1 + 5 \times 8 = 39$

The order of the resultant matrix is:

(number of rows in matrix 1) × (number of columns in matrix 2).

The two outer values of the orders of the matrices, when they are written as a product, give the order of the resultant while the two inner values satisfy the condition for multiplication if they are equal.

Example 20

Investigate the product $\begin{pmatrix} 4 & 6 & 1 \end{pmatrix} \begin{pmatrix} 5 \\ 7 \end{pmatrix}$.

The order of the first matrix is 1×3.

The order of the second matrix is 2×1.

This multiplication is not possible since the number of columns in the first matrix is not equal to the number of rows in the second.

Writing the product in terms of the orders of the matrices, $(1 \times 3)(2 \times 1)$, shows that the two inner numbers are not equal so the necessary condition for multiplication is not satisfied.

9.3.4 Matrix notation, commutativity and equality

Matrix notation and commutativity

Matrices are usually denoted by capital letters, for example, $\mathbf{A} = \begin{pmatrix} 2 & 3 \\ 4 & 5 \end{pmatrix}$, $\mathbf{B} = \begin{pmatrix} 3 & 1 \\ 1 & 6 \end{pmatrix}$.

Generally, multiplication of two matrices is not commutative: $\mathbf{A} \times \mathbf{B} \neq \mathbf{B} \times \mathbf{A}$.

Key fact

In summary, you can only multiply a matrix of order $m \times n$ by a matrix of order $p \times q$ if $n = p$.

Example 21

$\mathbf{A} = \begin{pmatrix} 2 & 3 \\ 4 & 5 \end{pmatrix}$ and $\mathbf{B} = \begin{pmatrix} 3 & 1 \\ 1 & 6 \end{pmatrix}$

Show that $\mathbf{A} \times \mathbf{B} \neq \mathbf{B} \times \mathbf{A}$.

Using the rules of matrix multiplication:

$$\mathbf{AB} = \begin{pmatrix} 2 & 3 \\ 4 & 5 \end{pmatrix}\begin{pmatrix} 3 & 1 \\ 1 & 6 \end{pmatrix} \qquad \mathbf{BA} = \begin{pmatrix} 3 & 1 \\ 1 & 6 \end{pmatrix}\begin{pmatrix} 2 & 3 \\ 4 & 5 \end{pmatrix}$$

$$= \begin{pmatrix} 6+3 & 2+18 \\ 12+5 & 4+30 \end{pmatrix} \qquad = \begin{pmatrix} 6+4 & 9+5 \\ 2+24 & 3+30 \end{pmatrix}$$

$$= \begin{pmatrix} 9 & 20 \\ 17 & 34 \end{pmatrix} \qquad = \begin{pmatrix} 10 & 14 \\ 26 & 33 \end{pmatrix}$$

So $\mathbf{A} \times \mathbf{B} \neq \mathbf{B} \times \mathbf{A}$.

Equal matrices

Two matrices are **equal** if they have the same order and the elements in corresponding positions are the same.

Key term

Equal matrices have the same order and the elements that are in corresponding positions are identical.

$$\begin{pmatrix} a & b \\ c & d \end{pmatrix} = \begin{pmatrix} j & k \\ l & m \end{pmatrix}$$

$2 \times 2 \qquad 2 \times 2$

These two matrices are equal.

They have the same order and corresponding elements are equal, which means that:

$a = j$, $b = k$, $c = l$ and $d = m$.

You can use the concept of equal matrices to solve problems.

Example 22

State the values of a, b and c from this matrix equation.

$$\begin{pmatrix} a & 3 & 1 \end{pmatrix} - \begin{pmatrix} 2 & b & 4 \end{pmatrix} = \begin{pmatrix} 5 & 6 & c \end{pmatrix}$$

Applying the rules for matrix subtraction:

$a - 2 = 5 \Rightarrow a = 7$

$3 - b = 6 \Rightarrow b = -3$

$1 - 4 = c \Rightarrow c = -3$

Example 23

State the values of p, q and r in this matrix multiplication.

$$\begin{pmatrix} -1 & 3 \\ r & 4 \end{pmatrix} \begin{pmatrix} p & q \\ 5 & 4 \end{pmatrix} = \begin{pmatrix} 9 & 0 \\ 7 & -10 \end{pmatrix}$$

Applying the matrix rules of multiplication:

$$\begin{pmatrix} -1 \times p + 3 \times 5 & -q + 3 \times 4 \\ pr + 20 & qr + 16 \end{pmatrix} = \begin{pmatrix} 9 & 0 \\ 7 & -10 \end{pmatrix}$$

$15 - p = 9 \Rightarrow p = 6$

$12 - q = 0 \Rightarrow q = 12$

$pr + 20 = 7 \Rightarrow 6r + 20 = 7$

$6r = -13 \Rightarrow r = -2\dfrac{1}{6}$

9.3.5 Creating transposed, adjoint and diagonal matrices

The transposed matrix

To **transpose** a matrix, you write the rows as columns and the columns as rows.

Key term

A **transposed** matrix is formed by interchanging the rows and columns of the matrix.

Example 24

Transpose the matrix $\mathbf{B} = \begin{pmatrix} 1 & -3 \\ 2 & 5 \end{pmatrix}$.

To transpose the matrix, make the rows into columns and the columns into rows.

$$\mathbf{B}^{\mathrm{T}} = \begin{pmatrix} 1 & 2 \\ -3 & 5 \end{pmatrix}$$

Adjoint and diagonal matrices

Some manipulations of matrices will require you to use the **adjoint** of a matrix, so you need to be able to calculate it.

If the adjoint matrix is multiplied by the original matrix, this produces the **diagonal** matrix.

Key term	Key term
The **adjoint** matrix of a 2 × 2 matrix is formed by interchanging the top left and bottom right elements and changing the signs on the top right and bottom left elements, for example, for matrix $\mathbf{P} = \begin{pmatrix} a & b \\ c & d \end{pmatrix}$, the adjoint is $\mathbf{P}^A = \begin{pmatrix} d & -b \\ -c & a \end{pmatrix}$.	The **diagonal** matrix of a 2 × 2 matrix is formed by multiplying a matrix by its adjoint, for example, for matrix $\mathbf{P} = \begin{pmatrix} a & b \\ c & d \end{pmatrix}$ and its adjoint $\mathbf{P}^A = \begin{pmatrix} d & -b \\ -c & a \end{pmatrix}$ their product $\mathbf{P}\mathbf{P}^A = \begin{pmatrix} h & 0 \\ 0 & h \end{pmatrix}$ where h is some number.

Example 25

Calculate the diagonal matrix for $\mathbf{P} = \begin{pmatrix} 1 & 3 \\ 5 & 4 \end{pmatrix}$.

The diagonal matrix is $\mathbf{P}\mathbf{P}^A$.

$$\mathbf{P}\mathbf{P}^A = \begin{pmatrix} 1 & 3 \\ 5 & 4 \end{pmatrix}\begin{pmatrix} 4 & -3 \\ -5 & 1 \end{pmatrix}$$

$$= \begin{pmatrix} 4 - 15 & -3 + 3 \\ 20 - 20 & -15 + 4 \end{pmatrix}$$

$$= \begin{pmatrix} -11 & 0 \\ 0 & -11 \end{pmatrix}$$

As you will see, both the adjoint matrix and the diagonal matrix are crucial in obtaining the inverse and unit matrices and subsequently in the solution of simultaneous equations.

9.3.6 The determinant and identity (unit), inverse and singular matrices

The determinant and identity matrices

Another useful quantity related to a matrix **A** is its **determinant**, detA, which is calculated from its elements.

Multiplying a diagonal matrix by the reciprocal of the determinant of the matrix produces an **identity matrix**.

Key term

The **determinant** of a 2 × 2 matrix is the difference between the product of the top left and bottom right elements and the product of the top right and bottom left elements. For the matrix $\begin{pmatrix} a & b \\ c & d \end{pmatrix}$ the determinant is $ad - bc$.

The general formula for obtaining a unit or identity matrix is, therefore:

$$\frac{1}{ad - bc}\begin{pmatrix} a & b \\ c & d \end{pmatrix}\begin{pmatrix} d & -b \\ -c & a \end{pmatrix}$$

Example 26

Given the matrix $\mathbf{A} = \begin{pmatrix} 2 & 4 \\ 6 & 8 \end{pmatrix}$, calculate:

(a) the transposed matrix   (b) the diagonal matrix   (c) the unit matrix.

(a) $\mathbf{A}^T = \begin{pmatrix} 2 & 6 \\ 4 & 8 \end{pmatrix}$

(b) The diagonal matrix, $\mathbf{B} = \mathbf{A} \times \mathbf{A}^A$

$$= \begin{pmatrix} 2 & 4 \\ 6 & 8 \end{pmatrix}\begin{pmatrix} 8 & -4 \\ -6 & 2 \end{pmatrix}$$

$$= \begin{pmatrix} 16 - 24 & -8 + 8 \\ -48 + 48 & -24 + 16 \end{pmatrix}$$

$$= \begin{pmatrix} -8 & 0 \\ 0 & -8 \end{pmatrix}$$

(c) The unit matrix $= \dfrac{1}{\det \mathbf{A}} \times$ diagonal matrix

$$= \frac{1}{16 - 24}\begin{pmatrix} -8 & 0 \\ 0 & -8 \end{pmatrix}$$

$$= -\frac{1}{8}\begin{pmatrix} -8 & 0 \\ 0 & -8 \end{pmatrix}$$

$$= \begin{pmatrix} 1 & 0 \\ 0 & 1 \end{pmatrix}$$

Note

Generally, a matrix multiplication is not commutative; however, in the case of the diagonal matrix, $\mathbf{A} \times \mathbf{A}^A = \mathbf{A}^A \times \mathbf{A}$ (commutative over multiplication).

$$= \begin{pmatrix} 2 & 4 \\ 6 & 8 \end{pmatrix}\begin{pmatrix} 8 & -4 \\ -6 & 2 \end{pmatrix} = \begin{pmatrix} 8 & -4 \\ -6 & 2 \end{pmatrix}\begin{pmatrix} 2 & 4 \\ 6 & 8 \end{pmatrix} = \begin{pmatrix} -8 & 0 \\ 0 & -8 \end{pmatrix}$$

$$\mathbf{A} \quad \times \quad \mathbf{A}^A \quad = \quad \mathbf{A}^A \quad \times \quad \mathbf{A} \quad = \quad \mathbf{D}$$

Inverse matrices

Unlike addition, subtraction and multiplication, the operation division is not associated with matrices. However, the reciprocal of a matrix **A**, denoted by **A⁻¹**, can be found. This reciprocal is known as the **inverse** matrix.

When an inverse matrix is multiplied by the original matrix, the result is the unit matrix. This is similar to the mathematical operation of multiplying a number by its reciprocal, resulting in 1; for example, $2 \times \frac{1}{2} = 1$ or $2 \times 2^{-1} = 1$, and in general, $n \times \frac{1}{n} = 1$ and $n \times n^{-1} = 1$.

> **Key term**
>
> The **inverse** of a matrix is the reciprocal of the matrix and is equal to the product of the reciprocal of the determinant and the adjoint matrix.

For any 2 × 2 matrix **A**, the product: $\mathbf{A} \times \mathbf{A}^{-1} = \mathbf{I}$ where **A** is the original matrix, **A⁻¹** is the inverse matrix and **I** is the unit matrix.

You have already seen that the unit matrix (**I**) can be determined by multiplying the original matrix by the reciprocal of the determinant and the adjoint matrix (**not** the transposed matrix). This means that the inverse matrix can be calculated from:

$$\mathbf{A}^{-1} = \frac{1}{\det \mathbf{A}} \times \mathbf{A}^{\mathrm{A}}$$

Example 27

Given the matrix $\mathbf{P} = \begin{pmatrix} 4 & 3 \\ 5 & 2 \end{pmatrix}$, calculate the inverse matrix **P⁻¹** and use it to obtain the unit matrix.

Inverse matrix $\mathbf{P}^{-1} = \dfrac{1}{\det \mathbf{P}} \times \mathrm{adj}\,\mathbf{P}$

$$= \frac{1}{8 - 15}\begin{pmatrix} 2 & -3 \\ -5 & 4 \end{pmatrix}$$

$$= -\frac{1}{7}\begin{pmatrix} 2 & -3 \\ -5 & 4 \end{pmatrix}$$

Unit matrix $= \mathbf{P}\mathbf{P}^{-1} = -\dfrac{1}{7}\begin{pmatrix} 2 & -3 \\ -5 & 4 \end{pmatrix}\begin{pmatrix} 4 & 3 \\ 5 & 2 \end{pmatrix}$

$$= -\frac{1}{7}\begin{pmatrix} 8-15 & 0 \\ 0 & -15+8 \end{pmatrix}$$

$$= -\frac{1}{7}\begin{pmatrix} -7 & 0 \\ 0 & -7 \end{pmatrix}$$

$$= \begin{pmatrix} 1 & 0 \\ 0 & 1 \end{pmatrix}$$

Singular matrices

Not every matrix has an inverse. Based on the formula used to calculate the inverse of matrix **B**, $\mathbf{B}^{-1} = \frac{1}{\det \mathbf{B}} \times \mathbf{B}^A$, so if det **B** = 0, since $\frac{1}{0}$ is undefined, the equation is invalid. In this case, the matrix is **singular**.

9.3.7 Using matrices to solve simultaneous linear equations

Matrices can be used to solve simultaneous equations by expressing these equations in the form of matrices and using inverse matrices to solve the resultant matrix equations.

Example 28

Use a matrix method to solve these simultaneous equations and verify the solution by using the substitution method.

$3x + 2y = 2$ **1**

$2x - y = 6$ **2**

Represent the equations in matrix form: the coefficients of x and y form a matrix (**A**).

Write the variables x and y in matrix form (**Z**) and the constant values in matrix form (**K**).

$$\begin{pmatrix} 3 & 2 \\ 2 & -1 \end{pmatrix} \begin{pmatrix} x \\ y \end{pmatrix} = \begin{pmatrix} 2 \\ 6 \end{pmatrix} \Rightarrow \begin{pmatrix} x \\ y \end{pmatrix} = \begin{pmatrix} 3 & 2 \\ 2 & -1 \end{pmatrix}^{-1} \begin{pmatrix} 2 \\ 6 \end{pmatrix}$$

$\mathbf{A} \times \mathbf{Z} = \mathbf{K} \Rightarrow \mathbf{Z} = \mathbf{A}^{-1} \times \mathbf{K}$

Z represents the unknown values (x, y) that satisfy the simultaneous equations, which are determined by solving the matrix equation.

$$\begin{pmatrix} x \\ y \end{pmatrix} = \frac{1}{\det \mathbf{A}} \times \mathbf{A}^A \times \mathbf{K}$$

det $\mathbf{A} = 3 \times (-1) - 2 \times 2 = -3 - 4 = -7$

$$\begin{pmatrix} x \\ y \end{pmatrix} = \frac{1}{-7} \begin{pmatrix} -1 & -2 \\ -2 & 3 \end{pmatrix} \begin{pmatrix} 2 \\ 6 \end{pmatrix}$$

$$= \frac{1}{-7} \begin{pmatrix} -2 - 12 \\ -4 + 18 \end{pmatrix}$$

$$= \frac{1}{-7} \begin{pmatrix} -14 \\ 14 \end{pmatrix}$$

$$= \begin{pmatrix} 2 \\ -2 \end{pmatrix}$$

Note

Based on the rules of matrix multiplication, it is not possible to calculate $k\mathbf{A}^{-1}$ since the orders (2 × 2 and 2 × 1) are not compatible; hence **Z** is written as $\mathbf{A}^{-1} \times k$.

1 Give two examples of everyday uses of matrices.

2 A farmer arranges his produce in four batches as shown here.

Batch	fruits	vegetables	eggs
A	100	150	240
B	200	200	120
C	300	350	180
D	250	100	300

Represent the information in a matrix.

3 Give one example of each of these types of matrix.

    a row   **b** column   **c** square   **d** diagonal   **e** null

    f unit   **g** rectangular

4 For each of these matrices, name the type and state the order.

    a $\begin{pmatrix} 3 & 0 \\ 5 & 0 \end{pmatrix}$ **b** $(11 \quad 1 \quad 8)$ **c** $\begin{pmatrix} 7 & 0 & 0 \\ 0 & 7 & 0 \\ 0 & 0 & 7 \end{pmatrix}$ **d** $\begin{pmatrix} 0 & 0 & 0 \\ 1 & 2 & 3 \end{pmatrix}$

5 Add the matrices.

    a $\begin{pmatrix} 9 & 6 \\ 15 & -2 \end{pmatrix} + \begin{pmatrix} 3 & -7 \\ 4 & 8 \end{pmatrix}$ **b** $\begin{pmatrix} 1 & 11 \\ 0 & 5 \\ 1 & -6 \end{pmatrix} + \begin{pmatrix} -1 & -2 \\ 13 & 4 \\ 15 & -7 \end{pmatrix}$

6 Subtract the matrices.

    a $(1 \quad 4 \quad 3) - (0 \quad 3 \quad -6)$ **b** $\begin{pmatrix} 8 \\ 0 \end{pmatrix} - \begin{pmatrix} 2 \\ -3 \end{pmatrix}$

7 Complete these scalar multiplications.

    a $7\begin{pmatrix} 3 & 0 \\ -1 & 5 \end{pmatrix}$ **b** $-11\begin{pmatrix} -1 \\ 0 \\ 3 \end{pmatrix}$

8 $A = \begin{pmatrix} 2 & 1 \\ 3 & 5 \end{pmatrix}$ $B = \begin{pmatrix} 9 & -1 \\ 6 & 5 \\ 4 & 0 \end{pmatrix}$ $C = (8 \quad 9 \quad 1)$ $D = \begin{pmatrix} 1 & 2 & 3 \\ 4 & 5 & 6 \\ 7 & 8 & 9 \end{pmatrix}$ $E = \begin{pmatrix} 9 & 1 & 0 \\ 3 & 5 & 4 \end{pmatrix}$ $F = (4)$

Complete these matrix multiplications where possible. Give reasons if the multiplication is **not** possible.

    a A^2 **b** $A \times B$ **c** $B \times D$ **d** $F \times C$ **e** $B \times E$

9 $P = \begin{pmatrix} 2 & 3 \\ 5 & 1 \end{pmatrix}$ and $Q = \begin{pmatrix} 6 & 4 \\ 8 & 9 \end{pmatrix}$. Show that $P \times Q \neq Q \times P$.

10 Study this matrix operation and calculate the values of x, y and z.

$$\begin{pmatrix} 4 & 3 \\ 5 & x \\ 1 & 2 \end{pmatrix} + \begin{pmatrix} y & 9 \\ 0 & -1 \\ 8 & z \end{pmatrix} = \begin{pmatrix} 8 & 12 \\ 5 & 6 \\ 9 & 7 \end{pmatrix}$$

11 Calculate the unknown values a, b and c from this matrix operation.

$$\begin{pmatrix} a & 1 \\ 2 & 3 \end{pmatrix} \times \begin{pmatrix} 7 & 4 & c \\ 0 & b & 6 \end{pmatrix} = \begin{pmatrix} 49 & 10 & -50 \\ 14 & -46 & 2 \end{pmatrix}$$

12 $\mathbf{P} = \begin{pmatrix} 2 & -1 \\ 4 & 9 \end{pmatrix}$

Calculate: **a** the transposed matrix **b** the adjoint matrix **c** the diagonal matrix.

13 $\mathbf{A} = \begin{pmatrix} -4 & 5 \\ 2 & 3 \end{pmatrix}$

a Calculate the determinant of **A**. **b** Obtain the unit matrix.

c Determine the inverse matrix ($\mathbf{A}^{-1}$).

14 **a** Give an example of a singular matrix and give reasons why it is singular.

b State whether each matrix is singular or not. Give reasons.

i $\begin{pmatrix} -3 & 6 \\ 2 & 4 \end{pmatrix}$ **ii** $\begin{pmatrix} 4 & -2 \\ -10 & 5 \end{pmatrix}$

15 Use matrices to solve these simultaneous equations. Verify your answers by using the elimination or other method.

a $2x - 3y = 9$ **b** $5x - 2y = 19$

$4x + 2y = 2$ $10x + y = 3$

9.4 Vectors, transformations and matrices

9.4.1 Vectors and matrices

You have already seen that a position vector or any other vector can be represented as a column

matrix. For the point P(a, b), the position vector $\mathbf{p} = \begin{pmatrix} a \\ b \end{pmatrix}$ and the distance of P from the origin, O, is

given by the magnitude of OP. This can be determined by using Pythagoras' theorem.

$OP = \sqrt{a^2 + b^2}$

The sum of two or more position vectors can be calculated by using matrix addition.

Example 29

Determine the resultant of two position vectors: $\mathbf{p} = \begin{pmatrix} 3 \\ 7 \end{pmatrix}$ and $\mathbf{q} = \begin{pmatrix} 4 \\ 2 \end{pmatrix}$.

The resultant $\mathbf{r} = \mathbf{p} + \mathbf{q}$

$\mathbf{r} = \begin{pmatrix} 3 \\ 7 \end{pmatrix} + \begin{pmatrix} 4 \\ 2 \end{pmatrix} = \begin{pmatrix} 7 \\ 9 \end{pmatrix}$

Example 29 continues on next page

The solution can be shown graphically.

R is the point (7, 9).

The direction of the resultant is found by using trigonometry:

$$\tan\theta = \frac{9}{7}$$

$$= 1.28$$

$$\tan^{-1} 1.28 = 52°$$

The magnitude or modulus of **r** is:

$$|\mathbf{r}| = \sqrt{7^2 + 9^2}$$

$$\sqrt{130}$$

$$|\mathbf{r}| = 11.4$$

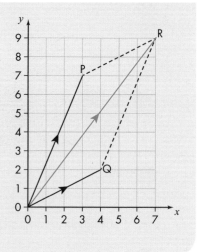

9.4.2 Transformations and matrices

Transformations were introduced in Chapter 4: **Geometry and trigonometry**. A transformation is defined as the mapping of an object onto an image. Chapter 4 introduced the transformations: translations, reflections, rotations, size transformations (enlargement and reduction), glide reflections, shears and stretches.

All of these transformations can be performed by using matrices.

Translations

A translation is a sliding, parallel movement of each point of an object through the same distance and in the same direction from one position to another.

If a point in the object is represented by the column matrix $\begin{pmatrix} x \\ y \end{pmatrix}$ and the

translation is denoted by another column matrix **T** $\begin{pmatrix} a \\ b \end{pmatrix}$ then the image is

found by matrix addition, $\begin{pmatrix} x \\ y \end{pmatrix} + \begin{pmatrix} a \\ b \end{pmatrix} = \begin{pmatrix} x' \\ y' \end{pmatrix}$.

The object and image triangles are shown in the diagram.

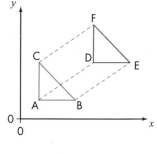

Example 30

The endpoints of the line PQ have coordinates P(2, 3) and Q(4, 7). Derive the coordinates of

the image of the line after a translation given by **T** $\begin{pmatrix} -5 \\ 1 \end{pmatrix}$.

The images P′ and Q′ of P and Q are found by adding matrix **T**.

$$P' = \begin{pmatrix} -5 \\ 1 \end{pmatrix} + \begin{pmatrix} 2 \\ 3 \end{pmatrix} = \begin{pmatrix} -3 \\ 4 \end{pmatrix} \text{ and } Q' = \begin{pmatrix} -5 \\ 1 \end{pmatrix} + \begin{pmatrix} 4 \\ 7 \end{pmatrix} = \begin{pmatrix} -1 \\ 8 \end{pmatrix}$$

The image of line PQ is the line P′Q′ through P′(–3, 4) and Q′(–1, 8).

Reflections

Reflection is the transformation that produces a mirror image of the object. A line of reflection separates the object from the image and each point on the object is the same perpendicular distance away from the line as its corresponding point on the image.

Reflection in the x-axis

Reflection of the point (x, y) is achieved through a matrix multiplication.

$$\begin{pmatrix} 1 & 0 \\ 0 & -1 \end{pmatrix}\begin{pmatrix} x \\ y \end{pmatrix} = \begin{pmatrix} x \\ -y \end{pmatrix}$$

Reflection in the y-axis

Reflection of the point (x, y) is achieved through a similar multiplication.

$$\begin{pmatrix} -1 & 0 \\ 0 & 1 \end{pmatrix}\begin{pmatrix} x \\ y \end{pmatrix} = \begin{pmatrix} -x \\ y \end{pmatrix}$$

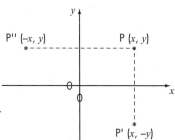

Example 31

Use matrices to determine the images of the point A(2, 3) after reflection in the x and y axes. Show your solution on a graph.

Reflection of A(2, 3) in the x-axis:

$$\begin{pmatrix} 1 & 0 \\ 0 & -1 \end{pmatrix}\begin{pmatrix} 2 \\ 3 \end{pmatrix} = \begin{pmatrix} 2 \\ -3 \end{pmatrix}$$

Reflection of A(2, 3) in the y-axis:

$$\begin{pmatrix} -1 & 0 \\ 0 & 1 \end{pmatrix}\begin{pmatrix} 2 \\ 3 \end{pmatrix} = \begin{pmatrix} -2 \\ 3 \end{pmatrix}$$

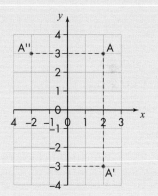

Reflection in the lines y = x and y = –x

The lines $y = x$ and $y = -x$ are shown in the diagram.

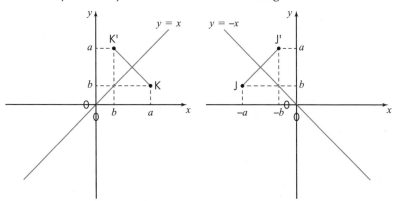

The reflection of any point K(a, b) in the line $y = x$, produces the image K′(b, a). The x and y coordinates are reversed. The transformation matrix which produces this reversal is $\begin{pmatrix} 0 & 1 \\ 1 & 0 \end{pmatrix}$.

Reflection in the line y = x is given by the matrix multiplication:

$$\begin{pmatrix} 0 & 1 \\ 1 & 0 \end{pmatrix}\begin{pmatrix} x \\ y \end{pmatrix} = \begin{pmatrix} y \\ x \end{pmatrix}$$

By a similar reasoning (see graph):

Reflection in the line $y = -x$ is given by the matrix multiplication

$$\begin{pmatrix} 0 & -1 \\ -1 & 0 \end{pmatrix} \begin{pmatrix} x \\ y \end{pmatrix} = \begin{pmatrix} -y \\ -x \end{pmatrix}$$

A general matrix multiplication for reflection of the point (x, y) in any line $ax + by + c = 0$

$$\frac{1}{a^2 + b^2} \begin{pmatrix} b^2 - a^2 & -2ab \\ -2ab & a^2 - b^2 \end{pmatrix} \begin{pmatrix} x \\ y \end{pmatrix} - \frac{2c}{a^2 + b^2} \begin{pmatrix} a \\ b \end{pmatrix} = \begin{pmatrix} x' \\ y' \end{pmatrix}$$

Where a and b are the coefficients of x and y respectively and c is the constant term.

Example 32

Find the coordinates of the point M(5, 2) after a reflection in the line $y = -x$.

For reflection in $y = -x$ use the matrix multiplication:

$$\begin{pmatrix} 0 & -1 \\ -1 & 0 \end{pmatrix} \begin{pmatrix} 5 \\ 2 \end{pmatrix} = \begin{pmatrix} -2 \\ -5 \end{pmatrix}$$

Rotations

If point $P(x, y)$ is given an **anticlockwise** rotation of magnitude θ about the origin, the matrix equation representing the transformation is given as:

$$\begin{pmatrix} \cos\theta & -\sin\theta \\ \sin\theta & \cos\theta \end{pmatrix} \begin{pmatrix} x \\ y \end{pmatrix} = \begin{pmatrix} x' \\ y' \end{pmatrix}$$

For a **clockwise** rotation one adjustment is made to the equation:

$$\begin{pmatrix} \cos\theta & \sin\theta \\ -\sin\theta & \cos\theta \end{pmatrix} \begin{pmatrix} x \\ y \end{pmatrix} = \begin{pmatrix} x' \\ y' \end{pmatrix}$$

For the centre of rotation not equal to the origin, for example, C(r, s), an anticlockwise rotation about C of magnitude θ:

$$\begin{pmatrix} \cos\theta & -\sin\theta \\ \sin\theta & \cos\theta \end{pmatrix} \begin{pmatrix} x \\ y \end{pmatrix} + \begin{pmatrix} 1-\cos\theta & \sin\theta \\ -\sin\theta & 1-\cos\theta \end{pmatrix} \begin{pmatrix} r \\ s \end{pmatrix} = \begin{pmatrix} x' \\ y' \end{pmatrix}$$

Example 33

A point M(2, 7) is rotated 60° anticlockwise about the origin. Determine the coordinates of the image of M.

$$\begin{pmatrix} \cos 60° & -\sin 60° \\ \sin 60° & \cos 60° \end{pmatrix} \begin{pmatrix} 2 \\ 7 \end{pmatrix} = \begin{pmatrix} 0.5 & -0.87 \\ 0.87 & 0.5 \end{pmatrix} \begin{pmatrix} 2 \\ 7 \end{pmatrix}$$

$$= \begin{pmatrix} -5.1 \\ 5.2 \end{pmatrix}$$

The image of M (M') = (−5.1, 5.2).

Size transformation or similarities

These are transformations that increase or decrease original lengths by a scale factor k.

$k > 1$ produces an enlargement and $k < 1$ produces a reduction.

All angles and shapes are maintained, hence a similar shape is produced.

The image in a size transformation, centre the origin, is obtained by multiplying the coordinates of

the object, written as a column matrix, by k or by the 2×2 matrix $\begin{pmatrix} k & 0 \\ 0 & k \end{pmatrix}$.

$$\begin{pmatrix} k & 0 \\ 0 & k \end{pmatrix}\begin{pmatrix} x \\ y \end{pmatrix} = k\begin{pmatrix} x \\ y \end{pmatrix}$$

$$= \begin{pmatrix} kx \\ ky \end{pmatrix}$$

$$= \begin{pmatrix} x' \\ y' \end{pmatrix}$$

Example 34

A triangle has vertices A(2, 3), B(6, 3) and C(6, 6). Determine the coordinates of the vertices of triangle ABC after an enlargement with centre the origin and scale factor 2. Show the position of the image by means of a sketch.

The image of triangle ABC is found by the matrix multiplication:

$$\begin{array}{cc} \text{A B C} & \text{A' B' C'} \\ \begin{pmatrix} 2 & 0 \\ 0 & 2 \end{pmatrix}\begin{pmatrix} 2 & 6 & 6 \\ 3 & 3 & 6 \end{pmatrix} = \begin{pmatrix} 4 & 12 & 12 \\ 6 & 6 & 12 \end{pmatrix} \end{array}$$

The graphical representation is shown in the diagram.

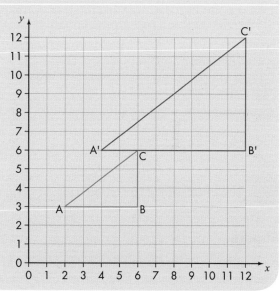

Size transformation with centre not equal to the origin with centre (h, k), the image of a point (x, y) after a size transformation scale factor f is derived as:

$$\begin{pmatrix} f & 0 \\ 0 & f \end{pmatrix}\begin{pmatrix} x \\ y \end{pmatrix} - f\begin{pmatrix} h \\ k \end{pmatrix} + \begin{pmatrix} h \\ k \end{pmatrix} = \begin{pmatrix} x' \\ y' \end{pmatrix}$$

Glide reflection

A glide reflection is a translation followed by a reflection.

Perform a glide reflection on the point (3, 5) given by the translation of $\begin{pmatrix} 7 \\ 2 \end{pmatrix}$ followed by a reflection in the y-axis.

Glide reflection: translation $\begin{pmatrix} 7 \\ 2 \end{pmatrix} + \begin{pmatrix} 3 \\ 5 \end{pmatrix} = \begin{pmatrix} 10 \\ 7 \end{pmatrix}$ followed by

Reflection $\begin{pmatrix} -1 & 0 \\ 0 & 1 \end{pmatrix}\begin{pmatrix} 10 \\ 7 \end{pmatrix} = \begin{pmatrix} -10 \\ 7 \end{pmatrix}$

Double transformations

In general a double transformation can be performed by first converting the two matrices into one matrix through multiplication, provided the multiplication is compatible, based on their orders. The sequence of the transformation must also be taken into consideration due to the non-commutative nature of most matrix multiplications.

The two matrices $L = \begin{pmatrix} 2 & 3 \\ 5 & 1 \end{pmatrix}$ and $M = \begin{pmatrix} 1 & 0 \\ 2 & 1 \end{pmatrix}$ represent two separate transformations on

a point P(2, 1). Given that the transformation M is performed first, find a single matrix that represents the transformation M followed by L and hence or otherwise find P′, the image of P.

$\begin{pmatrix} 2 & 3 \\ 5 & 1 \end{pmatrix}\begin{pmatrix} 1 & 0 \\ 2 & 1 \end{pmatrix} = \begin{pmatrix} 8 & 3 \\ 7 & 1 \end{pmatrix}$ which is the single matrix.

To find the image, multiply point (2, 1) by the single matrix.

$\begin{pmatrix} 8 & 3 \\ 7 & 1 \end{pmatrix}\begin{pmatrix} 2 \\ 1 \end{pmatrix} = \begin{pmatrix} 19 \\ 15 \end{pmatrix}$

The image of P(2, 1) is P′(19, 15).

Shears

A shear is a transformation that maps parallel lines onto parallel lines. There is also an invariant line (shown in the diagram). Area is preserved in this transformation but shape can change.

The matrix used to perform a shear with the invariant line

AB ($y = 0$) is $\begin{pmatrix} 1 & k \\ 0 & 1 \end{pmatrix}$.

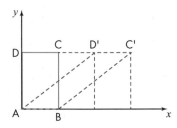

For a shear with the invariant line as $x = 0$ (y-axis), the matrix is $\begin{pmatrix} 1 & 0 \\ k & 1 \end{pmatrix}$.

One-way and two-way stretches

One-way stretches

A one-way stretch is an enlargement in one direction, with the y-axis or x-axis being the invariant line, as shown in the diagrams.

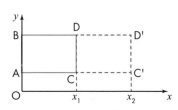

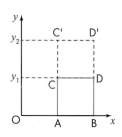

The y-axis is the invariant line. The image point is found by multiplying by the matrix $\begin{pmatrix} k & 0 \\ 0 & 1 \end{pmatrix}$.

The x-axis is the invariant line. The image point is found by multiplying by the matrix $\begin{pmatrix} 1 & 0 \\ 0 & k \end{pmatrix}$.

Two way stretches

A two-way stretch is a combination of two one-way stretches consisting of two scale factors (h and k), one for each axis, as shown in the diagram.

The matrix representing the two-way stretch is $\begin{pmatrix} h & 0 \\ 0 & k \end{pmatrix}$.

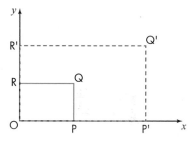

Using inverse transformations to obtain the original point from the image

The original point or object can be obtained by multiplying the image by the inverse of the transformation matrix.

If the image is (x', y'), the original point or object is (x, y) and the transformation matrix is $\begin{pmatrix} a & b \\ c & d \end{pmatrix}$. The object can be derived from the matrix equation:

$$\begin{pmatrix} x \\ y \end{pmatrix} = \begin{pmatrix} a & b \\ c & d \end{pmatrix}^{-1} \begin{pmatrix} x' \\ y' \end{pmatrix}$$

Example 37

A transformation given by the matrix $\begin{pmatrix} 2 & -3 \\ 1 & -1 \end{pmatrix}$ produces an image at (4, 1). Determine the coordinates of the object.

$$\begin{pmatrix} x \\ y \end{pmatrix} = \begin{pmatrix} a & b \\ c & d \end{pmatrix}^{-1} \begin{pmatrix} x' \\ y' \end{pmatrix}$$

$$= \begin{pmatrix} 2 & -3 \\ 1 & -1 \end{pmatrix}^{-1} \begin{pmatrix} 4 \\ 1 \end{pmatrix}$$

$$= \frac{1}{2 \times -1 - 1 \times -3} \begin{pmatrix} -1 & 3 \\ -1 & 2 \end{pmatrix} \begin{pmatrix} 4 \\ 1 \end{pmatrix}$$

$$= \frac{1}{1} \begin{pmatrix} -1 \\ -2 \end{pmatrix}$$

$(x, y) = (-1, -2)$

Exercise 9D

1 Calculate the sum of the position vectors $\begin{pmatrix} 18 \\ 12 \end{pmatrix} + \begin{pmatrix} -12 \\ 5 \end{pmatrix}$.

2 Calculate the sum of the three-dimensional position vectors $\begin{pmatrix} -2 \\ 3 \\ -14 \end{pmatrix} + \begin{pmatrix} 7 \\ 12 \\ 9 \end{pmatrix}$.

3 Determine the scalar magnitude and direction of the position vector $\begin{pmatrix} 8 \\ 10 \end{pmatrix}$.

4 Determine the image of the point P(3, 5) after a translation $\mathbf{T} \begin{pmatrix} -2 \\ 3 \end{pmatrix}$, showing the solution graphically.

5 Using the appropriate matrix multiplication, perform a reflection of the point P(2, 8):

   a in the y-axis   **b** in the x-axis   **c** in the line $y = x$ **d** in the line $y = -x$.

   Draw your answers on a coordinate grid.

6 Using the appropriate matrices, perform a glide reflection on the point (–4, 5) consisting of a translation $\mathbf{T} \begin{pmatrix} 9 \\ -3 \end{pmatrix}$ followed by a reflection in the y-axis. Show your solution graphically.

7 A point R(5, 1) was given a double transformation represented by the matrix $\begin{pmatrix} 2 & 0 \\ 1 & 3 \end{pmatrix}$ followed by the matrix $\begin{pmatrix} 2 & 1 \\ 1 & 2 \end{pmatrix}$. Write one matrix to represent the double transformation and use that matrix to find R′, the image of R. Show your solution graphically.

8 Using matrices perform a rotation of 90° in an anti-clockwise direction, of the point R(2, 6) with the origin as the centre of rotation.

9 Using matrices, perform a size transformation of magnitude $k = 2$, centre the origin, on the line AB, with endpoints A(1, 5) and B(3, 8). Show your solution graphically.

10 Perform a one-way stretch of magnitude $k = 3$, on rectangle PQRS, with vertices P(5, 0), Q(10, 0), R(10, 4) and S(5, 4), using the x-axis as the invariant line.

11 Perform a two-way stretch, given by the matrix $\begin{pmatrix} 3 & 0 \\ 0 & 4 \end{pmatrix}$, on triangle ABC, with vertices A(2, 0), B(4, 0) and C(3, 5). Show your solution graphically.

12 Perform a shear given by the matrix $\begin{pmatrix} 1 & 4 \\ 0 & 1 \end{pmatrix}$, on the rectangle PQRS, with vertices P(3, 3), Q(7, 3), R(7, 11) and S (3, 11). Show your solution graphically.

13 Derive the coordinates of the image of triangle DEF, with vertices D(–4, 2), E (–8, 1) and F(–6, 8), after a translation $\mathbf{T} \begin{pmatrix} 10 \\ 4 \end{pmatrix}$. Show the image graphically.

14 Rectangle ABCD, with vertices A (1, 2), B (6, 2), C (6, 6), D (1, 6), is reflected:

  a in the y-axis   **b** in the x-axis   **c** in the line $y = x$ **d** in the line $y = -x$.

  Using the appropriate matrix, determine the coordinates of the image of the rectangle in each of the four lines of reflection and show each solution graphically.

15 Triangle LMN, with vertices L(1, 7), M(3, 4) and N(6, 6), is given a glide reflection

  consisting of a translation $\mathbf{T}\begin{pmatrix} 5 \\ -6 \end{pmatrix}$ followed by a reflection in the y-axis. Derive the

  coordinates of its image and show the solution graphically.

16 A triangle PQR is given a clockwise rotation of 120° about the origin. The coordinates of the vertices of the triangle are P(3, –5), Q(8, –5) and R(8, –1). Use a matrix multiplication to determine the coordinates of the image. Show your solution graphically.

17 The coordinates of the vertices of a triangle are H(2, 3), K(8, 3) and L(5, 9). Given that it receives a size transformation of magnitude 5 centred on the origin, use the appropriate matrix to determine the coordinates of the image of triangle HKL.

18 Derive the coordinates of the images of the points P(3, 1) and Q(2, 8) after reflection in the line $2x + 3y + 1 = 0$.

19 A triangle ABC with vertices A(1, 1), B(6, 1) and C(6, 6) undergoes an enlargement of magnitude $k = 3$ with centre (–4, –5). Use a matrix multiplication to determine the coordinates of its image and show the object and image on a diagram.

20 A transformation matrix $\begin{pmatrix} 2 & 1 \\ 1 & 0 \end{pmatrix}$ produces an image at (3, 2). Determine the coordinates of the object.

End of chapter summary

Concepts associated with vectors

- A **scalar** quantity is a quantity that has magnitude only, without direction. A **vector** quantity is a quantity with both magnitude and direction; for example, a force of 20 N acting horizontally or a velocity of 10 m/s acting south.

- The **magnitude** of a quantity is its size.

- Scalar multiplication increases or decreases the magnitude of a vector by a factor.

- Two vectors are parallel if they act in the same direction, for example, $\overrightarrow{AB} \parallel \overrightarrow{LM}$.

- Two vectors are equal if they are parallel and have the same magnitude, for example,

- Two vectors are **inverse** if they have the same magnitude but opposite directions, for example, $\overrightarrow{HK}$ and $\overrightarrow{SR}$ are inverse vectors. Generally the inverse of any vector $\overrightarrow{AB}$ is $-\overrightarrow{AB} = \overrightarrow{BA}$.

- Vector algebra involves the addition and subtraction of vectors and multiplication of a vector by a scalar.

- The general rule for adding two or more vectors is that the vectors must be aligned head to tail while maintaining their original direction. The combined vector is the resultant, often represented by **r**.

- Subtracting one vector from another is equivalent to adding the inverse of the subtracted vector to the first vector.

- The **resultant** of two or more vectors is the vector that 'results' from the effects of the individual vectors.

- The **direction** of the resultant AC is given by the angle θ between the resultant and the northerly direction, AB.

- The **magnitude** of AC is evaluated by applying Pythagoras' theorem to the right-angled triangle.

- The **triangle law** states that if two adjacent sides of a triangle, drawn in order, represent the magnitudes and directions of two vectors acting from the same point, then the third side of the triangle represents the resultant of the two vectors in magnitude and direction.

- The **parallelogram law** states that if two vectors acting at a point are represented in magnitude and direction by two adjacent sides of a parallelogram, drawn from the same point, then the resultant of the two vectors is represented in magnitude and direction by the diagonal of the parallelogram, drawn from the same point.

Position vectors

- A **unit vector** is a vector with a magnitude of 1. It is sometimes called a **direction vector**.

- A **position vector**, sometimes called a location vector, is a vector that represents the position of a point P in space in relation to a fixed reference point called the origin (O) and is denoted by an arrow from the origin

- A position vector can be two-dimensional $\begin{pmatrix} x \\ y \end{pmatrix}$ or three-dimensional $\begin{pmatrix} x \\ y \\ z \end{pmatrix}$.

- For the two-dimensional representation, the unit vectors are **i** along the *x*-axis and **j** along the *y*-axis; for the three-dimensional representation the unit vectors are **i** along the *x*-axis, **j** along the *y*-axis and **k** along the *z*-axis.

- The **magnitude** of a position vector is a scalar quantity represented by the distance from the origin to the point and can be found by using Pythagoras' theorem.

- It is possible to convert any vector into a unit vector by dividing the vector by its magnitude.

- The **unit vector** $\bar{\mathbf{a}}$ associated with vector **a** is calculated as $\bar{\mathbf{a}} = \dfrac{\mathbf{a}}{|\mathbf{a}|}$ where $|\mathbf{a}|$ is the magnitude of **a**.

- **Collinear vectors** are vectors that act along the same line. Vectors **a** and **b** are collinear if $\mathbf{b} = k\mathbf{a}$ where k is a scalar constant.

- Another way of representing a position vector is as a **column vector**. The *x* and *y* components of a column vector are listed vertically, in a single column, with the upper value representing the *x* component and the lower value the *y* component, for example, for the point P(*x*, *y*), $\overrightarrow{OP} = \begin{pmatrix} x \\ y \end{pmatrix}$.

Matrices

- A **matrix** is a square or rectangular array of numbers, symbols or expressions arranged in rows and columns.

- The **order** of a matrix gives the number of rows and the number of columns. In the matrix $\begin{pmatrix} 1 & 3 & 4 & 8 \\ 5 & 0 & 9 & 1 \\ 3 & 6 & 8 & 7 \end{pmatrix}$, the order is three by four, written as 3×4.

- There are several types of matrix.

  - A **row matrix** has only one row, for example, $(2 \quad 3 \quad 1)$.

  - A **column matrix** has only one column, for example, $\begin{pmatrix} 7 \\ 1 \end{pmatrix}$.

  - A **null matrix** has zero as every element, for example, $(0 \quad 0 \quad 0)$.

  - A **square matrix** has equal numbers of rows and columns, for example, $\begin{pmatrix} 9 & 3 \\ 5 & 2 \end{pmatrix}$.

  - A **diagonal matrix** is a square matrix with at least one element of its leading or principal diagonal being non-zero but all other elements being zero, for example, $\begin{pmatrix} 3 & 0 \\ 0 & 4 \end{pmatrix}$.

  - An **identity** or **unit matrix** is a square matrix in which all the leading diagonal elements equal 1 and all others are 0, for example, $\begin{pmatrix} 1 & 0 \\ 0 & 1 \end{pmatrix}$.

  - A **rectangular matrix** is one in which the numbers of rows and columns are different and the elements are arranged in a rectangular shape, for example, $\begin{pmatrix} 3 & 8 \\ 4 & 2 \\ 0 & 1 \end{pmatrix}$.

- Matrices can be added and subtracted provided that they have the same order. Corresponding elements of the matrices are added or subtracted.
- To multiply a matrix by a scalar you simply multiply each element of the matrix by the scalar quantity outside the brackets.
- Two matrices can be multiplied provided that the number of columns in the first matrix is equal to the number of rows in the second (the two inner values in the multiplication of the orders).
- To multiply two matrices, you successively multiply the rows of the first matrix by the columns of the second.

$$\begin{pmatrix} a & b \\ c & d \end{pmatrix} \times \begin{pmatrix} p & q \\ r & s \end{pmatrix} = \begin{pmatrix} ap + br & aq + bs \\ cp + dr & cq + ds \end{pmatrix}$$

- You can only multiply a matrix of order $m \times n$ by a matrix of order $p \times q$ if $n = p$.
- Matrices are usually denoted by capital letters, for example, $\mathbf{A} = \begin{pmatrix} 2 & 3 \\ 4 & 5 \end{pmatrix}$, $\mathbf{B} = \begin{pmatrix} 3 & 1 \\ 1 & 6 \end{pmatrix}$.
- Generally, multiplication of two matrices is not commutative: $\mathbf{A} \times \mathbf{B} \neq \mathbf{B} \times \mathbf{A}$.
- **Equal** matrices have the same order and the elements that are in corresponding positions are the same.
- A **transposed** matrix is formed by interchanging the rows and columns of the matrix, for example, if $\mathbf{B} = \begin{pmatrix} 1 & -3 \\ 2 & 5 \end{pmatrix}$ then $\mathbf{B}^{\mathsf{T}} = \begin{pmatrix} 1 & 2 \\ -3 & 5 \end{pmatrix}$.
- The **adjoint** matrix of a 2×2 matrix is formed by interchanging the top left and bottom right elements and changing the signs on the top right and bottom left elements, for example, for matrix $\mathbf{P} = \begin{pmatrix} a & b \\ c & d \end{pmatrix}$, the adjoint is $\mathbf{P}^{\mathsf{A}} = \begin{pmatrix} d & -b \\ -c & a \end{pmatrix}$.
- The **diagonal** matrix of a 2×2 matrix is formed by multiplying a matrix by its adjoint, for example, for matrix $\mathbf{P} = \begin{pmatrix} a & b \\ c & d \end{pmatrix}$ and its adjoint $\mathbf{P}^{\mathsf{A}} = \begin{pmatrix} d & -b \\ -c & a \end{pmatrix}$ their product $\mathbf{P}\mathbf{P}^{\mathsf{A}} = \begin{pmatrix} h & 0 \\ 0 & h \end{pmatrix}$ where h is some number.
- The **determinant** of a 2×2 matrix is the difference between the product of the top left and bottom right elements and the product of the top right and bottom left elements. For the matrix $\begin{pmatrix} a & b \\ c & d \end{pmatrix}$ the determinant is $ad - bc$.
- An **identity** or **unit** matrix is a square matrix in which every element along the leading diagonal is 1 and all other elements are 0, for example, $\begin{pmatrix} 1 & 0 \\ 0 & 1 \end{pmatrix}$.
- The general formula for obtaining a unit or identity matrix is: $\dfrac{1}{ad - bc}\begin{pmatrix} a & b \\ c & d \end{pmatrix}\begin{pmatrix} d & -b \\ -c & a \end{pmatrix}$.
- The **inverse** of a matrix is the reciprocal of the matrix and is equal to the product of the reciprocal of the determinant and the adjoint matrix.
- A **singular** matrix is one in which the determinant is zero and as a result it has no inverse.
- Matrices can be used to solve simultaneous equations by expressing these equations in the form of matrices and using inverse matrices to solve the resultant matrix equations.

Vectors, transformation and matrices

- A position vector or any other vector can be represented as a column matrix. For the point P(a, b), the position vector $\mathbf{p} = \begin{pmatrix} a \\ b \end{pmatrix}$ and the distance of P from the origin, O, is given by the magnitude of OP, which can be determined by using Pythagoras' theorem, $OP = \sqrt{a^2 + b^2}$.

- The sum of two or more position vectors can be calculated by using matrix addition.

- All of the transformations (translations, reflections, rotations, size transformations (enlargement and reduction), glide reflections, shears and stretches) can be performed by using matrices.

- For a translation of $\begin{pmatrix} x \\ y \end{pmatrix}$ through $\mathbf{T}\begin{pmatrix} a \\ b \end{pmatrix}$, $\begin{pmatrix} x \\ y \end{pmatrix} + \begin{pmatrix} a \\ b \end{pmatrix} = \begin{pmatrix} x' \\ y' \end{pmatrix}$

- For a reflection:

  - in the x-axis, $\begin{pmatrix} 1 & 0 \\ 0 & -1 \end{pmatrix}\begin{pmatrix} x \\ y \end{pmatrix} = \begin{pmatrix} x \\ -y \end{pmatrix}$

  - in the y-axis, $\begin{pmatrix} -1 & 0 \\ 0 & 1 \end{pmatrix}\begin{pmatrix} x \\ y \end{pmatrix} = \begin{pmatrix} -x \\ y \end{pmatrix}$

  - in the line $y = x$, $\begin{pmatrix} 0 & 1 \\ 1 & 0 \end{pmatrix}\begin{pmatrix} x \\ y \end{pmatrix} = \begin{pmatrix} y \\ x \end{pmatrix}$

  - in the line $y = -x$, $\begin{pmatrix} 0 & -1 \\ -1 & 0 \end{pmatrix}\begin{pmatrix} x \\ y \end{pmatrix} = \begin{pmatrix} -y \\ -x \end{pmatrix}$

- For a rotation:

  - anticlockwise through angle θ, $\begin{pmatrix} \cos\theta & -\sin\theta \\ \sin\theta & \cos\theta \end{pmatrix}\begin{pmatrix} x \\ y \end{pmatrix} = \begin{pmatrix} x' \\ y' \end{pmatrix}$

  - clockwise through angle θ, $\begin{pmatrix} \cos\theta & \sin\theta \\ -\sin\theta & \cos\theta \end{pmatrix}\begin{pmatrix} x \\ y \end{pmatrix} = \begin{pmatrix} x' \\ y' \end{pmatrix}$

  - with centre of rotation not equal to the origin, for example, C(r, s), an anticlockwise rotation about C of magnitude θ,

  $$\begin{pmatrix} \cos\theta & -\sin\theta \\ \sin\theta & \cos\theta \end{pmatrix}\begin{pmatrix} x \\ y \end{pmatrix} + \begin{pmatrix} 1-\cos\theta & \sin\theta \\ -\sin\theta & 1-\cos\theta \end{pmatrix}\begin{pmatrix} r \\ s \end{pmatrix} = \begin{pmatrix} x' \\ y' \end{pmatrix}$$

- For a size transformation (enlargement or reduction):

$$\begin{pmatrix} k & 0 \\ 0 & k \end{pmatrix}\begin{pmatrix} x \\ y \end{pmatrix} = k\begin{pmatrix} x \\ y \end{pmatrix}$$

$$= \begin{pmatrix} kx \\ ky \end{pmatrix}$$

$$= \begin{pmatrix} x' \\ y' \end{pmatrix}$$

- A glide reflection is a translation followed by a reflection.

- In general a double transformation can be performed by first converting both matrices into one matrix through multiplication provided the multiplication is compatible based on the order.

- For a shear with the invariant line AB ($y = 0$) the matrix is $\begin{pmatrix} 1 & k \\ 0 & 1 \end{pmatrix}$.

- For a shear with the invariant line as $x = 0$ (y-axis), the matrix is $\begin{pmatrix} 1 & 0 \\ k & 1 \end{pmatrix}$.

- For one-way stretches:

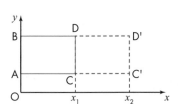

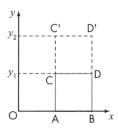

The y-axis is the invariant line. The image point is found by multiplying by the matrix $\begin{pmatrix} k & 0 \\ 0 & 1 \end{pmatrix}$.

The x-axis is the invariant line. The image point is found by multiplying by the matrix $\begin{pmatrix} 1 & 0 \\ 0 & k \end{pmatrix}$.

- For two-way stretches consisting of two scale factors (h and k), the matrix is $\begin{pmatrix} h & 0 \\ 0 & k \end{pmatrix}$.

- The original point or object can be obtained by multiplying the image by the inverse of the transformation matrix.

Examination-type questions for Chapter 9

1 Matrices **A** and **B** are given as $\mathbf{A} = \begin{pmatrix} 2 & 3 \\ 1 & 4 \end{pmatrix}$ and $\mathbf{B} = \begin{pmatrix} 0 & 5 \\ 7 & 6 \end{pmatrix}$.

   a Show by matrix multiplication that $\mathbf{AB} \neq \mathbf{BA}$.

   b Determine the inverse of **B**.

   c Determine the product $\mathbf{BB}^{-1}$.

   d Use a matrix method to solve the simultaneous equations:

      $3x - 2y = 4$

      $x - y = 1$

2 Two position vectors are given as $\overrightarrow{OP} = \begin{pmatrix} 5 \\ 10 \end{pmatrix}$ and $\overrightarrow{OF} = \begin{pmatrix} 8 \\ 16 \end{pmatrix}$.

   A point R is on OP such that $OP = 5OR$ and a point L is on OF such that OL is $\frac{1}{4}OF$.

   a Draw a sketch of each position vector, showing points R and L.

   b Write $\overrightarrow{OR}$ and $\overrightarrow{OL}$ as column vectors in the form $\begin{pmatrix} x \\ y \end{pmatrix}$.

   c Determine the vector sum of $\overrightarrow{PF}$ and $\overrightarrow{RL}$.

3 The matrix $\mathbf{W} = \begin{pmatrix} a & 1 \\ a & b \end{pmatrix}$ maps (2, 3) onto (5, 8).

   a Find the values of a and b.

   b Determine the image of (6, –5) under the transformation **W**.

4 The graph shows the points A, B and C in relation to the origin O.

   a Write the position vectors $\overrightarrow{OA}$, $\overrightarrow{OB}$ and $\overrightarrow{OC}$ in the form $\begin{pmatrix} x \\ y \end{pmatrix}$.

   b Determine the magnitude and direction of the position vector $\overrightarrow{OB}$.

   c Write the vector $\overrightarrow{BC}$ as a column vector.

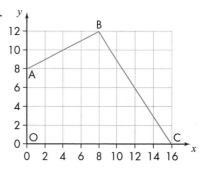

5 PQRS is a trapezium with sides PQ and SR parallel.

   $\overrightarrow{PQ} = \mathbf{t}$ and $\overrightarrow{SR} = 2\mathbf{t}$. $\overrightarrow{QR} = \mathbf{v}$.

   a State two parallel vectors.

   b Write $\overrightarrow{PQ} + \overrightarrow{SR}$ in terms of **t**.

   c Determine $\overrightarrow{SQ}$ in terms of **t** and **v**.

   d Write in terms of **t** and **v** the vector $\overrightarrow{PS}$.

   e Write in terms of **t** and **v** the vector $\overrightarrow{PR}$.

   f Write in terms of **t** and **v** the resultant of $\overrightarrow{PQ} - \overrightarrow{SR}$.

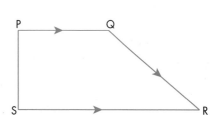

10 The School-Based Assessment in Mathematics

Introduction

The School-Based Assessment (SBA) is a practical, effective form of assessment used by many learning institutions. It has been part of the assessment used by the Caribbean Examination Council (**CXC**®) since it was first introduced in 1979.

SBAs help to monitor your performance over a period of time. They are meaningful assessments that help to create greater motivation. You actively participate in your own learning as well as in your own assessment. Through the SBA project activities, you are presented with various ways of demonstrating your knowledge, skills and attitudes.

One of the main reasons for introducing the SBA for Mathematics in 2018 was to help apply the mathematics you do in schools to everyday or real-world mathematics. Through the collection, representation and analysis of data the theoretical side of mathematics is now linked to real-world applications.

In the process, you will practise various problem-solving strategies such as:

- learning to define the problem in a personal way
- clarifying the problem through discussion with teachers, family, peers and resource persons in the community
- developing your own strategies, using words, charts, tables, symbols, models, algorithms, diagrams and figures
- evaluating the results
- developing a positive attitude towards mathematics and its use
- extending the mathematical processes to the understanding of other subjects in the curriculum.

The SBA project activities are linked to the syllabus and the marks obtained through these activities form part of your course assessment, which is then included in your final grade. It is therefore important that you score as highly in your SBA project as you can. You and your teacher can decide together on the project that you want to work on.

Managing the project

Your Mathematics School-Based Assessment (SBA) consists of **one project**, which has to be marked by the teacher, according to **CXC®** guidelines.

The use of projects is an ideal way of applying mathematical concepts, skills and procedures and to investigate and solve everyday mathematical problems, such as planning an outing, shopping, preparing a meal, building a simple structure or competing in a sporting event.

The role of the teacher is to:

- provide guidance to students in terms of choosing their project topics
- provide guidance and assistance throughout the SBA, ensuring that project content is also in line with the syllabus
- help maintain the continuity of the projects within scheduled classes and outside class time
- assess the projects, using the **CXC®** mark scheme provided and, together with students, keep both soft and hard copies.

Possible project activities include:

- an explanation of mathematical ideas
- carrying out practical tasks, using mathematical instruments such as rulers, compasses and protractors
- drawing, constructing, measuring, looking for patterns, counting, weighing, etc.
- the use of technological devices such as calculators and computers
- performing calculations mentally, with paper and pencil, using calculators, etc.
- oral responses to mathematics questions asked by teachers, peers and other interested parties
- identification of sections of the project to be done within normal class activities (supervised by the teacher) and outside normal class (in students' own time).

Specific activities for students

- State the task (nature, scope and focus).
- Devise a plan (what to do, how to do it), identifying materials needed, procedures to follow.
- Carry out the plan, procedure and/or activities.
- Record activities, stating what was done, how it was done, why it was done, using mathematical words and phrases, diagrams, tables, figures or charts.
- Reach a conclusion, including describing findings, comments, suggestions for improvement of findings, putting findings to greater use.

Devising a project topic

The topic that you choose for your SBA project should meet the following criteria:

1 It should allow you to use mathematics and mathematical problem-solving strategies to solve a problem, test a proposition, or investigate a hypothesis.

2 The topic – and the problem you are solving, the proposition you are testing or the hypothesis you are investigating – must be something that will give you data that will allow you to draw a conclusion that could have a meaningful impact on yourself, your family or friends, or your community. Ask yourself: what will I learn from this SBA that will have a positive impact on how I, my family and friends and community, live, act and make decisions? It is important that what you learn is meaningful and applicable.

3 The topic must be both *possible* and *achievable*: that is, it must be possible for you to get the data you need for your study, and it should be achievable in the time frame and with the resources that you have. The SBA should not require technical data that you are not going to be able to obtain.

Here are some examples of topics with notes on whether they meet the criteria for the SBA.

a What is the best angle at which to throw a javelin to achieve the maximum distance?

1 Can I use mathematics to investigate this?
Yes: I need to study angles and distances and to plot this information in charts.

2 Will I learn something from this SBA that I can use to improve how I act?
Yes: I will learn which angles give the best distance and will be able to share this information with athletes to help them improve their performance.

3 Will it be possible for me to get the data I need for this?
Yes: I will be able to work with an athlete in my school to collect the data I need.

4 Overall assessment: yes, this is a good topic for the SBA project and meets the criteria.

b A survey of the average amount of pocket money received by students per week/month.

1 Can I use mathematics to investigate this?
Yes: I need to collect data and put it into tables.

2 Will I learn something from this SBA that I can use to improve how I act?
No: I will learn how much pocket money students are given but it doesn't tell me very much that will have an impact. Parents will be able to see if they are more or less generous than the average, but this isn't very meaningful because parents will be limited in the pocket money they give their children because of economic factors.

3 Will it be possible for me to get the data I need for this?
Yes: I will be able to work with students in my school to collect the data.

4 Overall assessment: this is not an ideal project; it meets some of the criteria but is not going to tell me anything that will have a meaningful impact on how parents make decisions about how much pocket money to give. Also, the mathematics needed is at a fairly basic level.

c A survey on whether the average student spends more time on social media (such as Facebook) than on actual studies.

1 Can I use mathematics to investigate this?
Yes: I need to collect data and put it into tables.

2 Will I learn something from this SBA that I can use to improve how I act?
No: it will give me data about how students use their time but this data will not be linked to academic performance, for example, so won't tell me anything that will help me or my friends in any way.

3 Will it be possible for me to get the data I need for this?
Yes: I will be able to work with students in my school to collect the data.

4 Overall assessment: This is not a good topic as it is currently formulated. It will simply tell me about how students use their time without telling me anything about the impact or importance of how they use their time. It does use mathematics in that I have to collect data and plot that data into tables, but this is fairly low-level mathematics, and may not score highly in the assessment.

However, this project could be reformulated as follows to meet the SBA criteria.

d A survey of how time spent on social media compared with time spent on studies impacts performance at school.

1 Can I use mathematics to investigate this?
Yes: I need to collect data and put it into tables.

2 Will I learn something from this SBA that I can use to improve how I act?
Yes: it will give me data about how students use their time and about how the use of their time may or may not affect how well they do at school.

3 Will it be possible for me to get the data I need for this?
Yes: I will be able to work with students in my school to collect the data.

4 Overall assessment: yes, this could be a good topic if done in a carefully controlled way with helpful students to work with. This project may also require some research, and the conclusions based on a judgement between the results obtained and the expert research that can be obtained from online sources or which the teacher can supply.

e Does the use of diet drinks produce more health problems than the use of ordinary soft drinks?

1 Can I use mathematics to investigate this?
Yes: I will need to measure how much of each type of drink people consume, and plot this into tables.

2 Will I learn something from this SBA that I can use to improve how I act?
Yes: it could provide useful data to help people make better decisions about what to drink.

3 Will it be possible for me to get the data I need for this?
No: I will be able to work out what people drink, but I will not be able to assess the impact of these drinks on their health without being able to carry out blood tests, or to monitor their health over a longer period of time than is possible for the SBA. There are also likely to be other variables that impact on the health of my subjects during the study, such as exercise and nutrition, which I will not be able to include in the study.

4 Overall assessment: no, this fails the 'Can I collect the data I need?' criterion.

f How can I work out which is the best cell phone to buy for my personal use?

1 Can I use mathematics to investigate this?
Yes: I can calculate which options enable me to spend the least amount of money.

2 Will I learn something from this project that I can use to improve how I act?
Yes: this would enable me to buy the best value cell phone for my personal pattern of cell phone usage.

3 Will it be possible for me to get the data I need for this?
Yes: I can obtain information on the monthly and annual costs of different purchase options from cell phone retailers.

4 Overall assessment: yes, this would be a suitable SBA project.

Applying these three questions to your ideas for your SBA project will help you identify a good project topic.

  a Can I use mathematics to investigate this topic?

  b Will I learn something useful from this project?

  c Will it be possible for me to collect the data I need?

Your teacher can also help you work out if your project ideas meet the criteria for the SBA project. The important thing is to identify a clear specific objective for your project, and to say how you are going to investigate it.

Collecting data

Once you have decided what your SBA project is going to be about, you need to consider how you will collect your data. Some data can be collected by researching shops and retail outlets, or newspaper and magazines. Other data may need to be collected from your fellow students or members of your family or local community. In some cases this data may need to be collected with repeated surveys over a period of time.

Get together the equipment and resources you will need to collect your data. You will certainly need a new notebook with pens or pencils. You may also need measuring tapes, timers, weighing balances or other measuring devices. Gather these all together and keep them in one place ready for your project to start.

In your notebook create tables to show what data you will be collecting. Your tables will of course be empty to start with, but putting in the headings for the columns and rows will help to make it clear how much data you have to collect. For example, the table for the project on the cost of cell phones might look like this.

Plan	Time period	Data	Calls	Cost
Digicel Freedom LTE 1GB				
Digicel Freedom LTE 2GB				
Digicel Freedom LTE 8GB				
Flow Prepaid 1 day plan				
Flow Prepaid 3 days plan				

It may be necessary to have different tables for different suppliers and retailers.

Ensure that the amount of data that you have set yourself to collect is manageable and can be achieved in the time you have been allocated for carrying out your SBA project.

Problem-solving strategies

These strategies are useful with a wide range of projects.

  1 Simplify the problem by using simpler numbers, subsequently putting back the original numbers once the procedure is understood.

> **Example**
>
> Find the distance around a rectangular field of dimensions 187 m and 145 m.
>
> Using smaller numbers (3 m and 5 m), the distance could be found by the algorithm:
>
> $$P = 2(3 + 5) = 2 \times 8 = 16 \text{ m}$$
>
> Putting back the original numbers, $d = 2(187 + 145) = 2 \times 332 = 664 \text{ m}$

2 Sketch a simple diagram – using geometrical representation to interpret the problem.

Find the area of a field in the shape of a trapezium with two right angles at one end and with one of the angles at the other end equal to 140°. The lengths of the parallel sides are 80 m and 150 m.

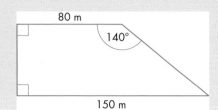

3 Make a table of results and then analyse it to discover any patterns.

This table represents the distances a ball is thrown by children aged 5–10 years.

Age (years)	Distance (m)				
	Throw 1	Throw 2	Throw 3	Throw 4	Throw 5
5	4	5	6	5	7
6	7	8	9	10	9
7	6	10	11	10	12
8	15	19	14	18	13
9	25	26	30	32	34
10	40	29	35	37	42

Various patterns can immediately be detected from the table, for example, *the greater the age the further the distance the ball is thrown.*

4 'Guess and check' consists of making some informed predictions or guesses about the solution to a problem, then checking the answer for accuracy.

A fruit basket contains three types of fruit: oranges, apples and pears. There are 55 fruits in the basket, comprising 15 oranges and three times as many pears as apples. How many of each of the three types of fruit are in the basket?

A 'guess and check' table can be devised as follows:

Number of			Total
apples	pears	oranges	
3	9	15	27
5	15	15	35
8	24	15	47
10	30	15	55

5 Look for patterns: these could be numerical or geometrical.

(a)   1, 2, 3, 3, 4, 5, 6, 6, 7, 8, 9, 9

(b)

6 Use algebraic symbols and equations to express ideas.

> An alternative method for solving example 4
>
> Let the number of apples be x.
>
> Number of fruits $= 15 + x + 3x$
>
> $$= 55$$
>
> $$\Rightarrow \quad 15 + 4x = 55$$
>
> $$\Rightarrow \quad 4x = 55 - 15$$
>
> $$= 40$$
>
> $$\Rightarrow \quad x = 10$$

7 Make full use of calculators (see chapter 12). This entails:

- exploration of number patterns, number ideas and number sense
- estimation skills associated with the arithmetic operations
- checking calculations.

This also allows more time to focus on the process of problem solving.

Project report

The project is presented in the form of a report. This should be made up of the following components:

1 Project title

2 Introduction

3 Method of data collection

4 Presentation of data

5 Analysis of data

6 Discussion of findings

7 Conclusion

You should write up your project carefully, producing a report that is no more than 1000 words long.

- Use proper grammar and carry out a spell check.
- Number your pages and provide a contents page.
- Include a cover page with personal and school information.
- Make an electronic copy and back up your document with both hard and soft copies.
- Place surveys, questionnaires, etc. in an appendix at the end of the document. Also provide a reference list or bibliography if you have cited sources in your introduction.

The project is marked out of a total of 20 and the marks will be allocated to each task and profile, as outlined below.

Project title

The title should be brief, clear and related to a real-world problem. It can take the form of a question or a precise and clear statement of intent – a hypothesis that indicates what you are trying to accomplish.

Introduction

The introduction consists of a comprehensive description of the project itself and should set the background for your intended course of action. It should include clearly stated and precise objectives.

Method of data collection

You must describe how you have collected your data. Methods may include *surveys, questionnaires, experiments* and *investigations*. The method chosen should be appropriate so that the data obtained is reliable, leading to reliable conclusions. In order to ensure that the method chosen is sound, advice can be sought from your teacher or other facilitator.

The tools used to collect data should also be stated: they may be blank tables with headings, survey questions, diagrams and general calculation formulae relating to what you plan to discuss.

Data collection was discussed on page 462.

Presentation of data

Data presentation should be well organised and accurate. The tools used to collect data (for example, surveys, tables) should be properly presented and clearly described and labelled. There should be at least one graph illustrating data.

The data may be represented in statistical graphs such as bar charts, pie charts, linear graphs and histograms. Every graph must have a concise but accurate title, for example, *A graph of…vs…*; axes must be clearly labelled and include any appropriate units.

Use of relevant software can be helpful in the plotting and analysis of graphs.

Analysis of data

The analysis of data refers to all the calculations that are done on your data and the values obtained. It will include values extracted from graphs and any other observations. The analysis of data must feature mathematical language, concepts and expressions and information should be presented in a logical, detailed but concise manner.

Discussion of findings

The discussion is based on what you have discovered from your analysis.

Consider these questions:

- What exactly is the data telling you?
- What patterns and trends are revealed?
- What comparisons are made? Include quantities, percentages and averages.

The analysis may or may not support your hypothesis. State findings clearly and precisely. Discussion must **not** be based on unsupported claims but should follow from your data and analysis.

Conclusion

The conclusion is basically a summary of what you have done for your project, but it must include the following key elements:

- Does the data that you have analysed and discussed confirm your hypothesis or does it disagree with the hypothesis?
- Can you make suggestions for improvements to your project investigation?
- Do your results enable you to make recommendations for changing your behaviour or the behaviour of others?

Project mark scheme

Italics indicate marks allocated for less well-executed criteria.

Project descriptors	Mark	Criteria
Title	1	• **Clear, concise and relates to a real-world problem – 1**
Introduction	4	• **Objectives clearly stated – 1** • **Comprehensive description of project – 2** • *Limited (concise) description of project – (1)* • **Detailed content page with page numbers – 1**
Data collection method	2	• **Method is clearly described, appropriate and flawless – 2** • *Method of data collection is clearly stated – (1)*
Presentation of data	5	• **Data is accurate and well organised – 2** • *Data is presented but is not well organised – (1)* • **Tables/graphs included, correctly labelled and used appropriately – 2** • *Tables/graphs included – (1)* • **Accurate use of mathematical concepts – 1**
Analysis of data	2	• **Detailed and coherent analysis – 2** • *Limited analysis of findings – (1)*
Discussion of findings	2	• **Statements of findings clear and follows from data collected – 2** • *Statement does not follow from data – (1)*
Conclusion	2	• **Based on findings and related to the objectives of the project – 2** • *Related to purpose/objectives of the project but not based on findings (1)*
Overall presentation	2	• **Information presented logically using correct grammar – 2** • *Information poorly presented (1)*

Sample Mathematics SBA

SBA title

What is the best angle at which to throw a javelin to achieve the maximum distance?

Introduction

Throwing the javelin is one of the most spectacular and anticipated field events in the athletic calendar, irrespective of the level at which the event is taking place: secondary, tertiary or international. It is an event that requires a lot of skill as well as strength, but someone who has mastered the technique can certainly make up for any lack of strength.

The javelin is, of course, a simple ballistic missile, that is an object moving through the air under the influences of gravity and friction only and without any form of power to make it fly further. A cricket ball, a discus, or a shot, etc. are other examples of ballistic missiles. In this SBA project, the specific objective is:

To discover whether there is indeed a specific angle of projection that will lead to a maximum horizontal distance achieved by the javelin.

Method of collecting data

The student conducting the SBA will seek the assistance of five athletes, possibly during PE classes, ensuring that they all throw the same javelin, and taking great care that these throws are made as close as possible to specific angles: 15°, 30°, 45°, 60° and 75°. A large classroom protractor, flat board and metre rule can be used to approximate the angles as closely as possible. Students may employ a short run up. (Note: 0°, representing a horizontal throw, and 90°, being vertical, would be too dangerous and as such will be omitted.)

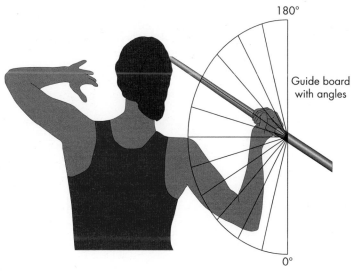

The set-up

With the above set-up, each student will be required to make one throw at each angle. The results will be tabulated and used for analysis.

Presentation of data

The data is presented in a table as shown below (Table 1): This table identifies each student and records the distance the javelin is thrown at each of the five angles.

Student	Angle of throw (°)				
	15	30	45	60	75
	Length of throw (m)				
A	31	39	42	36	26
B	28	36	39	35	28
C	33	43	44	36	29
D	29	35	38	33	24
E	29	37	37	35	23

Table 1

The information in Table 1 can be used to find the mean distance thrown at each of the five angles. These results are represented in a separate table (Table 2).

Angle of throw (°)	Average distance (m)
15	30
30	38
45	40
60	35
75	26

Table 2

Analysis of data

The data in Tables 1 and 2 is best analysed by drawing the bar chart and the line graph shown here. The bar chart shows the angles of throw of the javelin and the average distance thrown at each angle (in metres).

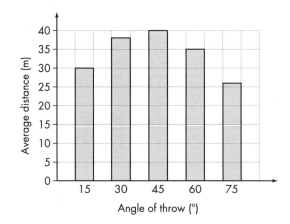

The line graph compares the angle of throw with the average distance thrown.

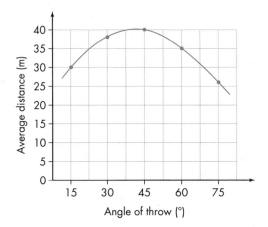

Discussion of findings

Table 1 shows that the distances thrown by each student increased for the first three angles: 15°, 30° and 45°, but dropped for the next two, 60° and 75°. An analysis of the mean distance for each angle also shows a similar trend. This is further confirmed by looking at the bar chart and line graph shown above.

The data shows that, on average, the distances thrown at 45° are the longest, while those thrown at angles of 15° or 75° are the shortest.

Conclusion

This SBA has proved, within the limits of experimental errors – and especially errors due to the maintaining of an accurate angle of throw – that a javelin thrower may be able to achieve best results by trying to maintain an angle of approximately 45° at the point of releasing the javelin. Scientifically, this fact has been proven for other missiles. The only consideration in the case of the javelin might be a difference in air resistance, due to the elongated shape of the missile, when compared to the round shape of a ball or even a discus. This may be investigated in a subsequent project. The results of this project, as relating to the objectives, have been generally satisfactory.

Examples of SBA topics

Shown below are some ideas for SBA project topics. Do discuss ideas for your topic with your teacher and try to come up with an idea that reflects your own interests. These topics can be reworded or adapted to suit individual students.

1. How does the amount of water given during germination affect the rate of growth of seedlings? This could be extended to investigate the effect of adding soluble plant food to the water.

2. What is the effect of regular watering, with plant nutrients added, on the amount of fruit produced by tomato plant seedlings?

3. Which is the most effective angle at which to throw:
   a a discus   b a cricket ball   c a shot?

4. From where in the field is a particular striker most effective at scoring goals: from close range, from the penalty spot, or from the outfield?

5. What is the range most effective for scoring goals on the football field/netball court/basketball court?

6. Does the width or weight of a cricket bat affect the number of runs scored by a particular batter?

7. Is typing faster than ordinary handwriting? This would need to take into account the time taken to correct typing mistakes.

8. Are spinners in the school cricket team more successful in taking wickets than fast bowlers?

9. What is the safest angle at which to rest a ladder on a wall? This topic would need to be investigated with proper regard for student safety.

10. Are the students who have taken the entrance exam at your school more successful at algebra than at trigonometry?

11. How does the distance a student lives from school affect their academic success?

12. Which is the best cell phone or other device to purchase that meets certain specifications using different purchasing schemes (consumer arithmetic)?

13. What is the best design for a bedroom, classroom, or school library with a given budget in which certain specifications must be met (measurement/linear programming)?

14. How can I budget my monthly allowance to be able to save a particular sum by the end of the year (consumer arithmetic)?

15. Who is the best batsman from the school's cricket team (statistics)?

CHAPTER

11 Examination Practice

Examination 1

Paper 1

(Duration 1½ hours)

Multiple choice questions

Each question has four suggested answers, A, B, C and D. Choose the correct one.

1 Find the value of 0.075 as a fraction in its lowest terms.

A $\dfrac{3}{4}$ **B** $\dfrac{7}{10}$ **C** $\dfrac{3}{40}$ **D** $\dfrac{3}{10}$

2 Write 0.0023 in standard form.

A 0.2×10^3 **B** 2.3×10^3 **C** 2.3×10^2 **D** 2.3×10^{-3}

3 Ann shared $48.00 among three children in the ratio 3 : 4 : 5. What was the largest share?

A $12.00 **B** $16.00 **C** $20.00 **D** $24.00

4 $3(5 - 7) - (2 + 8) =$

A 6 **B** 36 **C** -26 **D** -16

5 The cost of a marbles was $$x$. What is the cost of 2 marbles?

A $$2x$ **B** $\$\dfrac{x}{a}$ **C** $\$\dfrac{2x}{a}$ **D** $\$\dfrac{x}{2a}$

6 A student buys a text book for $80.00 and sells it at a 20% loss. The selling price is:

A $16.00 **B** $60.00 **C** $75.00 **D** $64.00

7 A shirt was bought for $120 and sold for $90.00. What was the percentage loss?

A 30% **B** 25% **C** 60% **D** 20 %

8 Tom borrowed $200.00 for 5 years, paying interest at 10% per annum. Calculate the interest.

A $1000.00 **B** $100.00 **C** $500.00 **D** $250.00

9 One notebook costs $$2x$ and 2 pens cost $$y$. Find the cost of 3 notebooks and 2 pens.

A $3x + 2y$ **B** $6x + 2y$ **C** $6x + y$ **D** $3x + y$

10 Given that $p * q = 2pq - p + q$, evaluate $2 * 3$.

A 13 **B** 12 **C** 7 **D** 9

11 Simplify $-5y(3 - 2y)$.

    A $-15y - 2y$ **B** $10y - 15y$ **C** $-15y + 10y^2$ **D** $15y - 10y^2$

12 Simplify $3ab(4a^2b^3)$.

    A $12a^2b^3$ **B** $12a^3b^3$ **C** $12a^4b^3$ **D** $12a^3b^4$

13 Find the range of values that satisfy the inequality $5x - 2 \leqslant 3$.

    A $x \leqslant 3$ **B** $x \leqslant \dfrac{1}{5}$ **C** $x \leqslant 1$ **D** $x \geqslant \dfrac{1}{2}$

14 Expand $5a(a - 3b) - 3b(4b - 2a)$.

    A $5a^2 - 3ab - 12b^2$ **B** $5a^2 - 12b^2 - 9ab$

    C $5a^2 - 15ab + 6ab$ **D** $5a^2 - 15ab - 12b^2 - 6ab$

15 Find the value of x in the equation $3(2x - 4) - 5(x - 2) = 1$.

    A $x = 0$ **B** $x = 3$ **C** $x = 23$ **D** $x = 1$

16 A square has sides of length 5 cm and a right-angled triangle has a base of 6 cm and a hypotenuse of length 10 cm.

    A The triangle has a greater area than the square.

    B The square has a greater perimeter than the triangle.

    C The sum of their areas is 49 cm².

    D The sum of their perimeters is 43 cm.

17 A right-angled triangle has sides of length $3x$, $4x$ and $5x$. Given that the perimeter of the triangle is 48 cm, find the value of x.

    A $x = 2$ cm   **B** $x = 3$ cm   **C** $x = 4$ cm   **D** $x = 8$ cm

18 Find the size of each of the interior angles of a regular pentagon.

    A 90°   **B** 108°   **C** 54°   **D** 98°

19 The area of a rectangle is 20 cm². Given that the length is five times the width, what is the perimeter?

    A 18 cm   **B** 24 cm   **C** 42 cm   **D** 20 cm

20 A box contains 8 red balls and 12 blue balls. If one ball is removed randomly, what is the probability that it is a blue ball?

    A $\dfrac{3}{5}$ **B** $\dfrac{8}{12}$ **C** $\dfrac{12}{8}$ **D** $\dfrac{8}{20}$

This frequency table shows the marks obtained by students in a test.
Use it to answer questions 21 and 22.

Marks	1	2	3	4	5	6	7	8	9	10
Number of students	8	5	6	4	12	16	10	3	2	1

21 The modal score is:

    A 1   **B** 5   **C** 16   **D** 6

22 The median score is:

    A 5   **B** 4   **C** 6   **D** 8

Use this arrow diagram to answer questions 23 and 24.

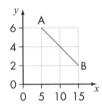

x f(x)

23 The relation shown in the arrow diagram can be described as:

    A a one-to-many function   **B** a many-to-one relation

    C a one-to-one function   **D** a many-to-many function

24 The function can be written as:

    A f(x) = x + 1   **B** f(x) = 2x **C** f(x) = x^2 **D** f(x) = 2x^2

Use this graph to answer questions 25 and 26.

25 The gradient of AB is:

    A $\dfrac{4}{10}$ **B** $\dfrac{10}{4}$ **C** $-\dfrac{2}{5}$ **D** $\dfrac{3}{4}$

26 The coordinates of the midpoint of AB are:

    A (10, 6)   **B** (10, 2)   **C** (10, 4)   **D** (9, 4)

27 The range of f: $x \rightarrow x^2$ for the domain {−3, −2, −1, 0, 1, 2, 3} is:

    A {−3, −2, −1, 0, 1, 2, 3}   **B** {−6, −4, −2, 0, 2, 4, 6}

    C {9, 4, 1, 0, 1, 4, 9}   **D** {18, 8, 2}

28 In triangle PQR, the value of sin R is:

    A 0.6   **B** 0.75   **C** $\dfrac{5}{3}$ **D** $\dfrac{4}{5}$

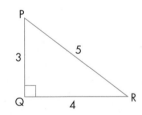

29 The image of a point (5, 3) under an enlargement of scale factor $k = 4$, with centre of enlargement the origin, is:

    A (20, 3)   **B** (5, 12)   **C** (9, 7)   **D** (20, 12)

30 In the diagram, TR is a tangent at R and PR is a chord.
The size of angle PQR is:

A 80°   B 90°   C 60°   D 70°

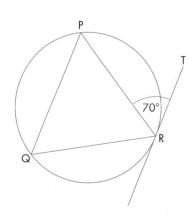

31 In the diagram, LM and RS are parallel; angle θ is therefore
equal to:

A 50°   B 130°   C 100°   D 160°

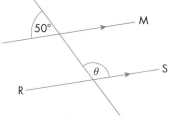

32 A boat set sail from a port K in an easterly direction, travelling for 20 miles, and then travels due
south for another 20 miles to a port J. The bearing of J from K is:

A 90°   B 045°   C 315°   D 135°

33 Write the mixed number $7\frac{1}{4}$ as a decimal, correct to 1 decimal place.

A 7.2   B 7.3   C 7.25   D 7

34 Write the number 29.005 correct to one decimal place.

A 29.0   B 3   C 30   D 30.0

35 Find the HCF of 36, 42 and 48.

A 4   B 8   C 6   D 2

36 Find the LCM of 36, 54, and 90.

A 18   B 4860   C 540   D 6

37 What is the place value of the 9 in the number 436.098?

A 90   B $\frac{9}{10}$ C 9   D $\frac{9}{100}$

38 Given that P = {factors of 18}, Q = {odd numbers between 8 and 14} and R = {even numbers from
10 to 18}, determine: $(P \cap R) \cup Q$.

A {9, 11, 13, 18}   B {18}

C {1, 2, 3, 6, 9, 10, 11, 12, 13, 14, 16, 18}   D {6, 8, 10, 12}

39 The diagram shows two intersecting sets A and B. The shaded
region represents:

A $A \cap B$ B $A \cup B$ C $A \cap B'$ D $B \cap A'$

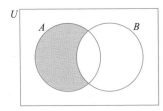

40 The volume of a cube of edge length 10 cm is:

    A 100 cm³   **B** 10 000 cm³   **C** 1000 cm³   **D** 30 cm³

41 Write the number 0.000 98 correct to 1 significant figure.

    A 0.0001   **B** 9.8   **C** 0.001   **D** 0.0009

Use this diagram to answer questions 42–44.

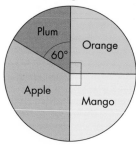

The pie chart shows the preference in fruits among 60 students.

42 The number of students who like oranges is:

    A 40   **B** 30   **C** 20   **D** 15

43 The number of students who prefer apples will be:

    A 50   **B** 15   **C** 20   **D** 30

44 What is the probability that a student chosen at random prefers plums?

    A $\dfrac{1}{3}$ **B** $\dfrac{2}{3}$ **C** $\dfrac{1}{2}$ **D** $\dfrac{1}{6}$

45 Which of these graphs represents a function?

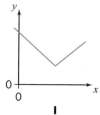 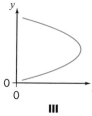 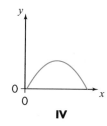

      I   II   III   IV

    A I only   **B** I and II only   **C** I and IV only   **D** I, II, III and IV

46 Which diagram **best** shows a reflection of the point P in the line $y = -x$?

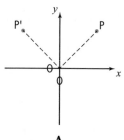 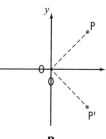 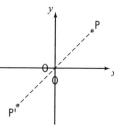 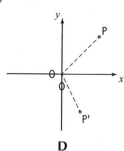

      A   **B**   **C**   **D**

47 Calculate the exact value of $0.0012 \div (6 \times 10^{-3})$.

    A 2   **B** 0.2   **C** 7.2   **D** 0.5

48 Evaluate $-\left(\dfrac{4}{9}\right)^{-\frac{1}{2}}$.

A $\dfrac{2}{3}$ **B** $\dfrac{3}{2}$ **C** $-1\dfrac{1}{2}$ **D** $-\dfrac{2}{3}$

49 In the diagram, sector OAB has a sector angle of 120° and radius OA = 3 m.

The area of the sector, in terms of π, is:

A 9π m² **B** $9\pi^2$ m² **C** 3π m² **D** 2π m²

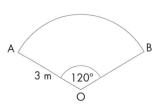

50 Ana has p beads, her friend Jane has three times that number and Tom has three-quarters as many as Ana and Jane together. If Tom loses 4 of his beads, how many is he left with?

A $p-4$ **B** $\dfrac{3}{4}p-4$ **C** $3p-4$ **D** $3p$

51 Written in standard form (or scientific notation), 0.00709×10^{-5} is:

A 7.09×10^{-7} **B** 0.7×10^{-7} **C** 7×10^7 **D** 7.09×10^{-8}

52 Water rates are charged as: $20.00 per month for rental of meter, $30.00 for the first 200 litres and $5.00 for each additional 20 litres. What is the total bill for 400 litres in one month?

A $200.00 **B** $150.00 **C** $100.00 **D** $55.00

53 Simplify $\dfrac{2x-4}{5} + \dfrac{x+1}{2}$.

A $\dfrac{3x-3}{10}$ **B** $\dfrac{5x-3}{7}$ **C** $\dfrac{9x-3}{10}$ **D** $\dfrac{11x-18}{7}$

54 Simplify $12m^3 \times 3m^4n^2 \times 2mn^4$.

A $72m^{12}n^8$ **B** $17m^7n^6$ **C** $30m^8n^6$ **D** $72m^8n^6$

55 ABCD is a trapezium in which AB = 10 cm, BC = 5 cm and CD = 6 cm. Find the area of the trapezium.

A 50 cm² **B** 30 cm² **C** 26 cm² **D** 40 cm²

56 What is the equation of the line that passes through the points P(0, 5) and Q(5, 0)?

A $y = x$ **B** $y = x + 5$ **C** $y = -x + 5$ **D** $y = -x$

57 The point P(−3, 2) is rotated 90° in an anticlockwise direction about the origin. The image of P is:

A P′(2, 3) **B** P′(−2, −3) **C** P′(3, −2) **D** P′(3, 2)

58 Write this expression as a fraction in its simplest form.

$$\dfrac{(2-4)^3}{4^3 - 4^2}$$

A $-\dfrac{8}{48}$ **B** $\dfrac{1}{8}$ **C** $\dfrac{8}{48}$ **D** $-\dfrac{1}{6}$

59 The dimensions of this cube are in centimetres. Calculate the volume, in terms of x, in the appropriate units.

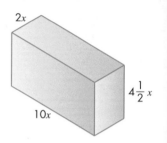

 A 16.5x^3 cm³   B 80x^2 cm³   C 120x^3 cm³   D 90x^3 cm³

60 Which of these represents a null set?

 A Multiples of 5 greater than 6   B Even numbers between 1 and 3
 C Factors of 6 from 7 to 12   D Prime numbers from 7 to 11

Paper 2

(Duration 2 hours 40 minutes)

Section 1

Answer all questions in this section.

1 a Using a calculator or otherwise, calculate:

    i $\dfrac{2\frac{3}{4} - 1\frac{1}{2} \times 5\frac{1}{3}}{5\frac{1}{2} + 2\frac{1}{4} - 4\frac{3}{4}}$

    giving your answer as a fraction in its lowest terms   (2 marks)

    ii the exact value of $(2.4)^2 \times \dfrac{2}{5} + \dfrac{4.5}{0.4}$ (2 marks)

b Give your answer for part **a ii**:

    i   correct to 2 decimal places   (1 mark)

    ii correct to 1 significant figure   (1 mark)

    iii in standard form.   (1 mark)

c A man invests $800.00 for 5 years at 5% simple interest.

    i   What is the amount he receives at the end of the period?   (1 mark)

    ii At what rate should he invest in order to make three times that amount?   (1 mark)

                                                            (Total 9 marks)

2 a Given that $p * q = 3p^{\frac{1}{3}} + 2q^{\frac{1}{2}}$, find:

    i $8 * 9$ (2 marks)

    ii $\dfrac{1}{8} * \dfrac{4}{9}$ in its simplest form.   (2 marks)

b Solve the inequality: $3x - 2 < 5(2x + 1)$, representing your solution on a number line.   (2 marks)

c One mango costs $x and 1 banana costs $y. 3 mangoes and 1 banana together cost $5.00 and 5 mangoes and 2 bananas cost $9.00.

    i   Write down two equations in x and y to represent these statements.   (1 marks)

    ii Solve the equations to find the cost of a banana and a mango.   (2 marks)

                                                            (Total 9 marks)

3 a In a youth sports clubs with 60 participating members, 35 play football, 32 play cricket and 25 play netball. Of these members 20 play both football and cricket, 10 play football and netball, 7 play cricket and netball and 10 play neither cricket nor netball.

    i   Represent the information in a Venn diagram and determine the number who play all three sports.   (3 marks)

    ii How many members play only one sport?   (2 marks)

b **i**   Using a pair of compasses, pencil and ruler only, construct triangle KLM in which KL = 10 cm, angle KLM = 150° and LM = 7 cm.   (2 marks)

    ii Join K to M and measure the length of KM.   (1 mark)

    iii Construct the point N such that KLMN is a parallelogram.   (1 mark)

                                                            (Total 9 marks)

4 a The equation of a straight line is given as $12x = 3y - 6$.

    i Write the equation in the form $y = mx + c$, hence find the gradient and the *y*-intercept.
                                    (1 mark)

    ii A point P(2, *h*) lies on the line. Find the value of *h*.   (1 mark)

    iii Find the equation of a line passing through P and perpendicular to the line $y = mx + c$.
                                        (2 marks)

b Represent these inequalities on the same graph.

    i $x \geqslant 2$ (1 mark)

    ii $x \leqslant 6$ (1 mark)

    iii $y \geqslant 0$ (1 mark)

    iv $y \leqslant 2x$ (1 mark)

c Shade the region that satisfies the four inequalities in part **b**.   (1 mark)

(Total 9 marks)

5 a In a regular polygon the interior angle is eight times the exterior angle.

    i Find the value of each interior and exterior angle of the polygon.   (2 marks)

    ii Find the number of sides of the polygon.   (1 mark)

    iii Given that the length of each side of the polygon is 10 cm, what is the perimeter?   (1 mark)

b A cylindrical water tank of internal diameter 1.8 m and internal depth 2.4 m is $\frac{7}{8}$ full of water. When an iron ball is gently lowered into the tank of water it immediately sinks to the bottom and the water level rises right up to the brim. Given that the ball displaces its own volume of water, determine:

    i the capacity of the tank, in litres   (2 marks)

    ii the volume of the ball   (1 mark)

    iii the diameter of the ball.   (2 marks)

(Total 9 marks)

6 a A survey was carried out in a supermarket to determine the sales of various items on a daily basis. The results were displayed in a pie chart, with angles of 30° for fruit, 50° for vegetables, 70° for flour, 90° for rice and 80° for meat. The amount taken for drinks was $400.

    i What angle would represent drinks?   (1 mark)

    ii How much money was taken for fruits and flour?   (2 marks)

    iii What were the total daily sales?   (1 mark)

    iv Make a sketch of the pie chart, showing all the information.   (1 mark)

b Find the image of triangle PQR after reflection in the line $y = 2$.

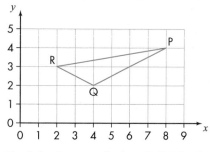

                                      (2 marks)

c Find the image of triangle PQR after an anti-clockwise rotation of 90° about the origin. Give the coordinates of P', Q' and R' after the rotation.   (2 marks)

d Draw a diagram to show the positions of the images formed in parts b and c.   (4 marks)

(Total 13 marks)

7 The diagram shows the first three shapes in a sequence made from squares of side 1 cm.

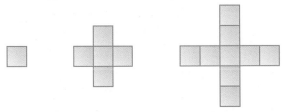

a Draw the fourth figure in the sequence. (2 marks)

b Each row in this table relates to a specific shape in the sequence. The table displays a pattern of numbers.

Shape	Number of squares	Perimeter of shape (cm)	
1	1	$1 \times 4 = 4$	
2	5	$3 \times 4 = 12$	
3	9	$5 \times 4 = 20$	
4			**i** (2 marks)
...			
		$11 \times 4 = 44$	**ii** (2 marks)
	29		**iii** (2 marks)
...			
32			**iv** (2 marks)

Complete the rows numbered (i), (ii), (iii) and (iv).

(Total 10 marks)

Section 2

Answer all questions in this section.

Relations, functions and graphs

8 a i Copy and complete this table for the function $2y + 4x^2 = 16x + 6$, by rearranging the equation and finding and inserting the missing values of y.

x	−1	0	1	2	3	4	5
$y =$	−7	3		11	9		

(3 marks)

ii Use your completed table to plot a graph of the function. (3 marks)

iii Draw a graph of the function $g(x) = 5$ on the same set of axes as the function in parts **ai** and **ii**, and hence or otherwise solve the equation $2x^2 - 8x + 2 = 0$. (2 marks)

b The velocity–time graph shows the motion of an object during a 14-second period.

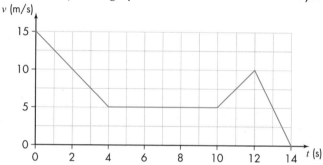

i During which period was the object decelerating most and what was the rate of this deceleration? (1 mark)

ii During which period was the object accelerating and what was the rate of this acceleration? (1 mark)

iii Find the total distance covered by the object in this 14-second interval. (2 marks)

(Total 12 marks)

Geometry and trigonometry

9 a A ship P is directly east of another ship Q, at a distance of 10 km from it. A third ship R is on a bearing of 160° from Q and is 14 km from Q.

i Make a sketch showing the positions of P, Q and R. (2 marks)

ii Calculate the distance PR and the bearing of R from P. (4 marks)

b The diagram (not drawn to scale) shows a circle, centre O, with a tangent PT. Angle QTS = 60°, RM = 4 cm and PM = 5 cm.

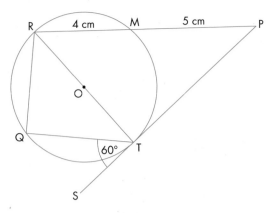

i   Calculate angle RQT.   (1 mark)
ii Calculate angle QRT.   (1 mark)
iii Calculate angle RTP.   (1 mark)
iv Given RM = 4 cm and PM = 5 cm, find the length of PT.   (3 marks)

(Total 12 marks)

Vectors and matrices

10 a The point P(2, 3) is mapped onto the point P′(x, y) by the transformation $\mathbf{T} = \begin{pmatrix} 1 & 5 \\ 0 & 1 \end{pmatrix}$.

    i Find the values of x and y.   (2 marks)

    ii P(2, 3) is mapped onto P″(r, s) by the transformation $\mathbf{W} = \begin{pmatrix} 1 & 0 \\ 5 & 1 \end{pmatrix}$.
    Find the values of r and s.   (2 marks)

    iii Describe fully both transformations, **W** and **T**, and state the differences between
    the images obtained by transformations **T** and **W**.   (2 marks)

    iv Find the matrix for the single transformation that represents the double
    transformation **WT**.   (2 marks)

    v Find the image of the point R(1, 6) under **WT**.   (1 mark)

b In trapezium ABCD, AB ∥ DC, AB = 2DC, $\overrightarrow{AB} = 2\mathbf{p}$ and $\overrightarrow{AD} = \mathbf{q}$.

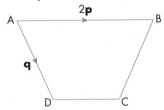

Express, in terms of **p** and **q**:

    i $\overrightarrow{BD}$ (1 mark)
    ii $\overrightarrow{AC}$ (1 mark)
    iii the sum of the vectors $\overrightarrow{BD} + \overrightarrow{AC}$.   (1 mark)

(Total 12 marks)

Examination 2

Paper 1

(Duration 1$\frac{1}{2}$ hours)

Multiple choice questions

Each question has four suggested answers, A, B, C and D. Choose the correct one.

1 Given that 70% of a number is 49, the number is:

   A 60   **B** 70   **C** 51   **D** 30

2 2.75 × 0.09 correct to 2 decimal places, is:

   A 0.24   **B** 0.20   **C** 0.25   **D** 0.23

3 $44.00 was shared among four students in the ratio: 1 : 2 : 3 : 5. What was the greatest share?

   A $20.00   **B** $12.00   **C** $40.00   **D** $11.00

4 Find the LCM of 8, 12 and 16.

   A 48   **B** 24   **C** 4   **D** 96

5 What is the largest prime number between 10 and 20?

   A 17   **B** 13   **C** 19   **D** 11

6 Given the subsets A = {prime numbers between 1 and 10}, B = {factors of 10} and C = {odd numbers from 1 to 10} then $A \cup B \cap C$ is:

   A {1, 2, 3, 5, 7, 10}   **B** {2, 5}   **C** {5}   **D** {1, 3, 5, 7}

7 Simplify $\left(\sqrt{169} - \sqrt{144}\right)^{\frac{1}{2}}$.

   A 5   **B** 1   **C** 25   **D** 12

8 Given the universal set U = {square numbers between 0 and 40}, $n(U)$ is:

   A 6   **B** 5   **C** 38   **D** 8

9 Find the HCF of 27, 36 and 45.

   A 3   **B** 540   **C** 9   **D** 15

10 Find 250% of 1000.

   A 500   **B** 4   **C** 2500   **D** 250

11 In the Venn diagram, $n(A)$ = 21, $n(B)$ = 15, $n(A \cap B)$ = 5, $n(A \cup B)'$ = 2.

Given that the universal set is represented by ε, find $n(\varepsilon)$.

   A 43   **B** 41   **C** 38   **D** 33

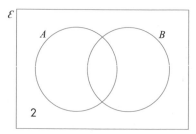

12 Find $2\frac{1}{2}$% of $10.00.

   A $2.50   **B** $25.00   **C** 25¢   **D** $15.00

13 John bought a bicycle for $210.00, including 5% sales tax. How much was the bicycle before tax was added?

    A $200.00   **B** $190.00   **C** $220.00   **D** $215.00

14 $(5x - 4)(2x - 6) =$

    A $10x^2 - 30x - 8x - 24$ **B** $10x^2 + 38x - 24$ **C** $10x^2 + 22x + 24$ **D** $10x^2 - 38x + 24$

15 A mother gives her son $y each day for his allowance (Monday to Friday). She decides to increase that allowance by $5.00 more than $2\frac{1}{2}$ times his original daily allowance. How much will the weekly amount increase by?

    A $2\frac{1}{2}y + 5$ **B** $12\frac{1}{2}y - 5y$ **C** $7\frac{1}{2}y + 5$ **D** $12\frac{1}{2}y + 25$

16 $-2y(3y^2 - 4y) =$

    A $6y^2 - 4y$ **B** $-6y^3 - 8y$ **C** $-6y^3 + 8y^2$ **D** $6y^3 - 8y^2$

17 $p * q = \dfrac{2pq - p^2q}{p + q}$, therefore 2 * 3 is:

    A $\dfrac{1}{5}$ **B** $\dfrac{2}{5}$ **C** $\dfrac{3}{5}$ **D** 0

18 Solve the equation $3(2x - 1) - 4x + 5 = 2(2x - 2)$.

    A $x = 1$ **B** $x = 2$ **C** $x = -1$ **D** $x = 3$

19 I think of a number x, I double the number, take away 5 from the doubled number and divide the result by 3, obtaining an answer of 1. What is the number x?

    A   2   **B**   1   **C**   4   **D**   5

20 Abigail takes out a loan of $1000.00 at 10% simple interest for 5 years. How much will she have to repay?

    A $5000.00   **B** $1500.00   **C** $2500.00   **D** $2000.00

21 Simplify $20x - 15y - 3(5x - 6y)$.

    A $5x + 3y$ **B** $35x - 21y$ **C** $5x - 33y$ **D** $5x + 9y$

22 Given that $1 * 4 = 10$ and $4 * 6 = 10$ then $p * q =$

    A $5q - 10p$ **B** $4q - 6p$ **C** $2q - \dfrac{1}{2}p$ **D** $3q - 2p$

23 Solve the simultaneous equations:

$x + 2y = 2$

$2x - 2y = 1$

    A $x = 1, y = \dfrac{1}{2}$ **B** $x = -1, y = 1\dfrac{1}{2}$ **C** $x = 2, y = 1$ **D** $x = 1, y = 1$

24 The area of this circle, in square units, is:

    A   147   **B**   154   **C**   196   **D**   148   7

25 A minor arc is subtended by an angle of 30° at the centre of a circle of diameter 20 cm. Find the length of the minor arc.

    A $\dfrac{10\pi}{3}$ **B** $1\dfrac{1}{2}\pi$ **C** $\dfrac{5\pi}{3}$ **D** 5π

26 The area of a rectangle is $30x^2$. If one side of the rectangle is $5x$, the perimeter is:

    A 20x   **B** 22x   **C** 11x   **D** 23x

27 The volume of a cube of side 1.1 cm is:

    A 13.31 cm³   **B** 1.331 cm³   **C** 0.1331 cm³   **D** 133.1 cm³

28 A cyclist travels 6 km in $1\frac{1}{2}$ hours, then a further 8 km in 2 hours. Find his average speed, in kilometres per hour.

    A 2 km/h   **B** 3 km/h   **C** 4 km/h   **D** 5 km/h

29 The circumference of a circular track is 400 m. What distance is covered by an athlete who runs straight across the track through the centre from one side to the other? Choose the closest answer.

    A 139 m   **B** 127 m   **C** 130 m   **D** 140 m

30 Find the area of a right-angled triangle in which the hypotenuse is $5x$ long and one other side is $3x$ long.

    A $15x^2$ **B** $7.5x^2$ **C** $10x^2$ **D** $6x^2$

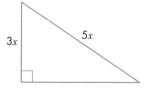

31 Given $5x - 3 \geqslant 2(x - 1)$, find the range of values for x.

    A $x \geqslant 3$ **B** $x \geqslant \dfrac{2}{7}$ **C** $x \geqslant \dfrac{1}{3}$ **D** $x \leqslant 3$

32 Simplify $\left(\dfrac{8}{27}\right)^{-\frac{2}{3}}$

    A $\dfrac{2}{3}$ **B** $\dfrac{3}{2}$ **C** $\dfrac{4}{9}$ **D** $\dfrac{9}{4}$

33 Convert 2.5 km to centimetres.

    A 2.5×10^5 cm   **B** 2500 cm   **C** 25 000 cm   **D** 250 cm

34 Mary leaves home for school at 07:00 and arrives at the bus stop at 07:35. The bus takes 20 minutes to reach the bus-stop that is closest to the school and Mary walks for another 5 minutes to get to school. If her average speed, including the bus ride, is 20 km/h, how far does Mary live from the school?

    A 20 km   **B** 50 km   **C** 35 km   **D** 40 km

35 Find the perimeter of this sector in terms of π.

    A 12π m   **B** 18 m   **C** 12π m   **D** $(12 + 2\pi)$ m

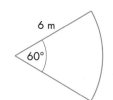

36 Work out the area of this shape.

    A 50 m²   **B** 65 m²   **C** 52 m²   **D** 60 m²

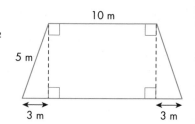

37 Find the mean of the data set 9, 8, 2, 1, 6, 3, 4, 1, 2, 2, 6.

    A 4   **B** 3   **C** 2   **D** 1

38 Given that the perimeter of an equilateral triangle is 12 cm, the height of the triangle, in surd form, is:

    A $3\sqrt{2}$ cm   **B** $2\sqrt{5}$ cm   **C** $5\sqrt{2}$ cm   **D** $2\sqrt{3}$ cm

39 A farmer selling his produce has on his table 20 apples, 120 bunches of grapes, 65 oranges, 100 bananas, 40 mangoes and 55 pears. What is the probability that the first fruit purchased randomly is an apple?

    A $\dfrac{1}{10}$ **B** $\dfrac{1}{20}$ **C** $\dfrac{1}{40}$ **D** $\dfrac{1}{4}$

40 Given the equation of a straight line is $5y - 10 + 15x = 0$, write down the gradient, m, and the y-intercept, c.

    A $m = 3, c = 10$ **B** $m = -3, c = 10$ **C** $m = -3, c = 2$ **D** $m = -15, c = -10$

41 These pairs of equations represent straight lines. Which pair are parallel?

    A $2x + y = 3$ **B** $2y + 4x = 6$ **C** $15y + 45x = 12$ **D** $x + y = 0$
       $5y = 4 - 10x$ $5x + 10y = 1$ $3y - 6x - 9 = 0$ $5x - 5y = 1$

42 What is the gradient of the line AB in this diagram?

    A $\dfrac{2}{5}$ **B** $-\dfrac{2}{5}$ **C** $\dfrac{3}{5}$ **D** $\dfrac{2}{3}$

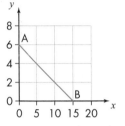

43 Given that $g(x) = \dfrac{2x - 4}{3x^3 + 5}$, find $g(-1)$.

    A -1 **B** $-\dfrac{1}{4}$ **C** -3 **D** $-\dfrac{3}{4}$

44 Which of these relations is many-to-one?

        A   **B**   **C**   **D**

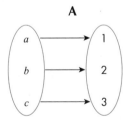 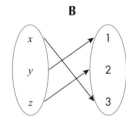 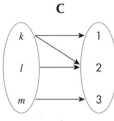 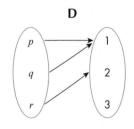

45 The sum of the exterior angles of a triangle is equal to:

    A $180°$ **B** $270°$ **C** $360°$ **D** $720°$

46 The table represents a function f.

x	1	2	3
f(x)	3	5	7

The function is:

    A $f(x) = x + 2$ **B** $f(x) = 4x - 1$ **C** $f(x) = 3x$ **D** $f(x) = 2x + 1$

47 A vehicle travels a distance of 5 km due north, from point A to B, then 8 km north-east to C, then 10 km due south to D. Which sketch best represents the vehicle's journey?

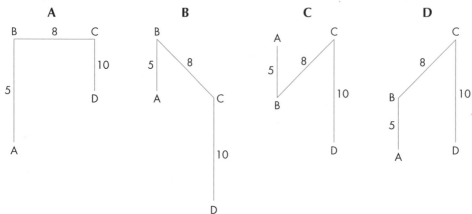

48 In the diagram, AB∥EC so angle ADC is equal to:

A 30°   **B** 60°   **C** 180°   **D** 150°

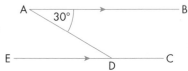

49 A quadratic function is expressed as $f(x) = -4x^2 - 3x - 1$. The graph of this function intercepts the y-axis at the point:

A (0, 1)   **B** (1, 0)   **C** (0, –1)   **D** (1, 1)

50 The individual letters of the word 'Mississippi' were placed in a bag and one was drawn at random. What Is the probability of drawing an s?

A $\dfrac{4}{11}$ **B** $\dfrac{2}{5}$ **C** $\dfrac{1}{3}$ **D** $\dfrac{4}{9}$

51 Fifty seedlings were measured to determine their growth rate after a specific period.

The table shows the results.

Height (cm)	1–5	6–10	11–15	16–20	21–25	26–30
Number of seedlings	5	10	16	9	6	4

The class boundaries of the first class are:

A 1 and 5   **B** 1.5 and 5.5   **C** 0.5 and 5.5   **D** –0.5 and 5.5

52 The pie chart shows the preferences of 720 students in a school for five different extra-curricular activities. The number of students who preferred singing is:

A 110   **B** 200   **C** 220   **D** 360

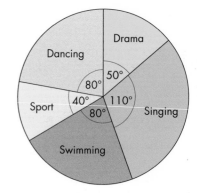

53 A'B' is the image of AB after a transformation.
What is the transformation?

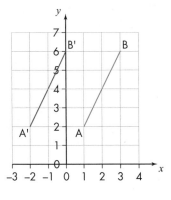

A reflection in the y-axis

B rotation about the point $(0, 2)$

C translation $\mathbf{T} = \begin{pmatrix} -3 \\ 0 \end{pmatrix}$

D reflection in the line $y = -1$

54 The triangle with vertices P(2, 1), Q(6, 1) and R(4, 4) undergoes an enlargement, centre (0, 0) and scale factor $k = 3$. What are the coordinates of the image P'?

A (8, 1)   **B** (4, 2)   **C** (6, 3)   **D** (6, 6)

55 In the right-angled triangle HJK, HK = 10 cm and JK = 8 cm.
What is the value of sin K?

A 0.8   **B** 0.6   **C** 0.75   **D** 1.25

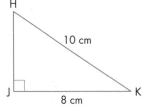

56 The volume of a solid cylinder of height 10 cm, with circular base of diameter 14 cm is:

A 140 cm³   **B** 21560 cm³   **C** 70π cm³   **D** 1540 cm³

Use this diagram to answer questions 57 and 58.

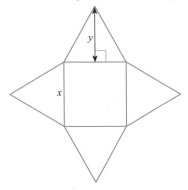

57 The diagram shows four equilateral triangles, each placed on a side of a square of side x cm.
The height of each triangle is y so the perimeter of the whole shape is:

A $4x$ cm   **B** $12x$ cm   **C** $8x$ cm   **D** $16x$ cm

58 The area of the shape is:

A $2xy + x^2$ **B** $x^2 + 4xy$ **C** $x^2 + 4y^2$ **D** $xy + x^2$

59 Find the volume of this prism, with the dimensions as shown.

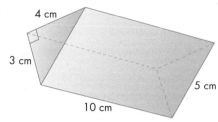

   A 100 cm³   **B** 75 cm³

   C 200 cm³   **D** 60 cm³

60 In this triangle (not drawn to scale), AC = 5, BC = 12, angle CAB = 80° and angle ABC = 40°.

The length of the side AB (y) can be found from the formula:

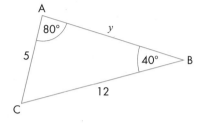

   A $\cos 40° = \dfrac{y}{12}$ **B** $12^2 = 5^2 + y^2$

   C $\sin 40° = \dfrac{5}{y}$ **D** $12^2 = 5^2 + y^2 - 2 \times 5 \times y \cos 80°$

Paper 2

(Duration 2 hours 40 minutes)

Section 1

Answer all questions in this section.

1 a i Using a calculator or otherwise, find the exact value of:

$$\frac{4\frac{1}{4} - 2\frac{1}{2} \times 1\frac{1}{2}}{2.56 \times 0.25 + 1.86}$$

(2 marks)

   ii Write the answer correct to 1 decimal place. (1 mark)

   iii Write the answer correct to 1 significant figure. (1 mark)

b A farmer purchases a plot of land priced at $20 000.00 by paying a deposit of 5% followed by monthly instalments of $1500.00 for 24 months.

   i What is the total amount to be repaid? (1 mark)

   ii What is the rate of interest paid? (1 mark)

c In a local school, electricity charges consist of a fixed fuel charge of 50 cents per kWh and an energy charge of 30 cents per kWh. The table gives the electricity consumption over the most recent billing period.

Meter reading (kWh)		Number of kWh used	Energy charges ($)	Fuel charges ($)
Last	Present			
20 452	44 632			

   i Find the total amount of kWh used. (1 mark)

   ii Determine the energy charge. (1 mark)

   iii Determine the fuel charge. (1 mark)

(Total 9 marks)

2 a Factorise these expressions.

   i $25x^2 - 10x + 1$ (2 marks)

   ii $81a^2 - 4b^2$ (2 marks)

   iii $3x^2y - 9xy^2$ (1 mark)

b Solve this equation.

$$\frac{5x + 2}{3} - \frac{3x - 1}{2} = \frac{3}{4}$$

(2 marks)

c Given the formula:

$$\sqrt{\frac{b^2 - a}{a + b}} = \frac{a}{x}$$

   i make x the subject of the formula (1 mark)

   ii find the value of x when $a = 1$ and $b = 2$. (1 mark)

(Total 9 marks)

3 a The universal set $\varepsilon = \{x: x \in \mathbb{W}, 1 \leqslant x < 20\}$ $E = \{$even numbers$\}$ and $F = \{$factors of 18$\}$.

    i List the elements of E.   (1 mark)

    ii List the elements of F.   (1 mark)

    iii Draw a Venn diagram to represent the universal set and the sets E and F.   (2 marks)

    iv List $E \cap F$.   (1 mark)

b i Using ruler and compasses only, construct a trapezium PQRS in which PQ = 8 cm, angle PQR = 120°, QR = 6 cm, RS = 12 cm and PQ is parallel to RS.   (2 marks)

    ii Measure the length of PS.   (1 mark)

    iii Measure angle QPS.   (1 mark)

(Total 9 marks)

4 a Given the functions $f(x) = 2x - 1$ and $g(x) = 4x + 1$:

    i evaluate $f(-1)$ (1 mark)

    ii evaluate $g^{-1}(-1)$ (2 marks)

    iii write down an expression for $fg(x)$ (1 mark)

    iv find $gf^{-1}(2)$.   (2 marks)

b On a Cartesian grid, plot the points A(1, 2) and B(6, 3). Find the equation of the line passing through AB.

Find the equation of the perpendicular bisector of AB.

Sketch AB and its perpendicular bisector on the same axes.   (3 marks)

(Total 9 marks)

5 a In the diagram (not drawn to scale), BC = 8 m, angle ADE = 70°, angle DAC = 35° and AB is parallel to DC.

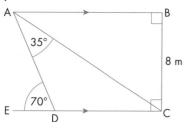

    i What sort of quadrilateral is ABCD?   (1 mark)

    ii What size is angle BAC?   (1 mark)

    iii What size is angle ACD?   (1 mark)

    iv Find the length of AC.   (1 mark)

    v Find the length of AD.   (1 mark)

    vi Calculate the perimeter of ABCD.   (1 mark)

b Triangle LMN, with vertices L(2, 6), M(5, 6) and N(5, 2) is mapped onto triangle L'M'N' by a

clockwise rotation of 90° about the origin. L'M'N' is now transformed by $\mathbf{T} = \begin{pmatrix} 5 \\ -6 \end{pmatrix}$ onto L"M"N".

Draw triangles LMN, L'M'N' and L"M"N" on a set of Cartesian axes.   (3 marks)

(Total 9 marks)

6 a The diagram shows the shape of a swimming pool, where OA and OB are radii, each of length 14 m.

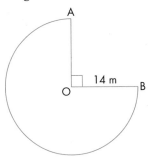

i   Use the information you have been given to find the surface area of the pool.   (2 marks)

ii Find the distance around the pool, walking from point O to A to B to O.   (2 marks)

iii Given that the depth of the water in the pool is 2.5 metres, calculate the volume of water it contains when completely filled.   (2 marks)

b A cylindrical drum with a base radius of 50 cm is $\frac{4}{5}$ full of water. The volume of water is 785 000 litres.

i   Find the height of water in the drum when it is $\frac{4}{5}$ full.   (1 mark)

ii Find the total height of the drum.   (1 mark)

iii Find the total capacity of the drum.   (1 mark)

(Total 9 marks)

7 The diagram shows the first three shapes in a sequence.

a Draw the fourth shape in the sequence.   (2 marks)

b Each row in this table relates to a specific shape in the sequence. The table displays a pattern of numbers.

Shape	Number of stars	Number of straight lines between the stars
1	3	1 × 3 = 3
2	5	2 × 3 = 6
3	7	3 × 3 = 9
i 4		
...		
ii		8 × 3 = 24
iii	27	
...		
iv 43		

Complete the rows numbered (i), (ii), (iii) and (iv).

(Total 10 marks)

Section 2

Answer all questions in this section.

Relations, functions and graphs

8 a A vehicle starts from rest and reaches a velocity of 40 km/h after 30 minutes. It maintains that velocity for $1\frac{1}{2}$ hours. The vehicle then increases its velocity to 60 km/h over a period of 30 minutes and continues at this new velocity for a further hour. Finally the vehicle slows to a stop over a period of 30 minutes.

    i Assuming that all changes in velocity take place at constant rates, draw a velocity-time graph to represent the motion of the vehicle. *(2 marks)*

    ii Find the acceleration of the vehicle, given by the gradient of the graph, during the first 30 minutes. *(1 mark)*

    iii During which period does the vehicle experience the greatest acceleration? *(1 mark)*

    iv Describe the motion of the vehicle during the last 30 minutes of the journey. Determine the gradient of the graph over this period and explain its significance. *(2 marks)*

    v Find the total distance travelled by the vehicle. *(2 marks)*

b Solve these simultaneous equations.

$$2xy + 2x + 4 = 0$$

$$y + 2x = 2$$

(4 marks)

(Total 12 marks)

Geometry and trigonometry

9 a A traveller sets off on a journey from village A towards another village B, 600 m away. The bearing of B from A is 060°. From B, he travels to another village C located 1 km east of B. Finally, he travels directly back to A.

    i Make a sketch of his journey, including all the distances and angles. *(2 marks)*

    ii Determine the bearing of C from A. *(3 marks)*

    iii Calculate the distance AC. *(2 marks)*

b In the diagram, O is the centre of the circle, angle QSU = angle QSR = 60° and RS is a tangent at S.

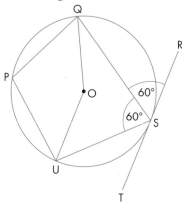

    i Calculate angle UPQ. *(2 marks)*

    ii Calculate angle UST. *(1 mark)*

    iii Calculate angle OQS. *(2 marks)*

(Total 12 marks)

Vectors and matrices

10 a The point P(2, 5) is mapped onto the point (*a*, *b*) by the transformation $\mathbf{T} = \begin{pmatrix} 2 & 0 \\ 0 & 2 \end{pmatrix}$.

  Find the values of *a* and *b* and describe fully the transformation **T**. (4 marks)

b Use a matrix method to solve these simultaneous equations.

$2x + 3y = 4$

$2x - 3y = 8$ (4 marks)

c In the diagram, $\overrightarrow{OP}$ and $\overrightarrow{OQ}$ are position vectors.

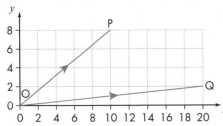

i Express both position vectors in the form $\begin{pmatrix} x \\ y \end{pmatrix}$. (1 mark)

ii Given that K is a point on OP such that OK = $\dfrac{1}{4}$OP, write position vector $\overrightarrow{QK}$ in the form $\begin{pmatrix} x \\ y \end{pmatrix}$. (1 mark)

iii Given that H is a point on OQ such that OH = $\dfrac{1}{2}$OQ, write position vector $\overrightarrow{QH}$ in the

  form $\begin{pmatrix} x \\ y \end{pmatrix}$. (1 mark)

iv Using the information given, express the vector $\overrightarrow{HK}$ in the form $\begin{pmatrix} x \\ y \end{pmatrix}$. (1 mark)

(Total 12 marks)

Examination 3

Paper 1

(Duration $1\frac{1}{2}$ hours)

Multiple choice questions

Each question has four suggested answers, A, B, C and D. Choose the correct one.

1 The number 5.98 correct to 1 decimal place is:

    A 5.1   **B** 6.1   **C** 6.0   **D** 5.9

2 The exact value of 0.098×0.05 is:

    A 0.05   **B** 0.005   **C** 0.49   **D** 0.0049

3 0.001 25 written in standard form is:

    A 0.125×10^2 **B** 1.25×10^2 **C** 1.25×10^{-3} **D** 1.25×10^{-2}

4 How many millimetres are there in 3.48 metres?

    A 348   **B** 34.8   **C** 3480   **D** 34800

5 Three children shared twenty-four plums in the ratio: 2 : 4 : 6. What was the largest share?

    A 8 plums   **B** 12 plums   **C** 24 plums   **D** 18 plums

6 A man donates 5% of his salary to charity. If he gives $200.00 to charity, what was his salary?

    A $4000.00   **B** $3800.00   **C** $2000.00   **D** $2800.00

7 The fraction $\frac{3}{8}$ written as a decimal is:

    A 0.35   **B** 0.375   **C** 0.75   **D** 2.67

8 What number is missing in the sequence: 0, 3, 8, …., 24?

    A 9   **B** 13   **C** 15   **D** 10

9 When a certain number is subtracted from $3\frac{7}{8}$ the result is $2\frac{1}{4}$. The number is:

    A $1\frac{5}{8}$ **B** $1\frac{3}{4}$ **C** $1\frac{2}{8}$ **D** $1\frac{1}{4}$

10 Simplify $\left(\frac{4}{9}\right)^{\frac{1}{2}} \times \frac{3}{4}$.

    A $\frac{1}{3}$ **B** $\frac{4}{27}$ **C** $\frac{2}{3}$ **D** $\frac{1}{2}$

11 The HCF of 2, 3 and 4 is:

    A 2   **B** 24   **C** 1   **D** 4

12 Find the exact value of $\dfrac{2.34 + 0.56 - 1.9}{1\frac{1}{2} \times \frac{1}{3}}$.

    A $\frac{1}{2}$ **B** 2   **C** $\frac{1}{4}$ **D** $\frac{1}{3}$

13 The largest prime number between 20 and 40 is:

    A 39   **B** 31   **C** 37   **D** 29

In this Venn diagram the universal set *U* represents the students in a class and the other sets represent their interests in different sporting events: athletics (*A*) and Basketball (*B*). Use it to answer questions 14 and 15.

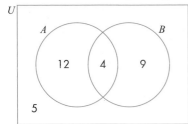

14 The number of students interested in only one of the two sporting events is:

    A 12   **B** 9   **C** 21   **D** 25

15 The total number of students in the class is:

    A 25   **B** 30   **C** 32   **D** 40

16 Matilda bought a book for $80.00 and sold it for $100.00.

    A Her percentage profit = 25%   **B** Her percentage loss = 20%

    C Her percentage profit = 20%   **D** Her percentage loss = 25%

17 How long will it take $2000.00, invested at a rate of 5 % per annum, to generate simple interest of $200.00?

    A 1 year   **B** 2 years   **C** 3 years   **D** 4 years

18 A woman selling newspapers is paid a commission of 25 cents on each newspaper sold. Given that each newspaper is sold for $2.50, find the total commission on sales of papers amounting to $150.00.

    A $25.00   **B** $20.00   **C** $15.00   **D** $60.00

19 Simplify the expression $3x - (-2)2y + (4)5x - (-1)3x$.

    A $11x - 4y$ **B** $26x - 4y$ **C** $26x + 4y$ **D** $23x + 7y$

20 Given that $(25)^2 = 625$, then $\sqrt{0.000\,625}$ is:

    A 0.025   **B** 0.25   **C** 0.0025   **D** 2.5

21 $\dfrac{12y \times 8}{4 + 6} =$

    A $4y$ **B** $\dfrac{48y}{5}$ **C** $12y$ **D** $\dfrac{6y + 4}{5}$

22 Solve the equation $4(3x - 1) + 5(4 - 2x) = 2(2 - x)$.

    A $x = 3$ **B** $x = -3$ **C** $x = 1$ **D** $x = 4$

23 Given that $x \in \mathbb{N}$, the solution set of $3x - 5 \leqslant 2(x - 2)$ is:

    A {0, 1}   **B** {0}   **C** {1}   **D** {0, 1, 2}

24 Given that $f(x) = \dfrac{4x - 1}{3}$, find f(–2).

    A 3   **B** 1   **C** $\dfrac{7}{3}$ **D** –3

25 $\dfrac{-3 + 2 \times 4}{-1(2 - 3) + 0 \div 4} =$

    A –5   **B** 5   **C** 4   **D** 20

26 Given that $h(x) = \dfrac{6x - 5}{2}$, find $h^{-1}(-2)$.

  A $\dfrac{1}{6}$ **B** $-\dfrac{17}{2}$ **C** $1\dfrac{1}{2}$ **D** $-1\dfrac{1}{2}$

27 Jane is $2x$ years old and her sister Beryl is 6 years younger. In 10 years' time their brother Thomas's age will be half the sum of the ages of Jane and Beryl. Given that Thomas is now 17 years old, what is Jane's age now?

  A 24   **B** 20   **C** 14   **D** 16

28 What is the gradient of the straight line $3x - 4y = 12$?

  A 3   **B** 4   **C** $\dfrac{3}{4}$ **D** $1\dfrac{1}{3}$

29 Given that a, b and c are constants and $a > 0$, then a possible equation for this graph is:

  A $y = ax^2 + bx + c$ **B** $y = -ax^2 + bx + c$
  C $y = ax^2 + bx$ **D** $y = ax^2 - bx + c$

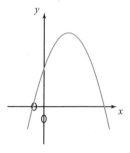

30 Find the volume of this cone.

  A 126 cm³   **B** 1386 cm³   **C** 462 cm³   **D** 693 cm³

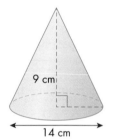

Use this diagram (not drawn to scale) to answer questions 31 and 32.

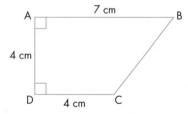

31 The area of ABCD is:

  A 16 cm²   **B** 28 cm²   **C** 22 cm²   **D** 44 cm²

32 The perimeter of ABCD is:

  A 15 cm   **B** 19 cm   **C** 22 cm   **D** 20 cm

33 In triangle PQR, sin P is:

  A $\dfrac{3}{4}$ **B** $\dfrac{4}{3}$ **C** $\dfrac{3}{5}$ **D** $\dfrac{4}{5}$

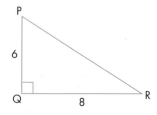

34 In the Venn diagram, the shaded area represents the set:

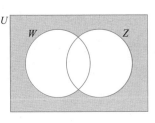

A $W \cup Z$ **B** $W' \cup Z$ **C** $(W \cap Z)'$ **D** $(W \cup Z)'$

35 Given $S = \{1, 2, 3, 4\}$, the number of subsets that can be made from S is:

A 4   **B** 8   **C** 16   **D** 32

36 Simplify $\dfrac{x+1}{y} - \dfrac{x-3}{2y}$.

A $\dfrac{xy + 5y}{2y}$ **B** $\dfrac{x+5}{2y}$ **C** $\dfrac{x-1}{2y}$ **D** $\dfrac{xy + 5y}{2y^2}$

37 Given that that $m * n = \dfrac{m+n}{m-n}$, work out $3 * 2$.

A 5   **B** $\dfrac{1}{5}$ **C** -5 **D** 1

38 In this circle, O is the centre of the circle and ABO is an equilateral triangle of sides 7 cm.

Taking π as $\dfrac{22}{7}$, the area of the shaded segment AB is approximately:

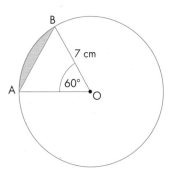

A 4 cm²   **B** 5 cm²   **C** 6 cm²   **D** 3 cm²

Use this triangle (not drawn to scale) to answer questions 39 and 40.

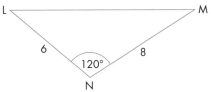

39 The side LM $= n$ can be calculated from the formula:

A $\dfrac{n}{\sin 120°} = \dfrac{8}{\sin L}$ **B** $LM^2 = 6^2 + 8^2$ **C** $n^2 = 6^2 + 8^2 - 2 \times 6 \times 8 \times \cos 120°$ **D** $\sin M = \dfrac{6}{n}$

40 The area of triangle LMN is:

A $\dfrac{1}{2} \times 6 \times 8$ **B** $\dfrac{1}{2} \times 6 \times 8 \times \sin 120°$ **C** $\dfrac{1}{2} \times 10 \times 8$ **D** $\dfrac{1}{2} \times 10 \times 6$

41 The number 1011_2 in base 10 is:

A 3   **B** 10   **C** 11   **D** 12

42 The image of the point P(2, 3) after an anticlockwise rotation of 90° about the origin is:

A $(-3, 2)$ **B** $(3, -2)$ **C** $(2, 3)$ **D** $(2, -3)$

The marks gained by 10 students in a test, marked out of 10 are 8, 9, 6, 3, 8, 9, 6, 5, 8, 1.

Use this information to answer questions 43–45.

43 The median mark is:

A 6   **B** 8   **C** 9   **D** 7

44 The modal mark is:

A 8   **B** 6   **C** 9   **D** 5

45 The mean mark is:

A 6.0   **B** 6.3   **C** 7.2   **D** 7.5

46 A square has rotational symmetry of order:

A 1   **B** 2   **C** 3   **D** 4

47 In the diagram, AB∥CD.

Angles X and Z are:

A complementary   **B** alternate

C vertically opposite   **D** supplementary

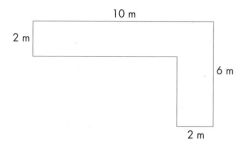

48 Given that $x = -2$, evaluate $2x^3 - x^2 + x$.

A −18   **B** 20   **C** −22   **D** −10

49 The shape in the diagram has dimensions as stated.

The perimeter is:

A 20 m   **B** 32 m

C 28 m   **D** 30 m

50 Find the radius of a sphere with volume 36π cm³.

A 6 cm   **B** 3 cm

C 6π cm   **D** 12 cm

51 The diagram (not drawn to scale) shows a cylinder of radius r and height 10 cm. Given that the volume of the cylinder is 1000 cm³, find the radius in terms of π.

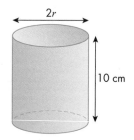

A $\dfrac{1000}{\pi}$ **B** $\dfrac{10}{\sqrt{\pi}}$ **C** 10π **D** 100π

52 Given that a varies inversely as the square of b and when $a = 2$, $b = -3$, find the value of a when $b = 5$.

A $\dfrac{50}{9}$ **B** $\dfrac{18}{25}$ **C** $\dfrac{3}{5}$ **D** $5\dfrac{1}{2}$

53 Simplify $\dfrac{5^x \left(25\right)^{2y}}{125^{\frac{1}{3}}}$ and write your answer as a power in base 5.

A 5^{2xy} **B** 5^{4xy-1} **C** 5^{x+4y-3} **D** 5^{x+4y-1}

54 $\dfrac{0.01}{10} =$

A 1% **B** 0.01% **C** 0.1% **D** 10%

55 Which shape is a quadrilateral with only two sides parallel?

A square **B** rectangle **C** parallelogram **D** trapezium

56 A school has a population of 300 girls, 400 boys, 60 teachers and 40 auxiliary staff members. What is the probability that if one member of the school population is chosen that person is not a student?

A $\dfrac{3}{40}$ **B** $\dfrac{1}{20}$ **C** $\dfrac{1}{8}$ **D** $\dfrac{7}{8}$

57 86 729 correct to 2 significant figures is:

A 86 729.00 **B** 87 **C** 87 000 **D** 867 000

This Venn diagram represents three different types of pets: cat (C), dog (D) and rabbit (R) owned by a number of households. Use it to answer questions 58 and 59.

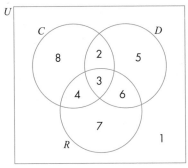

58 How many households owned at least two pets?

A 12 **B** 15 **C** 20 **D** 1

59 How many households owned only one pet?

A 20 **B** 1 **C** 3 **D** 32

60 What is the least number of grapes that can be shared equally among 12, 16 or 24 children?

A 4 **B** 48 **C** 96 **D** 144

Paper 2

(Duration 2 hours 40 minutes)

Section 1

Answer all questions in this section.

1 a i Using a calculator or otherwise. find the exact value of $\dfrac{0.028\,(7.3 + 1.5)}{0.09 \times 82 - 0.98}$.

(2 marks)

    ii Give your answer correct to 2 significant figures. (1 mark)

b A man wishes to fill a drum of diameter 0.8 m and height 1.8 m with water. He uses a cylindrical bucket of diameter 0.4 m and height 0.5 m. How many buckets of water will be required to fill the drum? (2 marks)

c Jane invests $2480.00 at 5% per annum simple interest.

    i Calculate the simple interest earned after $1\frac{1}{2}$ years. (1 mark)

    ii Find the total amount Jane has after five years. (1 mark)

    iii After how many years will the total amount invested reach $3720.00? (2 marks)

(Total 9 marks)

2 a i Solve the inequality $3(2x - 1) \leqslant 5(2 - x)$. (1 mark)

    ii Show the solution on a number line. (1 mark)

b Simplify the expression $\dfrac{2x - 1}{3x + 5} - \dfrac{x + 2}{3x - 5}$. (2 marks)

c Simplify the expression $\dfrac{2x^3 \times 4x^3 \times 6x^4}{48x^5 \times \frac{1}{4}x^3}$. (2 marks)

d Factorise the expression $2x^2 - 8x + 6$. (3 marks)

(Total 9 marks)

3 a 80 students were interviewed about their favourite colour(s), from red (R), white (W) and blue (B). The Venn diagram shows the results.

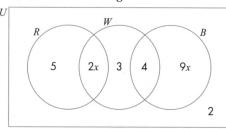

    i Find the value of x. (1 mark)

    ii How many students like red? (1 mark)

    iii How many students like white? (1 mark)

    iv How many students like blue? (1 mark)

    v How many students like two colours? (1 mark)

b i Using ruler and compasses only, construct parallelogram PQRS in which PQ = 6 cm, angle PQR = 120° and QR = 7 cm. (2 marks)

  ii Measure angle SPQ. (1 mark)

  iii Measure the diagonal PR. (1 mark)

(Total marks 9)

4 a This table displays the results of a survey that was done in a certain town, to assess the numbers of children in various families.

Number of children	1	2	3	4	5	6	7	8	9
Number of families	10	20	18	12	7	5	3	2	1

  i Draw up a frequency polygon to represent the information shown in the table. (2 marks)

  ii Find the median number of children in the families. (1 mark)

  iii Find the modal number of children in the families. (1 mark)

  iv Find the mean number children in the families. (1 mark)

  v What is the probability that a family chosen at random will have 5 children? (1 mark)

b The scores of 10 players in a cricket match are 10, 20, 40, 60, 90, 80, 1, 35, 55 and 0. Find the range, median and interquartile range of the scores. (3 marks)

(Total 9 marks)

5 a The diagram (not drawn to scale) shows the angles of depression from a point P on a wall to two points R and S on the level ground. These angles are given as 50° (P to R) and 30° (P to S).

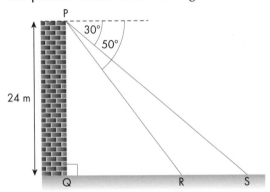

Given that the height of the wall is 24 m, find, correct to one decimal place:

  i the distance QS (1 mark)

  ii the distance QR (1 mark)

  iii angle PRS. (2 marks)

b In the graph, A'B'C'D' is the image of ABCD under a transformation **T**.

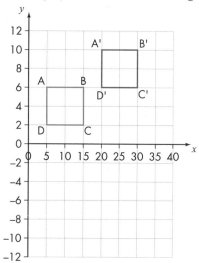

   i Write down the coordinates of the vertices of ABCD and A'B'C'D'.   (2 marks)

  ii Describe fully the transformation that maps ABCD onto A'B'C'D'.   (1 mark)

 iii Draw A"B"C"D", which is the image of ABCD after reflection in the *x*-axis.   (2 marks)

(Total 9 marks)

6 A group of young people was surveyed to determine the number of 'friends' they each had on Facebook. The table shows the results.

Number of friends on Facebook	Number of young people (frequency)
0–50	2
51–100	5
101–150	8
151–200	10
201–250	15
251–300	20
301–350	18
351–400	12
401–450	6
451–500	4

a Draw up a cumulative frequency table to represent the information shown.   (1 mark)

b Using an appropriate scale, draw a cumulative frequency curve.   (3 marks)

c Use your graph to estimate the median.   (2 marks)

d Find the number of young people with the greatest number of friends on Facebook.   (1 mark)

e What is the probability that a young person chosen at random will have 325 friends or more on Facebook?   (2 marks)

(Total 9 marks)

7 The diagram shows the first three shapes in a sequence. Each shape consists of a number of 2 cm × 3 cm rectangles.

a Draw shape 4 of the sequence. (3 marks)

The table shows the number of rectangles and the total area for each shape.

b Complete the table by inserting the missing values in rows numbered (i) to (iv).

Shape	Number of rectangles (R)	Total area of the shape (cm²)	
1	4	24	
2	7	42	
3	10	60	
i	4			3 marks
...			
ii	31		2 marks
...			
iii		276	2 marks
...			
iv	62		2 marks

(Total 12 marks)

Section 2

Answer all questions in this section.

Relations, functions and graphs

8 a i Using the information given in the diagram, find the equations of AD and BC.

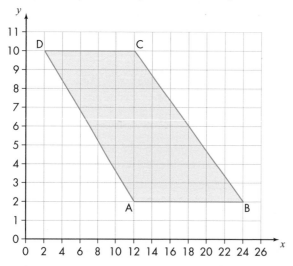

(4 marks)

ii State the four inequalities that represent the shaded area. (2 marks)

The function $F = 6x - 5y + 3$ satisfies the solution set represented by the shaded region.

iii Give the coordinates of the point where F has a maximum value, and where F has a minimum value. (2 marks)

b Given functions $f(x) = \dfrac{5x^2 - 3x}{4}$ and $g(x) = \dfrac{2x + 3}{4}$, find:

i $f(-1)$ (1 mark)

ii $g(-3)$ (1 mark)

iii $g^{-1}(2)$ (2 marks)

(Total 12 marks)

Geometry and trigonometry

9 a In the diagram (not drawn to scale), O is the centre of the circle, DE is a tangent to the circle at E and DB is a secant.

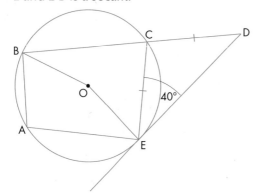

Given that DC = CE and angle DEC = 40°, calculate:

   i ∠BCE (2 marks)

   ii ∠BAE (1 mark)

   iii reflex angle BOE (2 marks)

   iv ∠OBC. (2 marks)

b Three vessels, A, B and C, are out at sea. Vessel A is due south of B and vessel C is on a bearing of 068° from B. The distance of A from B is 26 km and C is 42 km from B.

   i Draw a diagram indicating the positions of A, B and C, including all the given measurements. (1 mark)

   ii Find the distance AC. (2 marks)

   iii Find the bearing of A from C. (2 marks)

(Total 12 marks)

Vectors and matrices

10 a Points P, Q and R have coordinates: P(3, 7), Q(–4, 5) and R(4, –5).

   i Write position vectors $\overrightarrow{OP}$, $\overrightarrow{OQ}$ and $\overrightarrow{OR}$ in the form $\begin{pmatrix} x \\ y \end{pmatrix}$. (3 marks)

   ii Write $\overrightarrow{PQ}$ and $\overrightarrow{PR}$ in the form $\begin{pmatrix} x \\ y \end{pmatrix}$. (4 marks)

   iii What is the relationship between $\overrightarrow{OQ}$ and $\overrightarrow{OR}$? (1 mark)

b Given the matrices $\mathbf{H} = \begin{pmatrix} 2 & 3 \\ -1 & 4 \end{pmatrix}$, $\mathbf{K} = \begin{pmatrix} 5 & 1 \\ 6 & 2 \end{pmatrix}$ and $\mathbf{M} = \begin{pmatrix} 9 & 0 \\ 8 & 3 \end{pmatrix}$:

   i evaluate $\mathbf{HK} + \mathbf{M}$ (2 marks)

   ii find $\mathbf{M}^{-1}$. (2 marks)

(Total 12 marks)

12 Using your calculator

Using the calculator in the mathematics classroom

In this modern world almost everyone can have access to a calculator, either hand-held or built into your phone or computer.

The arrangement of the number keys is normally as shown in the photograph. It might be different or the same on your phone or computer.

The most important keys on the calculator

• The **addition** key – used for addition	+
• The **subtraction** key – used for subtracting	−
• The **multiplication** key – used to multiply	× ∗
• The **division** key – used for division and fraction to decimal conversion	÷
• The **equals** key – used to find the answer	=
• The **decimal point** key – used to place a decimal point	•
• The **+/− key** – used to change positive to negative and negative to positive	+/−
• The **percentage** key – divides the number by 100%	%

Additional keys on the scientific calculator

• The **inverse** key – changes the number to its reciprocal (inverts the number or places 1 over the number)	$1/x$
• The **square, cube, square root and cube root** keys – inputs the square, cube, square root or cube root of a number	x^2 x^3 $\sqrt{}$ $\sqrt[3]{}$
• The **exponent** key – allows input of numbers in scientific notation	10^x
• The **power** key – expresses the input number to a specific power, for example, $2^9 = 512$	x^y
• The **nth root** key – expresses the nth root of the input number	$\sqrt[n]{\square}$
• The **fraction key** – used to input fractions, such as $2\frac{1}{3}$	$a\,^b/_c$

In addition to these keys, which are generally the most frequently used, there are other important keys on the scientific calculator that are used in more advanced mathematical calculations.

• The **trigonometrical functions** keys – for inputting sine, cosine, tangent, $\sin^{-1}$, $\cos^{-1}$, $\tan^{-1}$, …	sin cos tan $\sin^{-1}$ $\cos^{-1}$ $\tan^{-1}$
• The **logarithm** keys – for inputting logarithms to the base e (ln) and to the base 10 (log)	ln log
• The **parentheses** keys – facilitate BOMDAS type questions that contain a string of different operations and brackets, for example, simplify: $3 - 4\,(5 \times 2) \div 8 + 7$	( )
• The π key – inserts the value of π	π
• The **shift** or **2nd function** – enables additional functions such as $\sin^{-1}$, and is situated at the top of the calculator, usually in a colour different from the normal keys	SHIFT
• The **mode** – changes between radians and degrees and other functions such as standard form	MODE
• The **nCr and nPr** keys – calculates combinations and permutations	nCr nPr
• The **factorial** key, calculates the product $x(x-1)(x-2)(x-3)\ldots$	$x!$

Examples of using a calculator for calculations and problem-solving

It is very important that you master the concepts and procedures before engaging in extensive calculator practice. The examples shown on the next five pages are intended to introduce you to the correct procedures. Of course, for multiple choice examinations, where calculators are not allowed, it is imperative that you know how to solve the problems that you will encounter without resorting to a calculator.

Number theory and computation

This is the topic that best lends itself to the use of calculators.

The four basic operations of addition, subtraction, multiplication and division are quite straightforward and are performed by pressing the appropriate keys.

For example: $15 + 23 \Rightarrow$ [1] [5] [+] [2] [3] [=]

Answer: 38

Combined operations can be completed by pressing the keys in the same order as the digits and operators appear in the calculation.

For example: $-1 + 3 \times 14 \div 7 - 8 \Rightarrow$ [1] [±] [+] [3] [×] [1] [4] [÷] [7] [−] [8] [=]

Answer: −3

BOMDAS type questions are done in a similar way. Great care must be exercised when using the open and close parentheses keys to ensure that for each open parenthesis there is a closed one.

For example: $12 \div 6(8 \times 2^{10}) \Rightarrow$ [1] [2] [÷] [6] [(] [8] [×] [2] [x^y] [1] [0] [)] [=]

Answer: 2.4×10^{-4}

If there are more than two sets of brackets, the number of opening brackets must be equal to the number of closing brackets.

For example: [2] [5] [(] [−] [3] [+] [6] [)] [+] [(] [9] [÷] [3] [)] [−] [1] [×] [4] [+] [8] [=]

Using the calculator with fractions

Some calculators have a fraction key [$a^{b}/_c$] while others treat fractions like decimals.

For example, for a calculator without a fraction key:

$$\frac{2}{5} + \frac{4}{5} \Rightarrow$$ [2] [÷] [5] [+] [4] [÷] [5] [=] 1.2

or [0] [.] [4] [+] [0] [.] [8] [=] 1.2

Answer: 1.2

The fraction operation is converted to a decimal operation.

For example, for a calculator with fraction key:

$$\frac{2}{5} + \frac{4}{5} \Rightarrow$$ [2] [$a^{b}/_c$] [5] [+] [4] [$a^{b}/_c$] [5] [=] $1\lrcorner1\lrcorner5$

Answer: This display is read as $1\frac{1}{5}$.

To express a mixed number, add the whole number part to the fractional part.

Using the calculator properly, it is possible to solve fraction problems as complex as:

$$2\frac{3}{4} + 1\frac{1}{2}\left(\frac{4}{5} - \frac{3}{8}\right)$$

$\Rightarrow$ [2] [+] [3] [a^b/c] [4] [+] [(] [1] [+] [1] [a^b/c] [2] [)] [×] [(] [4] [a^b/c] [5] [−] [3] [a^b/c] [8] [)] [=]

$3\lrcorner3\,1\lrcorner80$

The answer is $3\frac{31}{80}$

Using the calculator with decimals

Using the calculator to carry out mathematical operations with decimals is quite straightforward but requires the use of the decimal key.

For example: $2.3 \times 4.5 \Rightarrow$ [2] [•] [3] [×] [4] [•] [5] [=] 10.35

Answer: 10.35

Note: It is sometimes appropriate to omit leading zeros, for example, to input 0.324 you can key [•] [3] [2] [4], but to input 5.02 it is essential to key [5] [•] [0] [2], since the 0 is a place holder.

Using the calculator with percentages

This is where the percentage key [%] is used.

For example: to find 5% of $20.00 $\Rightarrow$ [2] [0] [×] [5] [SHIFT] [%] [=] 1

Answer: $1.00

Note that for this particular calculator, you input the number ($20) first and press the [%] key last.

To find percentage change (increase or decrease) after the percentage is calculated add it to (or subtract it from) the original amount.

For calculators without a [%] key, you can use decimals to find percentages since n% of an amount is the same as $\frac{n}{100}$ of it.

For example: to find 15% of $60.00 $\Rightarrow$ [6] [0] [×] [1] [5] [÷] [1] [0] [0] [=] 9

or [6] [0] [×] [•] [1] [5] [=] 9

Answer: $9.00

Some calculators have a special feature that allows you to calculate the percentage change in one single operation, for example, to add a 10% sales tax on $120.00, the operation will be [1] [2] [0] [+] [1] [0] [%].

Using the calculator with standard form, significant figures, exponents or powers, squares and square roots

Different calculators may have slightly different procedures. Taking as an example the CASIO fx–260 solar, which is quite popular, you enter **scientific notation and significant figures** by pressing [MODE] followed by [8] and entering the number of significant figures required.

For example: to write the number 0.003 59 in standard form, correct to 3 significant figures, input the number 0.00359, press [MODE] followed by [8], then [3].

Answer: 3.59×10^{-3}

Note: On the CASIO fx–260 solar, the [ON] key is used to switch between scientific and normal data entry.

You can use [EXP] key to enter data in **exponential form**.

For example: to calculate $2.5 \times 10^9 \times 8 \times 10^7$ on the scientific calculator

$\Rightarrow$ [2] [•] [5] [EXP] [9] [×] [8] [EXP] [7] [=] 2^{17}

Answer: 2^{17}

Operations can also be carried out between data in exponential and ordinary form, such as $9.2 \times 10^6 + 938\,936 = 10\,138\,936$.

For example: to calculate $8 \times 10^2 + 200$ $\Rightarrow$ [8] [EXP] [2] [+] [2] [0] [0] [=] 1000

Answer: 1000

You can find the **square** of any number by pressing [x^2] after the number.

For example: to calculate $(2.3)^2$ $\Rightarrow$ [2] [•] [3] [x^2] [=] 5.29

Answer: 5.29

You can **raise a number to any power** by using the [x^y] key.

For example: to calculate 3^{13} $\Rightarrow$ [3] [x^y] [1] [3] [=] 1.59×10^6

Answer: 1.59×10^6

To find the **square root**, use [SHIFT] the [√] key.

For example: to calculate the square root of 6.25 $\Rightarrow$ [6] [•] [2] [5] [√] [=] 2.5

Answer: 2.5

Using the calculator in trigonometry

You can find the trigonometrical functions sine, cosine and tangent on the calculator. Always take care, however, that the calculator is in the right mode.

One way to ensure that the calculator is in the correct mode is to find the sine, cosine or tangent of a known angle, such as sin 30° = 0.5 or cos 60° = 0.5.

The procedure (using a calculator) to find cos 60° may be: [cos] [6] [0] [=] or [6] [0] [cos] [=] 0.5

Once the ratio is known, the angle can also be found by using the arccos, arcsin and arctan functions.

For example: given cos $\theta = 0.7$, then
$\theta = \cos^{-1} 0.7$ so $\theta = 45.6°$.

On the calculator: [•] [7] [cos⁻¹] [=] 45.6

Answer: $\theta = 45.6°$

> **Note**
>
> If the calculator has a [cos⁻¹] key, there is no need to press the [SHIFT] key.

To convert from degrees to radians and vice versa (on the CASIO fx-260), use the [SHIFT] and [MODE] keys.

For example: to convert 180° to radians, begin in degree mode

$\Rightarrow$ [1] [8] [0] [SHIFT] [MODE] [5] [=] 3.142

Answer: 3.142 (or π) radians

For example: to convert 1.5π radians to degrees, begin in radian mode

$\Rightarrow$ [1] [•] [5] [×] [π] [SHIFT] [MODE] [4] [=] 270

Answer: 270°

Note that with some calculators [sin], [cos] and [tan] are pressed before the numbers representing the angles.

In trigonometrical and many other types of computation it is good practice to estimate the answer before calculating, in case an error arises due to incorrect data entry or pressing the incorrect key, resulting in a major deviation from the estimated value.

Using the calculator in statistics

The factorial function ($n!$) as well as the combination (nCr) and permutation (nPr) functions have applications in statistics. These can be used to help determine the number of combinations and permutations (ways of arranging objects).

For example: $_5C_2 \Rightarrow$ `5` `nCr` `2` `=` 10

$_5P_2 \Rightarrow$ `5` `nPr` `2` `=` 20

$8! \Rightarrow$ `8` `SHIFT` `x!` `=` 40320

Calculators with the random `RAND` key can be used to generate random decimal and whole numbers, simulating the numbers obtained by throwing dice, drawing a card from a pack or tossing a coin, presenting data for experimental purposes.

For example: `SHIFT` `RAND` `1`   or `SHIFT` `RAND` `0`

More advanced uses of the calculator

Solution of complex equations and functions

Calculators, especially those with more advanced features, can be used to solve equations consisting of several levels of brackets and complex functions.

Some advanced calculators contain a `MENU` key which when pressed, reveals different types and categories of equation. For example:

- equations involving many brackets and operations, such as $2 \times 3^4 - (5 \, (8 \div 4) + 6^{-3}) - 9!$
- complex functions such as factorials and negative fractional indices.

Drawing graphs based on equations

Such a calculator will have a `y` button near the top of the key pad.

The graphing procedure

- Press the `y` button, to reveal different y-values representing different graphs, such as Y1, Y2, and Y3.
- Key in the equation and press enter – the equation will appear to the right of Y1.
- For the x-part of the equation, press the `x`, `T`, `n` or `-` key.
- Press the `Graph` key – the graph will appear on the screen.

Representing coordinates of a point in space in different forms

Cartesian coordinates expressed in the form, P(x, y) can be rewritten as polar coordinates in the form (r, θ) by pressing `→rθ`.

Polar coordinates in the form (r, θ) are converted to Cartesian coordinates (x, y) by pressing `→xy`.

`←··→` displays (r, θ) and (x, y).

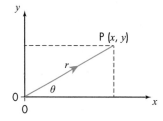

For example: to write the Cartesian coordinates P(x, y) as polar coordinates

`8` `SHIFT` `2` `SHIFT` `→rθ` $r=$ `8` `·` `2`

`SHIFT` `←··→` $\theta=$ `1` `4`

Using the calculator with number bases

More advanced calculators can convert between numbers in binary, pental, octal, decimal and hexadecimal. The four basic operations can also be performed between these bases.

The logical operations AND, OR, NOT, NEG, XOR, and XNOR can also be carried out on binary, pental, octal, and hexadecimal numbers.

Pressing these keys perform the following conversions.

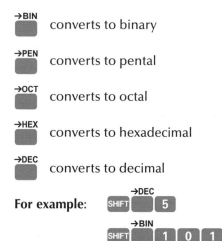

→BIN　converts to binary

→PEN　converts to pental

→OCT　converts to octal

→HEX　converts to hexadecimal

→DEC　converts to decimal

For example:　SHIFT →DEC 5

SHIFT →BIN 1 0 1

Analysing statistical data

Some advanced scientific calculators have a statistics function, which allows data to be entered for statistical calculations and data to be cleared and corrections carried out.

For example, a table of values as shown below can be entered for analysis:

Number	1	2	3	4	5	6	7	8	9
Marks	10	20	30	40	50	60	70	80	90
Number of students	3	5	9	16	12	11	10	4	

Keys:

DATA　　CD

Data　Clears data

Mathematical exercises, problems and investigations

One of the most effective teaching and learning approaches in mathematics is to allow students to experience a wide variety of mathematical activities. These may include paper and pencil exercises in the classroom, mathematical discussions, practical work such as model-making and even song, drama and dance.

Generally, such learning activities can be classified into three types:

- exercises
- problems
- investigations.

A **mathematical exercise** is a repetitive task that is understood by the student and for which there exists a known procedure or algorithm for carrying it out.

Example

Find the area of a rectangle of length 6 m and width 4 m.

$A = l \times w$

$\quad = 6 \times 4$

$\quad = 24 \text{ m}^2$

The ability to perform mathematical exercises may be developed after plenty of practice. The exercises are excellent for sharpening basic computation skills in addition, subtraction, multiplication and division.

In a **mathematical problem**, students know what is being asked and understand the task, but they do not have a direct way of doing it. Mathematical problems are therefore more difficult than exercises and some creative skills are normally required to solve them.

Example

A man is now twice as old as his son. Six years ago the man was 40 years old; how old is his son now?

The man's present age is $40 + 6 = 46$.

The son's age is $\frac{1}{2} \times 46 = 23$.

Mathematical problems are used at two crucial stages in the teaching and learning of mathematics:

- at the beginning of a lesson to help develop the understanding of a concept
- at the end of the lesson as a means of assessment and reinforcement of the concept.

These problems help develop critical thinking skills and students learn how to analyse, generalise, recognise patterns and make conjectures.

The most important aspect of learning to solve problems is that the skills and procedures acquired are used to solve problems encountered in everyday life.

In a **mathematcial investigation**, students do not know what is being asked (they do not understand the task) and do not have a way of finding a solution.

For example, students may be presented with a figure, such as this compound shape, and asked to investigate it.

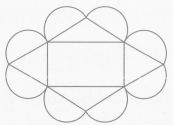

Students can then choose their own problem, based on the shape of the figure.

Some may choose to describe the shape or shapes, find areas or perimeters or seek some practical application. The student can thus create their own problem and present their solution to the problem.

If they are given a sequence of shapes, as shown here, they can investigate the shape of the *n*th shape as well as the number of lines making up the *n*th shape.

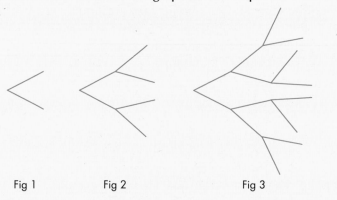

| Fig 1 | Fig 2 | Fig 3 | Fig *n* |

In some cases they could also find the perimeter and area of the *n*th shape.

Mathematical investigations are normally carried out over a longer period of time than exercises or problems and findings can be presented in a report form – oral or written.

Each of the mathematical tasks described (exercises, problems and investigations) plays an important role in students' mathematical development. The exercises, as mentioned before, provide a good starting place, introducing inexperienced students to basic computation skills and laying the foundation for more challenging mathematics to come.

The mathematics problems help develop higher-order thinking skills and the investigations assist students not only to improve their problem-solving skills but also to create and develop their own mathematical understanding.

An interesting observation is that any one investigation can be used to reinforce all three mathematical tasks, depending on the level of the questions asked.

In the sequence of shapes, for example, simple exercises can be found, such as the number of lines in the first three shapes. One problem based on the sequence could be a requirement to draw the fourth or *n*th figure in the sequence and students can be challenged to investigate any aspect of the sequence.

Exercises

An exercise may be described as a routine problem, The solutions are generally straightforward, requiring the use of basic computation skills and familiar algorithms.

Routine examples of exercises

Here are some routine examples of exercises.

1 $45 \div 5$

2 13×7

3 $235 + 17$

4 $24 - 13$

5 If one box weighs 12 kg, find the mass of 14 boxes.

6 18 loaves of bread together cost $54.00. What is the price of one loaf?

7 $72.00 is shared equally among 8 boys. How much does each receive?

8 There are 29 ripe mangoes on a tree. Then some boys picked 14. How many mangoes remained on the tree?

9 One man performs a job in 25 days. In how many days will five men of equal ability together perform the same job?

10 A single piece of string is cut into smaller pieces of lengths 12 cm, 14 cm, 22 cm, 8 cm and 19 cm. What was the length of the original piece?

11 Simplify $\frac{3}{4} + \frac{1}{4} + 1\frac{1}{2}$.

12 Work out $\frac{7}{8} - \frac{1}{2}$.

13 Work out $\frac{2}{5} \times 1\frac{1}{5}$.

14 Work out $4\frac{1}{2} \div \frac{3}{4}$.

15 Find 20% of 40.

16 Three girls are aged 14 years, 12 years and 7 years. What is their mean age?

17 Work out:
   a $123.78 + 82.93$ b $11.7 - 8.8$ c 2.35×1.2 d $12.48 \div 1.2$.

18 Work out:
   a $(12)^2$ b $\sqrt{81}$.

19 Share 15 plums in the ratio 2 : 3.

20 Given the universal set $U = \{$whole numbers up 20$\}$, write down:
   a {even numbers}   b {odd numbers}   c {prime numbers}.

21 Find the LCM of 12, 9 and 24.

22 Find the HCF of 48, 84 and 126.

23 Find the area and perimeter of each shape.

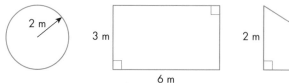

Problems: examples, strategies and solutions

Unlike simple, one-step routine problems, non-routine problems are not generally solved by simple calculations or estimation: they are usually solved in two or more steps. Not only are there different problem-solving strategies but one problem may present more than one solution.

Solving non-routine problems is beneficial to students in many ways.

- They become familiar with the problem-solving process and the various strategies that can be used to solve everyday problems.
- Through problem-solving, students are also able to practise their basic computation skills.
- With teacher guidance and modelling, students are encouraged and build confidence in the process.
- They are also trained in the step-by-step approach that is also vital in the solution of everyday problems.

One important aspect of problem solving is group work; it is recommended that two or more students work together to discuss, explain, receive feedback and encourage one another.

Non-routine problems

Here are some examples of non-routine problems.

1 Write down all the words you can think of that describe:

   a a trapezium   **b** a circle.

2 A man's present age is a multiple of 6, in two years' time it will be a multiple of 11. What is his present age?

3 Write down all the numbers that can be made from the digits 3, 4, 5 and 6.

4 Suppose the letters of the alphabet, from a to z, are worth $1.00 to $26.00 respectively.

   a How much is the name 'Jonathan' worth? What is your name worth?

   b Find a name that is worth $20.00.

5 Joanna decides to do her exercises by running up and down a flight of 30 steps. Starting from the first step she runs up half the flight then returns to the first step. She then runs right to the top of the flight then runs down to step number 5, after which she runs up to step 28. How many steps did she run up and down, during her exercise routine?

6 Two adult tickets and 3 child tickets for a movie cost $33.00. The price of one child ticket is $\frac{1}{4}$ the price of an adult ticket. The total money taken for a showing was $480.00 and 64 children attended the movie. How many adults attended the movie?

7 A ladder with rungs 30 cm apart is 24 times as long as it is wide. If the ladder has a total of 20 rungs, what is its width?

8 A total of 124 students need to be transported to a function by buses. If each bus can hold 18 students how many buses are needed?

The problem-solving process

The problem-solving process consists of three fundamental steps:

- understanding the problem
- solving the problem
- checking the answer.

Understanding the problem is crucial if there is to be any progress in solving it. Students must read or listen to the problem very carefully, to ensure that the task is clear. It is often necessary for them to listen to or read the problem more than once, reflect or think deeply about it, discuss it with others and even rewrite it in their own words, using symbols and diagrams.

Solving the problem involves the application of different strategies, choosing those that are most effective. These strategies include:

- using concrete materials or manipulatives
- using diagrams or sketches to help visualise the problem
- 'guessing and checking' – making assumptions then using data to prove whether you are right or wrong
- organising the data into lists, tables, charts, graphs – thus bringing clarification and order
- looking for a pattern – it is much easier to detect patterns, whether numerical or geometrical, in well-organised data
- dividing up the problem into smaller parts, for example, wages received at the end of a week could be broken up into daily wages received
- simplifying the problem by using the same algorithm or procedure with smaller, more manageable numbers
- working backwards, by applying the steps but in the reverse order; operations are also reversed.

Checking back, or **checking answers**, entails reading the question again, to ensure that the actual question asked has been answered, and double-checking calculations and answers to make sure that the answers make sense.

Example

Sandra thinks of a number, multiplies that number by 4, subtracts 2 from the result, adds 20 and then divides by 3. She ends up with the number 7. What was the number she thought of?

Use the strategy of working backwards.

Begin with the number she ends up with and work backwards to the number she thought of; the **operations are reversed** in the process.

She ends up with 7

$\Rightarrow$ the number before was $7 \times 3 = 21$

$\Rightarrow$ the number before adding 20 was $21 - 20 = 1$

$\Rightarrow 2 + 1 = 3$

$\Rightarrow 3 \div 4 = \dfrac{3}{4}$

The number she thought of was $\dfrac{3}{4}$.

Math Investigations give students opportunities to solve open-ended problems, create their own learning and make their own mathematical discoveries. Through this process, students are able to understand and retain the information much more easily.

Key stages in the investigation process

These include:

- exploration – students explore various aspects of the area under investigation to gather useful information
- collection of data – the information obtained is organised and recorded
- examination of the data to detect patterns
- making guesses (conjectures) and testing them to obtain information that will be used to make important predictions – there are times when it will be necessary to make amendments and retest
- discussion and explanation of results
- generalising – applying the solution to other cases
- presenting a conclusion based on findings.

Investigation tasks are therefore very diverse and varied. Students can be presented with a complicated figure, as discussed above, a practical problem, a mathematical statement or a numerical, algebraic or geometric sequence – or even a social issue. Students may also be given the option to devise their own investigations. For each investigation (open-ended or restricted), a report can be submitted, based on the precedures during the investigative process.

Number sequences that can be investigated

Arithmetic sequences

An arithmetic sequence is formed by repeatedly adding the same value to the previous value.

The constant value added is the **common difference**.

For example, for the sequence 5, 9, 13, 17, …, the common difference is 4 and the next three terms in the sequence are 21, 25, 29.

The common difference may also be negative.

For example, for the sequence 12, 9, 6, 3, …, the common difference is –3 and the next three terms in the sequence are 0, –3, –6.

Geometric sequences

A geometric sequence is formed by multiplying each consecutive term by the same value. The value in this case is the **common ratio**.

For example, for the sequence 1, 2, 4, 8, 16, … the common ratio is $r = 2$ and the next three terms are 32, 64, 128.

The common ratio may also be negative or less than 1.

For example, for the sequence 48, 24, 12, 6, … the common ratio is $r = \dfrac{1}{2}$ and the next three terms are 3, 1.5, 0.75.

Special number sequences

The triangular numbers

This sequence may be represented by dots arranged in the shape of triangles.

1   3   6

The triangular number sequence is 1, 3, 6, 10, 15, ...

The square numbers

This sequence may be represented by dots arranged in the shape of squares, or by simply squaring consecutive whole numbers, $(0)^2$, $(1)^2$, $(2)^2$, $(3)^2$.

The sequence of square numbers is 0, 1, 4, 9, ...

The cube numbers

The sequence of cube numbers is formed by cubing consecutive numbers.

The sequence of cube numbers is 1, 8, 27, 64, ...

The Fibonacci numbers

This sequence is formed by consistently adding the two numbers immediately before each term: 0, 1, 1, 2, 3, 5, 8, ...

Other sequences

Students may even come up with their own sequences, for example, 1, 3, 2, 6, 4, 9 ...

Sequences based on geometric shapes and figures

Sequences can also be formed from arrangements of shapes and lines, for example,

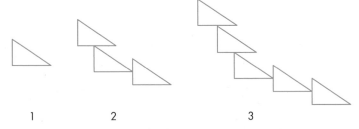

1   2   3

In each case, the investigation may be open-ended or restricted by specific questions. For example, for the series af shapes, the instructions could be:

- draw the next shape (shape 4) in the sequence
- count the triangles or lines in a particular term
- find the area or perimeter of a particular (or *n*th) term.

Alternatively students may simply be instructed to 'investigate' the sequence, and thus they will describe their own findings.

Examples of investigations

1 Investigate the arithmetic sequence 1, 8, 15, 22, …

2 Investigate the geometric sequence 80, –20, 5, …

3 Investigate the sequence 1, 2, 4, 8, 32, …

4 Create your own number sequence and investigate it.

5 Investigate Pythagoras' theorem.

6 Investigate the relation between the horizontal distance covered and the height of a projectile.

7 Investigate the difference in perimeters of successive squares: 1^2, 2^2.

8 Investigate the difference in volumes of successive cubes: 1^3, 2^3.

9 Investigate particular shapes with the same perimeter and area.

10 Investigate the addition of all the digits of consecutive even numbers.

11 Investigate the sequence formed by these coins.

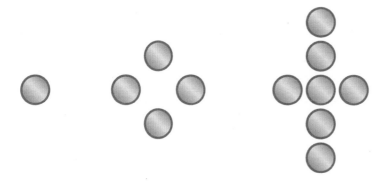

Continue the sequence and try to derive a formula.

12 Investigate the number that appears the fewest times when a dice is thrown.

13 Investigate the number appearing most times when a dice is thrown.

14 Given any four-digit number, rearrange the digits to make the biggest number and rearrange to make the smallest number, subtract the smallest from the biggest. Investigate.

15 Investigate the relationship between the perimeter and area of the same right-angled triangle.

16 Investigate the relationship between the sum and product of the sides of the same triangle.

17 Investigate the relationship between the perimeter and area of a square.

18 Investigate the difference between the perimeters of a square and a rectangle of the same area.

19 Investigate the statement: 'The attitude of the mathematics teacher and student greatly affect the teaching and learning of mathematics.'

20 Investigate the statement: 'Drinking cold water in the morning is more effective at weight loss than drinking hot water.'

Example of an open-ended investigation

Open-ended investigation: *The differences in the areas of successive squares*

Exploration

Areas of successive squares are checked, starting from the first whole-number square.

- side = 1, area = 1 = 1^2
- side = 2, area = 4 = 2^2

The difference is the result of subtracting the smaller from the larger (4 – 1).

These values represent the data to be analysed for patterns.

Collection of data

A table is drawn up, recording a list of successive numbers and the corresponding square numbers, starting from 1, and the differences between consecutive square numbers.

Consecutive whole numbers	Consecutive square numbers	Difference between successive squares
1, 2, 3, 4, 5, 6, 7, 8, 9, 10	1, 4, 9, 16, 25, 36, 49, 64, 81, 100	3, 5, 7, 9, 11, 13, 15, 17, 19

Examination of the data to detect patterns

The data in column 3 is carefully observed and a pattern is observed.

The differences between successive squares produces a sequence with a common difference of 2.

Making guesses and testing them

It is safe to make a conjecture that the differences between successive squares result in an arithmetic progression in which any term can be found through the formula $a + (n – 1)d$, where a is the first term, d is the common differnce and n is the term number.

To test this, the ninth term can be found as:

$$a + (n – 1)d = 3 + (9 – 1)2$$
$$= 3 + 16$$
$$= 19$$

Conclusion

The conclusion of this investigation is pretty straightforward. When successive squares are subtracted (the smaller from the larger) the result is an arithmetic progression.

Unlike a more complex and challenging investigation, there is not much scope for discussion or generalisation.

Directed investigation

A directed investigation should result in a report explaining the entire investigation process, consisting of three parts: **Introduction**, **Mathematical process** and **Conclusion**. A directed investigation is not open-ended but presents a specific problem to be solved.

- **The introduction** presents an outline of the problem, its objectives and the strategies implemented to accomplish these objectives. Strategies will include mathematical processes, technology and data-gathering methods and instruments, and ways of recording, such as tables and graphs.

- **The mathematical process** or middle part of the report lays out in detail the strategies, methods, instruments and mathematical calculations used to acquire and manipulate data appropriately to solve the problem.

- **The conclusion** presents a summary of the main findings of the second part of the report (mathematical process) and a suggestion for how these findings can be used in other cases (generalisations).

The use of calculators

Contrary to the beliefs of some, the use of calculators in the classroom can actually enhance students' mathematics learning rather than detract from it. Students become better at problem solving which is one of the main objectives of learning mathematics.

In support of the use of calculators

Here are some of the ways use of calculators can actually contribute to the teaching and learning of mathematics.

- **It can reduce the time spent doing mathematics.**
  Some basic computation, such as basic operations, percentages, averages, especially involving big numbers, can be performed much faster with the use of calculators.

- **It can assist students in developing certain mathematical concepts.**
  Concepts such as long division, decimal places and square roots can become quite tedious without the use of calculators, especially when decimals are involved. For example, with the use of a calculator, the square root of 7.84 is quickly found to be 2.8. Without the use of the calculator it is much more difficult.

- **It plays an essential role in drill and practice.**
  Students can quickly check the accuracy of their working and answers, especially when they need to complete a number of exercises in a short time. This can actually help students to develop concepts quickly and become more confident in their mathematical ability as they have more time to practise.

- **It contributes to growth in mental computation.**
  Students' mental mathematical ability can be tested against the speed of the calculator. It might be much easier to compute 800×40 than 842×24 but, with practice, assisted by the calculator, speed of mental computation can increase.

- **It improves students' problem-solving ability.**
  The use of calculators affords students the luxury of working with practical, more realistic values without having to worry about difficult computation of large numbers. In this way students have more time to focus on understanding the problem and applying the appropriate strategy to solve it.

Criticisms of the use of the calculator

Students may become over dependent on the calculator and fail to practise basic mathematical computation. This concern can be addressed by the teacher, who needs to ensure that basic skills are mastered through the use of small numbers. For example, to help students grasp the concept of addition, setting the example: $3 + 5 = 8$ might be more effective than $387 + 451$, which can be used later with the calculator once the concept is mastered.

Manipulatives and teaching aids

Manipulatives are any real objects that students can use to enhance learning. These may be:

- existing objects, utensils, tools, toys, blocks, any object found in the environment or in the home
- specially created objects or models, such as cubes, spheres and other structures.

Examples of very useful manipulatives include:

- cardboard boxes that can be used to find surface areas, nets, volumes of real objects
- balls, funnels, cylindrical cans, dice, cones, spoons, measuring cups, any flat or curved surface, marbles, matches
- a geoboard consisting of a flat board with nails placed at equal and regular distances apart, which could be used with rubber bands to investigate geometrical concepts.
  Among its many diverse uses, the geoboard can be used to find areas, perimeters and angles and investigate or study the properties of various polygons.

- Other examples include devices to estimate height and sectioned models of objects of various shape.

Teaching aids are normally classified as anything that may be used to teach a particular concept in mathematics. Manipulatives are a special type of teaching aid in their own right but may include charts, graphs, pictures, diagrams, computer images, which are not three-dimensional but can aid with the visual understanding of a concept.

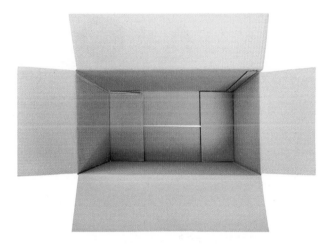

A cardboard box – a real, everyday object

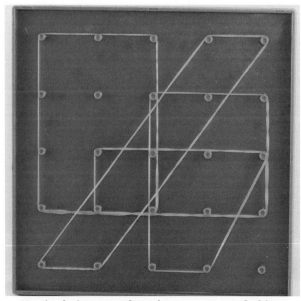

Manipulative: a geoboard – a constructed object

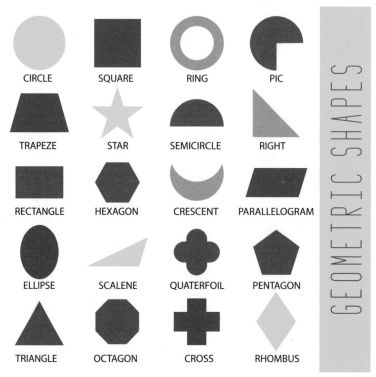

Example of a teaching aid – chart on shapes

Manipulatives and teaching aids are extremely useful in allowing students to receive that hands-on practical experience which leaves a more lasting impression on the mind and is an invaluable tool in concept development.

Answers

Chapter 1 — Number theory and computation

Exercise 1A

1 a XII **b** LXXXIII

2 a 19 **b** 95

3

4

5

6 a XXIX **b** CL

7 a 103 **b** 140

8

9 MMXVIII

10 a 186 **b** 2610

11 a 2905 – Hindu Arabic numerals
 b MMCMV - Roman numerals
 c

Exercise 1B

1 a 1 (natural number)
 b 0 (whole number)
 c –7 (integer)
 d $\frac{3}{4}$ (rational number)
 e $\sqrt{2}$ (irrational number)
 f 8 (real number)
 g $\sqrt{-6}$ (imaginary number)

2 Natural numbers: 1, 2, 3, 4, 5, 6, 7, 8, 9

3 Whole numbers: 0, 1, 2, 3, 4, 5, 6, 7, 8, 9, 10

4 Integers: –10, –9, –8, –7, –6, –5, –4, –3, –2, –1, 0, 1, 2, 3, 4, 5

5 Ascending order: 0.001, 10^{-2}, $\frac{2}{100}$, 1.000001, $(100)^{\frac{1}{2}}$

6 Descending order: 27 000, $\sqrt{1000}$, $\frac{92}{3}$, 3(10.2), $\sqrt[3]{3^3}$

7 a Integers: –1, 100, 0
 b Natural numbers: 100
 c Real numbers: –1, $\sqrt{9}$, 1.5, $1\frac{3}{4}$, $\sqrt{\frac{1}{4}}$, $\sqrt{3}$, 100, 0, –0.45, $\frac{14}{3}$
 d Irrational numbers: $\sqrt{3}$
 e Imaginary numbers: $\sqrt{-4}$

8 Integers: 2 and 3

9 Number of whole numbers: 101

10 The set of prime numbers is not closed. Examples may vary.

11

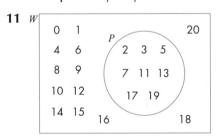

12 Largest set: Real numbers

13 No largest integer – tend to infinity; smallest positive integer is 0

14 No largest whole number – tend to infinity; smallest whole number is 0.

15 The triangular numbers between 1 and 20 are: 3, 6, 10, 15

16 Rectangular numbers between 1 and 20: 2, 6, 8, 9, 10, 12, 14, 15, 16, 18.

17 a i Largest prime number: 19

    ii Largest composite number: 18

    iii Largest odd number: 19

    iv Largest even number: 18

  b Sum: 74   **c** Product 116 964

Exercise 1C

1 3   **2** 6

3 3 and 300   **4** 9 and 9000

5 400 and 70; digit 4 has the greater place value.

6 '235' has the digit with the smallest place value (5), place value of 5.

7 a Digit with highest place value is 9; Digit with lowest place value is 1
  b Sum = 10

8 Place value of the digit 4: 40 000 000.

9 Place value table for 1 021 508 and 9 723 456.

Millions	HTH	TTH	TH	H	T	U
1	0	2	1	5	0	8
9	7	2	3	4	5	6

10 900 000 000 + 80 000 000 + 7 000 000 + 000 000 + 00 000 + 0000 + 200 + 10 + 3

11 Place value of 9 = 9000; face value of 4 = 4; 9000 × 4 = 36 000

12 Sum of the face values: 20

13 Product of place values: 0

Exercise 1D

1 a 11   **b** 5   **c** 24   **d** 2.67

2 5 mangoes

3 a 1222   **b** 235   **c** 30 015   **d** 89

4 a 2763   **b** 29 964   **c** 1 695 708   **d** 42

5 $10 043.00

6 Number of males: 6666

7 43 bottles   **8** 47 804

9 1092 km   **10** 74 808

11 a 88 966   **b** 1279   **c** 393 577 008   **d** 122

12 968 769

13 824 893 women

14 100 000 marbles

15 $29 559 575

16 1 375 000 grains

Exercise 1E

1 a 100; additive identity
  b 40; multiplicative identity

2 $\frac{4}{3}$

3 a 0   **b** 2   **c** $-1\frac{1}{9}$

4 12   **5** 7   **6** 19

7 a 35   **b** −36

8 a $20 \div 5 = 4$ and $5 \div 20 = \frac{1}{4}$, which shows the commutative law does not apply
  b $15 - 6 = 9$ and $6 - 15 = -9$, which shows the commutative law does not apply

9 $\frac{3}{2}$ or $1\frac{1}{2}$ **10** 1

11 a 9   **b** −15

12 −92.9   **13** $-\frac{13}{38}$

Exercise 1F

1 39   **2** 20

3 $21\frac{1}{3}$ **4** 2^{11}

5 10^3 **6** $2^{11} \times 3^4$

7 1

8 a Square number: 36   **b** Square number: 49

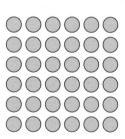

9 $2^5 \times 3^3$ **10** 1

Exercise 1G

1 Factors of 12: 1, 2, 3, 4, 6, 12

2 Prime factors of 20: 2, 5

3 Multiples of 3; 12, 15, 18

4 HCF = 4   **5** LCM = 30

6 a 225   **b** 400   **c** 1024

7 a 9   **b** 11   **c** 15

8 $-\dfrac{8}{7}$

9 Largest prime = 7; smallest prime = 2; product = 14

10 $3 \times 5 \times 7$

11 10, 20, 30, 40, 50, 60, 70, 80, 90

12 $\dfrac{1}{2}$ **13** LCM = 48

14 30 seconds

15 a Largest square: 4 cm²
   b Number of squares = 60

Exercise 1H

1 a $\dfrac{1}{7}$ **b** $\dfrac{1}{6}$

2 $\dfrac{1}{8}$ is shaded

3 a $\dfrac{3}{4}$ – proper fraction

   b $\dfrac{5}{3}$ – improper fraction

   c $1\dfrac{1}{2}$ – mixed number

4 a   **b**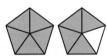

   c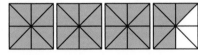

5 Examples of four fractions equivalent to $\dfrac{3}{5}$:
   $\dfrac{6}{10}; \dfrac{12}{20}; \dfrac{30}{50}; \dfrac{36}{60}$

6 a $\dfrac{2}{3}$ **b** $\dfrac{1}{3}$

7 $\dfrac{5}{8}$

8 a $\dfrac{1}{7}$ **b** Proper/unit fraction

9 Fraction of circle NOT shaded: $\dfrac{23}{24}$

10 $\dfrac{1}{1\,209\,600}$

Exercise 1I

1 $\dfrac{1}{3}$ **2** $\dfrac{2}{5}$

3 $1\dfrac{4}{15}$ **4** $-\dfrac{1}{8}$

5 $\dfrac{1}{4}$ **6** $\dfrac{5}{6}$

7 $8\dfrac{1}{4}$ **8** $1\dfrac{1}{2}$

9 8   **10** $9\dfrac{3}{8}$

11 $7\dfrac{1}{5}$ **12** 2

13 $7\dfrac{1}{2}$ **14** $1\dfrac{13}{20}$ acres

15 $\dfrac{10}{127}$ **16** $1444\dfrac{19}{20}$ ft

17 $\dfrac{19}{48}$ yards   **18** $5\dfrac{7}{10}$

19 1 cake   **20** $\dfrac{3}{10}$

21 $\dfrac{3395}{105\,003}$

22 a $1\dfrac{1}{10}$ km²   **b** $\dfrac{433}{20\,000}$ km²   **c** $1\dfrac{1567}{20\,000}$ km²

Exercise 1J

1 Five examples of a decimal number: 0.8; 1.3; 145.2; 0.000 08; 1.0102

2 a Smallest: 0.001   **b** Largest 1.1

3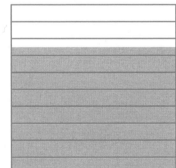

4.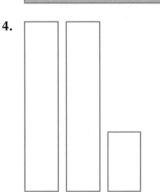

5 **a** 0.1 **b** 0.01 **c** 0.02 **d** 0.003
  e 0.015 **f** 0.4 **g** 0.16 **h** 1.4
  i 0.375

6 **a** $\dfrac{7}{10}$ **b** $\dfrac{95}{100}$ **c** $1\dfrac{9}{1000}$ **d** $\dfrac{3}{10\,000}$

  e $157\dfrac{6}{100}$

7 $123.56 = 100 + 20 + 3 + \dfrac{5}{10} + \dfrac{6}{100}$

8 **a** Place value of: 2: $\dfrac{2}{1000}$; 8: 8000; 0: $\dfrac{0}{10}$;

  6: 60; 5: $\dfrac{5}{100\,000}$

  b The face values of the: 2, 8, 0, 6 and 5 are respectively: 2, 8, 0, 6 and 5

9 1.03 can be represented by this diagram.

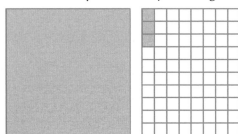

10 0.075

11 $\dfrac{101}{12\,500\,000}$

12 $4350.892 = 4000 + 300 + 50 + 0 + \dfrac{8}{10} + \dfrac{9}{100} + \dfrac{2}{1000}$

13 8070.195

14 $300\,000 + 50\,000 + 6000 + 800 + 70 + 2 +$

  $\dfrac{0}{10} + \dfrac{1}{100} + \dfrac{9}{1000} + \dfrac{4}{10\,000}$

15 A geometric shape is divided into 1000 equal parts. One of these parts together with one whole shape represents 1.001

16 0.025 **17** 2 010 872.016

18 Six identical geometric shapes are chosen. One of these is divided into 1000 identical parts. Five of these parts together with the remaining five whole shapes represent 5.005.

Exercise 1K

1 **a** 6.69 **b** 1.327

2 **a** 12.84 **b** 42.3

3 56.44 **4** 132.77

5 87.6 **6** 0.0403

7 543.237 **8** 316.774

9 44.6 **10** 83.1

11 117.68 **12** 214

13 2086.752 **14** 0.04

15 8.16×10^4 **16** 3.56×10^{-1}

17 **a** 1.4 **b** 1.37

18 **a** -8.75×10^{-1} **b** -8.75×10^{-1}

19 **a** 40.776 m² **b** 1095.63 m² **c** 94860 tiles

20 2 sweets

Exercise 1L

1 **a** 50% **b** 75% **c** 420% **d** 75%
  e 182% **f** 0.63% **g** 140% **h** 200%

2 **a i** $\dfrac{2}{5}$ **ii** 0.4

  b i $1\dfrac{1}{5}$ **ii** 1.2

  c i $\dfrac{1}{500}$ **ii** 0.002

  d i $\dfrac{21}{2000}$ **ii** 0.0105

  e i $5\dfrac{13}{20}$ **ii** 5.65

3 $12.00 **4** 500 students

5 $\dfrac{1}{5}$

6 Garden: 60% planted in vegetables

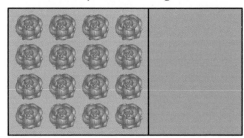

7 25%

8 Ascending order: $\dfrac{18}{25}$; $\dfrac{3}{4}$; $\dfrac{39}{50}$; $1\dfrac{1}{5}$; $1\dfrac{1}{2} = 150\%$; $1\dfrac{3}{5}$

9 New salary: $12\,000.00

10 133.3 cm **11** 95 marbles

12 James: $28.00; Tommy: $84.00; Phillip: $28.00

13 Fraction remaining is $\dfrac{6}{100}$ or $\dfrac{3}{50}$.

14 $\dfrac{1}{6048}$% **15** $3333.33

16 a More singers than dancers and more dancers than musicians ($0.01 > 0.0085 > 0.005$)
  b Number of musicians = 25
  c Total population = 500 000

17 a Percentage not voting: about 6%
  b Total number of voters: about 100
  c Winner: C

18 Final price: $155.82

Exercise 1M

1 a 14 500 m **b** 2×10^{-1} m or 0.2 m

2 a 205 mg **b** 10^5 mg

3 2580 cc **4** 60 minutes

5 8.3 metres in 1 second

6 1.7 m **7** 453 600 g

8 2 400 000 litres

9 0.000 041 m in 1 second

10 6×10^{-8} m³ **11** 64 857 bottles

12 Combined mass: 171.8 lbs

Exercise 1N

1 Ratio in its simplest form: 2 : 3 : 4 : 5

2 $10 : $20 : $30 **3** $9.00

4 10 days **5** Average = 6 years

6 200 km **7** 3.24 s

8 $9 : $15 : $21 **9** $100.00

10 30 mangoes **11** 5 years

12 168 gallons per day

13 a No, the total mass of 593 kg will exceed the maximum load
  b 143 lbs

14 The sixth number is 11

15 1 : 4 : 8 : 80

16 Six tailors take the same 3 days

17 41 **18** 84.6

19 1.44 m **20** 66 ears

21 34 524 words **22** $26 \times 4 \times 100$

23 a Group B had the longest combined jump
  b Difference $1\frac{1}{2}$ m

24 14.7

Exercise 1O

1 15, 18

2 a Rule: Add $(0.2 + 0.n)$, $(n \in W)$
    W = {whole numbers}
  b 2.8, 3.5

3 2, 5, 10, 17 **4** $0, \frac{1}{2}, 1, 1\frac{1}{2}, 2$

5 $x = 61$

6 a Rule: Multiply by 10
  b Next two terms: 10^5, 10^6

7 Next two terms: $6\frac{1}{2}$, 8

8 1, 2, 3, 4, 5

9 Rule: Add two to each consecutive whole number and add two to the denominator of each consecutive fraction.

10 Next two terms: –2, 1

11 Rule: Subtract n^2 from each consecutive term ($n \in$ {natural numbers})

12 $A = 1$

13 Next three terms: $\frac{1}{6}, 7, \frac{1}{8}$

14 Sequence: –0.01, –0.02, –0.03, –0.04, –0.05

Exercise 1P

1 $87\,569_{10} = 9 + 60 + 500 + 7000 + 80\,000$

2 $21_{10} = 10\,101_2$

3 Four examples of binary numbers = 1, 101, 11, 110

4 a base 10 **b** base 3
  c base 5 **d** base 7
  e base 8

5 22_{10} **6** $111\,111_2$

7 1000_2 **8** $1\,001\,101_2$

9 11_2 **10** 3223_4

11 63_{10} **12** 2123_4

13 223_4 **14** $323\,010_4$

15 30_4 **16** 1734_8

17 470_{10} **18** 735_8

19 675_8 **20** $207\,442_8$

21 3_8 **22** 66_{10}

23 $12\,015_{10}$ **24** $10\,212\,111_4$

25 $1\,000\,110\,111_2$ **26** $1\frac{4}{27}_8$

Examination-type questions for Chapter 1

1 **a** $\frac{1}{3}$ **b** 0.82

2 **a** 9.22 **b** 9.22

3 **a** **i** 30 boys **ii** 50 students

    b **i** Total: 30 sweets

        ii 2-year-olds: 4 sweets; 3-year-olds: 6 sweets; 4-year-olds: 8 sweets; 6-year-olds: 12 sweets

4 **a** 5th innings score: 330
    b **i** Length of the trip is 250 km
        ii Time $= 3\frac{4}{7}$ hours

5 **a** $K = 4^3 = 64$ **b** 50067_8 **c** 110010_4

6 Percentage increase = 8% (to the nearest 1%)

Chapter 2 — Consumer Arithmetic

Exercise 2A

1 $190.00 **2** $14.25

3 $16\frac{2}{3}$ % **4** $85.50

5 $86.96 **6** $221.16

Exercise 2B

1 $64.20 **2** $2280.00

3 50% **4** $48.08

Exercise 2C

1 Loss 65¢ **2** Loss $12.00

3 Loss $40.00 **4** Loss = 25%

5 Profit = 12.5% **6** Loss = 11.7%

7 Original price: $24.29 **8** $5189.50

Exercise 2D

1 25% **2** 25%

3 66.7% **4** 4.87 billion people

5 221.1 lbs

6 $2123.21, difference $78.21

Exercise 2E

1 $30.00 **2** $3000.00

3 $1666.67 **4** $833.33

5 $7.30 **6** $1\frac{1}{3}$ years

7 $2\frac{1}{7}$% per annum **8** $6083.26

9 $99 800 **10** $590.49

11 10.5 years **12** 7.2% per annum

13 **a** $71 963.44 **b** 60%

Exercise 2F

1 $2070.00

2 Monthly instalment: $79.20

3 Interest rate: 9.8%

4 Marked price: $2900.00; period: 8 months.

5 Interest: $120.00; monthly instalment: $153.33

6 **a** $5000.00 **b** $20 000.00 **c** 120
    d $24 000.00 **e** $29 000 **f** $4000
    g 20%

Exercise 2G

1 Deposit = $10 000.00; mortgage = $90 000.00

2 5%

3 **a** $45 500.00 **b** $304 500.00
    c $885 500.00

4 **a** $9600.00 **b** $70 400.00
    c $16 000.00 **d** $96 000.00

5 **a** $17 600.00 **b** $202 400.00
    c $1777.78 **d** $117 600.00
    e $320 000.00

6 **a** Deposit = $20 100.00
    b Cost of house = $100 500.00
    c Number of monthly instalments = 180
    d Interest $9600.00
    e Total amount payable = $110 100.00

Exercise 2H

1 $1800.00 **2** 0.175 or 17.5%

3 $1454.55 **4** 0.5 or 50%

5 $10 820.00

6 Rates payable = $1104.00; $96.00 more

7 Income from rates = $1600; new rate = 0.105 or 10.5%

Exercise 2I

1 2.88 kWh

2 144 kWh in 30 days

3 60 kWh (units)

4 280 kWh (units)

5 168 kWh (units)

6 31.25¢ per kWh

7 $180.00   **8** $257.13

9 **a** 540 kWh used
    b Energy charge: $91.80
    c Fuel charge $216.00
    d Total amount $322.80
    e After discount: $284.06

Exercise 2J

1 $91.20   **2** $4583.00

3 Water bill highest in 2017 by $79.25

4 Total amount: $6014.87

Exercise 2K

1 $69.75   **2** $224.25

3 Number of local calls: 200
    a $70.00   **b** $20.00
    c 95.00 (70 × 0.50 + 80 × 0.75)
    d $34.53   **e** $29.34
    f $314.07

Exercise 2L

1 **a** 503¢   **b** $200.85

2 $125.80 = $100 + $20 + $5 + 3 × (25¢) + 5¢
    $125.80 = 6 × $20 + $5 + 8 × (10¢)
    $125.80 = 2 × ($50) + 5 × ($5) + 16 × (5¢)

3 **a** $448.25   **b** $205.55

4 $31.01   **5** $4788.00

6 $12.90

Exercise 2M

1 **a** US$92.59   **b** US$69.59
    c BDS$442.00   **d** US$76.69
    e US$47   **f** GUY$9452.00
    g CNY1019.7

2 **a** TT$63.98   **b** BDS$3.39
    c EC$8.30   **d** EC$263.56
    e US$529.19

3 US$153.76

4 The different currencies remaining are:
    TT$1168.46; EC$187.50; BDS$69.44

Exercise 2N

1 $33 600.00   **2** $4000.00

3 **a** $3086.00   **b** $40 800.00
    c $3768.00   **d** $37 032.00

4 $340.00   **5** $13.50 per hour

6 50 hours   **7** $200.00

8 **a** $122.00   **b** $322.00

9 $455.00   **10** $488.00

11 $1030.05

12 **a** $11 per hour
    b Total overtime: 36.36 hours

Exercise 2O

1 $4464.00 per annum

2 $2727.27

3 **a** $4155.00   **b** $2308.33

4 The company charging $600 insurance premium

5 $100 per year

6 **a** $19 620.00   **b** $6365.00
    c $15 096.00

Examination-type questions for Chapter 2

1 **a** $1240.00   **b** $7590.00
    c $759.00

2 $25 624.45   **3** $52 610.00

4 **a** $26 800.00   **b** $11 200.00   **c** 41.8%

5 **a** P = $70.00; Q = 50; R = $0.75; S = $78.00;
        W = 14; Y = $377.20; Z = $75.44.
    b Loss = $64.64; percentage loss = 14.3

6 **a** $1086.00   **b** $411.00   **c** 62.15%

7 **a** BDS$ = 1640.00   **b** TT$715.00; EC$204.29

8 **a** Total number of local calls = 105
    b Charge on 55 local calls: $19.25
    c Tax on overseas calls = $44.53
    d Total charge = $254.18 (including arrears)

Chapter 3 — Algebra

Exercise 3A

1 **a** 6   **b** –9   **c** –9
    d 8   **e** 0

2 82   **3** –116

4 8   **5** –150

6 33 degrees   **7** –59

8 43   **9** 581

10 80 620   **11** 21 648 515

Exercise 3B

1 a 12   b 15   c –9   d –12

2 161   3 119

4 –210   5 –276

6 1665   7 770 cm

8 168 168   9 98 256

10 5 711 130   11 –26 545 266

Exercise 3C

1 a 4   b 5   c –1   d –3

2 4   3 4

4 –5   5 13 remainder 200 cm

6 0.5   7 –51

8 4010   9 –1

Exercise 3D

1 –4   2 –72

3 –15   4 –17

5 41   6 $-\dfrac{51}{8}$

7 –4   8 $-\dfrac{5}{4}$

Exercise 3E

1 $x, 5y, x^{-2}$

2 5 is a coefficient; x and y are variables; power is –2

3 $(a + r)$

4 Pairs of like terms: $2ax, 12ax$; $5x, 100x$; $3x^2, 10x^2$; $12xy, 16xy$; $24xy^2, 6xy^2$; unlike terms: $2x^2y, 3ax^2$

5 Total cost ($) = $21\dfrac{1}{2}x + 15y$

Exercise 3F

1 $a + b + c$ 2 $x - y$

3 $lmnq$ 4 $\dfrac{p}{h}$

5 $(a + b) \div 2(a - b)$

6 $4(a + b + c + d) \times (abcd)^2$

7 $\dfrac{3}{4}\left(\dfrac{a}{b}\right) \times \dfrac{1}{2}ab$

8 $\sqrt{a + b + c + d + e} \div (abcde)^3$

9 $\left(\dfrac{l}{m}\right)^2 \times \sqrt{\dfrac{1}{2}(l - m)} + 5(lm - (l + m))$

Exercise 3G

1 a p^9 b q^{-1} or $\dfrac{1}{q}$ c r^3

2 $8x^6 + 5x^{12}$

3 $\dfrac{4}{x^9} - \dfrac{5}{4}x^4$ 4 $2\dfrac{x^3}{y^4}$

5 $\dfrac{ab^2}{4}$ 6 10

Exercise 3H

1 $10x$ 2 $43x^2 + 26x$

3 a $28x$ b $25ay$

4 $13x^2$ cm^2 5 $2x$

6 $ab + 3ab^2 - 4a^2b$

7 $12 + 2x^2 - 3x + bx - ax$

8 $6n^2 - 5n$ 9 $\dfrac{9\pi r^2 h}{256}$

Exercise 3I

1 xyz 2 $6a^3$

3 $24abc$ 4 $6y^3$ units^3

5 $a^3b^3c^2$ 6 $400p^5q^4$

7 $5600r^4$ 8 $24a^3$ units^3

9 a $x^3 + 36x^2 - 22x - 24$

   b $3ab^2cd^2 + 3ab^2c - 15a^2bc^2 + 25ab^3c^3 - bc - abc$

Exercise 3J

1 $2b$ 2 r

3 $2ab^2c^3$ 4 $\dfrac{x^2y + 1}{5}$

5 $2p^2$ 6 $\dfrac{3}{4}lbh$

7 $\dfrac{25x^2y^3}{2z}$ 8 $\dfrac{y^{24}z^3}{27}$

Exercise 3K

1 $30y^2 + 11y$ 2 $14a - 10ab$

3 $11p^2 - 4p$ 4 $\dfrac{8qr + r^2 - 2q}{5qr}$

5 m^2n^3vw 6 $\dfrac{2b^2}{acx^2y^2z}$

Exercise 3L

1 For example, $\dfrac{1}{2}x$; $\dfrac{a}{b}$; $\dfrac{21}{3m}$; $\dfrac{2x-y}{3x+y}$; $\dfrac{9x^2+y}{5r^2+2r-1}$

2 $\dfrac{5}{12y^3}$ **3** $\dfrac{5a^2}{b^3}$

4 $\dfrac{yz+2xz+3xy}{xyz}$ **5** $\dfrac{2ax^2+3bc-4c}{x^3}$

6 $\dfrac{75a-80b+12c}{60y}$ **7** $\dfrac{9xy+4x}{(2y-3)(y+2)}$

8 $\dfrac{(a+2)(y+2)(y+3)-(b+3)(y+1)(y+3)+(2b+5)(y+1)(y+2)}{(y+1)(y+2)(y+3)}$

Exercise 3M

1 9   **2** 11

3 −330   **4** 3

5 $\dfrac{30}{17}$ **6** 28

Exercise 3N

1 1   **2** 8

3 5   **4** $\sqrt{-11}$

5 19   **6** 8

7 4229   **8** $\dfrac{31}{63}$

Exercise 3O

1 $2a^2-4ab$

2 $12w^2-8w^3$

3 $6x^3-12x^2-15x$

4 $2x^2+5x-12$

5 $2x^3-13x^2+22x-8$

6 $12x^4+15x^3y-4x^2y^2-5xy^3$

7 $\dfrac{6x^2+2xy-20y^2}{x^2-4y^2}$ or $\dfrac{6x-10y}{x-2y}$

Exercise 3P

1 $3ax$ **2** $5a$

3 **a** $6x(3x^2-2x+4)$ **b** $5p^2r^3(pr+2)$

4 $ab\left(\dfrac{2b^4}{x^3y^3}+\dfrac{6a}{5x^2y^2}-\dfrac{9}{10}\right)$

5 $tp(x^2-y^3)(a-p)$

---

6 $\dfrac{2a+b}{3x-5}-\dfrac{x(2a+b)}{\left(3x^2-5\right)}=(2a+b)\left[\dfrac{1}{3x-5}-\dfrac{x}{\left(3x^2-5\right)}\right]$

A fuller answer:

$\dfrac{2a+b}{3x-5}-\dfrac{3x(2a+b)}{3\left(3x^2-5\right)}$

$\quad =(2a+b)\left[\dfrac{1}{3x-5}-\dfrac{x}{\left(3x^2-5\right)}\right]$

$\quad =(2a+b)\left[\dfrac{\left(3x^2-5\right)-x(3x-5)}{(3x-5)\left(3x^2-5\right)}\right]$

$\quad =\dfrac{5(2a+b)(x-1)}{(3x-5)\left(3x^2-5\right)}$

Exercise 3Q

1 $x=12$ **2** $x=2$

3 $x=10$ **4** $x=-3$

5 $x=11$ **6** $x=10$

7 $x=8$ cm   **8** 21 cm^2

9 $x=\dfrac{14}{13}$ **10** $x=5\dfrac{5}{6}$

11 $x=\dfrac{1}{60}$ **12** $x=46\dfrac{2}{3}$ degrees

13 Adam: \$40, Bill: \$80; Total: \$180

14 $x=4\dfrac{3}{32}$ **15** $x=-\dfrac{17}{36}$

16 a Vegetable market: $3y$; clothes store: $3y+80$
   b $y=55.71$; fish: \$55.71; vegetables: \$167.14; clothes: \$247.14

Exercise 3R

1 $x<2$ **2** $x>9$

3 length $\leqslant\sqrt{40}\leqslant 6.3$ cm

4 4 cm $(w\leqslant 4$ cm$)$ **5** $x\leqslant 1$

6 $x\geqslant\dfrac{40}{63}$ or $x\geqslant 0.63$ **7** $x\leqslant\dfrac{44}{27}$

8 $r>1.95$ **9** $x\leqslant\dfrac{65}{18}$

10 $x<\dfrac{65}{294}$

11 $h<0.36$ cm

Exercise 3S

1 $m = \dfrac{F}{a}$

2 $t = \dfrac{v - u}{a}$

3 $\dfrac{-2u + \sqrt{4u^2 + 8as}}{2a}$

4 $n = \dfrac{PF}{kmL}$

5 $T = \sqrt{\dfrac{4\pi^2 r^2}{G(M_1 + M_2)}}$

6 $m = \dfrac{RLagp}{bhq}$

Exercise 3T

1 **a** $x = 3; y = 1$ **b** $x = 1; y = -\dfrac{1}{2}$
   c $a = 5$ and $b = -2$

2 Exercise book = $1, pencil = $1

3 $x = 3, y = \dfrac{1}{2}$

4 $x = \dfrac{27}{23}, y = -\dfrac{17}{23}$

5 $x = 6\dfrac{11}{19}, y = 2\dfrac{16}{19}$

6 $x = -15, y = -6$

7 $b = \$1.05, M = \$1.14$

Exercise 3U

1 $(k + r)(a + b)$ **2** $(m^3 - n^2)(a - b)$

3 $(m + n)(y^2 - x^2)$ **4** $(p^2 l + 1)(m - n)$

5 $(6x + 3y)(x - 2y)$ **6** $(x + y)(x - y)$

7 $(4a + 5b)^2$ **8** $(3x + 4y)(2a + b - 2)$

9 $(3x + y)(2x - y + 2)$

Exercise 3V

1 **a** $x^2 + 2x + 3$ **b** $5x^2 - 4x + 3$
   c $x^2 + 9$ **d** $5x^2 - 4x$
   e $9x^2 + 30x + 25$ **f** $4a^2 - 4$

2 **a** $(x + 5)(x + 1)$ **b** $(2x + 1)(x + 1)$
   c $(2x + 2)^2$ **d** $(5a - 3)(5a + 3)$

3 $12x(x - 2)$ **4** $(x + 2)^2 - 1$

5 $x(x + 1) = 12; 3, 4$ or $-3, -4$

6 $4(x - 1)^2 + 9$ **7** $-5(x - 1)^2 + 8$

8 $8.4(x - \dfrac{1}{4})^2 - 5.725$

Exercise 3W

1 **a** $x = -9, -1$ **b** $x = -\dfrac{2}{3}; x = -1$

   c $x = \dfrac{9}{8}, x = -\dfrac{9}{8}$ **d** $x = 1$

2 $x = -4; x = 2$ **3** $x = 4, 2$

4 $x = 1, 2$ **5** $1.18, -0.61$

6 length = 1

7 Dimensions: 3 cm by 6 cm

8 $x = 3.93, 0.073$ **9** $x = 2.43, -0.927$

10 2 or $-\dfrac{2}{3}$

Exercise 3X

1 $x = 1; y = 1$ **2** $x = 1, y = -1$

3 **a** The first is an equation which is true for only one value of x: $x = \dfrac{5}{7}$. The second is an identity. When you expand the LHS it becomes the RHS.

   b Equations: $5x = 14 + x^2, 2a^2 - ab + b = b - 3$; identities: $2x(2x - 3)^2 = 2x(4x^2 - 12x + 9)$, $2x - 3(x + 4) = -12 - 3x + 2x$

4 $x = -1, y = -1$

5 $x = 5, y = -1$ and $x = -1, y = 5$

6 Expand and simplify LHS to get RHS

7 $x = 1, y = 0$ and $x = 1.4, y = -0.8$

8 $a = 12, b = -11, c = -1$

9 $x = 0.521, y = -1.96$ and $x = 2.09, y = 1.18$

Exercise 3Y

1 $k = \dfrac{4}{3}, B = 7\dfrac{1}{2}$

2 $k = 20, x = 2\dfrac{1}{2}$

3 $P = 0.8$

4 $V = 0.53$ units^3

5 Number bacteria, $M = 4^5 = 1024$

6 $H = 0.26$

Examination-type questions for Chapter 3

1 **a** **i** 17   **ii** 225

   b $\dfrac{3s + 32}{12}$

2 a i $(700 - x)$ ml

  ii $(700 - x - 6y)$ ml

  b 1 cricket bat costs $40, 1 ball costs $80

3 a $(5x + 2y)(2x - y)$ **b** $2a(2a^2 + 6a - 5)$

  c $(5x + 2)(x - 1)$ **d** $(2y - 8)(2y + 8)$

4 a $x = 6$

  b $x = 3.53$ and $x = 0.4$, $y = 7.58$ and $y = -1.58$

5 a i $\dfrac{57}{4}$ **ii** $\dfrac{40\,593}{580}$

  b a^2/b^2

6 a $v = 5$, $w = 3\dfrac{3}{4}$ **b** $4(x - \dfrac{3}{2})^2 - 4$

7 a $F = \dfrac{9 + 12P}{4 + 6P}$;   **b** $P = -\dfrac{11}{16}$

8 a $\$2x$ **b** $\$(120 - x)$

  c 20 large, 20 small

9 a $2a(2a + 1) - b(b - 1)$

  b $\dfrac{-2x^3 + 8x - 3}{x^2 - 4}$

Chapter 4 — Geometry and Trigonometry

Exercise 4A

10 A cuboid has: 12 edges, 6 faces, 8 vertices

11 A tin of orange juice: 2 edges

12 Ice-cream cone: 1 edge

15 A triangular prism, 6 vertices, 9 edges.

Exercise 4B

4 a 90°   **b** 10°

5 $2\dfrac{1}{2}$ revolutions = 900°

8 141°

9 0.7 revolution = 252°

10 1260° = 3.5 revolutions

11 210°

13 120°

14 $\dfrac{1}{12}$ revolution

15 43 right angles

Exercise 4C

13 EF = 10 cm

14 Angle WXY = 82°

15 JL = $16\dfrac{1}{2}$ cm

16 AC = 19.1 cm

17 QT = 18.2 cm, angle QTS = 33°

18 UY = 7.3 cm, angle VWY = 120°

Exercise 4D

14 a lines of symmetry: 1; order of rotational symmetry: 1

  b lines of symmetry: 2; order of rotational symmetry: 2

  c lines of symmetry: 1; order of rotational symmetry: 1

16 a lines of symmetry: 0; order of rotational symmetry: 8

  b lines of symmetry: 2; order of rotational symmetry: 2

Exercise 4E

1 90°; 0.25 revolution; ¼ turn

2 About 180°; 0.5 revolution

3 42°: acute angle; 90°: right angle; 99°: obtuse angle; 200°: reflex angle; 180°: straight angle; 360°: one revolution.

4 a For example 30° and 60°

  b For example 130° and 50°

6 $x = 130°$, $w = 130°$, $p = 50°$, $q = 50°$, $r = 130°$, $u = 130°$, $v = 50°$

7 a x and w are vertically opposite

  b w and q are co-interior

  c q and r are adjacent

  d p and q are alternate

  e p and v are corresponding.

8 a smallest angle PQR

  b largest side PQ

  c Angle RQS = R + P

  d $\angle$PRQ + $\angle$RPQ + $\angle$PQR = 180°

  e Angle RQS = exterior angle

14 Students' own answers

15 a $x = 8$;   **b** $h = 3$

16 a $a = 5$ cm; $b = 4.47$ cm; $c = 10$ cm

  b Perimeter = 34.5 cm

17 b $a = 23.3$, $l = 5$, $r = 5.83$

  c Area ratio = 144 : 9 = 16 : 1

18 Thread length = perimeter = 121.5 cm, GA = 50 cm; RP = 30 cm

19 a $x = 110°$; $y = 70°$
   b Exterior angles: L = 120°; M = 80°; N = 90°
   c Sum of exterior angles = 360°

20 a Pentagon: interior angle 108°; exterior angle 72°
   b Hexagon: interior angle 120°; exterior angle 60°
   c Heptagon: interior angle 128.6°, exterior angle 51.4°
   d Octagon: interior angle 135°, exterior angle 45°
   e Nonagon: interior angle 140°, exterior angle 40°
   f Decagon: interior angle 144°, exterior angle 36°

21 a hendecagon,
   dodecagon,
   tridecagon,
   tetradecagon,

Exercise 4F

2 $x = 50°$; $y = 100°$; $a = 50°$; $w = 90°$; $p = q = 90°$; $y = 130°$; $r = 92°$; $k = 88°$; $\theta = 60°$; $i = h = 90°$

3 a $x = 90°$ **b** $y = 90°$ **c** $m = 125°$

4 PQ = 13 cm; $\angle$QOP = 60°

5 KT = 9.8 cm

6 a 4 cm **b** 8 cm

7 a = 90°; b = 80°; c = 80°

8 c = 90°, d = 150°

9 $\angle$PRT = 20°, $\angle$RTS = 20°, $\angle$STQ = 70°

10 $\angle$ADC = $\angle$BCD = 100°; $\angle$BAD = $\angle$ABC = 80°

11 $P = 60°$, BDC = 120°

12 a $\angle$ABO = 65° **b** $\angle$AOR = 25°
   c $\angle$ROT = 90° **d** $\angle$SOK = 90°
   e $\angle$DOC = $\angle$AOB = 50°

13 a $\angle$LOA = 70° **b** $\angle$OAB = 70°

14 a $\angle$BCD = 100° **b** $\angle$CBD = 60°
   c $\angle$ADB = 60° **d** $\angle$BCH = 20°

15 $a = b = 120°$, $y = 60°$, $x = 60°$, $z = 120°$, $r = 60°$

Exercise 4G

1 $A'\begin{pmatrix} 4 \\ 4 \end{pmatrix}$; $B'\begin{pmatrix} 8 \\ 8 \end{pmatrix}$

2 P'(−2, 1), Q'(−5, 5)

3 The images are both congruent to the object triangle, but the first is flipped, to produce a mirror image, the second is flipped again to produce an identical image.

4 The images are both congruent to the object triangle, but the first is flipped, to produce a mirror image, the second is flipped again to produce an identical image but has been rotated.

5 W′(3, −1), K′(8, −5)

6 $A' = (4, 8)$; $B' = 2\begin{pmatrix} 6 \\ 4 \end{pmatrix} = (12, 8)$;
   $C' = 2\begin{pmatrix} 6 \\ 8 \end{pmatrix} = (12, 16)$; $D' = 2\begin{pmatrix} 2 \\ 8 \end{pmatrix} = (4, 16)$

7 $H'' = \begin{pmatrix} 4 \\ -10 \end{pmatrix}$; $K'' = \begin{pmatrix} 9 \\ -8 \end{pmatrix}$; $T'' = \begin{pmatrix} 7 \\ -4 \end{pmatrix}$

8 $A' = (0, 4)$, $B' = (4.8, 1.6)$, $C' = (3.2, 8)$

9 Centre of rotation about (−3.8, 3)

10 Centre (−11, −9), scale factor 2

11 J′(−2.5, −1.5); F′(1.5, −1.5); K′(2.5, 0.5); E′(−2.5, 0.5)

12 N′(−10.3, −4.0); I′(−13.3, 1.2); P′(−16.8, −0.8); K′(−13.8, −6.0)

Exercise 4H

1 $q = 12$ cm; $r = 10.4$ cm; $R = 60°$

2 $n = 8.9$; $N = 48.2°$; $L = 41.8°$

3 Tree: 38 ft or 11.5 m

4 30°

5 a sin 60° **b** −cos 20° **c** −tan 60°

6 a Area = 17 units² **b** Area = 35.5 units²

7 $A = 90°$; $B = 37°$; $C = 53°$

8 $r = 8.7$ cm; $P = 23.4°$; $Q = 36.6°$

9 $K = 13°$; $l = 11.4$ mm; $k = 3.8$ mm

10 $V = 50.9°$; $U = 25.1°$; $u = 10.9$ mm

11 AD = 140 km; bearing of D from A = 164.8°

12 Distance from Harry to Paul = 404 m; bearing of John from Paul = 208.4°

16 XY = 103.9 cm
   Angle of XY with the base = 35.3°

Examination-type questions for Chapter 4

2 $\angle KTJ = 40°$; $\angle TJK = 50°$; $\angle JRT = 90°$;
   $\angle OTR = 30°$; $\angle RJO = 60°$

3 **b** HG = 17.3 cm
   c FH = 20 cm; area FGH : area NMH = 25 : 9

4 One-way stretch followed by translation:
   $P(x, y) \Rightarrow P'(\frac{3}{2}x + 5, y)$

5 **a** A to B: 7 cm; B to D: 6.5 cm; A to C: 7.7 cm
   b A to D: 32 km; B to C: 11.75 km
   c Area of map ≈ 69 cm²;
      area of island ≈ 431 km²

6 **b** 159°   **c** 240°
   d BS = 241.5 m; PS = 152.3 m

Chapter 5 — Measurement

Exercise 5A

1 **a** 2 m   **b** 8000 m   **c** 0.01 m

2 **a** 0.002 km   **b** 4 000 000 cm
   c 200 cm   **d** 5 km
   e 500 mm

3 **a** 5×10^3 mm   **b** 4.5×10^3 mm
   c 4.58×10^{-4} km   **d** 2.3×10^4 cm

4 **a** 2.5×10^3 mm   **b** 8×10^{-7} cm
   c 2.59×10^8 m

5 1.451×10^{17} s (using 365 days per year)

6 **a** 1.7×10^{-6} ag   **b** 3.995×10^{-10} pg

7 1.599×10^3 Em   **8** 10^{-36} Em

Exercise 5B

1 **a** 2 kg   **b** 0.3 kg
   c 5×10^3 kg   **d** 0.0001 kg

2 **a** 2 g   **b** 2×10^{-7} Mg   **c** 7×10^{-4} Kg

3 **a** 2.3×10^4 µg   **b** 2.35×10^{11} mg
   c 2.55×10^{-7} Mg   **d** 1×10^{-17} Mg

4 **a** 2×10^3 pg   **b** 2×10^{19} cg
   c 4×10^{-22} Gg   **d** 8×10^9 ng

Exercise 5C

5 30 cm   **6** 1.56 cm

14 5938 g or 5.94 kg   **15** 13.4 cm

Exercise 5D

1 **a** 3.3 minutes   **b** 10 hours
   c 2 days   **d** 3 weeks

e 9 months   **f** $2\frac{1}{2}$ years
g 8 decades   **h** 20 centuries

2 **a** 1200 seconds   **b** 9000 seconds
   c 259 200 seconds

3 3 June   **4** 07:53

5 **a** 14:50   **b** 09:00

6 **a** 81 days $(2 \times 30 + 21)$ = 1944 hours
               = 116 640 minutes
               = 6 998 400 seconds
   b 0.28 hours

7 4 September   **8** 4:10 p.m.

9 **a** 00:15   **b** 23:59

10 **a** 3.17×10^{-7} years   **b** 1.75×10^7 hours

11 5 April   **12** 68 hours

Exercise 5E

1 1750 N   **2** 20 kg

3 The crate is heavier   **4** 5 m s⁻¹

5 3 g cm⁻³   **6** 2.5 km⁻¹, 0.694 m s⁻¹

7 $1333\frac{1}{3}$ seconds or 22 minutes

8 250 cm³   **9** 70 seconds   **10** 18 kg

Exercise 5F

1 **a** P = 25.6 cm; A = 30 cm²
   b P = 28 m; A = 45 m²
   c P = 37.7 mm; A = 113.1 mm²
   d P = 36 cm; A = 100 cm²
   e P = 36.5 cm; A = 35.6 cm²

2 A = 4800 m²; P = 280 m   **3** D = 628 m

4 x = 4 cm, y = 10 cm; P = 40 cm; A = 80 cm²

5 P = 28.7 cm; A = 51.3 cm²

6 P = 44 m; A = 84 m²   **7** 12 cm²

8 P = 36 m; cost: $1080

9 A = 575 m² or 5 750 000 cm²
   Number of tiles = 115 000

10 P = 84.9 cm; A = 168 cm²

Exercise 5G

1 12 cm

2 Greatest error: 0.005; 0.005; 0.05; 0.005
   Smallest error: −0.005; −0.005; −0.05; −0.005

3 Smallest area = 2043.9775 m²
   Greatest area = 2056.0275 m²

4 Perimeter = 130.64 m
Greatest error = 0.02 m
Smallest error = –0.02 m

5 **a** h = 17.3 cm
b Perimeter = 94 cm, area = 465 cm²
c **i** Perimeter:
Maximum error = 0.2 cm
Minimum error = –0.2 cm
ii Area
Maximum error = 2.5 cm²
Minimum error = –2.5 cm²

6 73.47 m²

7 Maximum volume = 2 500 604.072 mm³
Minimum volume = 2 491 171.286 mm³
They both agree to 2 s.f: 2 500 000 mm³

Exercise 5H

1 **a** 8.8 km   **b** 8.4 km

2 Area of Nevis is roughly 93 km²

3 Area of St Kitts is roughly 168 km²

4 The perimeter of Nevis is roughly 34 km
The perimeter of St Kitts is roughly 73 km

5 11 km

6 The perimeter of Dominica is 148 km and its area is 750 km². Check students' answers are close to this.

7 17.5 km

8 Dominica is approximately 47 km long and 26 km wide. Check students' answers are close to this.

9 **a** 15.2 m²   **b** 2 windows, 1 door
c Window: 10 cm
Door: 4.5 cm
d Window: 2 m
Door: 0.9 m
e 182 tiles (13 × 14)

Exercise 5I

1 0.576 m³

2 1177.5 ml

3 V = 554 cm³

4 V = 216 cm³

5 **a** 780 cm³
b 156 cm²

6 **b** 0.1 m   **c** 5.3037 m³
d 24.696 m³ or 24 696 litres

Exercise 5J

1 **a** 24 000 cm³   **b** 5.24 m³

2 849 cm³   **3** 48 cm³

4 85.3 m³

5 **a** 1026.4 cm³   **b** 340 cm³

7 2093.3 ml

Exercise 5K

1 523.3 cm³   **2** 56 520 m³

3 Volume of cone: 1071.8 cm³
Volume of sphere: 2143.6 cm³

4 **b** 3349.3 m³   **5** 66 987 cm³

6 **a** 226.1 cm³
b They hold the same amount.
c Difference is zero.

7 Number of balls = 25 (when packed in layers)
Remaining space = 70 766 cm³

Exercise 5L

1 **a** 202 cm²   **b** 174 cm²
c 835 cm²   **d** 716 cm²

2 r = 5.6 cm

3 **a** Paper not large enough.
b 104 cm²

4 25 cm²   **5** h = 14 cm

6 290 cm²   **7** **b** 226 cm²

8 **a** cuboid; triangular-based prism; rectangular-based pyramid
b **i** Surface area = 683.4 cm²
ii Volume = 458.7 cm³

Examination-type questions for Chapter 5

1 **a** 10:45
b $203\frac{1}{3}$ km
c 5 hours 45 minutes
d 35.4 km h⁻¹
e 2.03 × 10⁵ m

2 **a** 1.5 cm   **b** 3 cm²
c 6 cm   **d** 15 km; 60 km
e 30 km   **f** 3 km
g Approximately 400 km²

3 Volume of water = 7.3 m³
Surface area = 21.4 m²

4 RV = 36 cm
volume = 2304 cm³

5 **a** $A = 558\ m^3$
  $B = 714.2\ m^3$
  $C = 1205.8\ m^3$
  b Tank C has the greatest value.

6 **a** $1186.9\ cm^3$ **b** $535.4\ cm^2$

7 Volume of wedge = $30\ 000\ cm^3$
  Volume of space = $60\ 000\ cm^3$

8 **a** Greatest perimeter = 90.4 m
   Smallest perimeter = 89.4 m
  b 89.9 m
  c Greatest area = $333.1\ m^2$
   Smallest area = $325.8\ m^2$
  d $329\ m^2$ **e** 164

Chapter 6 — Sets

Exercise 6A

1 Non-examples are **b** and **c**

4 **a** A set of types of transportation
  b A set of vegetables
  c A set of dining utensils
  d A set of stationery
  e A set of farm animals

5 **b** $c \notin \mathbb{N}$
  c Mango $\in$ {Fruits}
  d Cabbage $\notin$ {Fruits}
  e 'A house is not a member of the set of vehicles'
  f 'Radio is a member of the set of electronic equipment'

6 **a** W = {apple, plum, orange, banana}
   $\rightarrow n(W) = 4$
  b $n(V) = 5$ **c** $n(P) = 26$
  d $n(G) = 2$ **e** $n(R) = 3$
  f $n(M) = \emptyset$ or { }

8 Equal sets:
  K = {1, 2, 3, 4} and H = {3, 4, 1, 2}
  Equivalent sets: K, P, J and H

9 **i** {1, 2}: { }, {1}, {2}, {1, 2}
  ii {a, b, c}: { }, {a, b, c}, {a, b}, {a, c}, {b, c}, {a}, {b}, {c}
  iii {dog, pig, cat, goat}: { },{d, c, p, g}, {d}, {c}, {p}, {g}, {d, c}, {d, p}, {p, c}, {d, g}, {p, g}, {c, g}, {d, c, p}, {d, c, g}, {p, d, g}, {p, c, g}

10 **a** Set of odd numbers from 1 to 9
  b Set of even numbers from 2 to 10
  c Set of prime numbers from 2 to 13
  d Set of powers of 10 from 0 to 4
  e Set of negative values of powers of 4 from 1 to 4

11 a = 2, b = 16, c = 32

12 Number of subsets = 64

13 $n(Z) = 128$

15 U = {alphabet}; D = {Letters from b to y};
  D' = {a, z}

16 A = {0, 1, 2, … 30}
  Q' = {0, 1, 2, 4, 5, 7, 8, 10, 11, 13, 14, 16, 17, 19, 20, 22, 23, 25, 26, 28, 29}

17 **a** Y = {9, –9}
  b M = {p, q, r, s, t, u, v, w, x, z}
  c E = {red, yellow, blue}

18 **a** I = {x: x is member the windward islands}
  b C = {x: x is a member of the OECS}
  c R = {x: x is a reptile}

19 **a** Finite, not well-defined
  b Not well-defined
  c Finite, not well-defined
  d Well-defined
  e Well-defined
  f Null or empty
  g Equivalent sets

Exercise 6B

1 **a** $I \cap W$ = {3}
  b $P \cap R$ = {e}
  c $A \cap B$ = {14, 15, 16, 17, 18}
  d $D \cap W$ = {plums}

2 **a** $I \cup W$ = {1, 2, 3, 4, 5}
  b $P \cup R$ = {a, b, c, d, e, f, g, h, j}
  c $A \cup B$ = {1, 2, 3, 4, 5, 6, 7, 8, 9, 10, 11, 12, 13, 14, 15, 16, 17, 18, 19, 20, 21, 22}
  d $D \cup W$ = {apples, mangoes, plums, grapes, oranges, cherries}

3 $S \cap T$ = {20, 30}
  $S \cup T$ = {10, 20, 30, 40, 50, 60}

4 **a** $F \cap G$ = {1, 3}
  b $M \cap K$ = {12}

5 **a** $F \cup G$ = {1, 2, 3, 4, 6, 9, 12}
  b $M \cup K$ = {3, 4, 6, 8, 9, 12, 15, 16}

6 $R \cap Z$ = {0, 1, 2, 3, 4, 5, 6, 7, 8, 9} or {x: $0 \leqslant x \leqslant 9$}
  $R \cup Z$ = {x: $-4 < x < 15$}

7 **a** H' = {0, 2, 4, 6, 8, 10, 12, 14, 16, 18, 20, 22, 24, 26, 28, 30}
  b $Y \cup F$ = {2, 4, 5, 6, 8, 10, 12, 14, 15, 16, 18, 20, 22, 24, 25, 26, 28, 30}
  c $H \cap F$ = {5, 15, 25}

8 **a** $K = \{0, 1, 2, 3, 4, 5, 6, 7, 8, 9, 10\}$
       $M = \{5, 6, 7, 8, 9, 10, 11, 12, 13, 14\}$
       $Z = \{-15, -14, -13, -12, -11, -10, -9, -8, -7, -6\}$
  b M & Z and K & Z are disjoint sets
  c $K \cap M = \{5, 6, 7, 8, 9, 10\}$
  d $K \cap Z = \emptyset$ or $\{\}$
  e $(K \cup M \cup Z)'' = \{-16, -17, -18, -19, -20, -5,$
      $-4, -3, -2, -1, 15, 16, 17, 18, 19, 20\}$

9 **a** Smallest universal set (U) containing O, P
      and E = {Natural numbers up to 10}
  b $P \cap E = \{2\}$
  c $O \cup P = \{2, 3, 5, 7, 9\}$
  d $O \cap P \cap E = \{\}$ or $\emptyset$

10 **a** P = {prime factors of 20} = $\{2, 5\}$
      M = {Multiples of 2 up to 20} = $\{2, 4, 6, 8, 10,$
      $12, 14, 16, 18, 20\}$
      F = {Factors of 20} = $\{1, 2, 4, 5, 10, 20\}$
  b $P \cap M \cap F = \{2\}$
  c $P \cup M \cup F = \{1, 2, 4, 5, 6, 8, 10, 12, 14,$
      $16, 18, 20\}$
  d $(P \cup F)' = \{6, 8, 12, 14, 16, 18\}$, assuming
      $U = \{x: 1 \leqslant x \leqslant 20\}$

12 **a** $E = \{e, n, c, y, c, l, o, p, e, d, i, a\}$
      $M = \{m, a, t, h, e, m, a, t, i, c, s\}$
      $F = \{f, i, g, u, r, a, t, i, v, e\}$
  b $E \cup M \cup F = \{e, n, c, y, l, o, p, d, i, a, m, t, h,$
      $s, f, g, u, r, v\}$
  c $(E \cup M \cup F)' = \{b, j, k, q, w, x, z\}$
  d $E \cap M \cap F = (a, i, e)$
  e $n(E \cup M \cup F) = 19$

13 **a** Universal set for the following:
      i Days of the week: One week
      ii Months of the year: One year
      iii Weeks of the year: One year
  b D = {Monday, Tuesday, Wednesday,
      Thursday, Friday, Saturday, Sunday}
  c M = {January, February, March, …
      December}
  d W = { x: x is a week of the year}
  e $D \cap M \cap W = \{\}$ or $\emptyset$
  f $n(D) = 7$; $n(M) = 12$; $n(W) = 52$

14 **a** $Q = \{1, 2, 4, 5, 10, 20, 25, 50, 100\}$
      $S = \{1, 4, 9, 16, 25, 36, 49, 64, 81, 100\}$
      $M = \{5, 10, 15, 20, 25, 30, 35, 40, 45, 50, 55, 60,$
      $65, 70, 75, 80, 85, 90, 95, 100\}$
      $P = \{2, 3, 5, 7, 11, 13, 17, 19, 23, 29, 31, 37, 41,$
      $43, 47, 53, 59, 61, 67, 71, 73, 79, 83, 89, 97\}$
  b $(Q \cap M \cap S) \cup P = \{2, 3, 5, 7, 11, 13, 17, 19,$
      $23, 25, 29, 31, 37, 41, 43, 47, 53, 59, 61, 67, 71,$
      $73, 79, 83, 89, 97, 100\}$
  c $(Q \cup M \cup P \cup S)' = \{6, 8, 9, 12, 14, 18, 21, 22,$
      $24, 26, 27, 28, 32, 33, 34, 38, 39, 42, 44, 46, 48, 51,$
      $52, 54, 56, 57, 58, 62, 63, 66, 68, 69, 72, 74, 76, 77,$
      $78, 82, 84, 86, 87, 88, 91, 93, 94, 96, 98, 99\}$

d $(Q \cup M \cup P) \cap S = \{1, 4, 25, 100\}$
e $(Q \cup M)' \cup (P \cup S)' = \{3, 6, 7, 8, 9, 10, 11, 12,$
    $13, 14, 15, 16, 17, 18, 19, 20, 21, 22, 23, 24, 26,$
    $27, 28, …, 98, 99\}$

15 **a** $M \subset E$; $H \subset E$
  b **i** 550   **ii** 400   **iii** 400
      iv 150   **v** 300
  c The universal set could be the set of all
      students in the school.

Exercise 6C

1 **b** **i** $A \cap B = \{10, 11, 12, 13, 14, 15\}$
      ii $(A \cup B)' = \{\}$ or $\emptyset$

2 **b** **i** $P \cup Q = \{a, b, c, d, e, f, g, h, l, j, k\}$
      ii $(P \cap Q)' = \{a, b, c, d, e, g, h, i, j, k, l, m\}$
      iii$P' = \{g, h, i, j, k, l, m\}$
      iv $Q' = \{a, b, c, d, e, i, l, m\}$

3

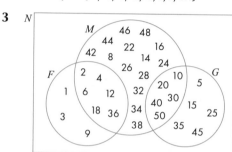

4 **a** F = {mango, apple, banana, cherry,
      tangerine, pear, melon, grape, plum, orange,
      berry, guava, sapodilla, grapefruit, peach}
      H = {grape, plum, orange, berry, guava}
      G = {sapodilla, grapefruit, peach, berry,
      guava}
  b $H \cap G$ = {berry, guava}
  c $H \cup G$ = {grape, plum, orange, berry,
      guava, sapodilla, grapefruit, peach}
  d $(H \cup G)'$ = {mango, apple, banana, cherry,
      tangerine, pear, melon}
  e $(H \cap G)'$ = {mango, apple, banana, cherry,
      tangerine, pear, melon, grape, plum,
      orange, sapodilla, graprefruit, peach}

5 **a** Number of students in class = 26
  b Number of students studying Maths = 10
  c Number of students studying English = 13
  d Number of students studying both = 4
  e Number of students studying neither
      Maths nor English = 7
  f Number of students studying Maths only = 9

6 **a** Number of sportsmen doing 3 sports = 3
  b Number of sportsmen doing 2 sports = 24
  c Number involved in 2 sports = 53
  d Number involved in none of the three
      sports = 12
  e Total = 92

7
  a Number in the class = 31
  b Number that like vanilla only = 8
  c Number that like chocolate only = 3

8
  a $n(O \cap P) = 8$
    Number of students that like only
    one fruit = 40

9 Number of students studying both Spanish
  and French = 2

10 Number that like dancing = 45
  Number that like singing = 25

11
  a Number of housewives surveyed = 52
  b **i** $n(P)$ only = 11
    ii $n(B)$ only = 4
    iii $n(C)$ only = 8

12 $n(P \cap C \cap i) = 6$

13
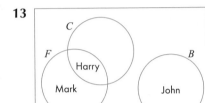

14 Number of athletes in field events only = 6

15 $n(A \cup B \cup C) = 101$

16
  a $P \cap Q \cap R = \{z\}$
  b $P \cap Q = \{i, j, z\}$
  c $Q \cap R = \{z, q\}$
  d $P \cap R = \{r, z\}$
  e $P = \{a, b, c, d, e, f, g, h, i, j, r, z\}$
  f $Q = \{i, j, k, l, m, n, o, p, q, z\}$
  g $R = \{r, q, s, t, u, v, z\}$
  h $P \cup Q \cup R = \{$ a, b, c, d, e, f, g, h, i, j, r, z, q, k, l, m, n, o, p, s, t, u, v$\}$
  i $(P \cup Q \cup R)' = \{w, x, y\}$

Exercise 6D

1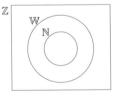

2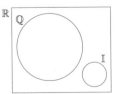

  $\mathbb{I}$ and $\mathbb{Q}$ are disjoint
  and $(\mathbb{I} \cup \mathbb{Q}) = \mathbb{R}$

3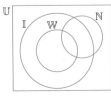

4 $x = 6$; $n(F)$ only = 2; $n(V)$ only = 4

7
  a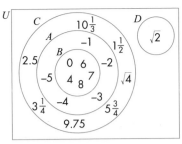

  b **i** $A \cap B \cap C = \{0, 4, 6, 7, 8\}$
    ii $(A \cap B) \cup D = \{0, 4, 6, 7, 8, \sqrt{2}\}$
    iii $(B \cap C) \cup A =$
      $\{0, 4, 6, 7, 8, -5, -4, -3, -2, -1\}$
    iv $D' = \{-5, -4, -3, -2, -1, 0, 1\frac{1}{2}, 2.5, 3\frac{1}{4}, 4,$
      $5\frac{3}{4}, 6, 7, 8, 9.75, 10\frac{1}{3}, \sqrt{4}\}$
    v $(A \cup B)' = \{2.5, 1\frac{1}{2}, 3\frac{1}{4}, 5\frac{3}{4}, 9.75, 10\frac{1}{3}, \sqrt{2},$
      $\sqrt{4}\}$
    vi $n(U) = 18$
    vii $C \cap D = \{\quad\}$

8
  a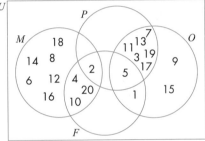

  b **i** $F = \{1, 2, 4, 5, 10, 20\}$
    ii $(M \cap F) = \{2, 4, 10, 20\}$
    iii $P \cap M = \{2\}$
    iv $(P \cup M \cup F) = \{1, 2, 3, 4, 5, 6, 7, 8, 10, 11,$
      $12, 13, 14, 16, 17, 18, 19, 20\}$
    v $(P \cup M)' \cup F = \{1, 2, 4, 5, 9, 10, 15, 20\}$

9 Check students' answers.

10
  a 4 students use all three social media
  b WhatsApp only = 3
  c Facebook only = 15
  d Instagram only = 25
  e Two forms of social media = 40

11
  a $x = 15$
    Total number of girls = 56
  b B or N only = 45

12
  a $x = 5$
    The number liking all three = 7
    The number preferring only 1 = 29
    The number liking none = 15
    The number liking 2 types = 9

13 a ℝ

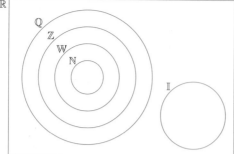

b i ℕ ∩ 𝕎 = ℕ
ii 𝕎 ∩ ℚ = 𝕎
iii ℚ ∩ 𝕀 = { }

14 U

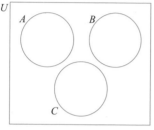

15 Check students' answers.

Examination-type questions for Chapter 6

1 a $10 + 5x - 3 + 2x + 3x - 7$
b $x = 4$
c $n(C \cap F) = 8$
d $n(C \cap F)' = 32$
e $n(C)$ only $= 17$.

2 a $n(U) = 29$
b $A = \{11, 13, 15, 17, 19\}$
c $B = \{1, 2, 4, 7, 14, 28\}$
d

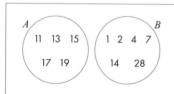

e $A \cup B = \{1, 2, 4, 7, 11, 13, 14, 15, 17, 19, 28\}$
$A \cap B = \{ \}$

3 a $x = 17$
b $n(U) = 57$
c Number of students not involved in sports $= 22$

4 a U

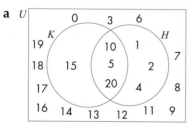

b i $K \cap H = \{5, 10, 20\}$
ii $K \cup H = \{1, 2, 4, 5, 10, 15, 20\}$
iii $(K \cup H)' = \{0, 3, 6, 7, 8, 9, 11, 12, 13, 14, 16, 17, 18, 19\}$

5 a i $U = \{a, b, c, e, f, g, h, i, j\}$
ii $P = \{e, f\}$
iii $Q = \{f, g, h\}$
iv $R = \{h, i, j\}$
v $P \cap Q = \{f\}$
vi $Q \cap R = \{h\}$
vii $P \cap R = \{ \}$
viii $P \cup Q \cup R = \{e, f, g, h, i, j\}$
ix $(P \cup Q \cup R)' = \{a, b, c\}$
x $n(U) = 9$

b

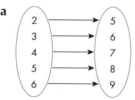

Shaded area: $(P \cap Q) \cup (Q \cap R)$

Chapter 7 — Relations, functions and graphs

Exercise 7A

1 Relation: (3, 4)
Non-example of relation: (3, 5, 7)

2 Domain $= \{2, 4, 6, 8\}$
Range $= \{5, 10, 15, 20\}$

3 Codomain $= \{$letters of the alphabet$\}$

4 a

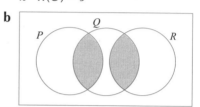

b

x	2	3	4	5	6
y	5	6	7	8	9

c

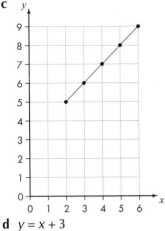

d $y = x + 3$

5 $y = 2x - 3$

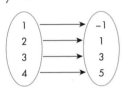

x	1	2	3	4	5
y	–1	1	3	5	7

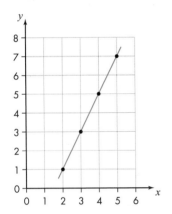

6 Function: Non-function:

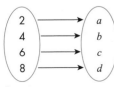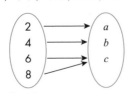

7 Function: F = {(1, 4) (2, 7) (3, 10)}
Non-example: P = {(1, 4) (2, 2) (3, 0) (1, 8)}

8

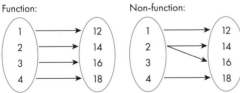

Function   Function
One-to-one   Many-to-one

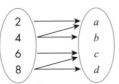

Relation, not a function
One-to-many

9

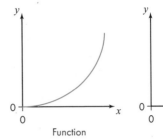

Function   Not a function

10 a 1   **b** –14

11 a f^{-1} = {(1, 1) (3, 2) (5, 3) (7, 4) . . . (17, 9) (19, 10)}

b

y	1	2	3	4	5	6	7	8	9	10
x	1	3	5	7	9	11	13	15	17	19

c

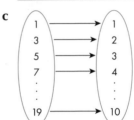

d

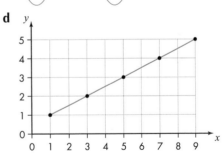

12 a 4   **b** $-\dfrac{1}{5}$

13 Domain = {2, 3, 4}; Range = {4, 9, 16}

14 a Domain = {0, 1, 2, 3, 4, 5}

Range = $\left\{0, \dfrac{2}{3}, \dfrac{4}{3}, \dfrac{6}{3}, \dfrac{8}{3}, \dfrac{10}{3}\right\}$

Codomain = {Rational numbers ⩾ 0}

b Domain = {0, 1, 2, 3, 4, 5, 6, 7}
Range = {1, 2, 5, 10, 17, 26, 37, 50}
Codomain = {Natural numbers}

15 a

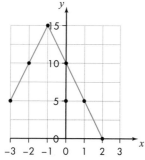

b A function because it is one-to-one.

16 $f^{-1}(x) = \dfrac{x+5}{3}$

17 $fg(x) = 6x - 17$; $fg(-1) = -23$

18 $f(a) = \left\{(1, -\dfrac{5}{8})\ (2, -\dfrac{1}{2})\ (3, -\dfrac{3}{8})\ (4, -\dfrac{1}{4})\right.$

$\left.(5, -\dfrac{1}{8})\ (6, 0)\ (7, \dfrac{1}{8})\ (8, \dfrac{1}{4})\ (9, \dfrac{3}{8})\ (10, \dfrac{1}{2})\right\}$

19 $y = 3x - 4$

20 $g^{-1}(x) = \dfrac{12y + 37}{8}$

21 $gf(x) = 3x - 8$; $gf(-4) = -20$

22 $ff^{-1}(x) = x$; $fg^{-1}(2) = -\dfrac{11}{5}$

23 a $fgh(x) = \dfrac{10x - 7}{4}$

b $gfh^{-1}(2) = \dfrac{91}{2}$

c $hfg^{-1}(-5) = \dfrac{19}{20}$

Exercise 7B

1 **a, c** and **d** are linear functions

2

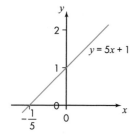

3

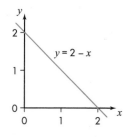

4 a

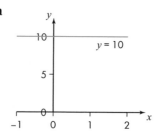

b

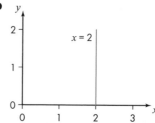

5 a, b

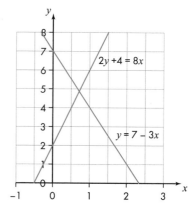

6 PQ = 6.4

7 **b** Right-angled triangle
c Perimeter = 24; Area = 24
d AC = 10; midpoint = (5, 5)

8 $y = \dfrac{2}{3}x - 2$; $m = \dfrac{2}{3}$, $c = -2$

9 $y = \dfrac{7}{2}x + \dfrac{3}{4}$ or $4y = 14x + 3$

10

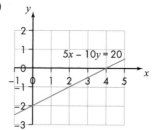

11 **b** Rectangle
c Perimater = 20; Area = 24
d PR = 7.21; midpoint = (5, 8)

12 a Neither
b Perpendicular
c Parallel

13 $y = \dfrac{7}{2}x + 1$

14 $y = 0.8x + 3.4$

15 $y = \dfrac{1}{2}x + 7$

16 $y = -\dfrac{2}{3}x + \dfrac{22}{3}$

17 $y = x$

18

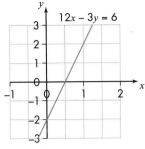

19 b Trapezium
   c Perimeter = 46.6
     Area = 126
   d TV = 16.64
   e (2.5, 1)

20 $y = \dfrac{-1}{2}x + 8$

21 a $y = -\dfrac{14}{9}x + \dfrac{4}{3}$

   b $y = \dfrac{9}{14}x - \dfrac{81}{14}$

22 $y = \dfrac{3}{7}x + \dfrac{6}{7}$

23 $x = 0$ **24** $y = 0$ **25** $y = -x$

26 a Quadrilateral
   b Perimeter = 34.1
     Area = 70
   c (1, −1)

27

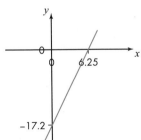

Exercise 7C

1 a Spain > Jamaica **b** $V_S < V_R$
   c $J \geqslant 12$ (years)   **d** $8 < h < 9$ (metres)
   e $B \leqslant 35$ ($B \in \mathbb{W}$)

2 a $x > 5$ **b** $x \leqslant \dfrac{3}{5}$

3 a $x = \{x: x \leqslant -2\}$
   b $x = \{x: x < 1\}, x \in \mathbb{R}$

4 a

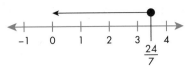

   b

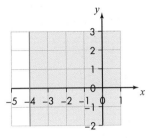

5 a

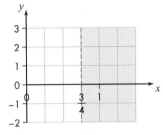

   b

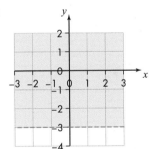

6 a

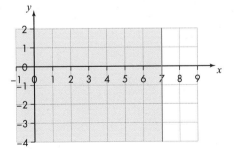

   b

   c

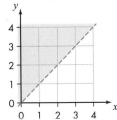

d

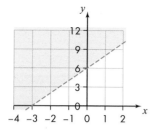

7 $x = \{x: x \geqslant 3\}, x \in \mathbb{W}$

8

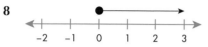

9

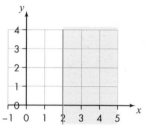

10

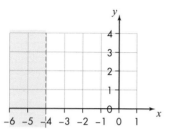

11

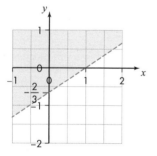

12

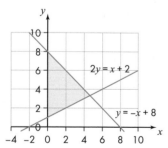

2y = x + 2
y = −x + 8

13 Maximum = 5
Minimum = −62.

14 $x = 4, y = 5$

15 $a = 3, b = 10$

16 $x = \varnothing$ (no solutions)

17

18

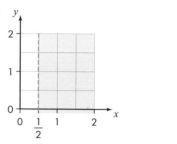

19

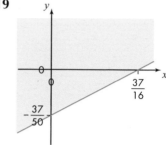

20

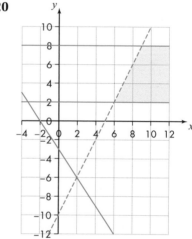

21 Maximum = 74, Minimum = −66

22 Line: $y = 2x - 2$, Perpendicular bisector:

$$y = -\frac{1}{2}x + \frac{11}{2}$$

23 $x = -1.6, y = 3.7$

Exercise 7D

1   **a** $a = 1$
    b $a = 2, c = -1$
    c $a = 9, b = -20$
    d $a = -3, b = -20, c = -2$

2   Answers may vary

3

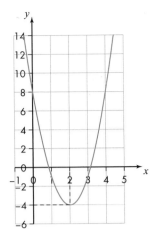

4

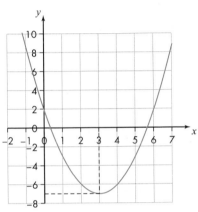

Axis of symmetry: $x = 3$

5   Axis of symmetry: $x = 0$

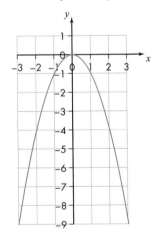

6

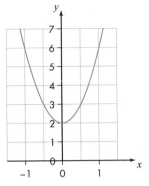

Gradient at $x = 2$ is 16

7   **a**   Minimum = –15 at $x = -1$
       b   Maximum = 18 at $x = 3$

8

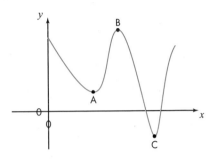

B is a maximum, the highest point on the graph
C is a minimum, the lowest point on the graph
(A is a turning point)

9 $x = 0.4$ and $x = 3.6$

10 $x = 1.2$ and $x = 2.8$

11   **a**   minimum is $f(x) = 0$ or $y = 0$
          root is $x = -1$
      b   maximum is $f(x) = 20$ or $y = 20$
          roots are $x = 0$ and $x = 4$

12   Axis of symmetry: $x = 10$
      minimum is $f(x) = 0$ or $y = 0$
      root is $x = 10$

13   **a**   minimum $f(x) = 4$
      b   minimum at $x = -1$
      The function has no roots

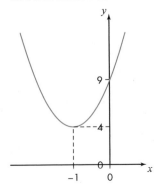

14 $x = \{x: x \leqslant -1\}$ or $x \geqslant 1$

15 axis of symmetry: $x = 4$

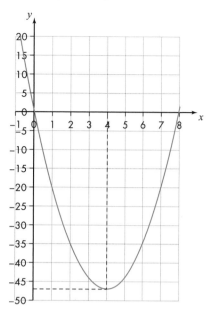

16

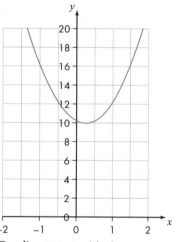

Gradient at $x = 1$ is 6

17 $x = \{x: x \in \mathbb{R}\}$, where $\mathbb{R}$ is the set of real numbers

18

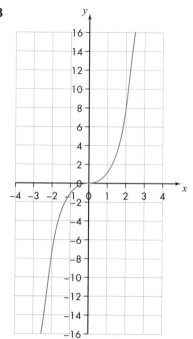

19

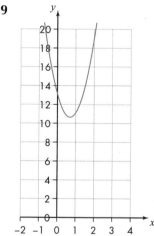

20

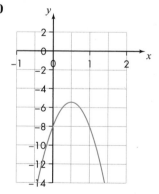

21 a

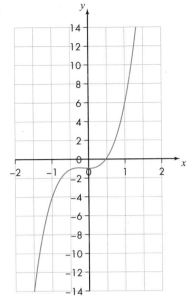

b Turning points: $(0, -1)$, $(-0.27, -0.95)$

22

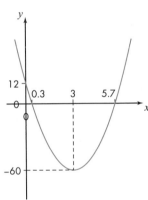

23

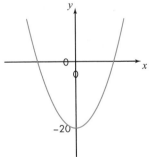

Axis of symmetry: $x = 0$ (y-axis)

24 $f(x) = 6x^2 - 8x + 10$ has no roots
$f(x) = 6x^2 + 20x - 3$ has roots $x = -2.2$ and $x = 0.23$

25

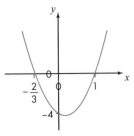

$x = \{x: -\dfrac{2}{3} \leqslant x \leqslant 1\}$

26

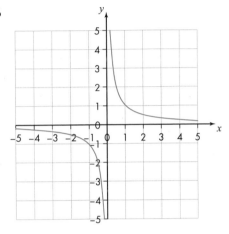

27

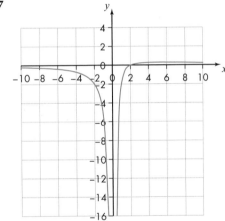

Examination-type questions for Chapter 7

1 a i $f(-2) = -3$
　　ii $x = -2$
　　iii Range $= \left\{f(x): -\dfrac{15}{4} < f(x) < -\dfrac{3}{2}\right\}$

　b $x \leqslant \dfrac{1}{2}$

c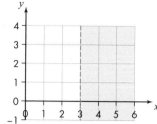

2 a $f(-1) = -21$; $g\left(\dfrac{1}{2}\right) = 3$

b $gf(-2) = -\dfrac{61}{2}$

c $f^{-1}(10) = \dfrac{19}{12}$

3 a $-\dfrac{2}{5}$

b $(4, 3)$ **c** $y = \dfrac{5}{2}x - 7$

4 a $y = \dfrac{k}{x}$ **b** $p = 1$, $q = 0.5$

c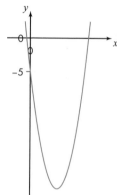

x-intercepts: $x = -0.4$,
$x = 6.4$
minimum $= -23$
at $x = 3$

d $x = -0.3$, $x = 6.8$

5 a Roots: $x = -3$, $x = 6$
b $c = -18$
c $b = -3$

d Axis of symmetry: $x = \dfrac{3}{2}$

minimum point is $\left(\dfrac{3}{2}, \dfrac{-81}{4}\right)$ or $(1.5, -20.25)$

6 a $y = -\dfrac{1}{2}x - \dfrac{3}{2}$

gradient $= -\dfrac{1}{2}$, *y*-intercept $= -\dfrac{3}{2}$

b $y = 2x - 1$

c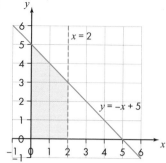

d Greatest value: vertex $(0, 0)$ or $(0, 5)$.
Least valve: vertex $(2, 0)$ or $(2, 3)$.

7 a 2 stops; longest was 3 hours because
CD > AB

b 15 hours; $2\dfrac{2}{15}$ or 2.13 km h^{-1}

c slowest: EH; speed $= 1\dfrac{1}{3}$ or 1.33 km h^{-1}

fastest: OA; speed $= 6$ km h^{-1}
d towards his destination
e 5 hours

Chapter 8 — Statistics

Exercise 8A

1 Examples may vary

2 Examples may vary

3 Examples may vary

4

Runs	Tally	Frequency
0–20	‖‖‖ ‖‖‖ ‖	11
21–40	‖‖‖	5
41–60	‖‖‖	3
61–80	‖‖	2
81–100	‖	1

5

Runs	Frequency
0–20	11
21–40	5
41–60	3
61–80	2
81–100	1

6 a i

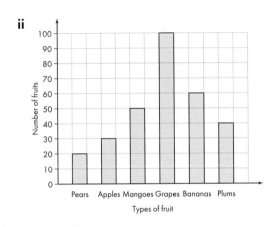

ii

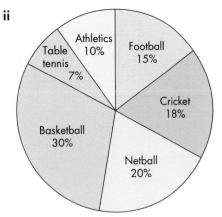

b Percentages are shown in the pie chart

b Pears 6.7%
Apples 10%
Mangoes 16.7%
Grapes 33.3%
Bananas 20%
Plums 13.3%

7 a i

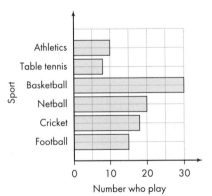

8

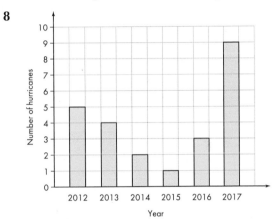

9

10 a

Mass	Tally	Frequency
51–60	‖	2
61–70	‖‖ ‖‖ ‖	11
71–80	‖‖ ‖‖ ‖	11
81–90	‖‖ ‖‖ ‖‖ ‖	16

b 60 and 51

c 9

11 a Vegetables   **b** $300

12 a 4

b

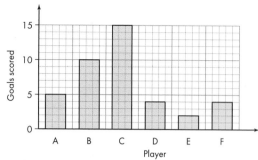

13

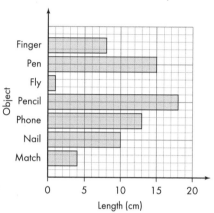

14

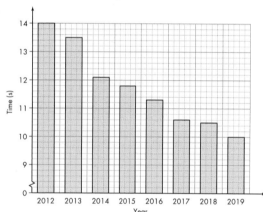

15

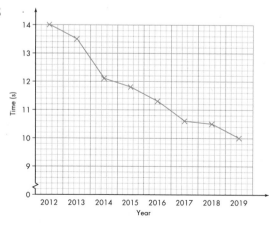

16

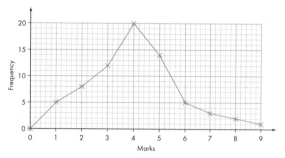

17 a Class boundaries: 50.5–60.5
                                60.5–70.5
                                70.5–80.5
                                80.5–90.5

b Class width: 10

c

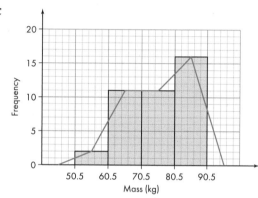

18 Class boundaries: 149–156, 156–163, 163–170, 170–177, 177–184, 184–191, 191–198, 198–205.

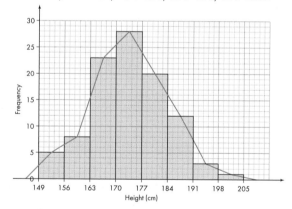

19

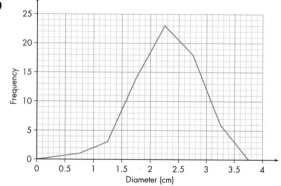

20

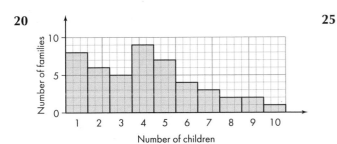

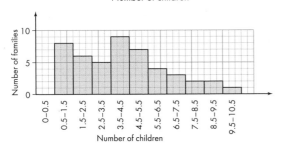

21

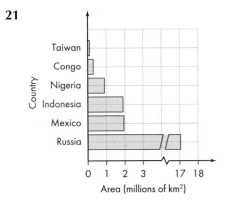

22

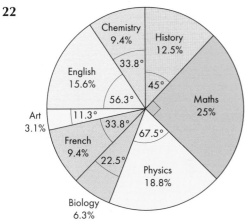

23 a 15°   **b** 2   **c** 48
  d grapes 27.1%, melons 4.2%, apples 37.5%, bananas 14.6%, cherries 16.7%.

24 a 22 000
  b 2004, 5000
  c 2001, 1000
  d i 2002, 3000; 2004, 2000; 2006, 1000; 2007, 1000.
    ii 2003, 1000; 2005, 3000.

25

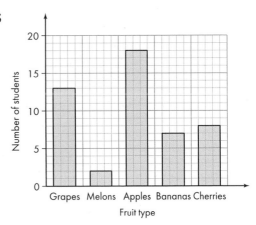

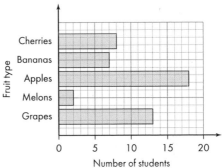

26 a

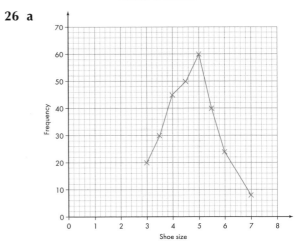

  b Discrete

27

Age (years)	Class limit	Class boundary	Frequency
10–19	10, 19	9.5–19.5	50
20–29	20, 29	19.5–29.5	60
30–39	30, 39	29.5–39.5	34
40–49	40, 49	39.5–49.5	18
50–59	50, 59	49.5–59.5	40
60–69	60, 69	59.5–69.5	54
70–79	70, 79	69.5–79.5	10

28

Number of wickets	Frequency
0–4	40
5–9	20
9–12	20
13–16	40
17–20	0

Exercise 8B

1 6   **2** 8   **3** 3

4 median: 4
mode: 3
mean: 4.44

5 median: 5.5
mode: 4 and 8
mean: 5.5

6 mean: 5.24
median: 5
mode: 6

7 9.8   **8** 9.8

9 a

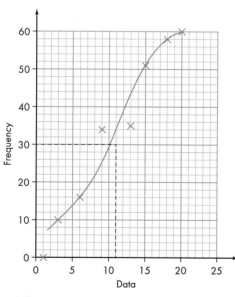

b median: 10.5
c 7–9

10 a $Q_1 = 5.8$, $Q_2 = 11$, $Q_3 = 12.8$
interquartile range = 7.0, semi-interquartile range = 3.5

11 a Range = 7, interquartile range = 4,
Semi-interquartile range = 2.
b Range: 10, interquartile range = 4
Semi-interquartile range = 2

12 a Standard deviation measures the spread of data
b Class B. They had a greater standard deviation.
c About 40

13 a 60   **b** $\frac{1}{15}$ **c** 60%

14 a 30   **b** 12   **c** 2

Exercise 8C

1 $\frac{1}{2}$ **2** $\frac{1}{52}$ **3** $\frac{1}{6}$

4 $\frac{2}{3}$ **5** 0   **6** $\frac{1}{13}$

7 1   **8** $\frac{3}{11}$ **9** $\frac{1}{3}$

10 $\frac{12}{17}$ **11** $\frac{3}{7}$

12 $U = \{$letters of the alphabet$\}$
Vowels, $V = \{a, e, i, o, u\}$
$V' = \{$consonants$\}$

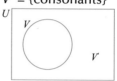

13 $U = \{1, 2, 3, 4, 5, 6, 7, 8, 9, 10\}$
Favourable, $F = \{1, 4, 9\}$
Unfavourable, $F' = \{2, 3, 5, 6, 7, 8, 10\}$

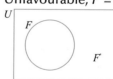

14 $\frac{1}{12}$, independent events   **15** $\frac{17}{52}$.

16 a $\frac{5}{17}$ **b** $\frac{3}{8}$ **c** $\frac{1}{3}$

17 a $\frac{6}{49}$ **b** $\frac{8}{49}$ **c** $\frac{6}{49}$

18 a

Youths	Boys	Girls	Totals
Facebook	40	60	100
Instagram	30	10	40
Totals	70	70	140

b $\frac{3}{7}$ **c** $\frac{6}{7}$

d The preference is dependent.

Examination-type questions for Chapter 8

1 a 18, 26   **b** 8

c

Distance thrown	Frequency	Cumulative frequency
0–8	1	1
9–17	3	4
18–26	6	10
27–35	12	22
36–44	8	30
45–53	4	34

d 10

e

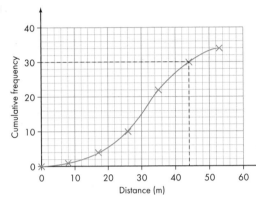

f line drawn on graph in (e)

g $\dfrac{12}{17}$

2 a i 30

   ii 16.5

   iii 74

b i

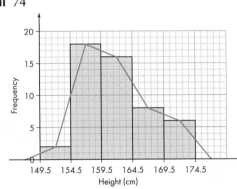

  ii 161.8

  iii 155–159

  iv $\dfrac{8}{25}$

3 a

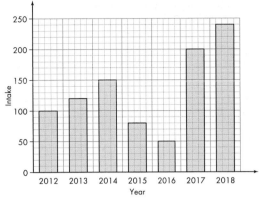

b i 2018

   ii 2016

c i 2016–2017

   ii 2014–2015

d 940

4 a 20°   **b** Vanilla

  c 25%   **d** 180

5 a

Number of fruits	Frequency
0	10
2	20
4	20
6	25
8	10
10	15
12	5

b Median = 6, mode = 6

c 5.33   **d** 105   **e** $\dfrac{2}{21}$

Chapter 9 — Vectors, matrices and transformations

Exercise 9A

1 a 5 N   **b** 15 m

  c 40 m s⁻¹   **d** 100 m s⁻²

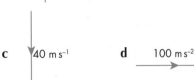

2 a 36 N   **b** 432 N **c** 171 N **d** 441 N

3 a 195 m **b** 45 m **c** 150 m **d** 367.5 m

4 a 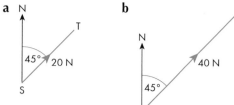 **b**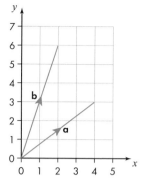

c 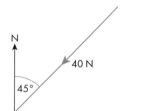 **d**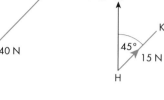

5 a 6a   **b** 2a   **c** −a   **d** 0

6   27.9 km

7   25 m at 53° to the horizontal

8   10 N at 53° anticlockwise from the direction of **p**

9   18.9 m on a bearing of 086.2°

10 $\sqrt{5}$ N at 26.6° anticlockwise from direction of 10 N vector.

11 13.2 N on a bearing of 059.5°

12 a 90.8 m
   b On a bearing of 300°

13 21 a

14 $\sqrt{a^2 + b^2 + c^2 + \sqrt{2}\,(\mathbf{bc} - \mathbf{ac})}$

15 a $\overrightarrow{AC} = \mathbf{a} + \mathbf{c}$

  b $\overrightarrow{BD} = \mathbf{b} - \mathbf{c}$

  c $\overrightarrow{AB} = \mathbf{a} + \mathbf{c} - \mathbf{b}$

  d $\overrightarrow{DQ} = \mathbf{c} - \dfrac{1}{2}\mathbf{b}$

  e $\overrightarrow{BP} = \mathbf{b} - \mathbf{c} - \dfrac{1}{2}\mathbf{a}$

  f $\overrightarrow{PC} = \dfrac{1}{2}\mathbf{a} + \mathbf{c}$

  g $\overrightarrow{PQ} = \dfrac{1}{2}\mathbf{a} + \mathbf{c} - \dfrac{1}{2}\mathbf{b}$

Exercise 9B

1 a $\overrightarrow{OP} = 3\mathbf{i} + 5\mathbf{j}$
     $\overrightarrow{OQ} = \mathbf{i} + 10\mathbf{j}$

  b $\overrightarrow{OP} = \begin{pmatrix} 3 \\ 5 \end{pmatrix}$

     $\overrightarrow{OQ} = \begin{pmatrix} 1 \\ 10 \end{pmatrix}$

2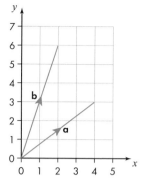

3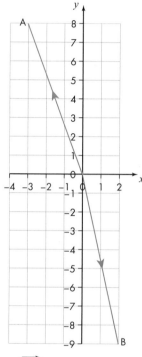

4 a $|\overrightarrow{OR}| = 5$ at 53°

  b $|\overrightarrow{OM}| = 10$ at 53°

5 a a, $\overrightarrow{OP}$; magnitude = 12.65 at 18.4°

6 $\dfrac{1}{14}(8\mathbf{i} - 12\mathbf{j})$

7 **b** − **a**

8 a 11a   **b** a

9 a $\overrightarrow{OW} = \begin{pmatrix} 2 \\ 6 \end{pmatrix}$, $\overrightarrow{OS} = \begin{pmatrix} 10 \\ 2 \end{pmatrix}$

  b $\begin{pmatrix} 12 \\ 8 \end{pmatrix}$ **c** $\begin{pmatrix} 8 \\ -4 \end{pmatrix}$

10 a $\overrightarrow{BC} = \mathbf{b} - \mathbf{a}$

  b $\overrightarrow{CD} = \mathbf{c} - \mathbf{b}$

  c $\overrightarrow{AB} = \mathbf{b} - \mathbf{a}$ or $\overrightarrow{AB} = \mathbf{c} - \mathbf{b}$

    $\overrightarrow{DE} = \mathbf{b} - \mathbf{a}$ or $\overrightarrow{DE} = \mathbf{c} - \mathbf{b}$

11 a $\overrightarrow{OP} = 5\mathbf{i} + 2\mathbf{j} + 4\mathbf{k}$

  b $|\overrightarrow{OP}| = 6.71$ **c** $41.8°$

12

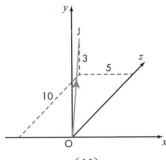

$|\overrightarrow{OJ}| = 11.58$

13 Resultant $= \begin{pmatrix} 11 \\ 6 \end{pmatrix}$

14 $\dfrac{100}{877}(-2\mathbf{i} + 3\mathbf{j} - 8\mathbf{k})$

15 a $\overrightarrow{OL} = \mathbf{p} + 2\mathbf{r}$;

    $\overrightarrow{OM} = 3\mathbf{p} + 6\mathbf{r}$
         $= 3(\mathbf{p} + 2\mathbf{r})$

    $\overrightarrow{ON} = -2\mathbf{p} - 4\mathbf{r}$
         $= -2(\mathbf{p} + 2\mathbf{r})$

    $\overrightarrow{OL}$, $\overrightarrow{OM}$ and $\overrightarrow{ON}$ are all multiples of $(\mathbf{p} + 2\mathbf{r})$, so they are parallel.

    They all have a common point, O, so L, M and N are colinear.

  b $\overrightarrow{LM} + \overrightarrow{MN} = -3(\mathbf{p} + 2\mathbf{r})$

  c

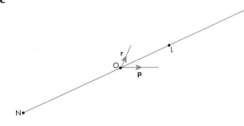

Exercise 9C

1 Examples may vary

2 $\begin{pmatrix} 100 & 150 & 240 \\ 200 & 200 & 120 \\ 300 & 350 & 180 \\ 250 & 100 & 300 \end{pmatrix}$

3 Examples:
  a $(2 \quad 5 \quad -3)$

  b $\begin{pmatrix} -8 \\ 5 \\ 3 \end{pmatrix}$ **c** $\begin{pmatrix} 3 & 5 \\ 0 & -9 \end{pmatrix}$

d $\begin{pmatrix} 0 & 0 & 3 \\ 0 & 16 & 0 \\ 12 & 0 & 0 \end{pmatrix}$ **e** $\begin{pmatrix} 0 & 0 \\ 0 & 0 \end{pmatrix}$

f $\begin{pmatrix} 1 & 0 & 0 \\ 0 & 1 & 0 \\ 0 & 0 & 1 \end{pmatrix}$ **g** $\begin{pmatrix} 3 & 9 & -5 \\ -1 & 6 & 0 \end{pmatrix}$

4 a Square, 2×2 **b** Row, 1×3
  c Diagonal, 3×3 **d** Rectangular, 2×3

5 a $\begin{pmatrix} 12 & -1 \\ 19 & 6 \end{pmatrix}$ **b** $\begin{pmatrix} 0 & 9 \\ 13 & 9 \\ 16 & -13 \end{pmatrix}$

6 a $(1 \quad 1 \quad 9)$ **b** $\begin{pmatrix} 6 \\ 3 \end{pmatrix}$

7 a $\begin{pmatrix} 21 & 0 \\ -7 & 35 \end{pmatrix}$ **b** $\begin{pmatrix} 11 \\ 0 \\ -33 \end{pmatrix}$

8 a $\begin{pmatrix} 7 & 7 \\ 21 & 28 \end{pmatrix}$

  b number of columns in **A** $\neq$ number of rows in **B**
  c Number of columns in **B** $\neq$ number of rows in **D**
  d $(32 \quad 36 \quad 4)$

  e $\begin{pmatrix} 78 & 4 & -4 \\ 69 & 31 & 20 \\ 36 & 4 & 0 \end{pmatrix}$

9 $PQ = \begin{pmatrix} 36 & 35 \\ 38 & 29 \end{pmatrix}$ $QP = \begin{pmatrix} 32 & 22 \\ 61 & 33 \end{pmatrix}$

  $PQ \neq QP$

10 $x = 7$, $y = 4$, $z = 5$

11 $a = 7$, $b = -18$, $c = -8$

12 a $\begin{pmatrix} 2 & 4 \\ -1 & 9 \end{pmatrix}$ **b** $\begin{pmatrix} 9 & 1 \\ -4 & 2 \end{pmatrix}$

  c $\begin{pmatrix} 22 & 0 \\ 0 & 22 \end{pmatrix}$

13 a -22

  b $\begin{pmatrix} 1 & 0 \\ 0 & 1 \end{pmatrix}$ **c** $\dfrac{1}{-22}\begin{pmatrix} 3 & -5 \\ -2 & -4 \end{pmatrix}$

14 a For example: $\begin{pmatrix} 4 & 8 \\ 3 & 6 \end{pmatrix}$ is singular because

$(4 \times 6) - (8 \times 3) = 0$

b i Not singular because $-12 - 12 \neq 0$
ii Singular because $20 - 20 = 0$

15 a $x = \dfrac{3}{2}, y = -2$
b $x = 1, y = -7$

Exercise 9D

1 $\begin{pmatrix} 6 \\ 17 \end{pmatrix}$

2 $\begin{pmatrix} 5 \\ 15 \\ -5 \end{pmatrix}$

3 12.8 at 51.3°

4 $\begin{pmatrix} 1 \\ 8 \end{pmatrix}$

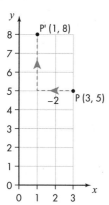

5

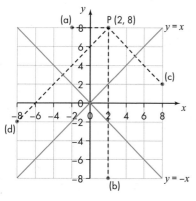

6

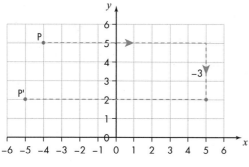

7 $\begin{pmatrix} 5 & 3 \\ 4 & 6 \end{pmatrix}\begin{pmatrix} 5 \\ 1 \end{pmatrix} = \begin{pmatrix} 28 \\ 26 \end{pmatrix}$

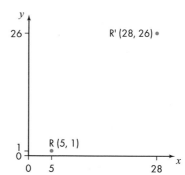

8 $\begin{pmatrix} 0 & -1 \\ 1 & 0 \end{pmatrix}\begin{pmatrix} 2 \\ 6 \end{pmatrix} = \begin{pmatrix} -6 \\ 2 \end{pmatrix}$

9

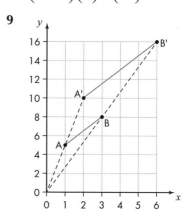

10 $P(5, 0) \rightarrow P'(5, 0)$
$Q(10, 0) \rightarrow Q'(10, 0)$
$R(10, 4) \rightarrow R'(10, 12)$
$S(5, 4) \rightarrow S'(5, 12)$

11

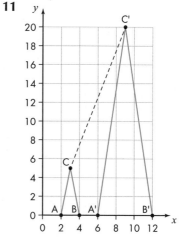

12

13

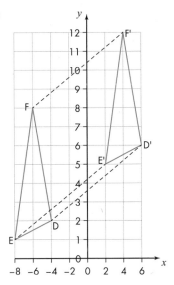

14 a A(1, 2) → A′(−1, 2)
B(6, 2) → B′(−6, 2)
C(6, 6) → C′(−6, 6)
D(1, 6) → D′(−1, 6)

b A → A′(1, −2)
B → B′(6, −2)
C → C′(6, −6)
D → D′(1, −6)

c A → A′(2, 1)
B → B′(2, 6)
C → C′(6, 6)
D → D′(6, 1)

d A → A′(−2, −1)
B → B′(−2, −6)
C → C′(−6, −6)
D → D′(−6, −1)

15

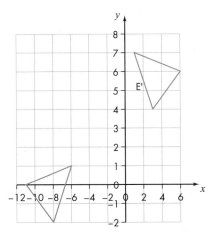

16

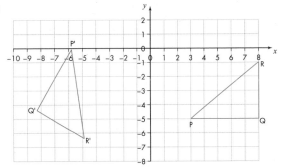

17 H′(10, 15), K′(40, 15), L′(25, 45)

18 P′(−0.08, −3.6), Q′(−6.9, −5.4) or

$$P'\left(\frac{-1}{13}, \frac{-47}{13}\right) \quad Q'\left(\frac{-90}{13}, \frac{-70}{13}\right)$$

19

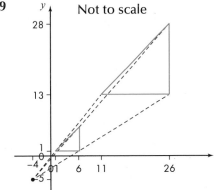

Not to scale

20 (2, −1)

Examination-type questions for Chapter 9

1 b $B^{-1} = -\dfrac{1}{31}\begin{pmatrix} 6 & -5 \\ -7 & 0 \end{pmatrix}$

c $BB^{-1} = \begin{pmatrix} 1 & 0 \\ 0 & 1 \end{pmatrix}$

d $x = 2, \quad y = 1$

2 a

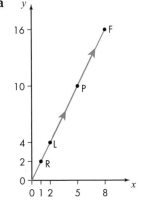

Not to scale

b $\overrightarrow{OR} = \begin{pmatrix} 1 \\ 2 \end{pmatrix}$, $\overrightarrow{OL} = \begin{pmatrix} 2 \\ 4 \end{pmatrix}$

c $\overrightarrow{PF} = \begin{pmatrix} 3 \\ 6 \end{pmatrix}$, $\overrightarrow{RL} = \begin{pmatrix} 1 \\ 2 \end{pmatrix}$

3 a $a = 1$, $b = 2$
  b $(1, -4)$

4 a $\overrightarrow{OA} = \begin{pmatrix} 0 \\ 8 \end{pmatrix}$, $\overrightarrow{OB} = \begin{pmatrix} 8 \\ 12 \end{pmatrix}$, $\overrightarrow{OC} = \begin{pmatrix} 16 \\ 0 \end{pmatrix}$

  b 14.4 at 56.3°

  c $\overrightarrow{BC} = \begin{pmatrix} 8 \\ -12 \end{pmatrix}$

5 a $\overrightarrow{PQ}$ and $\overrightarrow{SR}$
  b $\overrightarrow{PQ} + \overrightarrow{SR} = 3t$
  c $\overrightarrow{SQ} = 2t - v$
  d $\overrightarrow{PS} = v - t$
  e $\overrightarrow{PR} = t + v$
  f $\overrightarrow{PQ} - \overrightarrow{SR} = -t$

Chapter 11 — Examination Practice

Examination 1

Paper 1

1	C	2	D
3	C	4	D
5	C	6	D
7	B	8	B
9	C	10	A
11	C	12	D
13	C	14	B
15	B	16	C
17	C	18	B
19	B	20	A
21	D	22	A
23	C	24	A
25	C	26	C
27	C	28	A
29	D	30	D
31	B	32	D
33	B	34	A
35	C	36	C
37	D	38	A
39	C	40	C
41	C	42	D
43	C	44	D
45	D	46	C
47	B	48	C
49	C	50	C
51	D	52	C
53	C	54	D
55	D	56	C
57	B	58	D
59	D	60	C

Paper 2

Section 1

1 a i $\dfrac{20}{9}$
  ii 13.554
  b i 13.55
  ii 10
  iii 1.3554×10^1
  c i $1000.00
  ii 55%

2 a i 12
  ii $\dfrac{17}{6} = 3\dfrac{1}{6}$
  b $x > -1$

  c i $3x + y = 5$ $5x + 2y = 9$
  ii Mangoes cost $1.00; bananas cost $2.00

3 a i

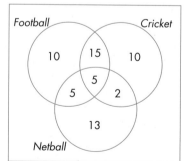

  5 people play all three sports

ii 10 people only play football; 10 people only play cricket and 13 people play only netball so 33 people play only one sport

b i & iii

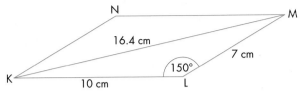

ii KM = 16.4 cm

4 a i $y = 4x + 2$; gradient = 4; intercept = 2
  ii $h = 10$
  iii $y = -\dfrac{x}{4} + \dfrac{21}{2}$

b i–iv & c

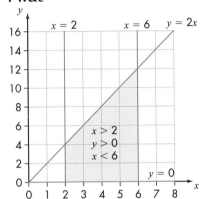

5 a i Interior angle 160° and exterior angle 20°
  ii 18 sides
  iii 180 cm
 b i 1944π l
  ii 243π l
  iii 0.57 m

6 a i 40°
  ii $1000
  iii $3600
  iv

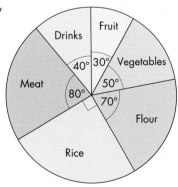

b P'(8, 0), Q'(4, 2), R'(2, 1)
c P'(−4, 8), Q'(−2, 4), R'(−3, 2)
d

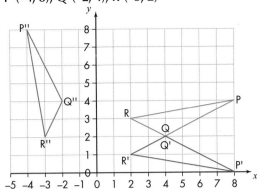

7 a

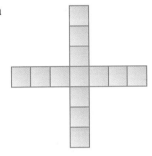

b i 4   13   7 × 4 = 28
  ii 6   21   11 × 4 = 44
  iii 8   29   15 × 4 = 60
  iv 32   125   63 × 4 = 252

Section 2

8 a i $y = 3 + 8x - 2x^2$

x	−1	0	1	2	3	4	5
y =	−7	3	9	11	9	3	−7

ii&iii

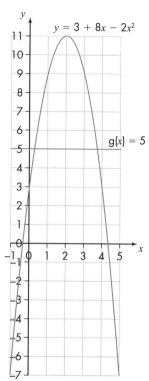

At the intersections of the graphs $3 + 8x - 2x^2 = 5$ which simplifies to $2x^2 - 8x + 2 = 0$ The solutions are the x values at the intersections: $x = 0.27$ and $x = 3.73$

b i 12 – 14 s; 5 m/s²

ii 10 – 12 s; 2.5 m/s²

iii 95 m

9 a i

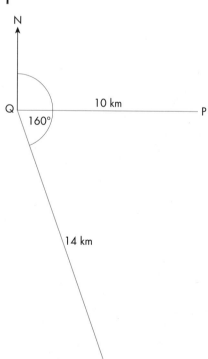

ii PR = 14.15 km; bearing of R from P is 158.4°

b i 90°

ii 60°

iii 90°

iv 6.7 cm

10 a i $x = 2, y = 13$

ii $r = 17, s = 3$

iii T moves point P parallel to the x-axis for a distance of five time the x coordinate W moves point P parallel to the x-axis for a distance of five times the y coordinate

iv $WT = \begin{pmatrix} 1 & 5 \\ 5 & 26 \end{pmatrix}$

v (31, 161)

b i $-2\mathbf{p} + \mathbf{q}$

ii $\mathbf{q} + \mathbf{p}$

iii $2\mathbf{q} - \mathbf{p}$

Examination 2

Paper 1

1	B	2	C	3	A
4	A	5	C	6	D
7	B	8	A	9	C
10	C	11	D	12	C
13	A	14	D	15	A
16	C	17	D	18	D
19	C	20	B	21	A
22	D	23	A	24	B
25	C	26	B	27	B
28	C	29	B	30	D
31	C	32	D	33	A
34	A	35	D	36	C
37	A	38	D	39	B
40	C	41	A	42	B
43	D	44	D	45	C
46	D	47	D	48	D
49	C	50	A	51	C
52	C	53	C	54	C
55	B	56	D	57	C
58	A	59	D	60	D

Paper 2

Section 1

1 a i 1.1
　ii 1
　b i \$37 000.00
　ii 85%
　c i 24 180 kWh
　ii \$7254.00
　iii \$12 090.00

2 a i $(5x - 1)^2$
　ii $(9a + 2b)(9a - 2b)$
　iii $3xy(x - 3y)$
　b $x = -2.5$
　c i $x = \sqrt{\dfrac{a^3 + a^2 b}{b^2 - a}}$
　ii $x = +1$ or -1

3 a i $E = \{2, 4, 6, 8, 10, 12, 14, 16, 18\}$
　ii $F = \{1, 2, 3, 6, 9, 18\}$
　iii

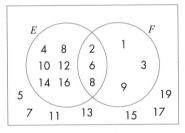

　iv $E \cap F = \{2, 6, 18\}$
　b i

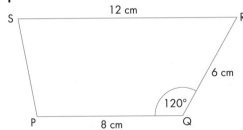

　ii PS = 6 cm
　iii Angle QPS = 120°

4 a i $f(-1) = -3$
　ii $g^{-1}(-1) = -\dfrac{1}{2}$
　iii $fg(x) = 8x + 1$
　iv $gf^{-1}(2) = 7$
　b

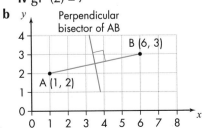

Equation of the line: $5y = x + 9$
Equation of the perpendicular bisector:
$y = -5x + 19$

5 a i Right-angled trapezium or right trapezoid
　ii 35°
　iii 35°
　iv 13.95 cm
　v 8.51 cm
　vi 36.45 cm
　b

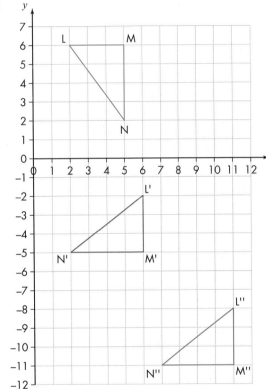

6 a i 147π m²
　ii $(21\pi + 28)$ m
　iii 367.5π m³
　b i 100 cm
　ii 120 cm
　iii 981 250 l

7 a

　b i 4　　9　　$4 \times 3 = 12$
　ii 8　　17　　$8 \times 3 = 24$
　iii 13　　27　　$13 \times 3 = 39$
　iv 43　　87　　$43 \times 3 \times 129$

Section 2

8 a i

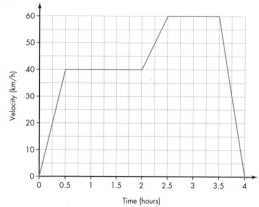

ii 80 km/h²

iii Between 2 – 2.5 hours after the start of the journey

iv Slowing down; gradient = –120 km/h²; rate of deceleration

v 170 km

b $x = 2$ and $y = -2$; $x = -\frac{1}{2}$ and $y = 3$

9 a i

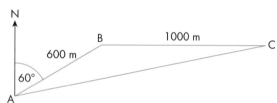

ii 079°

iii 1549 m

b i 120°

ii 60°

iii 30°

10 a $a = 4$, $b = 10$; enlargement scale factor 2

b

$$\begin{pmatrix} 2 & 3 \\ 2 & -3 \end{pmatrix}\begin{pmatrix} x \\ y \end{pmatrix} = \begin{pmatrix} 4 \\ 8 \end{pmatrix}$$

$$\frac{-1}{12}\begin{pmatrix} -3 & -3 \\ -2 & 2 \end{pmatrix}\begin{pmatrix} 4 \\ 8 \end{pmatrix} = \begin{pmatrix} 3 \\ -\frac{2}{3} \end{pmatrix}$$

$x = 3$, $y = -\frac{2}{3}$

c i $P = \begin{pmatrix} 10 \\ 8 \end{pmatrix}$ $Q = \begin{pmatrix} 20 \\ 2 \end{pmatrix}$

ii $\begin{pmatrix} 2.5 \\ 2 \end{pmatrix}$ **iii** $\begin{pmatrix} 10 \\ 1 \end{pmatrix}$ **iv** $\begin{pmatrix} -7.5 \\ 1 \end{pmatrix}$

Examination 3

Paper 1

1	C	**2**	D	**3**	C
4	C	**5**	B	**6**	A
7	B	**8**	C	**9**	A
10	D	**11**	C	**12**	B
13	C	**14**	C	**15**	B
16	A	**17**	B	**18**	C
19	C	**20**	A	**21**	B
22	B	**23**	C	**24**	D
25	B	**26**	A	**27**	B
28	C	**29**	D	**30**	C
31	C	**32**	D	**33**	D
34	D	**35**	C	**36**	C
37	A	**38**	A	**39**	C
40	B	**41**	C	**42**	A
43	D	**44**	A	**45**	B
46	D	**47**	D	**48**	C
49	B	**50**	B	**51**	B
52	B	**53**	D	**54**	C
55	D	**56**	C	**57**	C
58	B	**59**	A	**60**	B

Paper 2

Section 1

1 a i 0.0385

ii 0.039

b 14.4 buckets

c i $186.00

ii $3100.00

iii 10 years

2 a i $x \leqslant \frac{13}{11}$

ii

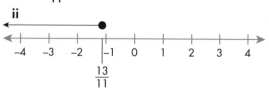

b $\dfrac{3x^2 - 24x - 5}{9x^2 - 25}$

c $4x^4$

d $(2x - 6)(x - 1)$

3 a i $x = 6$

   ii 17

   iii 19

   iv 58

   v 16

b i

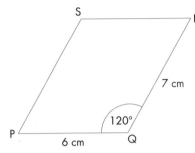

   ii Angle SPR = 27.5°

   iii PR = 11.3 cm

4 a i

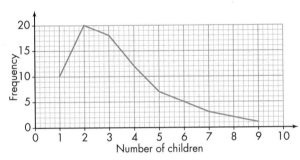

   ii 3

   iii 2

   iv 3.37

   v 7/78

b Range = 90; median = 37.5; interquartile range = 50

5 a i QS = 41.57 m

   ii QR = 20.14 m

   iii Angle PRS = 130°

b i A(5, 6), B(15, 6), C(15, 2), D(5, 2)

    A′(20, 10), B′(30, 10), C′(30, 6), D′(20, 6)

   ii Position of the x coordinate increases by +15 and the y coordinate increases by +4

iii

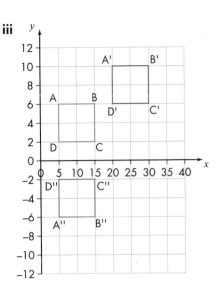

6 a

Number of friends on Facebook	Number of young people (frequency)	Cumulative frequency
0–50	2	2
51–100	5	7
101–150	8	15
151–200	10	25
201–250	15	40
251–300	20	60
301–350	18	78
351–400	12	90
401–450	6	96
451–500	4	100

b

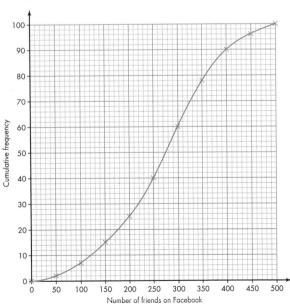

c Median = 275 friends on Facebook
d 4
e 0.3

7 a

b i 4 13 78
  ii 10 31 186
  iii 15 46 276
  iv 62 187 1122

Section 2

8 a i Equation of AD: $y = \dfrac{-4x}{5} + \dfrac{58}{5}$

Equation of BC: $y = \dfrac{-2x}{3} + 18$

ii $y \geqslant 2$; $y \leqslant 10$;

$y \geqslant \dfrac{-4x}{5} + \dfrac{58}{5}$

$y \leqslant \dfrac{-2x}{3} + 18$

iii Maximum (24, 2), minimum (2, 10)

b i $f(-1) = 2$

ii $g(-3) = \dfrac{-3}{4}$

iii $g^{-1}(2) = \dfrac{5}{2}$

9 a i 80° **ii** 100°
  iii 200° **iv** 30°

b i

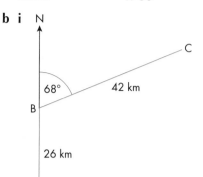

ii AC = 57.08 km

iii Bearing of A from C = 248°

10 a i $\overrightarrow{OP} = \begin{pmatrix} 3 \\ 7 \end{pmatrix}$; $\overrightarrow{OQ} = \begin{pmatrix} -4 \\ 5 \end{pmatrix}$; $\overrightarrow{OR} = \begin{pmatrix} 4 \\ -5 \end{pmatrix}$

ii $\overrightarrow{PQ} = \begin{pmatrix} -7 \\ -2 \end{pmatrix}$; $\overrightarrow{PR} = \begin{pmatrix} 1 \\ -12 \end{pmatrix}$

iii They are collinear / parallel

b i $HK + M = \begin{pmatrix} 37 & 8 \\ 27 & 10 \end{pmatrix}$

ii $M^{-1} = \dfrac{1}{27}\begin{pmatrix} 3 & 0 \\ -8 & 9 \end{pmatrix}$

Index

Acknowledgements

The publishers wish to thank the following for permission to reproduce photographs. Every effort has been made to trace copyright holders and to obtain their permission for the use of copyright materials. The publishers will gladly receive any information enabling them to rectify any error or omission at the first opportunity.

Cover Kudryashka/Shutterstock, p.2 SiBerry/Shutterstock, p.2 Tom Shaw/Staff/Getty, p.2 Teerasak Ladnongkhun/Shutterstock, p.2 Kattika Seemork/Shutterstock, p.3 ronstik/Shutterstock, p.3 pogonici/Shutterstock, p.3 Brian A Jackson/Shutterstock, p.3 AVS-Images/Shutterstock, p.3 BEPictured/Shutterstock, p.3 sirtravelalot/Shutterstock, p.3 Komkrit Noenpoempisut/Shutterstock, p.3 Lack-O'Keen/Shutterstock, p.3 Pixel Embargo/Shutterstock, p.4 James. Pintar/Shutterstock, p.6 Rick Strange / Alamy Stock Photo, p.40 photosounds/Shutterstock, p.42 freie kreation/Shutterstock, p.45 SHUTTER TOP/Shutterstock, p.57 Monkey Business Images/Shutterstock, p.60 INSAGO/Shutterstock, p.63 Bruce Johnstone/Shutterstock, p.66 Art Konovalov/Shutterstock, p.67 Mitch Gunn/Shutterstock, p.78 oksana2010/Shutterstock, p.85 Draftangle/Shutterstock, p.86 Horizons WWP / TRVL / Alamy Stock Photo, p.86 Svineyard/Shutterstock, p.86 Chutima Chaochaiya/Shutterstock, p.87 oasis15/Shutterstock, p.87 Andy Vinnikov/Shutterstock, p.87 Albert Nowicki/Shutterstock, p.87 Brian A Jackson/Shutterstock, p.87 Black Jack/Shutterstock, p.87 Visual Generation/Shutterstock, p.87 Song_about_summer/Shutterstock, p.87 Shyamalamuralinath/Shutterstock, p.87 carolgaranda/Shutterstock, p.87 Priceless-Photos/Shutterstock, p.87 Rawpixel.com/Shutterstock, p.87 Josep Suria/Shutterstock, p.87 Kenishirotie/Shutterstock, p.87 Andrey_Popov/Shutterstock, p.88 AXL/Shutterstock, p.89 Africa Studio/Shutterstock, p.90 Jacob Lund/Shutterstock, p.92 Mike Booth / Alamy Stock Photo, p.93 Imagedb.com/Shutterstock, p.93 Josep Suria/Shutterstock, p.97 S-F/Shutterstock, p.99 Valentin Valkov/Shutterstock, p.100 Andy Dean Photography/Shutterstock, p.104 LightField Studios/Shutterstock, p.104 Kokoulina/Shutterstock, p.112 IndustryAndTravel/Shutterstock, p.114 phortun/Shutterstock, p.115 kuzmaphoto/Shutterstock, p.115 Anton_Ivanov/Shutterstock, p.115 Pearl-diver/Shutterstock, p.115 Anton_Ivanov/Shutterstock, p.115 Linda Bestwick / Shutterstock, p.115 Oleg_Mit/Shutterstock, p.116 nik wheeler / Alamy Stock Photo, p.118 UfaBizPhoto/Shutterstock, p.119 Andrey_Popov/Shutterstock, p.126 lucadp/Shutterstock, p.126 Vectorforjoy/Shutterstock, p.126 Macrovector/Shutterstock, p.126 David R. Frazier Photolibrary, Inc. / Alamy Stock Photo, p.127 Steve Mann/Shutterstock, p.127 sirtravelalot/Shutterstock, p.127 Pegaz / Alamy Stock Photo, p.131 Foto by KKK/Shutterstock, p.133 Metta Image/Alamy Stock Photo, p.134 japansainlook/Shutterstock, p.134 Samuel Borges Photography/Shutterstock, p.151 Fokke baarssen/Shutterstock, p.178 vectorfusionart/Shutterstock, p.178 mkarco/Shutterstock, p.178 Yatra/Shutterstock, p.179 iceink/Shutterstock, p.179 inter reality/Shutterstock, p.179 Rudmer Zwerver/Shutterstock, p.179 vectorfusionart/Shutterstock, p.182 ScofieldZa/Shutterstock, p.182 Author's own, p.182 michaeljung/Shutterstock, p.183 Thorsten Spoerlein/Shutterstock, p.183 attaphong/Shutterstock, p.183 dsom/Shutterstock, p.184 ivn3da/Shutterstock, p.184 ivn3da/Shutterstock, p.184 Author's own, p.195 fivespots/Shutterstock, p.195 Mr.Prasit BOONMA/Shutterstock, p.195 StudioSmart/Shutterstock, p.195 Kotelnikov Andrii/Shutterstock, p.195 Lightspring/Shutterstock, p.195 Aleksey Fefelov/Shutterstock, p.195 skyearth/Shutterstock, p.205 John S. Sfondilias/Shutterstock, p.256 yevtushenko serhii/ Shutterstock, p.256 Sergieiev/Shutterstock, p.256 madtom/ Shutterstock, p.257 golf bress/Shutterstock, p.257 Maxisport/ Shutterstock, p.257 Andrey_Popov/ Shutterstock, p.257 Andrey_Popov/ Shutterstock, p.257 jarabee 123/ Shutterstock, p.257 txking/ Shutterstock, p.264 ATGImages/ Shutterstock, p.264 Sirocco/ Shutterstock, p.266 monticello/ Shutterstock, p.266 Justdoit777/ Shutterstock, p.283 dikobraziy/ Shutterstock, p.284 ashkabe/ Shutterstock, p.284 MaKars/ Shutterstock, p.291 MSPhotographic/ Shutterstock, p.293 MariishkaJon/ Shutterstock, p.293 Sashkin/ Shutterstock, p.304 Maryna Shkvyria/Shutterstock, p.304 Neveshkin Nikolay/Shutterstock, p.304 logonesia/ Shutterstock, p.304 MicroOne/Shutterstock, p.304 Kostikova Natalia/Shutterstock, p.304 Lilkin/Shutterstock, p.304 Monkey Business Images/Shutterstock, p.312 Chones/Shutterstock, p.312 Peter Hermes Furian/Shutterstock, p.320 Viktor1/Shutterstock, p.330 Gabor Havasi/Shutterstock, p.330 J. NATAYO/Shutterstock, p.330 wavebreakmedia/Shutterstock, p.331 Hollygraphic/Shutterstock, p.331 OpturaDesign/Shutterstock, p.331 WoodysPhotos/Shutterstock, p.331 Anthony Correia/Shutterstock, p.350 alexaldo/Shutterstock, p.376 Africa Studio/Shutterstock, p.376 Number1411/Shutterstock, p.376 BigKhem/Shutterstock, p.377 ojal/Shutterstock, p.377 Macrovector/Shutterstock, p.377 Allies Interactive/Shutterstock, p.377 graphixmania/Shutterstock, p.418 JeremyWhat/Shutterstock, p.418 MISS TREECHADA YOKSAN/Shutterstock, p.418 AndrewSproule/Shutterstock, p.418 pixelparticle/Shutterstock, p.419 Pulsmusic/Shutterstock, p.419 alri/Shutterstock, p.419 impromptuwitz/Shutterstock, p.419 ra2studio/Shutterstock, p.458 Wavebreak Media ltd/Alamy Stock Photo, p.458 John Birdsall/Alamy Stock Photo, p.458 John Birdsall/Alamy Stock Photo, p.506 Janine Wiedel Photolibrary / Alamy Stock Photo, p.506 photostock1/Alamy Stock Photo, p.506 Vitya_M/Shutterstock, p.513 R. MACKAY PHOTOGRAPHY, LLC/Shutterstock, p.513 Burlingham/Shutterstock, p.523 Denis Rozhnovsky/Shutterstock, p.523 Jesse Davis/Shutterstock, p.524 Irina Danyliuk/Shutterstockp.

Notes

Notes

Notes

Notes

Notes

Notes

Notes

Notes

Notes

Notes

Notes

Notes

Notes

Notes

Notes